HIGH ENERGY PHENOMENOLOGY

HIGH ENERGY PHENOMENOLOGY

Proceedings of the Forty Second Scottish
Universities Summer School in Physics,
St Andrews, August 1993.

A NATO Advanced Study Institute.

Edited by

K J Peach – University of Edinburgh

L L J Vick – University of Edinburgh

Series Editor

P Osborne – University of Edinburgh

Copublished by
Scottish Universities Summer School in Physics &
Institute of Physics Publishing, Bristol and Philadelphia

Copyright © 1994

The Scottish Universities Summer School in Physics

No Part of this book may be reproduced in any form
by photostat, microfilm or any other means without
written permission from the publishers.

British Library Catalogue-in-Publication Data:

*A catalogue record for this book is available
from the British Library*

    ISBN 0-7503-0326-3

Library of Congress Cataloguing-in-Publication Data are available

Copublished by

SUSSP Publications
The Department of Physics, Edinburgh University,
The King's Buildings, Mayfield Road, Edinburgh EH9 3JZ, Scotland

and

Institute of Physics Publishing, wholly owned by
The Institute of Physics, London

Institute of Physics Publishing, Techno House, Redcliffe Way, Bristol BS1 6NX, UK
US Editorial Office: Institute of Physics Publishing, The Public Ledger Building,
Suite 1035, Independence Square, Philadelphia, PA 19106, USA

Printed in Great Britain by J W Arrowsmith Ltd, Bristol

SUSSP Proceedings

1	1960	Dispersion Relations
2	1961	Fluctuation, Relaxation and Resonance in Magnetic Systems
3	1962	Polarons and Excitons
4	1963	Strong Interactions and High Energy Physics
5	1964	Nuclear Structure and Electromagnetic Interactions
6	1965	Phonons in Perfect and Imperfect Lattices
7	1966	Particle Interactions at High Energy
8	1967	Methods in Solid State and Superfluid Theory
9	1968	Physics of Hot Plasmas
10	1969	Quantum Optics
11	1970	Hadronic Interactions of Photons and Electrons
12	1971	Atoms and Molecules in Astrophysics
13	1972	Properties of Amorphous Semiconductors
14	1973	Phenomenology of Particles at High Energy
15	1974	The Helium Liquids
16	1975	Non-linear Optics
17	1976	Fundamentals of Quark Models
18	1977	Nuclear Structure Physics
19	1978	Metal Non-metal Transitions in Disordered Solids
20	1979	Laser-Plasma Interactions: 1
21	1980	Gauge Theories and Experiments at High Energy
22	1981	Magnetism in Solids
23	1982	Laser-Plasma Interactions: 2
24	1982	Lasers: Physics, Systems and Techniques
25	1983	Quantitative Electon Microscopy
26	1983	Statistical and Particle Physics
27	1984	Fundamental Forces
28	1985	Superstrings and Supergravity
29	1985	Laser-Plasma Interactions: 3
30	1985	Synchrotron Radiation
31	1986	Localisation and Interaction
32	1987	Computational Physics
33	1987	Astrophysical Plasma Spectroscopy
34	1988	Optical Computing

/continued

SUSSP Proceedings (continued)

35	1988	Laser-Plasma Interactions: 4
36	1989	Physics of the Early Universe
37	1990	Pattern Recognition and Image Processing
38	1991	Physics of Nanostructures
39	1991	High Temperature Superconductivity
40	1992	Quantitative Microbeam Analysis
41	1992	Spatial Complexity in Optical Systems
42	1993	High Energy Phenomenology
43	1994	Determination of Geophysical Parameters from Space
44	1994	Simple Quantum Systems
45	1994	Laser-Plasma Interactions: 5
46	1995	General Relativity
47	1995	Perspectives in Lasers
48	1996	High Power Electromagnetic Radiation

1. Christian Völcker
2. Siegfried Bethke
3. Andreas Nyffeler
4. Rocky Kolb
5. Jose Pelaez
6. Maria Chamizo
7. Jose Illana
8. Songhoon Yang
9. Stefania Ricciardi
10. Aaron Grant
11. Sarah Durston Johnson
12. Christine Beeston
13. Ravdandorj Togoo
14. Harry Blundell
15. Lance Vick
16. Ian Aitchison
17. James Stirling
18. Yong-Yeon Keum
19. Peter Negus
20. Pieter Rijken
21. Fani Lebessis
22. Orhan Toker
23. Henry Jaqaman
24. Sandro Ambrosanio
25. Antonio Di Domenico
26. Gulay Bahadiroglu
27. Doreen Wackeroth
28. Johan Rathsman
29. Rupert Seidlein
30. Geza Gyuk
31. Robert Rylko
32. David Saxon
33. Ayse Kuzucu
34. Alan Walker
35. Simona Rolli
36. Johannes Steuerer
37. Eda Eskut
38. Laura Bertolotto
39. Gabriella Sciolla
40. Yvonne Negus
41. Mary Finlay
42. Cecilia Gerber
43. Michael Earnshaw
44. Nick Willis
45. Jacques Bloch
46. Marcus Englert
47. Castulus Kolo
48. Mike Smith
49. Philip Reeves
50. Roland Bernet
51. Dietrich Lehner
52. Alick Macpherson
53. Krzysztof Piotrzkowski
54. Brian Dolan
55. Alain Blondel
56. Giovanni Abbiendi
57. Dimitris Fassouliotis
58. Roger Finlay
59. Pedro Teixeira-Dias
60. Roberto Ugoccioni
61. Sergio Lupia
62. Victor Elvira
63. Angela Benelli
64. Beatriz De Carlos
65. Jose Espinosa
66. Paul Maley
67. Remus Nicolaescu
68. Shigeo Yazaki
69. Thomas Hambye
70. David Delepine
71. Thomas Dignan
72. Philip Page
73. Lucia Bartolini
74. Antonio Riotto
75. Victor Uros Corrales
76. Alberto De Min
77. Bostjan Golob
78. Nikolaos Konstantinidis
79. Giancarlo Brugnola
80. Yakov Shnir
81. Tony Doyle
82. Pierre Bourdon
83. Stefano Moretti
84. Denis Comelli
85. Alessandro Culatti
86. Krzysztof Muchorowski
87. Jürgen Steegborn
88. Raimund Ströhmer
89. Marco Paganoni
90. Roman Friedrich
91. Patrick Van Esch
92. Christoph Schwick
93. Markus Schmidt
94. Andre Maul
95. Federico Farchioni
96. Ken Peach

Executive Committee

Prof D H Saxon	University of Glasgow	*Director*
Mr A Walker	University of Edinburgh	*Secretary*
Dr P J Negus	University of Glasgow	*Treasurer*
Dr K J Peach	University of Edinburgh	*Co-Editor*
Dr L L J Vick	University of Edinburgh	*Co-Editor*
Dr A T Doyle	University of Glasgow	*Steward*
Prof J F Cornwell	University of St Andrews	

International Organising Committee

Prof D H Saxon	University of Glasgow
Prof P Darriulat	CERN Geneva
Dr J R Ellis	CERN Geneva
Prof A Salam	ICTP Trieste
Prof C Jarlskog	University of Stockholm
Prof A Wagner	DESY Hamburg
Prof G Giacomelli	University of Bologna
Dr F Gilman	SSC Laboratory, Dallas

Lecturers

Ian J R Aitchison	University of Oxford
Siegfried Bethke	Universität Heidelberg
Alain Blondel	Ecole Polytechnique
Peter Jenni	CERN Geneva
Edward W Kolb	Fermilab
Roberto D Peccei	University of California at Los Angeles
W James Stirling	University of Durham
Günter Wolf	DESY Hamburg

Seminars given by:

Frank Close	Rutherford-Appleton Laboratory
Richard Kenway	University of Edinburgh
Chris Llewellyn-Smith	University of Oxford

Preface

The forty-second Scottish Universities Summer School in Physics had 82 students from five continents and courses of lectures (45 in all) from eight authoritative speakers. The school was supported by NATO and the Scottish Universities. We also acknowledge support for travel and fees from the UK Science and Engineering Research Council, the US National Science Foundation, the UK Institute of Physics and the British Council.

We would like to thank Hamilton Rentals and Digital Equipment Corporation for help with computing for the school.

The aim of the school was to provide a strong foundation of understanding, technique and interpretation to people active and becoming active in High Energy Phenomenology. The lecture courses mentioned above are reproduced in these proceedings. In addition we benefitted from expert seminars on lattice QCD as a phenomenological tool, on hadrons beyond the quark model and on the future of particle physics by R D Kenway (Edinburgh), F E Close (Rutherford) and C H Llewellyn-Smith (Oxford). The lively poster sessions by the students showed the vigour and range of their work, and on at least one occasion illuminated a question session in a lecture. The less formal interactions amongst lecturers and students, whether on points of the summation of series and renormalisation, or on Monte-Carlo methods of putting on the Himalayas Course, added greatly to the depth of the content. Some of us played the bagpipes and all of us learned Scottish dancing at the Ceilidh that concluded the school.

The success of a school depends on the early establishment of mutual confidence. The key element is the effort, and the welcoming style of it, put in by the committee. Let me thank Tony Doyle, Peter Negus, Ken Peach, Lance Vick and Alan Walker for their enthusiasm and dedication, ably supported by Mary Finlay and Yvonne Negus. The staff of John Burnet Hall handled the residential aspects with their customary efficiency and politeness.

D H Saxon
St Andrews, 1993

Contents

Standard Model Foundations 1
Ian J. R. Aitchison

    QED: a renormalisable U(1) gauge theory   11
    Global and local non-Abelian symmetries   36
    Spontaneous symmetry breaking   49
    The electroweak theory   61

Hadronic Physics in Electron-Positron Annihilation 79
Siegfried Bethke

    A short historical review   81
    Accelerators and detectors   84
    Basics of QCD and hadron production   86
    Hadronic event shapes   93
    Jet physics   96
    Tests of basic quantum numbers of quarks and gluons   103
    Measurements of α_s 106
    Differences between quark jets and gluon jets   127

HERA Physics 135
Günther Wolf

    The HERA collider   136
    The experiments   139
    Running conditions   147
    Photoproduction   171
    DIS and structure functions of the proton   191
    Final states in DIS   201
    Production of events with large rapidity gaps   210

Higgs Phenomenology 225
W. J. Stirling

    The Standard Model Higgs boson   226
    Higgs at LEP   237
    Higgs at hadron colliders   240
    Higgs and Supersymmetry   250

Physics and Experimental Challenge of Future Hadron Colliders 271
Peter Jenni

LHC and SSC — goals and experiments	272
Calorimetry	280
Electron identification	289
Inner detector	290
Muon detection	296
Trigger and data acquisition	300
Simulated physics	306

Beyond the Standard Model 319
Roberto Peccei

The question of mass	321
The SUSY alternative	329
Dynamical symmetry breaking	338
The question of forces	345
The question of matter	350

Particle Physics and Cosmology 361
Edward W. Kolb

A quick look at the Universe	362
The formation of structure	372
Inflation	388
Cosmological phase transitions	400

Precision Tests of the Standard Electroweak Model at LEP 417
Alain Blondel

What we knew before LEP started	419
A synopsis of the measured quantities	423
The Z line-shape	426
Partial widths into specific flavours	440
Helicity effects in Z production and decays	443

List of Acronyms 465

Participant's addresses 469

Index 477

Standard Model Foundations

Ian J R Aitchison

Department of Physics
University of Oxford

1 Introduction

'Foundations' is a big word. In one sense, this course of eight lectures is intended as a foundational one, upon which other courses at this School can build. Yet in another sense, as physicists we must always ask ourselves what are the foundations of the subject we are studying. For the electroweak theory, a classic answer was given by Weinberg (1980), in his Nobel lecture entitled 'Conceptual Foundations of the Unified Theory of Weak and Electromagnetic Interactions'. For Weinberg, the foundations of this theory were

- **local relativistic quantum field theory,**

subject to

- **symmetry principles,**

and a

- **principle of renormalisability.**

Further, the symmetry principles could be classified in two ways

- **manifest** (or **unbroken**) and **hidden** (or **spontaneously broken**)

as against

- **global and local.**

In this course, we shall have these principles very much in mind. The Standard Model (SM) of particle physics refers to the three quantum gauge theories which describe strong, weak, and electromagnetic interactions of quarks and leptons. All three theories are renormalisable, and are based on the following symmetries:-

- strong interactions: unbroken local SU(3)
- weak interactions: spontaneously broken local SU(2)×U(1)
- electromagnetic interactions: unbroken local U(1).

These are all gauge theories, the strong and weak interactions being described by non-Abelian generalizations of the Abelian local U(1) symmetry group of quantum electrodynamics (QED). We shall follow the obvious pedagogical path, therefore, of discussing QED first in some detail in Section 2; we stress, in particular, what is meant by a local quantum field theory, a local U(1) symmetry, and a renormalisable (and a non-renormalisable) theory, and give some basic results in QED at the 'one-loop' level. It may perhaps seem over-ambitious, or unnecessary, to deal with questions of renormalisation in a 'phenomenology' School such as this—but of course it is essential, since these questions arise in quantum field theory as soon as we try to calculate anything beyond the lowest order in perturbation theory, and already the accuracy of current electroweak experiments is such that higher order effects are clearly observable. The analysis of renormalisation at the one loop level in QED is, in fact, relatively simple.

Having described the dynamics of the simplest gauge theory, we then turn to symmetries, in the next two sections. Section 3 mainly deals with local unbroken non-Abelian symmetries, and leads up to QCD, the theory of the strong interactions between quarks and gluons, based on local SU(3) symmetry. Section 4 introduces the idea of spontaneously broken symmetry, the Goldstone and Higgs models, and the non-Abelian generalisation of the Higgs model which is a part of the electroweak theory. All the conceptual elements are now in place for the electroweak theory, which is described in Section 5; we include some discussion of theoretical possibilities concerning the Higgs sector, and of one loop radiative corrections, so as (we hope) to provide links to other courses.

Before proceeding, I want to surprise you perhaps, and stimulate you, by referring again to Weinberg, this time to his 1992 Dallas Conference Summary (Weinberg 1993). Among other things, he says

> "Non-renormalisable theories are just as renormalisable as renormalisable theories"

and

> "... we now believe that the field theories of which we are so proud, quantum electrodynamics, quantum chromodynamics, even general relativity for that matter, are not what they are because they are truly fundamental theories, but simply because they are the only way of satisfying the requirements of symmetries and S-matrix theory."

We shall touch on the first point in Section 2.4. As regards the second, that is perhaps best left for the discussion sessions. The trouble—and the fascination—of foundations is that they have a tendency to shift ... .

2 QED: a renormalisable U(1) gauge theory

2.1 Lagrangian and Feynman rules

The Lagrangian for QED consists of three parts:

$$\mathcal{L}_{\text{QED}} = \mathcal{L}_\psi + \mathcal{L}_A + \mathcal{L}_{\text{int}} \tag{2.1}$$

where $\mathcal{L}_\psi$ is the Dirac Lagrangian for the free field ψ (which destroys electrons and creates positrons), $\mathcal{L}_A$ is the Maxwell Lagrangian for the electromagnetic vector potential field A_μ (which creates and destroys photons), and $\mathcal{L}_{\text{int}}$ is the interaction between the photons and the electrons and positrons. We shall look at each of these in turn.

The Dirac Lagrangian for the free complex spinor field ψ is

$$\mathcal{L}_\psi = \bar{\psi}(i\slashed{\partial} - m)\psi \tag{2.2}$$

where as usual (Aitchison and Hey 1989) $\bar{\psi} = \psi^\dagger \beta \equiv \psi^\dagger \gamma^0$, $\slashed{\partial} = \gamma^\mu \partial_\mu$, and m is the electron/positron mass. The Euler-Lagrange equation for ψ is

$$\partial_\mu \left(\frac{\partial \mathcal{L}}{\partial(\partial_\mu \psi)} \right) - \frac{\partial \mathcal{L}}{\partial \psi} = 0 \tag{2.3}$$

and this easily leads to the Dirac equation for $\bar{\psi}$, from which taking the Hermitean conjugate we get the Dirac equation for ψ:

$$(i\slashed{\partial} - m)\psi = 0. \tag{2.4}$$

$\bar{\psi}$ and ψ in all the above are of course field *operators*. $\psi(x)$ can be expanded in the form

$$\psi(x) = \sum_{s=1,2} \int \frac{d^3 k}{\mathcal{N}_e} \left[c_s(k) u(k,s) e^{-ik.x} + d_s^\dagger(k) v(k,s) e^{ik.x} \right] \tag{2.5}$$

where s labels the two independent spin states of the spinors u and v satisfying

$$(\slashed{k} - m)u(k,s) = 0 \tag{2.6}$$
$$(\slashed{k} + m)v(k,s) = 0 \tag{2.7}$$

with $k^2 \equiv (k^0)^2 - \underline{k}^2 = m^2$ ($c = 1$). $\mathcal{N}_e$ is a normalisation factor which will not concern us as we shall not be doing detailed calculations. The conjugate field $\bar{\psi}$ has a similar expansion involving $c_s^\dagger$ and d_s. For each k, s the pairs $(c_s^\dagger(k), c_s(k))$ and $(d_s^\dagger(k), d_s(k))$ are like the familiar $a^\dagger$ and a which create quanta of the simple harmonic oscillator in quantum mechanics: they are *operators* which create and destroy quanta corresponding to specific momentum (and spin) components of the quantum field $\psi(x)$; $\psi(x)$ itself destroys an electron or creates a positron *at the point* x. However, there is of course one fundamental difference between $(c^\dagger, c)$ (and $(d^\dagger, d)$) and $(a^\dagger, a)$—the former pairs satisfy *anti*-commutation relations, unlike the latter which obey the commutation relation $[a, a^\dagger] = 1$; this correctly implements the exclusion principle for the fermionic quanta.

The Maxwell Lagrangian for the electromagnetic field is

$$\mathcal{L}_A = -\frac{1}{4} F_{\mu\nu} F^{\mu\nu} \tag{2.8}$$

where
$$F_{\mu\nu} = \partial_\mu A_\nu - \partial_\nu A_\mu. \tag{2.9}$$

Remembering that (2.8) (and (2.1)) is in fact a Lagrangian *density*, the action S being the integral of $\mathcal{L}$ over all space-time
$$S = \int \mathcal{L} d^4x, \tag{2.10}$$
we can do some partial integrations to show that (2.8) is equivalent to
$$\mathcal{L}_A = \frac{1}{2} A^\mu (g_{\mu\nu} \Box - \partial_\mu \partial_\nu) A^\nu. \tag{2.11}$$

The Euler-Lagrange equation for A_μ is then found to be
$$(g^{\mu\nu} \Box - \partial^\mu \partial^\nu) A_\nu = 0. \tag{2.12}$$

A_μ has a mode expansion of the form (2.5):–
$$A_\mu(x) = \sum_{\lambda=\pm 1} \int \frac{d^3k}{\mathcal{N}_\gamma} \left[a_\lambda(k) \epsilon_\mu(k,\lambda) e^{-ik\cdot x} + a_\lambda^\dagger(k) \epsilon_\mu^*(k,\lambda) e^{ik\cdot x} \right] \tag{2.13}$$

where $\mathcal{N}_\gamma$ is another normalisation constant, and the sum is over the *two* independent helicity states $\lambda = \pm 1$ (Aitchison and Hey 1989). The *polarisation vectors* ϵ_μ play the role of the Dirac spinors here, and can be chosen to satisfy (Aitchison and Hey 1989)
$$\epsilon \cdot k = 0 \quad \text{and} \quad \epsilon^2 = -1. \tag{2.14}$$

In (2.13), the bosonic operators $a_\lambda(k)$ and $a_\lambda^\dagger(k)$ destroy and create photons in the indicated states. From (2.12) and the first of (2.14) we have $k^2 = (k^0)^2 - \underline{k}^2 = 0$, for all k in (2.13), telling us that the photon is massless. $A(x)$ creates and destroys photons at the point x.

Of course, the electron/positron and photon particles are not free—they interact via the interaction Lagrangian
$$\mathcal{L}_{\text{int}}(x) = -e\bar{\psi}(x)\gamma_\mu \psi(x) A^\mu(x) \tag{2.15}$$

where $e > 0$ and $-e$ is the electron's charge. We shall have more to say about $\mathcal{L}_{\text{int}}$, but for the moment the essential point is that it is *local*: interactions are thought of in terms of processes in which particles are created or destroyed *at a point*. Thus (2.15) includes for example a process in which, at the point x, an electron is destroyed and another electron and one photon are created (Figure 1).

In between these interaction points in space-time, the particles are pictured as moving freely, so that their space-time histories consist of mostly free propagation punctuated by occasional local interaction events (Figure 2). The basis for this picture is really *perturbation theory*—that is, the idea that the interactions are a weak perturbation on the free motion.

Such a space-time picture is very appealing intuitively, but it does not actually correspond to the usual physical situation in, say, a scattering process. There, we think

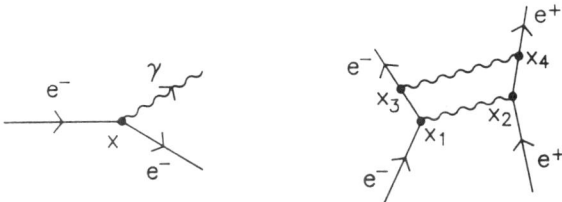

Figure 1.   Figure 2.

of the participating particles not as being localised in space but, on the contrary, as having definite energies and momenta; in other words, we work in momentum space rather than coordinate space. For example, the amplitude for the ψ-particles to travel between the space-time points x and y is the *propagator* $S(x-y)$ defined by

$$(i\partial\!\!\!/_x - m)S(x-y) = i\delta(x-y) \tag{2.16}$$

which in momentum space reads

$$(k\!\!\!/ - m)S(k) = i, \tag{2.17}$$

to which the solution is

$$S(k) = \frac{i}{k\!\!\!/ - m} = \frac{i(k\!\!\!/ + m)}{k^2 - m^2}. \tag{2.18}$$

Note that the denominator of (2.18) vanishes when $(k^0)^2 = \underline{k}^2 + m^2$, which is called the 'on-shell' condition and of course applies to free *real* ψ-particles. The propagator (2.18) includes the more general case $k^2 \neq m^2$, when the particles are free but *'virtual'*.

When (2.18) is Fourier transformed to get back to $S(x-y)$, one has to know how to deal with the singularity in (2.18) at $k^2 = m^2$: this is handled by giving m a small negative imaginary part (the 'Feynman $i\epsilon$ prescription') which correctly incorporates *causality*. The overall factor i is conventional.

The action corresponding to $\mathcal{L}_\psi$ is

$$S_\psi = \int d^4x \, \bar{\psi}(i\partial\!\!\!/ - m)\psi = \int d^4k \, \bar{\psi}(k)(k\!\!\!/ - m)\psi(k) \tag{2.19}$$

where we have introduced the Fourier-transformed fields. The Dirac propagator is therefore just (proportional to) the inverse of the matrix $(k\!\!\!/ - m)$ appearing in this free action, which is quadratic in the fields. For the photon, the corresponding free action is

$$S_A = \int d^4x \left\{ -\frac{1}{4}F_{\mu\nu}F^{\mu\nu} \right\} = -\frac{1}{2}\int d^4k \, A^\mu(k)(k^2 g_{\mu\nu} - k_\mu k_\nu)A^\nu(k) \tag{2.20}$$

so we expect the photon propagator to be proportional to the inverse of

$$P_{\mu\nu} = (k^2 g_{\mu\nu} - k_\mu k_\nu).$$

Unfortunately, however, this matrix does not have an inverse! This can be seen immediately from the fact that it has an eigenvector with eigenvalue zero, namely k^ν:

$$P_{\mu\nu}k^\nu = 0. \tag{2.21}$$

The origin of this lies in the invariance of $\mathcal{L}_A$ under *gauge transformations*

$$A_\mu(x) \to A'_\mu(x) = A_\mu(x) + \partial_\mu \chi(x). \tag{2.22}$$

Although for $\mu = 0,1,2,3$ we appear to have four degrees of freedom in $A_\mu(x)$, we can always use (2.22) to transform one of them to zero: $P_{\mu\nu}$ projects onto a three-dimensional subspace. Classically, of course, the vector potential is similarly subject to (2.22), and has therefore considerable arbitrariness. This is usually dealt with by imposing a 'gauge condition', a constraint which reduces the number of independent degrees of freedom to *two* (both transverse). *Quantizing* such a field theory with a constraint proves to be an awkward technical point, which we shall not delve into here. We merely note that the conventional way out is to introduce a 'gauge-fixing term' in $\mathcal{L}_A$:—

$$\mathcal{L}_A \to \mathcal{L}_{\text{Ag.f.}} = -\frac{1}{4} F_{\mu\nu} F^{\mu\nu} - \frac{1}{2\xi} (\partial \cdot A)^2 \tag{2.23}$$

where ξ is an arbitrary parameter. Clearly $\mathcal{L}_{\text{Ag.f.}}$ is no longer invariant under (2.22). On the other hand, the corresponding action is

$$S_{\text{Ag.f.}} = -\frac{1}{2} \int d^4k\, A^\mu(k) \left[k^2 g_{\mu\nu} - k_\mu k_\nu + \frac{k_\mu k_\nu}{\xi} \right] A_\nu(k), \tag{2.24}$$

and now the operator in [ ] *does* have an inverse, leading to the photon propagator

$$D_{\mu\nu} = \left[-g_{\mu\nu} + (1-\xi) \frac{k_\mu k_\nu}{k^2} \right] \frac{i}{k^2 + i\epsilon}. \tag{2.25}$$

A particularly convenient form is obtained when $\xi = 1$, which is called the Feynman gauge. It is not at all self-evident why the results of using $\mathcal{L}_{\text{Ag.f.}}$ should be physically correct, since it breaks the Maxwell symmetry (2.22); nor why, in particular, physical results are independent of the 'gauge parameter' ξ. Nevertheless, such is in fact the case: an essential point is that since the photon couples to a conserved current (see below) the $k_\mu k_\nu$ part of (2.25) never contributes to physical amplitudes.

Equations (2.18) and (2.25) give us the momentum-space amplitudes for free electron/positron and photon propagation; to complete the QED story we need to know the amplitude corresponding to an interaction. For example, the amplitude for the process in which an electron of momentum p and spin s is destroyed, and an electron of momentum p' and spin s', and a photon of momentum q and polarisation λ are created (Figure 3), is given in lowest-order perturbation theory by

$$i \langle p', s'; q, \lambda | \int d^4x\, \mathcal{L}_{\text{int}}(x) | p, s \rangle \tag{2.26}$$

where

$$|p, s\rangle = c_s^\dagger(p) |0\rangle, \qquad |q, \lambda\rangle = a_\lambda^\dagger(q) |0\rangle \tag{2.27}$$

and $|0\rangle$ is the vacuum state satisfying $c_s(p)|0\rangle = d_s(p)|0\rangle = a_\lambda(p)|0\rangle = 0$ for all p, s, λ. It is a simple exercise to insert the expansions (2.5) and (2.13) into (2.26), and use the commutation relations of the operators to show that (2.26) becomes

$$\epsilon_\mu^*(q,\lambda) \bar{u}(p', s')(-ie\gamma^\mu) u(p, s)(2\pi)^4 \delta^4(p' + q - p). \tag{2.28}$$

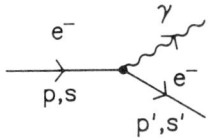

Figure 3.

Actually, it proves impossible to satisfy the δ-function condition for on-shell particles (except for the pathological case of zero-energy photons). Nevertheless, (2.28) does in fact contain the correct factors for the *interaction vertex* (and for incoming and outgoing particles) for processes which *do* occur. Namely, we identify the factors $u, \bar{u}$ and ϵ^* in (2.28) with the ingoing electron, the outgoing electron, and the outgoing photon respectively, and the factor $-ie\gamma^\mu$ with the interaction vertex '$e^- \to e^-\gamma$'; finally, there is four-momentum conservation at the vertex.

We now have all the pieces necessary to write down momentum-space amplitudes for QED processes in perturbation theory—the rules for doing so are the famous *Feynman rules* (which we have of course not derived—see any quantum field theory textbook). These rules associate a graphical element with each factor in the amplitude, so that to a given amplitude there corresponds a certain *Feynman graph* (or diagram), and vice versa. For an internal electron/positron we have

$$\frac{i}{\not{k} - m + i\epsilon}$$

(the arrow follows the 'current', with the e^- being conventionally a positive current); for an internal photon

$$\frac{-ig_{\mu\nu}}{k^2 + i\epsilon} \quad \text{(in Feynman gauge)};$$

and for an e-e-γ vertex

$$-ie\gamma^\mu.$$

Factors of $u, \bar{u}, v, \bar{v}$ correspond to ingoing or outgoing electrons or positrons, and ϵ, ϵ^* to ingoing or outgoing photons; four-momentum is conserved at every vertex. Since perturbation theory is conducted in powers of e, we effectively have an expansion in the number of vertices.

As an example, we just write down the lowest order ($O(e^2)$) expansion for $e^+e^- \to \mu^+\mu^-$ which is given by the diagram of Figure 4; the amplitude is

$$\bar{u}(p_3, s_3) ie\gamma^\nu v(p_4, s_4) \frac{-ig_{\mu\nu}}{(p_1 + p_2)^2 + i\epsilon} \bar{v}(p_2, s_2) ie\gamma^\mu u(p_1, s_1). \qquad (2.29)$$

It is most important to realize that, despite appearances, Feynman diagrams are most emphatically *not* a picture of a space-time process: as we have stressed, they correspond to momentum-space formulae such as (2.29).

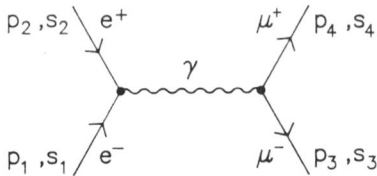

Figure 4.

2.2 Global and local U(1) symmetries

Consider the free Dirac Lagrangian

$$\mathcal{L}_\psi = \bar\psi(i\partial\!\!\!/ - m)\psi. \qquad (2.30)$$

Clearly this is invariant under the transformation

$$\psi \to \psi'(x) = e^{-i\alpha}\psi(x) \qquad (2.31)$$

where α is a constant. This is called a *global* invariance because the parameter α is the same for all space-time points; in a moment we will introduce the corresponding *local* transformation in which the phase angle in (2.31) depends on the space-time point x. To investigate the consequences of the invariance (2.31), it is convenient to consider an infinitesimally small transformation

$$\psi \to \psi' = (1 - i\epsilon)\psi \equiv \psi + \delta\psi. \qquad (2.32)$$

Since $\mathcal{L}_\psi$ is invariant under this transformation, we have

$$\delta\mathcal{L}_\psi \equiv \mathcal{L}_{\psi+\delta\psi} - \mathcal{L}_\psi = 0. \qquad (2.33)$$

However, we can also write

$$\delta\mathcal{L}_\psi = \frac{\partial\mathcal{L}}{\partial\psi}\delta\psi + \frac{\partial\mathcal{L}}{\partial(\partial_\mu\psi)}\delta(\partial_\mu\psi), \qquad (2.34)$$

which becomes

$$\delta\mathcal{L}_\psi = \partial_\mu\Big(\frac{\partial\mathcal{L}}{\partial(\partial_\mu\psi)}\Big)\delta\psi + \frac{\partial\mathcal{L}}{\partial(\partial_\mu\psi)}\partial_\mu(\delta\psi) \qquad (2.35)$$

after using (2.3) and the fact that $\delta(\partial_\mu\psi) = \partial_\mu(\delta\psi)$. Thus

$$\delta\mathcal{L}_\psi = \partial_\mu\Big[\frac{\partial\mathcal{L}}{\partial(\partial_\mu\psi)}\delta\psi\Big] = \epsilon\partial_\mu[\bar\psi\gamma^\mu\psi]. \qquad (2.36)$$

But $\delta\mathcal{L}_\psi$ is in fact zero, so that (2.36) tells us that (cancelling the constant parameter ϵ) we have the *current conservation* condition

$$\partial_\mu j^\mu_\psi = 0 \qquad (2.37)$$

where
$$j^\mu_\psi = \bar\psi\gamma^\mu\psi. \tag{2.38}$$

If we write $j^\mu_\psi = (\rho_\psi, \underline{j}_\psi)$ then (2.37) is

$$\frac{\partial\rho_\psi}{\partial t} = -\nabla\cdot\underline{j}_\psi. \tag{2.39}$$

Integrating (2.39) over all space, using the divergence theorem, and assuming that $\underline{j}_\psi$ falls off sufficiently fast that the surface integral can be dropped, we find the *conservation law*

$$\frac{dN_\psi}{dt} = 0, \tag{2.40}$$

where
$$N_\psi = \int \rho_\psi \, d^3x = \int \psi^\dagger\psi \, d^3x. \tag{2.41}$$

Thus from the symmetry (2.31) has followed a conservation law (2.40); this is a special case of Noether's theorem. Note that j^μ_ψ and N_ψ are *operators*; in the present case, if we insert the expansion (2.5) into (2.41) we find (for suitable normalisation $\mathcal{N}_e$)

$$N_\psi = \int \frac{d^3k}{(2\pi)^3} \sum_s \left[c^\dagger_s(k)c_s(k) - d^\dagger_s(k)d_s(k)\right] \tag{2.42}$$

where $c^\dagger c$ and $d^\dagger d$ are *number operators* whose eigenvalues are the numbers of electrons and positrons (respectively) in the indicated states. Thus the total number of electrons minus the total number of positrons in a state is a constant.

The transformation (2.31) can also be written as

$$\psi'(x) = U(\alpha)\psi(x) \tag{2.43}$$

where $U(\alpha)$ obviously satisfies

$$U(\alpha_1)U(\alpha_2) = U(\alpha_1 + \alpha_2); \tag{2.44}$$

that is, the product of any two such transformation factors is also one. The quantities $U(\alpha)$ (for all α) form the elements of a *group*, called the U(1) group ('U' = unitary, '1' = 1 × 1). Later we will meet the U(N) group of unitary N × N matrices, and SU(N) which is the group of all such matrices with determinant +1. In the case of U(1), a unitary 1 × 1 matrix is just a phase factor. Thus (2.30) is invariant under global U(1) transformations. It is worth noting that if we had two independent sorts of fermions (*e.g.* e's and μ's), so that we generalised the free Lagrangian (2.30) to

$$\bar\psi_1(i\slashed\partial - m)\psi_1 + \bar\psi_2(i\slashed\partial - m_2)\psi_2 \tag{2.45}$$

then we could have two independent U(1) transformations on ψ_1 and ψ_2.

What happens when we include the interaction term (2.15)? Clearly this is still invariant under the global transformation (2.31), so we still have the consequential number conservation law(s). However, more remarkably, the Lagrangian

$$\mathcal{L}_\psi + \mathcal{L}_{\text{int}} = \bar\psi(i\slashed\partial - m) - e\bar\psi\slashed{A}\psi \tag{2.46}$$

is invariant under the *local* version of (2.31), in which α becomes a function of the space-time point x:

$$\psi(x) \to \psi'(x) = e^{-i\alpha(x)}\psi(x). \qquad (2.47)$$

We check this important property as follows. Clearly, the term $\bar{\psi}m\psi$ remains invariant under (2.47). On the other hand, $\bar{\psi}i\slashed{\partial}\psi$ is not invariant, since

$$\begin{aligned}
\bar{\psi}i\slashed{\partial}\psi \to \bar{\psi}'i\slashed{\partial}\psi' &= \bar{\psi}(x)e^{i\alpha(x)}i\slashed{\partial}(e^{-i\alpha(x)}\psi(x)) \\
&= \bar{\psi}(x)e^{i\alpha(x)}[(\slashed{\partial}\alpha(x))e^{-i\alpha(x)}\psi(x) + e^{-i\alpha(x)}i\slashed{\partial}\psi(x)] \\
&= \bar{\psi}(x)i\slashed{\partial}\psi(x) + \bar{\psi}(x)(\slashed{\partial}\alpha(x))\psi(x). \qquad (2.48)
\end{aligned}$$

However, we can arrange for the Lagrangian (2.46) to be invariant under (2.47) if we agree that, when $\psi(x)$ undergoes (2.47), the potential $A_\mu(x)$ changes by

$$A_\mu(x) \to A'_\mu(x) = A_\mu(x) + \frac{1}{e}\partial_\mu\alpha(x), \qquad (2.49)$$

since then the unwanted second term on the right hand side of (2.48) is cancelled. But (2.49) is precisely a gauge transformation of the type (2.22), introduced before (identifying $\chi(x)$ with $\frac{1}{e}\alpha(x)$), which is itself a symmetry of the Maxwell Lagrangian $\mathcal{L}_A = -\frac{1}{4}F_{\mu\nu}F^{\mu\nu}$. Hence the full QED Lagrangian of (2.1) is invariant under the local U(1) transformation (2.47).

For future reference, we add some more comments regarding the local U(1) invariance of $\mathcal{L}_{\text{QED}}$. First, there is a very convenient general way of seeing how to construct a locally U(1) invariant Lagrangian from one that is *globally* U(1)-invariant. Consider the quantity

$$D_\mu\psi \equiv (\partial_\mu + ieA_\mu)\psi, \qquad (2.50)$$

where D_μ is called the 'U(1) covariant derivative'. Under the transformations (2.47) and (2.49) it is easy to check that $D_\mu\psi$ transforms as

$$D_\mu\psi \to D'_\mu\psi' = e^{-i\alpha(x)}(D_\mu\psi), \qquad (2.51)$$

from which it is obvious that $\bar{\psi}\slashed{D}\psi$ is invariant. In general, all terms without derivatives are invariant under (2.47) if they are under (2.31), and if any term involving a ∂_μ acting on a ψ is invariant under (2.31), it will be invariant under (2.47) if ∂_μ is replaced by D_μ. Thus merely by replacing ∂_μ by D_μ in a *free* matter Lagrangian we seem to have a unique way of introducing the electromagnetic interaction! (Actually this is not true for charged vector fields, as we shall see in Section 3.2.)

Secondly, we might wonder if a new conservation law follows from the (more restrictive) symmetry of (2.47) and (2.49). The answer is no—the symmetry current for electron/positron number conservation is still (2.38) and there is no new one. However, if we look at the equation of motion for the A_μ field, which turns out to be

$$(g^{\mu\nu}\Box - \partial^\mu\partial^\nu)A_\nu = e\bar{\psi}\gamma^\mu\psi = ej_\psi^\mu, \qquad (2.52)$$

we find a remarkable thing: the current which is the driving term on the RHS of the inhomogeneous Maxwell equations (2.52) is (e times) the global U(1) symmetry current (2.38). This is a key feature of 'gauge theories': the *dynamical* currents (RHS of field equations) are also *symmetry* currents. Indeed, the local symmetry (2.47), if accepted as a principle, essentially *fixes* the dynamics. This is why accepting it as a principle (the 'gauge principle') is so attractive theoretically.

Thirdly, we can consider adding a standard mass term for the A_μ field:–

$$\mathcal{L} \to \mathcal{L}_{\text{QED}} + \frac{1}{2} M_\gamma^2 A_\mu A^\mu. \tag{2.53}$$

The equation of motion for A_μ is then, using (2.11),

$$[g^{\mu\nu}(\Box + M_\gamma^2) - \partial^\mu \partial^\nu] A_\nu = e j_\psi^\mu. \tag{2.54}$$

(2.53) is *not* invariant under (2.49), so that one is inclined to associate gauge invariance with masslessness of the vector field. We shall see, however, that this connection can be evaded.

2.3 Loops and renormalisation

As we have said, Feynman diagrams represent terms in a perturbation theory expansion of physical amplitudes, where the expansion parameter is e—or, better, the fine structure constant $\alpha = e^2/4\pi \approx 1/137$, in the units which are conventional in high energy physics. Equivalently, this is an expansion in terms of the number of vertices appearing in the diagrams. A little experimentation should convince you that, for a given physical process, the lowest order diagrams (with the fewest vertices) are those in which each vertex is connected to every other one by just one internal line; these are called *tree diagrams*. Remembering that one of the Feynman rules is that 4-momentum is to be conserved at every vertex, more experimentation will convince you that, in tree diagrams, the 4-momenta of the single internal lines is defined uniquely in terms of the physical 4-momenta of the external particles: for example, in the formula (2.29) corresponding to Figure 4, the photon's 4-momentum is $p_1 + p_2$. (Note that since it is conserved at every vertex, 4-momentum 'flows through' a diagram, and the final total 4-momentum is automatically equal to the initial total 4-momentum, so in this example $p_1 + p_2 = p_3 + p_4$). Thus, in lowest order for any process, the rules already given produce a perfectly definite amplitude, and all we have to do to calculate the corresponding cross-section is go through the routine struggle with Dirac tracing, phase space, *etc*.

But tree diagrams will only give us the lowest order contribution to the amplitude. As soon as we go beyond this approximation, we meet *loops*—a concept easy to understand graphically, see for example Figure 5, which shows the $O(e^4)$ or $O(\alpha^2)$ diagrams for $e^+ e^- \to \mu^+ \mu^-$. Admittedly, since α is rather small, such corrections seem to be very tiny—but, as you will all know, there are some fantastically accurate measurements of certain quantities, such as the anomalous magnetic moments of the electron and muon (see below), which challenge theorists to calculate higher order processes in QED.

More to the point for this School, current experiments at LEP and other laboratories already have an accuracy sensitive to one loop corrections to the Standard Model. Hence an understanding of such corrections is now essential for phenomenology.

So consider the $O(e^4)$ diagrams for $e^+ e^- \to \mu^+ \mu^-$, shown in Figure 5. We can distinguish two types of loop here: ones which appear as modifications to propagators

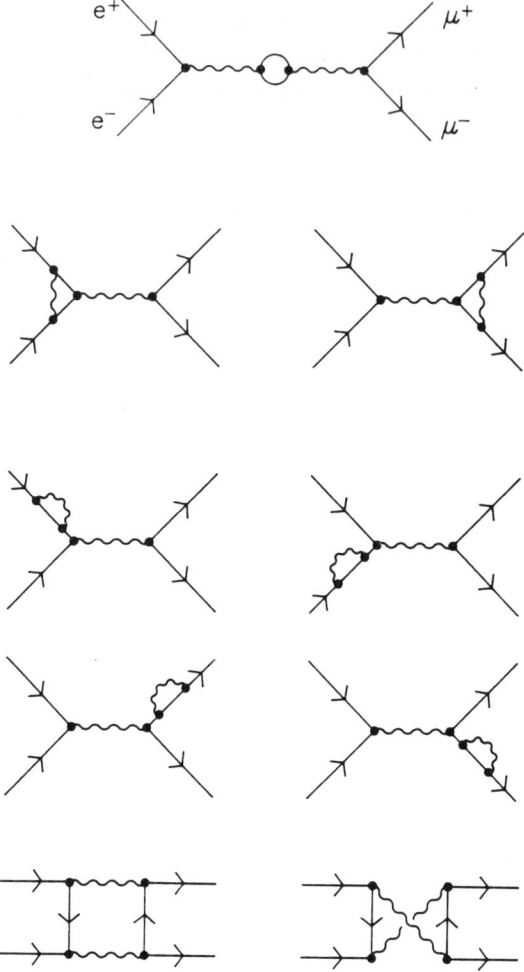

Figure 5.

or vertices, and ones which do not. In the latter class are the two 'box' diagrams at the bottom of Figure 5, which we shall come back to later. In the class of modifications (or 'insertions'), which we consider first, we have Figures 6(a) and 6(b) which modify the fermion and photon propagators respectively, and Figure 6(c) which modifies the vertex. For all loops, it is clear that the 4-momentum of the particles inside the loop is not uniquely defined: for example, in Figure 6(a) all that is required is that the 4-momenta of the internal fermion and γ add up to the 4-momentum of the external fermion. Thus there is one free momentum for each loop. Since this momentum is completely arbitrary, we have to integrate over it, with a standard normalisation factor $(2\pi)^{-4}$. Also, it turns out that for a fermion loop, as in Figure 6(b), we need to calculate

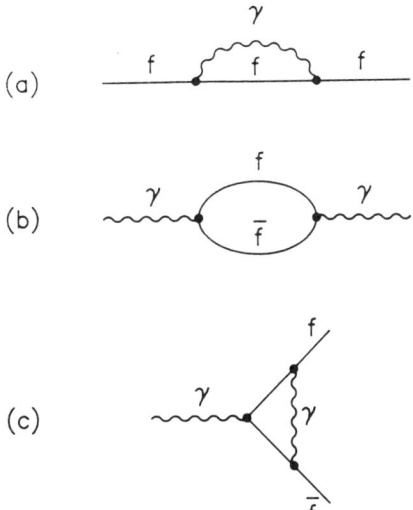

Figure 6.

the trace over whatever γ-matrices appear, and multiply by -1 as a consequence of the anticommuting character of the fermion fields.

We shall now outline the calculation of each of Figures 6(a)–6(c); this will lead us straight into the question of *renormalisation*, and the concept of a '*running coupling constant*'.

Insertions into propagator lines, such as Figures 6(a) and 6(b), are generically called 'self energies'. The reason for this name becomes clear when we consider a whole string of repeated insertions of Figure 6(a) into the fermion propagator, as shown in Figure 7.

Figure 7.

Calling this insertion $-i\Sigma$, the series is

$$\frac{i}{\not{p}-m} + \frac{i}{\not{p}-m}(-i\Sigma)\frac{i}{\not{p}-m} + \frac{i}{\not{p}-m}(-i\Sigma)\frac{i}{\not{p}-m}(-i\Sigma)\frac{i}{\not{p}-m} + \cdots \quad (2.55)$$

which sums up to

$$\frac{i}{\not{p}-m-\Sigma}, \quad (2.56)$$

from which we guess that, physically, Σ plays the role of a correction to the original mass m. (As we shall see, it is slightly more complicated than this). This correction is associated with the fermion's *self* interactions with the photons, since there is only the 'one' fermion present in Figure 7. Indeed there are no external particles in play at all in these processes: these self interactions are due to quantum excitations from the vacuum. The vacuum is therefore analogous to a condensed matter 'environment' through which, say, an electron propagates and which causes (in many cases) the electron to have an 'effective mass', different from its free space value. In (2.56), of course, propagation is precisely in the free space vacuum, and still this shift is occurring. It is an essentially quantum field theoretic effect. Let us see how large it is.

The quantity $-i\Sigma$ is normalised so that the amplitude for Figure 6(a) (which suppresses the external lines) is $\dfrac{i}{\not{p}-m}(-i\Sigma)\dfrac{i}{\not{p}-m}$ where

$$-i\Sigma = \int(-ie\gamma^\nu)\cdot\frac{-ig_{\mu\nu}}{k^2}\cdot\frac{i}{\not{p}-\not{k}-m}\cdot(-ie\gamma^\mu)\frac{d^4k}{(2\pi)^4} \qquad (2.57)$$

$$= -e^2\int\frac{1}{k^2}\gamma_\mu\frac{\not{p}-\not{k}+m}{(p-k)^2-m^2}\gamma^\mu\frac{d^4k}{(2\pi)^4}. \qquad (2.58)$$

The immediate difficulty with (2.58) is, of course, that it diverges for large values of the integration momentum k, as is evident from the fact there are only three powers of k in the denominator (we ought also to worry about the fact that it may diverge at the small k end of the integration—but this 'infra-red' problem is quite separate from the high-k or 'ultraviolet' divergence which is our present interest; we shall ignore the infra-red problem altogether). This seems to be disastrous, since what was supposed to be a small (mass) correction is actually infinite!

However, it is important, right from the outset, to ask whether we seriously believe our theory (in this case QED of e^-'s, μ^-'s $\cdots$) to remain literally unmodified for arbitrarily high virtual momentum scales—or, equivalently, for arbitrarily short distances. Indeed, we know very well that 'new physics' will come into play which is not at all contained in the original $\mathcal{L}_{\text{QED}}$ (for example, electroweak phenomena, physics at grand unified or Planck scales, $\cdots$). Thus it is surely more reasonable to take the view that we should only believe the theory we started with, $\mathcal{L}_{\text{QED}}$, up to some physical cut-off energy M, say (imposing a single sharp cut off like this is perhaps rather a crude way of representing the 'new physics at scale M', but it is good enough for our purpose). Thus we should not be too alarmed by the formal divergence as $M \to \infty$.

But we would certainly like our theory at 'low energies' ($E \ll M$) to be insensitive to what is going on at the scale M. Phrased in coordinate rather than momentum space, this means that we'd like our theory at relatively 'large' distances to be insensitive to unknown 'short' distance physics (i.e. to physics on scales less than or of order M^{-1}). To take an extreme example, most people think that quantum field theory as we know it will fail at distances of order the Planck length ($\ell_P = [G_N\hbar/c^3]^{\frac{1}{2}} \sim 10^{-33}$cm), where perhaps we shall find something like string theory. If we needed to know all the details of, for example, Planck-scale physics before being able to say something useful about experiments at the current energy scale of some hundreds of GeV (lengths $\sim 10^{-18}$cm), the prospects for phenomenology would be poor. Pursuing this line of thinking, we can reflect upon the fact that, for most purposes, atomic physicists manage to get along quite well without knowing details of nuclear structure, needing only certain gross parameters such as nuclear charge, mass, spin, magnetic moment, *etc*. Nuclear physicists, in turn, explain the values of some of these parameters in terms of nucleon-

nucleon interactions, using other parameters such as nucleon charges, masses, spins, *etc.*, but without needing to know much about quarks (perhaps rather surprisingly). Hadron physicists explain hadrons in terms of quarks ... and so we arrive at the idea that Nature may have a kind of 'hierarchy of scales', such that the physics at any one scale is successfully described in terms of certain 'effective degrees of freedom', with properties characterised by a smallish number of parameters which have to be empirically determined at that scale, but which might be calculable in terms of physics at a 'deeper' scale.

Returning to QED—and to the Standard Model of which it is a part—we would not be surprised to meet one or more new 'levels' even before, in principle, we got to the Planck scale; in fact, reasons will be discussed in this School to expect something to happen either quite soon (a few hundred GeV) or at least in the TeV region. For the immediate discussion of (2.58), however, we will simply think in terms of some finite energy cut-off M, which represents the scale at which 'new physics' enters. According to the 'hierarchy of scales' philosophy, then, we expect our low energy 'effective QED' theory to be characterized by a few parameters—charge, mass, ... — which depend on the physics at the (energy) scale M. Thus in the present case, the 'm' in 'the Lagrangian' should be thought of as $m(M)$, and the 'e' as $e(M)$ (the 'ψ' also needs an M-dependent normalisation, as we shall see). In particular, e and m in (2.58) are now $e(M)$ and $m(M)$.

From this viewpoint, (2.56) tells that the process of Figure 6(a) shifts the 'original' mass $m(M)$ to $m(M) + \Sigma(M)$, where the M-dependence of Σ is due to $e(M), m(M)$ and the explicit M-dependence of the loop integral in (2.58). Now we do not know anything about the M-dependence of $m(M)$ and $e(M)$. But the idea may suggest itself that perhaps these unknown M-dependences can be chosen so that the 'corrected' mass $m(M) + \Sigma(M)$, which we might interpret as the 'real' mass (to this order in e^2), is very weakly dependent on M (the 'unknown physics' scale)– and would even be finite if we went all the way and took $M \to \infty$. Indeed, if we write

$$m(M) = m_r - \Delta m(M) \tag{2.59}$$

where m_r is independent of M, we can imagine suitably adjusting $\Delta m(M)$ so that it cancels off the most significant M-dependence of the 'vacuum-corrected' mass $m_r - \Delta m(M) + \Sigma(M)$, and even renders this expression finite as $M \to \infty$.

To implement this idea we need to calculate $\Sigma(M)$. The complete expression is complicated, but for the present purpose it is enough to know that

$$\Sigma(M) = m(M)\frac{e^2(M)}{4\pi^2}\ln M^2 - \not{p}\frac{e^2(M)}{16\pi^2}\ln M^2$$
$$+ M\text{-independent terms} + O\left(\frac{1}{M^2}\right) \tag{2.60}$$

where an expansion in powers of m^2/M^2 has been made. Note that the apparent linear divergence of (2.58) as $M \to \infty$ does not in fact occur, since integration over an odd number of k's in the numerator of (2.58) gives zero; thus the divergence as $M \to \infty$ is only logarithmic. Note also that from (2.58) Σ obviously depends on p, though we have not indicated that on the left hand side of (2.60). Evidently, if we choose

$$\Delta m(M) = \frac{m(M)e^2(M)}{4\pi^2}\ln M^2, \tag{2.61}$$

the $\not{p}$-independent part of $m_r - \Delta m(M) + \Sigma(M)$, at least, will be weakly dependent on M, and will be finite as $M \to \infty$.

We can use (2.59) to write the free Dirac part of the Lagrangian

$$\bar{\psi}(i\not{\partial} - m(M))\psi \tag{2.62}$$

as

$$\bar{\psi}(i\not{\partial} - m_r)\psi + \Delta m(M)\bar{\psi}\psi . \tag{2.63}$$

In momentum space (c.f. (2.19)), the '$i\not{\partial}$' term would be just '$\not{p}$', which suggests that the way to deal with the $\not{p} \ln M^2$ term in (2.60) might be to allow the normalisation of ψ in (2.62) to depend on M:

$$\psi(x) \to \psi_M(x) = Z_2^{\frac{1}{2}}(M)\psi_r(x) \tag{2.64}$$

in the conventional notation. (This step is usually called 'wave function renormalisation', since (2.64) is equivalent to changing the normalisation of u and v in (2.5).) Then,

$$\bar{\psi}i\not{\partial}\psi = \bar{\psi}_r i\not{\partial}\psi_r + (Z_2(M) - 1)\bar{\psi}_r i\not{\partial}\psi_r \tag{2.65}$$

and the $\not{p}\ln M^2$ term in (2.60) can be cancelled if we choose

$$Z_2(M) = 1 - \frac{e^2(M)}{16\pi^2} \ln M^2 . \tag{2.66}$$

In this way the 'Lagrangian at scale M' has been rewritten in terms of quantities ψ_r and m_r, which are weakly dependent on M (and finite as $M \to \infty$), together with 'counter terms' which cancel off the unwanted large M-dependence of the loop.

At this point there are a number of comments to be made. First, it might be said (rightly, so far!) that this is an empty exercise: for each unwanted large M-dependence we merely postulate an appropriate cancelling dependence in some unknown parameter. But the remarkable fact is this: it is possible, by successively adjusting the M-dependence of the quantities appearing in $\mathcal{L}_{\text{QED}}(M)$, to eliminate all $\ln M^2$ terms *at each order of perturbation theory* (we have only looked at one $O(e^2)$ correction so far), so that as $M \to \infty$ all physical amplitudes are finite and depend only on the finite quantities m_r and e_r, which are to be taken from experiment in some suitably defined way (a point to which we shall return). This is what is meant by the statement that 'QED is renormalisable'.

Secondly there is clearly some arbitrariness in how we define quantities like $\Delta m(M)$: might we not, for instance, absorb some or all of the 'M-dependent terms' in (2.60) into $m(M)$? This raises some questions. For example, does it mean that our whole theory is arbitrary; and, how are the 'real mass' and 'real charge' defined? Here we meet a key aspect of the renormalisation procedure: the definition of the 'real physical parameters' at our scale $E \ll M$ is indeed (to a large extent) arbitrary, but some definition is necessary; however, observable physical quantities will be independent of the choice of definition (which is usually called a 'scheme'). Different definitions lead to finite shifts in the defined quantities, but these shifts will all conspire to cancel out in observable quantities (provided, of course, we do a complete calculation to all orders!).

In the case of the electron mass in QED, for example, a common definition is that the fully corrected electron propagator should have the form, as $M \to \infty$,

$$\frac{i}{\not{p} - m - \Sigma_r(p,m)}\text{''} \tag{2.67}$$

where 'm' (dropping the subscript r) is the conventional electron mass ($m \approx 0.51$ MeV), and where

$$\Sigma_r(p,m) = (\not{p} - m)^2 \times \text{finite quantity} + \text{higher powers in } (\not{p} - m) \tag{2.68}$$

for $\not{p} - m \approx 0$. This is called the 'on shell' definition of m, and means that the pole of the (momentum space) propagator is at $\not{p} = m$ or equivalently $p^2 = m^2$, which is indeed the free particle energy-momentum relation $(p^0)^2 = \underline{p}^2 + m^2$. Obviously, in this case, the energy scale at which the finite parameter m is defined is m itself, which seems most natural.

We now move on to consider the analogous quantity for the A_μ field, the photon 'self-energy' of Figure 6(b). Using the Feynman gauge, $\xi = 1$ (c.f. (2.25)), the amplitude for Figure 6(b) is

$$\frac{-ig^{\nu\rho}}{p^2}(-i\Sigma^\gamma_{\rho\sigma}(p))\frac{-ig^{\sigma\mu}}{p^2} \tag{2.69}$$

where

$$-i\Sigma^\gamma_{\rho\sigma}(p) = (-1)\text{Tr}\int \frac{d^4k}{(2\pi)^4}\frac{i}{\not{p}+\not{k}-m}(-ie\gamma_\rho)\frac{i}{\not{k}-m}(-ie\gamma_\sigma)$$

$$= -e^2\int \frac{d^4k}{(2\pi)^4}\frac{\text{Tr}[(\not{p}+\not{k}+m)\gamma_\rho(\not{k}+m)\gamma_\sigma]}{[(p+k)^2-m^2][k^2-m^2]}. \tag{2.70}$$

Although (2.70) is complicated, we can make useful progress by analysing its tensor character and noting that it has dimensions (mass)^2. The only tensors available to us after doing the integrals and the trace in (2.70) are $g_{\rho\sigma}$ and $p_\rho p_\sigma$, so $\Sigma^\gamma_{\rho\sigma}$ must have the form

$$\Sigma^\gamma_{\rho\sigma}(p) = g_{\rho\sigma}\{\text{terms in } M^2, m^2 \ln M^2, p^2 \ln M^2, \frac{m^4}{M^2}, \cdots\}$$

$$+ p_\rho p_\sigma\{\text{terms in } \ln M^2, \frac{m^2}{M^2}, \cdots\} \tag{2.71}$$

where we are interested in large values of $M(\gg m)$. There are no mM terms because the trace over the terms which are linear in m in (2.70) gives zero. Note that we seem to have a *quadratic* divergence as $M^2 \to \infty$, associated with the $\not{k}\not{k}$ term in the numerator of (2.70).

Consider a term of the form $\Sigma^\gamma_{\rho\sigma} = -Ag_{\rho\sigma}$ where $A \sim M^2$. When this is placed in (2.69), and the whole string of these 'fermion bubbles' (analogous to Figure 7 for the electron propagator) is summed up, the effect (c.f. (2.56)) is to replace the original photon propagator $-ig^{\nu\mu}/p^2$ by

$$\frac{-ig^{\nu\mu}}{p^2 - A} \tag{2.72}$$

where (c.f. 2.71) A can be strongly dependent on M. From (2.72) we see that the 'on-shell' point for the photon, which was originally correctly at $p^2 = 0$, has been shifted to $p^2 = A$; i.e. the photon has got a (strongly M-dependent) mass! Further, unlike the electron case, we have no '$M_\gamma(M)$' parameter in $\mathcal{L}_{\text{QED}}(M)$ which we can adjust to cancel this unwanted M-dependence.

So we have a problem. At this point we have to remember why there is no '$+\frac{1}{2}M_\gamma^2 A_\mu A^\mu$' term in $\mathcal{L}_{\text{QED}}$—it was because such a term violated the gauge invariance under (2.49). Indeed it can be shown that provided a gauge-invariant cut-off is used (which we will not specify further), no such M_γ^2 term arises. In fact, gauge invariance implies the constraints

$$p^\rho \Sigma^\gamma_{\rho\sigma} = p^\sigma \Sigma^\gamma_{\rho\sigma} = 0 \tag{2.73}$$

which in turn imply that $\Sigma^\gamma_{\rho\sigma}$ has the general structure

$$\Sigma^\gamma_{\rho\sigma} = (g_{\rho\sigma} p^2 - p_\rho p_\sigma) \Pi^\gamma(p^2). \tag{2.74}$$

Thus, *provided* $\Pi^\gamma(p^2)$ *has no pole at* $p^2 = 0$, there can be no term like $g_{\rho\sigma} M^2$ in $\Sigma^\gamma_{\rho\sigma}$ (the significance of the proviso will become clear in Section 4.2 when we discuss the U(1) Higgs model, in which a photon mass *is* generated). The quantity Π^γ (the 'photon vacuum polarisation') is then dimensionless, and contains at worst terms in $\ln M^2$.

Can we absorb these $\ln M^2$ terms by adjusting some M-dependence of $\mathcal{L}_{\text{QED}}$? The answer is yes: although we have no '$A_\mu A^\mu$' term, we do have the Maxwell term which in momentum space was given in (2.20). Comparing (2.20) with (2.74), we see that this $\ln M^2$ behaviour in $\Sigma^\gamma_{\rho\sigma}$ is like the $\slashed{p} \ln M^2$ behaviour in $\Sigma(M)$ of (2.60): as in the latter case, the former $\ln M^2$ behaviour can be absorbed into a renormalisation of a field, in this case the A_μ (photon) field:

$$A_\mu(x) \to A_{\mu,M}(x) = Z_3^{\frac{1}{2}}(M) A_{\mu,r}(x), \tag{2.75}$$

where (c.f. (2.66)) $Z_3 = 1 + O(e^2 \ln M^2)$. Once again there is a question of how much else of $\Pi^\gamma(p^2)$ we might absorb into $Z_3^{1/2}$; we shall return to this in a moment.

Before continuing with the interpretation of Π^γ, we turn to the last loop of Figure 6, the vertex correction Figure 6(c). Without writing it out in full (see (2.96) below), we can see by counting powers of k that it is dimensionless, and contains a new leading M-behaviour of the $\ln M^2$ type. However, we have one last card to play! We have not yet considered the interaction term in $\mathcal{L}_{\text{QED}}(M)$, namely

$$-e(M)\bar\psi_M(x)\slashed{A}_M(x)\psi_M(x). \tag{2.76}$$

We have used up the possible renormalisations of the ψ and A fields, but we are still free to explore the M-dependence of $e(M)$. Indeed, with an eye on (2.64) and (2.75), if we introduce a renormalised 'e' by

$$e(M) = Z_1(M) e_r / Z_2(M) Z_3^{\frac{1}{2}}(M), \tag{2.77}$$

we can rewrite (2.76) as

$$-e_r \bar\psi_r \slashed{A}_r \psi_r - e_r(Z_1 - 1)\bar\psi_r \slashed{A}_r \psi_r. \tag{2.78}$$

The first term in (2.78) generates the 'real' $e\text{-}e\text{-}\gamma$ vertex, while $Z_1(M)$ in the second can be chosen to cancel the $\ln M^2$ piece of Figure 6(c).

When one goes through the calculation of Figure 6(c) one discovers a remarkable coincidence, that the $\ln M^2$ behaviour of $Z_1(M)$ which is required for this cancellation to work is exactly the same as that of $Z_2(M)$, equation (2.66). This is actually a general result—in fact a *Ward identity* (Ryder 1985), which is (another) consequence of gauge invariance. Looking at (2.77) we see that this has the important physical effect that the corrections to the charge parameter are associated only with the photon factor $Z_3^{1/2}$, and are independent of the fermion wavefunction and vertex renormalisations. This means that for *all* charged fermions a *universal* charge renormalisation by a factor $Z_3^{1/2}$ as in (2.77) is sufficient. This is absolutely essential to preserve (under renormalisation) the universality of the replacement $\partial_\mu \to D_\mu = \partial_\mu + ieA_\mu$ (c.f. (2.50)), regardless of what particular particle with charge e (e^+, μ^+, p,...) is involved. Indeed, $e(M)A_{\mu,M}(x)$ is the same as $e_r A_{\mu,r}(x)$ if use is made of $e(M) = e_r/Z_3^{1/2}(M)$ and of (2.75).

We have tried to make plausible the idea that, at least at the one loop level, we can consistently suppose that the parameters and fields in $\mathcal{L}_{\text{QED}}(M)$ depend on M in such a way that all the $\ln M^2$ behaviour arising from loops can be absorbed, leaving physical amplitudes weakly dependent on M^2, and independent of M^2 as $M^2 \to \infty$. It is now time to gather together what we have learned about all the graphs in Figure 5, and examine the resulting 'physical amplitude' more closely.

We first note that when the external fermion-antifermion lines are stuck back onto Figure 6(b), we recreate the corresponding term in Figure 5, which we show separately in Figure 8.

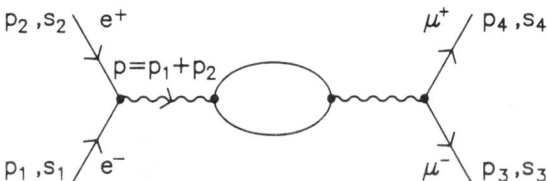

Figure 8.

The resulting amplitude will be (2.69) multiplied by factors of

$$\bar{u}(p_3, s_3)ie\gamma^\nu v(p_4, s_4) \quad \text{and} \quad \bar{v}(p_2, s_2)ie\gamma^\mu u(p_1, s_1) \tag{2.79}$$

at each end (c.f. (2.29)). Referring then to (2.74), the $p_\rho p_\sigma$ term in $\Sigma^\gamma_{\rho\sigma}$ will appear as

$$\bar{u}(p_3, s_3)ie\slashed{p}v(p_4, s_4)\bar{v}(p_2, s_2)ie\slashed{p}u(p_1, s_1), \tag{2.80}$$

where the momentum p flowing along the photon line is $p_1 + p_2$. So the second '$\bar{v}u$' part in (2.79), for example, is

$$ie\bar{v}(p_2, s_2)(\slashed{p}_1 + \slashed{p}_2)u(p_1, s_1) \tag{2.81}$$
$$= ie\bar{v}(p_2, s_2)(m - m)u(p_1, s_1) = 0, \tag{2.82}$$

using (2.6) and (2.7). Actually, this is a general result, since the photon is always coupled to a conserved current (c.f. (2.37) which becomes $p_\mu j^\mu = 0$ in momentum space, where p^μ is effectively the photon momentum). Thus we need only bother with the $g_{\rho\sigma}$ part of (2.74), which makes (2.69) much easier to handle.

Next, consider the last two 'box' graphs on Figure 5. It turns out that their sum is actually *finite* as $M^2 \to \infty$, so that for finite M they have only a weak dependence on M. This is of crucial significance, since it seems intuitively clear that, if they had contained a large M-dependent piece, it would *not* have been possible to absorb it into a redefinition of the parameters and fields in $\mathcal{L}_{\text{QED}}(M)$; it would be a new divergence (as $M \to \infty$), and we have no cards left to deal with it. Indeed, we would have to absorb it into a new term of the form

$$G\bar{\psi}\psi\bar{\psi}\psi \tag{2.83}$$

which would, in lowest order in G, correspond to an interaction in which the four fermions interact at a point (Figure 9). But this is no longer QED! We reserve further discussion of (2.83)—which is conventionally classed as a 'non-renormalisable interaction'—for later (in connection with the Fermi theory), noting here only that G would have to have *dimensions* of (mass)$^{-2}$, since $\bar{\psi}\psi$ has dimension (mass)^3, and the action $\int d^4x\, \mathcal{L}$ is dimensionless.

Figure 9.

So, remembering that the large-M dependences associated with the vertex part (Figure 6(c)) and the fermion self-energy (Figure 6(a)) cancel each other, large-M renormalisation for Figure 5 really only involves the vacuum polarisation term (2.69) which we have to add to the 'Born term' (2.29). Suppressing (as in (2.69)) the '$\bar{v}u\bar{v}u$' external leg factors, but keeping the overall $e^2(M)$ factor, and ignoring the $p_\rho p_\sigma$ terms in (2.74), we obtain for the sum of the two terms

$$-ig^{\mu\nu}\frac{e^2(M)}{p^2}\left(1 - \Pi^\gamma(p^2)\right) \tag{2.84}$$

$$\approx \frac{-ig^{\mu\nu}}{p^2}\frac{e_r^2}{Z_3(M)(1+\Pi^\gamma(p^2))} \tag{2.85}$$

$$\approx \frac{-ig^{\mu\nu}e_r^2}{p^2(Z_3(M)+\Pi^\gamma(p^2))} \tag{2.86}$$

where in (2.85) we have used (2.77) with $Z_1 = Z_2$, and in (2.85) and (2.86) the fact that $Z_3 = 1 + O(e^2)$ and $\Pi^\gamma = O(e^2)$, so that the replacements are all consistent to $O(e^2)$. (It does not matter what 'e' we use in Z_3 and Π^γ, since any difference would only show up in the next order). From (2.86) we learn again (as we already knew from (2.74) and (2.75)) that we need to choose Z_3 to absorb the large M-dependence of $\Pi^\gamma(p^2)$. However, (2.86) shows us that different choices of the *finite* part of Z_3 will amount to varying the 'effective charge'. So we have to decide, finally, what we are going to *mean* by 'the charge e'; i.e. the 'physical' one such that $e^2/4\pi \approx \frac{1}{137}$. The answer is that *this* 'charge' is defined via the limit of (2.86) as the photon 4-momentum p tends to zero; namely, if we define

$$Z_3 = 1 - \Pi^\gamma(0) \tag{2.87}$$

then

$$\frac{-ig^{\mu\nu}e_r^2}{Z_3 p^2}\left(1 - \Pi^\gamma(p^2)\right) \approx -ig^{\mu\nu}\frac{e_r^2}{p^2}\left(1 + \Pi^\gamma(0) - \Pi^\gamma(p^2)\right) \tag{2.88}$$

and e_r^2 can be identified as *the* e (dropping the subscript r) such that $e^2/4\pi \approx \frac{1}{137}$.

The result (2.88) can therefore be written as

$$\frac{-ig^{\mu\nu}}{p^2}e^2\left(1 - \hat{\Pi}^\gamma(p^2)\right) \tag{2.89}$$

where now the 'subtracted vacuum polarisation' $\hat{\Pi}^\gamma$ is entirely finite as $M \to \infty$, and vanishes as $p^2 \to 0$. This quantity is of vital significance for high precision electroweak experiments, and much else besides as we shall see. We therefore end this section with some discussion of its properties (for $M \to \infty$):–

1. We have introduced $\hat{\Pi}^\gamma(p^2)$ via the process shown in Figure 8, which contributes to $e^+e^- \to \mu^+\mu^-$ in the annihilation channel. In this case, p^2 (the invariant mass $(p_1 + p_2)^2$ of the e^+e^- pair) is necessarily greater than zero, indeed greater than $4m_\mu^2$. For $p^2 > 4m^2$ (m = electron mass), $\hat{\Pi}^\gamma(p^2)$ develops an imaginary part, which is proportional to the 'phase space factor' ($\sim (p^2 - 4m^2)^{\frac{1}{2}}$) for the e^+e^- pair. However, $\hat{\Pi}^\gamma$ is well-defined for all p^2—for example, it will appear with $p^2 < 0$ in the e^-e^- scattering diagram shown in Figure 10.

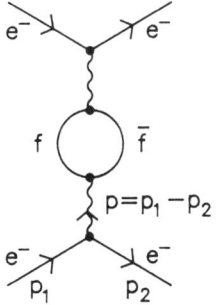

Figure 10.

2. All charged $f\bar{f}$ pairs can contribute to $\Pi^\gamma(p^2)$, and their contributions simply add up. Thus for instance we can immediately consider $\Pi^\gamma(p^2, m_e) + \Pi^\gamma(p^2, m_\mu)$, and so on.

3. For $|p^2| \ll m_f^2$, where m_f is the mass of the particular fermion in $\Pi^\gamma(p^2, m_f)$, one finds

$$\Pi^\gamma(p^2, m_f) \approx \frac{\alpha}{3\pi} \frac{p^2}{5m_f^2} + \text{higher powers of } \left(\frac{p^2}{m_f^2}\right) \qquad (2.90)$$

which illustrates the notion of 'decoupling': the contribution of heavier particles is suppressed relative to that of the lighter ones by a factor $(m_{\text{light}}/m_{\text{heavy}})^2$, which is about 10^{-5} for the electron and muon, for instance.

4. For $|p^2| \gg m_f^2$, one has

$$\text{Re } \hat{\Pi}^\gamma(p^2, m_f) \approx \frac{\alpha}{3\pi}\left(\frac{5}{3} - \ln\left(\frac{|p^2|}{m_f^2}\right)\right) + O\left(\frac{1}{|p^2|}\right). \qquad (2.91)$$

From (2.91) we see that as $|p^2|$ continues to grow, the product $\alpha \ln(|p^2|/m_f^2)$ will not remain small, and our perturbation theory will fail. It is a most important result (following from 'renormalisation group' arguments, which we cannot enter into here) that the effect of all large $\ln(|p^2|/m_f^2)$ terms is correctly given by replacing (2.89) by

$$\frac{-ig^{\mu\nu}}{p^2} \frac{e^2}{\left(1 + \hat{\Pi}^\gamma(p^2)\right)} \qquad (2.92)$$

with

$$\text{Re } \hat{\Pi}^\gamma(p^2) \approx -\sum_f Q_f^2 \frac{\alpha}{3\pi} \ln\left(\frac{|p^2|}{m_f^2}\right) \qquad (2.93)$$

for each $f\bar{f}$ pair, where Q_f is the charge of the fermion in units of e. Equation (2.92) allows us to introduce the idea of the *running charge* $e^2(p^2)$, where

$$e^2(p^2) = \frac{e^2}{1 + \hat{\Pi}^\gamma(p^2)}, \qquad (2.94)$$

which measures the change in e^2 away from the $p^2 = 0$ value, as p^2 varies. In current experiments this shift can be quite substantial: for Z^0 physics we are interested in $p^2 = m_Z^2$, for which one finds (Consoli *et al.* 1989)

$$\text{Re } \hat{\Pi}^\gamma(m_Z^2) \approx -0.060 \pm 0.001, \qquad (2.95)$$

corresponding to an 'α at m_Z^2' of about $1/128.8$ rather than $1/137$ (see Section 5.5 below).

An interesting aspect of (2.94) is the behaviour implied at large $|p^2|$. Evidently, the running charge increases with $|p^2|$: as we go to shorter distances, the interaction strength gets larger, while at larger distances the charge decreases. This is similar to what happens in a polarisable dielectric medium; here, the medium is the vacuum, and it is the e^+e^- pairs in the loop of Figure 6(b) which are 'screening' the charge (see Section 9.3 of Aitchison and Hey 1989). In QCD, as we shall see, the effect of the gluon self-interactions is exactly the opposite.

2.4 $(g-2)$, effective Lagrangians, and non-renormalisability

We end this section on QED by discussing briefly two further one-loop calculations which (like the box graph contribution in Figure 5) contain no $\ln M^2$ pieces, and are finite as they stand, even if $M \to \infty$. The first is the contribution of the vertex correction of Figure 6(c) to the anomalous magnetic moment of the fermion (e^- or μ^-, for example); the second is the γ-γ scattering process of Figure 11.

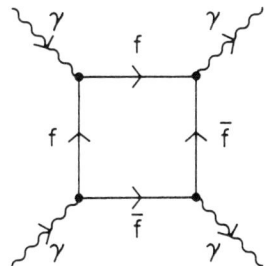

Figure 11.

The vertex correction modifies the lowest order vertex $-ie\gamma^\mu$ by the addition of $-ie\Gamma^\mu$ where (see Figure 12)

$$-ie\Gamma^\mu = \int \frac{d^4k}{(2\pi)^4} (-ie\gamma^\nu) \cdot \frac{-ig_{\nu\lambda}}{k^2} \cdot \frac{i}{\slashed{p}_2 - \slashed{k} - m} \cdot (-ie\gamma^\mu) \cdot \frac{i}{\slashed{p}_1 - \slashed{k} - m} \cdot (-ie\gamma^\lambda). \quad (2.96)$$

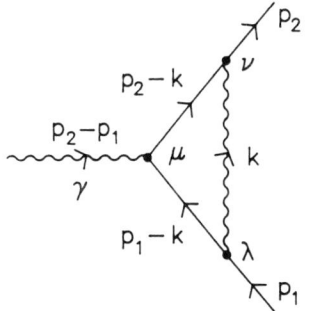

Figure 12.

After rationalising the fermion propagators, inspection shows that (2.96) contains a $\ln M^2$ divergence associated with a $\slashed{k}\slashed{k}$ term in the numerator, while the part without these numerator factors is convergent. The whole '$\slashed{k}\slashed{k}$' piece is proportional to γ^μ, and can be absorbed into the multiplicative factor Z_1, introduced earlier. In any physical process, the vertex will be sandwiched between spinors $\bar{u}(p_2)\ldots u(p_1)$, and the result of the calculation of the convergent part yields

$$-ie\bar{u}(p_2)\Gamma_c^\mu u(p_1) = -ie\bar{u}(p_2)\frac{\alpha}{2\pi} i\sigma^{\mu\nu} \frac{(p_2-p_1)_\nu}{2m} u(p_1), \quad (2.97)$$

where $\sigma_{\mu\nu} = \frac{1}{2}i(\gamma_\mu\gamma_\nu - \gamma_\nu\gamma_\mu)$. To interpret this result, we write the lowest order vertex contribution in the form

$$-ie\bar{u}(p_2)\gamma^\mu u(p_1) = -ie\bar{u}(p_2)\left[\frac{(p_1+p_2)^\mu}{2m} + i\sigma^{\mu\nu}\frac{(p_2-p_1)_\nu}{2m}\right]u(p_1), \qquad (2.98)$$

using the Dirac equation (2.4) and

$$\gamma^\mu\gamma^\nu + \gamma^\nu\gamma^\mu = 2g^{\mu\nu}. \qquad (2.99)$$

It is a familiar exercise to show that the second term in (2.98) is associated with the magnetic moment of the particle, and implies $g = 2$ (the Dirac value, of course). The effect of the correction (2.97) is therefore to modify the $g = 2$ result, producing a so-called 'anomalous' (*i.e.* non-Dirac) contribution: this is usually expressed in terms of the quantity 'a' defined by

$$a = \frac{1}{2}(g-2) \qquad (2.100)$$

so that (2.97) yields

$$a = \alpha/2\pi, \qquad (2.101)$$

as first obtained by Schwinger (1948).

There are a number of further comments to be made. First, suppose that the contribution (2.97) had involved a $\ln M^2$ piece. Then, as with the box graphs of Figure 5, we would have had to introduce a new term, not in the original $\mathcal{L}_{\text{QED}}$, in order to cancel it (note that the correction Z_1, which would contribute via (2.98) to a $\sigma_{\mu\nu}$-type term, is already 'used up'). This new term would have the form

$$G_a\bar{\psi}(x)\sigma_{\mu\nu}\psi(x)(\partial^\mu A^\nu(x) - \partial^\nu A^\mu(x)), \qquad (2.102)$$

as can be seen by considering the lowest order matrix element of (2.102). Unlike a photon mass term, (2.102) is gauge invariant. But it is a 'non-renormalisable' interaction, the meaning (and consequences) of which we shall return to later. Note, meanwhile, that G_a has to have dimension $(\text{mass})^{-1}$.

Secondly, the anomalous contribution a for the electron is one of the most accurately measured numbers in physics (PDP 1992):

$$a_e = 1,159,652,193(10) \times 10^{-12}, \qquad (2.103)$$

and for the muon the value is (PDP 1992)

$$a_\mu = 11,659,230(84) \times 10^{-10}. \qquad (2.104)$$

The quantity $\alpha/2\pi$ only gets us as far as the first two or three digits in (2.103) and (2.104). The calculation of higher and higher order corrections (in α) to 'a' is a continuing challenge to theorists. So far no discrepancy has been established. Nevertheless, the accuracy of (2.103) and (2.104) raises several interesting questions: for example, now that we know that QED belongs inside the electroweak theory (see below), and cannot be regarded as existing on its own apart from the weak interactions, do weak interactions make a contribution to 'a', and if so how big is it?

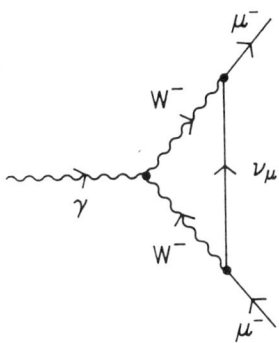

Figure 13.

Indeed, the diagram of Figure 13 does contribute an amount (Jackiw and Weinberg 1972)

$$a_\mu^W = \frac{Gm_\mu^2}{6\pi^2\sqrt{2}} \left\{ \frac{5}{2} + 3(g_W - 2)\ln\left(\frac{M}{M_W}\right) \right\} \qquad (2.105)$$

where G is the Fermi constant, g_W is the g-factor for the W (see Section 3.2 below) and M is an ultraviolet cut-off. (The similar quantity a_e^W involves m_e^2 rather than m_μ^2 and is therefore very much smaller). We shall learn in Section 3.2 that the electroweak theory predicts $g_W = 2$, so that the new divergence in (2.105) is actually absent! (This begins to illustrate the intricate interdependence of these interactions.) In that case the magnitude of (2.105) is about 4×10^{-9}, which is within the margin of experimental error in (2.104), but getting close to being observable. An experiment at the Brookhaven AGS (May 1988; Hughes 1989) is expected to achieve an accuracy of order 4×10^{-10}. Of course, other contributions (from the Z^0 and the Higgs particle) also have to be included for consistency, as was first done by Jackiw and Weinberg (1972).

Returning to (2.105), we shall see later that the electroweak theory relates the Fermi constant G to the W mass via a relation of the form $G \sim e^2/M_W^2$, so that (2.105) gives $a_\mu^W \sim \alpha(m_\mu^2/M_W^2)$. This is an important order of magnitude estimate. Compared with (2.101), it is down by a factor of m_μ^2/M_W^2. The m_μ^2 factor has its origin in the weak $(1 - \gamma_5)$ couplings of the W (see below), which for massive leptons introduce suppression factors proportional to the mass (as in the suppression of $\pi^- \to e^-\bar{\nu}_e$ relative to $\pi^- \to \mu^-\bar{\nu}_\mu$); this must clearly be balanced by a (mass)$^{-2}$ factor, which can be interpreted as 'setting the scale of the new physics'—in this case the electroweak scale $\sim M_W$. We are meeting here for the first time the important idea that *precision measurements at low energies can test for 'new physics' at high mass scales*. For a_μ this idea is pursued in a recent paper by Méry et al. (1990), for example.

Finally, we turn to the γ-γ process of Figure 11. This amplitude must have the form

$$\epsilon_3^{\lambda*}\epsilon_4^{\sigma*}\Pi_{\mu\nu\lambda\sigma}\epsilon_1^\mu\epsilon_2^\nu \qquad (2.106)$$

so that this graph contributes to a fourth rank tensor (the vacuum polarisation tensor $\Sigma_{\mu\nu}^\gamma$ had only external two photons). Simple counting of powers of k in the integral of $\Pi_{\mu\nu\lambda\sigma}$ would imply that it contains a $\ln M^2$ factor—again one that cannot be cancelled from within $\mathcal{L}_{\text{QED}}(M)$, since the latter contains no 'fundamental' $\gamma\gamma \to \gamma\gamma$ vertex, Figure 14. It is certainly possible to construct such 'A_μ^4' vertices—even gauge invariant ones; if we go no higher than four derivatives we can construct two:

$$G_\gamma^{(1)}(F_{\mu\nu}F^{\mu\nu})^2 \quad \text{and} \quad G_\gamma^{(2)}(F_{\mu\nu}\tilde{F}^{\mu\nu})^2 \tag{2.107}$$

where $\tilde{F}^{\mu\nu} = \epsilon^{\mu\nu\lambda\sigma}F_{\lambda\sigma}$. We have met '$F.F$' before in (2.20); it can be written in terms of the $\underline{E}$ and $\underline{B}$ fields as $2(\underline{B}^2 - \underline{E}^2)$. The $F.\tilde{F}$ product equals $4\underline{B}.\underline{E}$, which is a *pseudoscalar*, and would therefore (in combination with $F.F$) violate parity. However, the squared form $(\underline{B}.\underline{E})^2$ is perfectly acceptable. Reckoning up dimensions, the G_γ's in (2.107) would have dimension (mass)$^{-4}$. The problem with (2.107), however, is that—like (2.83) and (2.102)—they are 'non-renormalisable' (see further below).

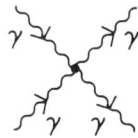

Figure 14.

Once again, gauge invariance saves us. As with $\Sigma_{\mu\nu}^\gamma$, it can be shown (Jauch and Rohrlich 1976) that imposing gauge invariance reduces the apparent divergence of $\Pi_{\mu\nu\lambda\sigma}$, in this case causing the coefficient of the $\ln M^2$ term to vanish, leaving a convergent expression which is clearly of order α^2. Although the exact expression is complicated, it simplifies greatly in the 'low-energy' limit, in which the γ-γ c.m.s. momentum is much smaller than m, the mass of the fermion in the loop. Such a limit corresponds precisely to the 'low number of derivatives' approach of (2.107) (p^μ is the momentum space version of $i\partial^\mu$). In fact, the leading contribution for $p \ll m$ can be expressed as an 'effective Lagrangian' (Jauch and Rohrlich 1976)

$$\mathcal{L}_{\text{eff},4\gamma} = \frac{2\alpha^2}{45m^4}\left(\left(\frac{1}{2}F.F\right)^2 + 7\left(\frac{1}{8}F.\tilde{F}\right)^2\right), \tag{2.108}$$

in the sense that the matrix element of $i\int \mathcal{L}_{\text{eff},4\gamma}\, d^4x$ between 2γ states (c.f. (2.26)) reproduces the amplitude correctly. We note that the coefficients in (2.108) are indeed of dimension (mass)$^{-4}$ as expected. As in the case of the weak contributions to a_μ, it is these denominator factors which 'set the scale'—of the momentum expansion, in this case.

We close with a few (initial) comments about the 'non-renormalisability' of terms like (2.102) and (2.107), supposing they were to be regarded as 'fundamental', and deserving of a place in $\mathcal{L}_{\text{QED}}$. Clearly, they give rise to no trouble (large M^2 behaviour from integration loops) at tree level. Problems begin to appear when we consider the behaviour of a loop which contains either (2.102) or (2.107) as an internal part, so that they function as 'effective vertices' (e-e-γ for (2.102), γ-γ-γ-γ for (2.107)). Both involve positive powers of the photon momentum, as compared to the situation for the dimensionless coupling of $\mathcal{L}_{\text{int}}$ of (2.15). Thus they seem very likely to cause new divergences in the 'primitive' amplitudes we have already considered in Figure 6, and which we have so carefully freed of the large-M^2 dependences (this causing of new divergences in these diagrams is exactly what higher order processes in $\mathcal{L}_{\rm QED}$ do not do, and is an important part of 'renormalisability'). Also, they will generate new types of divergences in amplitudes other than those of Figure 6. For example, if we consider the 4-γ vertex of (2.107) at the one loop level (Figure 15), and add it to the tree level process of Figure 14, we obtain something like

$$\sim \frac{p^4}{M_s^4}\left(1 + \frac{1}{M_s^4}\int\frac{d^4k}{k^2k^2}k^4\right) \tag{2.109}$$

where M_s is the 'scale-setting' mass, required on dimensional grounds, and the k^4 in the numerator also follows from dimensional analysis. The integral in (2.109) diverges as M^4 where M is the cut off scale—a very sharp dependence on high-M physics.

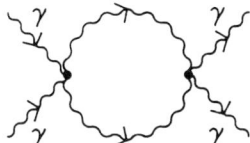

Figure 15.

Now we can of course imagine absorbing such an infinity into the 'original' vertex of Figure 14—actually there will be two independent such vertices, as in (2.107). That is, the numerical coefficients of these vertices will be regarded as parameters whose M-dependence can cancel the M^4 divergences, leaving (let us suppose) no strong M-dependence in the sum of Figures 14 and 15. The theory will now contain two new parameters, namely the finite parts of these two vertex parameters after the large-M^2 cancellations; the value of these parameters will have to be fixed by experiment.

This may not seem too bad—but unfortunately there is worse to come. Further analysis shows that there is no end to this process—higher orders generate ever new divergences in more and more complicated diagrams, involving arbitrarily many photons and fermions, and we ultimately need to introduce an infinite number of phenomenological parameters. This is what is traditionally called a 'non-renormalisable' theory—and it does sound pretty bad!

However, let us take another look at (2.109). The crucial point is that this kind of non-renormalisable interaction involves powers of some mass (here M_s) in the denominator—it is this which 'causes' the bad k^2 behaviour in (2.109). Yet if M_s is itself actually quite 'large' (compared with 'the current energy scale', for instance), the correction integral in (2.109) may be quite small for cut-offs $M \ll M_s$, and for energies $E \ll M_s$. Thus 'non-renormalisable' interactions really only lead to the 'infinite number of constants' disease when we ask them to make sense at cut-offs or energies which are of order their 'intrinsic' mass scale. In QED itself, the only natural mass scale is the mass of the electron, which is really why we would *not* want to deal with a 'fundamental' interaction of the type (2.102) or (2.107) in QED.

In fact, the situation is rather better than might be thought, in the sense that we can even consider $M \gg M_s$. Consider again (2.108) for definiteness; used at tree level this vertex gives the $O(p/m)^4$ contribution to $\gamma\gamma$ scattering. If we are content to work at this order in (p/m), all $\gamma\gamma$ processes (e.g. $\gamma\gamma \to 4\gamma$, $\gamma\gamma \to 6\gamma$, ...!) are correctly calculated by using (2.108) at tree level (Weinberg 1993; see also Weinberg 1979). Further, if we want to go to order $(p/m)^8$, we can get rid of large-M^2 behaviour in *all* processes involving just external γ's at the expense of introducing just the two counter terms given in (2.108); *i.e.* two new physical parameters. Of course, for QED itself the full theory copes with all processes and all energies with *no* further arbitrary constants. It's a fine theory! (But actually incomplete, as we now know).

In general, then, for a 'non-renormalisable' interaction with an intrinsic scale M_s, we can arrange for it to be insensitive to the cut-off M if we restrict ourselves to a given order of (p/M_s). At each higher order, more parameters will be needed to define the theory—and as the energy scale approaches M_s we shall be back with the 'infinite number of arbitrary constants' problem. Before then, however, we can have interesting physics, which may even be able to tell us something about what we are in for as we go up in energy.

3 Global and local non-Abelian symmetries

After this review of the dynamics of QED, it is time to introduce the gauge theories which describe the weak and strong forces between quarks and leptons. These both involve a somewhat more complicated type of local symmetry (called 'non-Abelian') than the simple Abelian U(1) symmetry of QED. Nevertheless, the approach followed in Section 2.2 can be generalised rather straightforwardly. We shall not give a general treatment however, preferring to limit the discussion to the particular ingredients of the Standard Model.

3.1 Global non-Abelian symmetry

Consider the Lagrangian for two free fermions of the same mass $m_1 = m_2 = m$

$$\mathcal{L}_2 = \bar{\psi}_1(i\slashed{\partial} - m)\psi_1 + \bar{\psi}_2(i\slashed{\partial} - m)\psi_2 ; \tag{3.1}$$

in terms of the 'doublet' field

$$\psi = \begin{pmatrix} \psi_1 \\ \psi_2 \end{pmatrix} \tag{3.2}$$

it can easily be rewritten as

$$\mathcal{L}_2 = \bar{\psi}(i\slashed{\partial} - m)\psi . \tag{3.3}$$

Note that although (3.3) looks formally like the single-field $\mathcal{L}_\psi$ of (2.30), it is of course quite different physically, representing two different sorts of particle (*e.g.* up and down quarks, and their antiparticles). Nevertheless, (3.3) is invariant under a symmetry rather like (2.31), namely the 2×2 unitary transformation

$$\psi \to \psi' = U\psi, \quad UU^\dagger = U^\dagger U = 1. \tag{3.4}$$

The U in (3.4) is a 2×2 matrix of numbers (not field operators) acting on the 2 components of ψ in (3.2), and they commute with the Dirac γ's. Such unitary 2×2 U's form a group, U(2). Since U in (3.4) does not involve x, we call (3.4) a global symmetry.

In general, two U's do not commute with each other, and it is called a non-Abelian symmetry.

From elementary properties of determinants we have

$$\det UU^\dagger = \det U.\det U^\dagger = \det U.\det U^* = |\det U|^2 = 1 \tag{3.5}$$

so that $\det U = e^{-2i\alpha}$, say. We can therefore write

$$U = e^{-i\alpha}\tilde{U} \tag{3.6}$$

where $\tilde{U}$ has determinant +1. Matrices of the form $\tilde{U}$ form the SU(2) group, where the S just means they have unit determinant. The phase factor in (3.6) corresponds to a simultaneous U(1) transformation of ψ_1 and ψ_2 (with the same phase angle) and leads, as in Section 2.2, to a conservation law of the total number of '1' particles and '2' particles. (For quarks this would be part of baryon number conservation). The new physics is contained in the $\tilde{U}$ part.

Groups such as SU(2) (and, later, SU(3)) have the important feature that their physically important properties can be found by studying infinitesimal transformations, of the form (*c.f.* (2.32))

$$\tilde{U} = 1 - i\xi \tag{3.7}$$

where ξ is a 2×2 matrix with small entries. The condition $\det U = 1$ gives $\mathrm{Tr}\xi = 0$ (neglecting terms of order ξ^2), while $\tilde{U}\tilde{U}^\dagger = 1$ reduces to $\xi = \xi^\dagger$. So ξ is a Hermitian traceless matrix. Such a thing depends on only three real parameters and can be written as

$$\xi = \underline{\epsilon}\cdot\underline{\tau}/2 \tag{3.8}$$

where $\underline{\epsilon} = (\epsilon_1, \epsilon_2, \epsilon_3)$ are the three parameters, and $\underline{\tau} = (\tau_1, \tau_2, \tau_3)$ are the Pauli matrices. Thus an infinitesimal SU(2) transformation on the doublet ψ is

$$\psi \to \psi' = (1 - i\underline{\epsilon}\cdot\underline{\tau}/2)\psi. \tag{3.9}$$

This should be compared with (2.32), from which it is clear that the 'ϵ' in that case becomes a *matrix* in (3.9). The form for a finite SU(2) transformation is

$$\psi \to \psi' = e^{-i\underline{\alpha}\cdot\underline{\tau}/2}\psi \tag{3.10}$$

which generalises (2.31) (note that for a matrix A, $\exp A = 1 + A + A^2/2! + \cdots$).

Since (3.9) or (3.10) are invariances of $\mathcal{L}_2$ we expect an associated conservation law. Indeed, since we have three independent transformations (using each of ϵ_i in turn) we expect three conservation laws. Following a generalisation of (2.33)–(2.36) one finds that the three quantities $T_1^\mu(x)$, $T_2^\mu(x)$, $T_3^\mu(x)$ defined by (*c.f.* (2.38))

$$T_i^\mu(x) = \bar{\psi}(x)(\tau_i/2)\gamma^\mu\psi(x) \tag{3.11}$$

satisfy

$$\partial_\mu T_i^\mu(x) = 0 \tag{3.12}$$

and are therefore symmetry currents. The corresponding 'charges'

$$T_i = \int \psi^\dagger(x)\frac{\tau_i}{2}\psi(x)\,d^3x \tag{3.13}$$

are conserved. These are the (field theoretic) 'isospin' operators, which have the very interesting property

$$[T_i, T_j] = i\epsilon_{ijk} T_k \tag{3.14}$$

as can be explicitly checked from (3.13) (using the proper commutation relations for the ψ fields). The relations (3.14) are of course exactly the commutation relations of the familiar angular momentum operators, which is why the name iso*spin* was coined; (3.14) is called the 'SU(2) algebra'.

In thinking about more complicated SU(2) multiplets than doublets (which we shan't need to do much) this angular momentum analogy is very helpful. The essential step is to find larger matrices which satisfy commutation relations of the form (3.14). For example, the three 3×3 matrices t_1, t_2 and t_3, defined by

$$(t_i)_{jk} = -i\epsilon_{ijk} \tag{3.15}$$

satisfy $[t_i, t_j] = i\epsilon_{ijk} t_k$. Then if we consider a triplet of three real degenerate fields (bosonic, say)

$$\underline{\phi} = \begin{pmatrix} \phi_1 \\ \phi_2 \\ \phi_3 \end{pmatrix} \tag{3.16}$$

with Lagrangian

$$\mathcal{L}_B = \frac{1}{2}\partial_\mu \underline{\phi} \cdot \partial^\mu \underline{\phi} - \frac{1}{2} m^2 \underline{\phi} \cdot \underline{\phi}, \tag{3.17}$$

$\mathcal{L}_B$ is invariant under

$$\underline{\phi} \to \underline{\phi}' = (1 - i\underline{\epsilon} \cdot \underline{t})\underline{\phi}. \tag{3.18}$$

Using (3.15), (3.18) is equivalent to

$$\underline{\phi}' = \underline{\phi} + \underline{\epsilon} \times \underline{\phi} \tag{3.19}$$

which should be familiar as the 'infinitesimal rotation' of an ordinary vector.

The SU(2) transformation of (3.4) can be generalised to the case of *three* degenerate fermion fields. If $\mathcal{L}_3$ is (3.1) with the addition of $\bar{\psi}_3(i\not{\partial} - m)\psi_3$, it too can can be written as in (3.3) where now

$$\psi = \begin{pmatrix} \psi_1 \\ \psi_2 \\ \psi_3 \end{pmatrix}. \tag{3.20}$$

Note particularly that unlike the ϕ's in (3.16), the ψ's in (3.20) are complex: each ψ_i contains c_i and $d_i^\dagger$ operators as in (2.5). $\mathcal{L}_3$ is invariant under $\psi \to \psi' = U\psi$ where U is now an x-independent 3×3 unitary matrix. Extracting the overall phase again, we are left with a global SU(3) transformation. An infinitesimal SU(3) matrix has the form

$$\tilde{U} = 1 - i\chi \tag{3.21}$$

where χ is a Hermitean traceless 3×3 matrix. Such a χ involves eight parameters and can be written as

$$\chi = \underline{\eta} \cdot \underline{\lambda}/2 \tag{3.22}$$

where $\underline{\eta} = (\eta_1, \ldots \eta_8)$ are the arbitrary parameters and the eight $\underline{\lambda}$'s are 3×3 Hermitean traceless matrices generalising the three τ's. They obey the commutation relations (the 'SU(3) algebra')

$$\left[\frac{\lambda_a}{2}, \frac{\lambda_b}{2}\right] = if_{abc}\frac{\lambda_c}{2} \tag{3.23}$$

where the f_{abc} are numbers characteristic of SU(3) (a, b, c all run from 1 to 8). If ψ_1, ψ_2, ψ_3 are taken to be the u, d, s quarks, this global SU(3) symmetry would be the SU(3) of strong interaction flavour symmetry (which however is not exact as m_u, m_d and m_s are not equal). Similarly, if we take 1, 2, 3 to be colour indices we have the exact SU(3)$_c$ colour symmetry of QCD, which we shall shortly see is a local symmetry. The currents corresponding to the SU(3) symmetry of $\mathcal{L}_3$ are (c.f. (3.11))

$$G_a^\mu(x) = \bar{\psi}(x)(\lambda_a/2)\gamma^\mu\psi(x) \tag{3.24}$$

and the associated eight 'charges'

$$G_a = \int \psi^\dagger(x)(\lambda_a/2)\psi(x)\, d^3x \tag{3.25}$$

generalise the three isospin operators, and obey the commutation relations

$$[G_a, G_b] = if_{abc}G_c. \tag{3.26}$$

As in the case of SU(2), larger multiplets are possible too. The key requirement is to find matrices which satisfy (3.26), since these commutation relations effectively define the group. For SU(3), the only larger multiplet in which we shall be interested is the octet, **8**, which is analogous to the triplet of SU(2). The matrices for the **8** are defined analogously to the t's of (3.15), namely $(F_a)_{bc} = -if_{abc}$ where the f's are as in (3.26). Notice that since there are eight 'charges' G_a, and all the indices a, b, c in (3.26) run from 1 to 8, the eight matrices F_a are each 8×8. In the same way, the three matrices t_i of (3.15) are each 3×3, since there are three SU(2) charges. This kind of pattern can be extended to arbitrary SU(N); the 'representation' in which the matrices are equal (with a factor of $-i$) to the 'structure constants' (the ϵ's and f's in (3.15) and (3.26)) is generally called the adjoint or regular representation.

3.2 Local SU(2) symmetry

Global symmetries and their associated (possibly approximate) conservation laws are certainly interesting, but they do not have the *dynamical* significance of local symmetries. We saw in Section 2.2 how the 'requirement' of local U(1) symmetry seemed to lead almost automatically to QED, with the symmetry current of the ψ matter fields now playing the role of the dynamical current on the RHS of the equation of motion for the A_μ field. A similar link between symmetry and dynamics follows if we generalise the preceding non-Abelian global symmetries to local ones. In this section we carry through the analysis for SU(2).

We begin by considering again a fermion doublet as in (3.3), without yet specifying exactly what the physical application will be. We want to extend the global SU(2) symmetry transformation (3.10) to the local one

$$\psi(x) \to \psi'(x) = e^{-ig\underline{\alpha}(x)\cdot\underline{\tau}/2}\psi(x) \tag{3.27}$$

by analogy with (2.47); note that we have slipped in a constant g in the exponent (c.f. the discussion after (2.49)). g is not a g-factor—it will be analogous to the charge e. Clearly, although the $\bar{\psi}m\psi$ part of (3.3) is still invariant under (3.27), the $\bar{\psi}i\not{\partial}\psi$ part is not—just as in the U(1) case (2.48), since the $\not{\partial}$ will pull down a $\not{\partial}\underline{\alpha}(x)$ factor. As in the U(1) case, we try to compensate this factor by introducing some vector field whose change under an appropriate transformation (accompanying (3.27)), exactly cancels this $\not{\partial}\underline{\alpha}(x)$ part. This time, since there are three $\underline{\alpha}(x)$'s ($\alpha_1(x)$, $\alpha_2(x)$, $\alpha_3(x)$) we immediately see that we need three vector (gauge) fields, called $W_1^\mu(x)$, $W_2^\mu(x)$, $W_3^\mu(x)$, or $\underline{W}^\mu(x)$ for short.

The key step in constructing the locally U(1) invariant Lagrangian of QED was the introduction of the 'covariant derivative' D_μ of (2.50): we saw that if ∂_μ was replaced by D_μ in the free Dirac $\mathcal{L}$ we obtained a locally U(1) invariant interaction automatically. The essential feature of $D_\mu\psi$ which ensured this was equation (2.51). In the present case, the required generalisation of $D^\mu\psi$ is

$$D^\mu\psi = (\partial^\mu + ig\underline{\tau}\cdot\underline{W}^\mu/2)\psi \tag{3.28}$$

when acting on an SU(2) doublet field such as ψ. The property required of (3.28) is that $D^\mu\psi$ should transform under the local symmetry (3.27) exactly as $\partial^\mu\psi$ does under the global one (3.10)—compare (2.51) and (2.47). In that case a term like $\bar{\psi}\not{D}^\mu\psi$ is automatically invariant under local SU(2).

This requirement on $D^\mu\psi$ determines the transformation law of the fields $\underline{W}^\mu$. The algebra is easier if we consider an infinitesimal transformation

$$\delta\psi = (-ig\underline{\epsilon}(x)\cdot\underline{\tau}/2)\psi(x); \tag{3.29}$$

we then require

$$\delta(D^\mu\psi) = (-ig\underline{\epsilon}(x)\cdot\underline{\tau}/2)D^\mu\psi. \tag{3.30}$$

It is a good exercise to verify that (3.30) implies that

$$\delta\underline{W}^\mu(x) = \partial^\mu\underline{\epsilon}(x) + g\underline{\epsilon}(x)\times\underline{W}^\mu(x), \tag{3.31}$$

which tells us how the $\underline{W}^\mu$'s must transform. The first term in (3.31) is the straightforward analogue of the infinitesimal version of (2.49), with $\alpha(x)\to e\epsilon(x)$. Comparing the second term of (3.31) with (3.19), we see that it implies that the three W-fields form the components of an SU(2) triplet. Thus *the W's carry SU(2) 'charge'*.

We now know the generalisation of (3.3) which makes it locally SU(2) invariant:

$$\mathcal{L}_{2W} = \bar{\psi}(i\not{D} - m)\psi = \bar{\psi}(i\not{\partial} - m)\psi - g\bar{\psi}\gamma^\mu\underline{\tau}/2\psi\cdot\underline{W}^\mu, \tag{3.32}$$

the last term being the generalisation of $\mathcal{L}_{\text{int}}$ in QED (equation (2.15)). We can immediately read off the ψ-ψ-W vertex factor as

$$-ig\frac{\tau_i}{2}\gamma^\mu. \tag{3.33}$$

In (3.33) 'i' refers to the SU(2) component of the W field quantum, and 'μ' to the Lorentz component of its polarisation vector.

We can easily generalise (3.25) to other SU(2) multiplets than doublets, by using appropriately larger matrices instead of the $\underline{\tau}/2$. For example, for an SU(2) triplet of fields $\underline{\phi} = (\phi_1, \phi_2, \phi_3)$, (3.28) becomes

$$D^\mu \phi_i = (\partial^\mu + ig\underline{t}\cdot\underline{W}^\mu)\phi_i \qquad (3.34)$$

where the three 3×3 matrices $\underline{t}$ are defined in (3.15). Under infinitesimal transformations, this changes by

$$\delta(D^\mu \phi_i) = (-ig\underline{\epsilon}(x)\cdot\underline{t})(D^\mu \phi_i) \qquad (3.35)$$
$$= (g\underline{\epsilon}(x) \times D^\mu \underline{\phi})_i \qquad (3.36)$$

(c.f. (3.18), (3.19), and (3.29)).

However, there is still an important part of the non-Abelian analogue of $\mathcal{L}_{\text{QED}}$ unaccounted for—namely the bit corresponding to the Maxwell-term $-\frac{1}{4}F.F$ for the gauge fields $\underline{W}^\mu$. Note that, as in the QED case, a simple mass term involving $\underline{W}^\mu.\underline{W}_\mu$ will violate invariance under (3.31), so these quanta are massless. Clearly we have a problem here in applying this 'local SU(2)'—as we eventually will—to weak interactions, which are very short ranged. This is where we will need the Higgs mechanism—see Section 4.3.

To get the non-Abelian '$F.F$' term, the obvious thing might be to consider

$$\partial^\mu A^\nu - \partial^\nu A^\mu \to D^\mu \underline{W}^\nu - D^\nu \underline{W}^\mu \qquad (3.37)$$

with D^μ given by (3.34), since the W's are an SU(2) triplet. The hope would be that by using the D's, $D^\mu \underline{W}^\nu - D^\nu \underline{W}^\mu$ would transform under local SU(2) transformations exactly as $\partial^\mu \underline{W}^\nu - \partial^\nu \underline{W}^\mu$ does under global ones—i.e. like (3.36). Then the 'dot product' $(D^\mu \underline{W}^\nu - D^\nu \underline{W}^\mu).(D_\mu \underline{W}_\nu - D_\nu \underline{W}_\mu)$ would be a locally invariant '$F.F$' term. Unfortunately it is not quite that simple. The problem is that the W's are a rather special triplet: whereas an ordinary triplet $\underline{\phi}$ would transform via only the second term in (3.31), the W's *also* have the first ('non-homogeneous') term as well. You can verify that in fact

$$\delta(D^\mu \underline{W}^\nu - D^\nu \underline{W}^\mu) \neq g\underline{\epsilon}(x) \times (D^\mu \underline{W}^\nu - D^\nu \underline{W}^\mu) \qquad (3.38)$$

so that the proposed '$F.F$' term will not work.

With the aid of some hindsight, we can be led to the right answer as follows. Consider, in the U(1) case, the quantity

$$(D^\mu D^\nu - D^\nu D^\mu)\phi \qquad (3.39)$$

where ϕ is any field of charge e and $D^\mu = \partial^\mu + ieA^\mu$. Evaluating (3.39) one finds

$$(D^\mu D^\nu - D^\nu D^\mu)\phi = ieF^{\mu\nu}\phi \qquad (3.40)$$

where $F^{\mu\nu} = \partial^\mu A^\nu - \partial^\nu A^\mu$. This suggests that we should look at the commutator of two covariant derivatives $[D^\mu, D^\nu]$. It does not matter whether we use the D from (3.28) or (3.34)—the result is essentially the same for all cases. Using the D^μ from (3.28) one finds

$$[D^\mu, D^\nu] = ig\underline{\tau}/2\cdot\underline{F}^{\mu\nu} \qquad (3.41)$$

where
$$\underline{F}^{\mu\nu} = \partial^\mu \underline{W}^\nu - \partial^\nu \underline{W}^\mu - g\underline{W}^\mu \times \underline{W}^\nu. \tag{3.42}$$

(Had we used (3.34) we would have got (3.41) with $\underline{\tau}/2 \to \underline{t}$.) When we now investigate the effect of the local SU(2) transformation (3.31) on $\underline{F}^{\mu\nu}$ we find

$$\delta \underline{F}^{\mu\nu}(x) = g\underline{\epsilon}(x) \times \underline{F}^{\mu\nu}(x) \tag{3.43}$$

precisely as desired (but not accomplished) in (3.38)—i.e. the inhomogeneous part in (3.41) has been got rid of. Thus $\underline{F}^{\mu\nu}$ does transform under local SU(2) transformations exactly as if it were an ordinary triplet under global SU(2) transformations and so the quantity

$$\mathcal{L}_W = -\frac{1}{4}\underline{F}_{\mu\nu} \cdot \underline{F}^{\mu\nu} \tag{3.44}$$

is indeed locally SU(2) invariant. This is the famous Yang-Mills (1954) term, the non-Abelian generalisation of the Maxwell term. $\underline{F}^{\mu\nu}$ is the non-Abelian field strength tensor.

The argument leading to (3.44) has been given in some detail since the result is of fundamental importance. Looking at (3.42) and (3.44) it is clear that, unlike the Maxwell term $\mathcal{L}_A$ of (2.8), the Yang-Mills term $\mathcal{L}_W$ of (3.44) *includes interactions between the gauge fields*—in addition, of course, to the expected 'free' part

$$-\frac{1}{4}(\partial_\mu \underline{W}_\nu - \partial_\nu \underline{W}_\mu) \cdot (\partial^\mu \underline{W}^\nu - \partial^\nu \underline{W}^\mu).$$

The free part leads to a W-propagator which is the same as (2.25), with a δ_{ij} factor to 'dot' the W's together. The interactions included in (3.44) are of two types: W-W-W (trilinear) and W-W-W-W (quadrilinear). This is quite unlike QED, where no fundamental γ-γ vertices are present (c.f. Section 2.4). It arises here because the W's both 'transmit' the gauge field force and feel it themselves since they are not SU(2) neutral (as the γ was U(1) neutral). Another important point to note is that these self-interactions among the W's come in with a coupling constant which is the same one as appears in the ψ-ψ-W vertex (3.33)—the W's 'couple universally'.

The physics application of all this is to the SU(2) of the weak interactions (see Section 5). There, the W_1^μ and W_2^μ fields correspond to the charged gauge bosons $W^{\pm\mu}$ (the combination $\frac{1}{\sqrt{2}}(W_1 - iW_2)$ destroys W^+ or creates W^-). As we shall see, the field W_3^μ is a linear combination of the photon γ and Z^0 fields:

$$W_3^\mu = \sin\theta_W A^\mu + \cos\theta_W Z^\mu \tag{3.45}$$

where θ_W is the 'weak angle', and the SU(2) gauge coupling constant g is related to e by

$$g\sin\theta_W = e. \tag{3.46}$$

We can then pick out the W-W-γ vertex from (3.44), and find that it is given by

$$ie\left[g_{\nu\lambda}(k_1 - k_2)_\mu + g_{\lambda\mu}(k_2 - k_\gamma)_\nu + g_{\mu\nu}(k_\gamma - k_1)_\lambda\right] \tag{3.47}$$

where the momenta and indices are as in Figure 16.

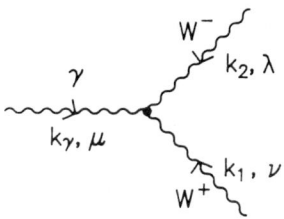

Figure 16.

The vertex (3.47) is to be compared with a general W-W-γ vertex given long ago by Lee and Yang (1962) which has the form (in our notation)

$$ie\left[g_{\nu\lambda}(k_1 - k_2)_\mu + g_{\lambda\mu}(k_2 - \kappa k_\gamma)_\nu + g_{\mu\nu}(k_\gamma - \kappa k_1)_\lambda\right]. \tag{3.48}$$

The Lee-Yang vertex arises from an interaction Lagrangian

$$\mathcal{L}_{W\gamma} = -\frac{1}{2}[(\partial_\mu - ieA_\mu)W_\nu]^\dagger [(\partial^\mu - ieA^\mu)W^\nu] - ie\kappa W_\mu^\dagger W_\nu F^{\mu\nu}. \tag{3.49}$$

The first term in (3.49) is what would follow from the standard replacement (2.50) for the field W_μ which destroys W^-'s; the second has the form of an 'anomalous' magnetic moment, as in (2.102), coupling to $F^{\mu\nu}$. Note, however, that unlike the G_a in (2.102), the κ in (3.49) is dimensionless, so that the coupling (3.49) is at least not obviously unrenormalisable. In fact, κ is sometimes called the 'ambiguous' moment, since if we had started from a free $W^\pm$ kinetic term of the form (2.11), and made the standard substitution $\partial_\mu \to D_\mu$, we would have encountered a problem in knowing whether to use $D^\mu D^\nu$ or $D^\nu D^\mu$, since they do not commute (c.f. (3.40)). Thus an arbitrary amount of 'WWF' can always (from this point of view) be added on, and the standard replacement $\partial_\mu \to D_\mu$ does not, in this case, lead to a single unique interaction. In the case of the W-W-γ vertex of (3.47), however, we have a perfectly definite *prediction* for κ, namely

$$\kappa = 1, \tag{3.50}$$

which corresponds (Lee and Yang 1962) to a g-factor for the W of

$$g_W = 1 + \kappa = 2. \tag{3.51}$$

This is just the value required to get rid of the $\ln M$ in (2.105), and it has arisen here precisely because, via (3.45), the photon takes part in a larger (SU(2)) local symmetry than just the electromagnetic U(1). Eventually we will see that the true symmetry is SU(2)×U(1): the parameter θ_W in (3.45) arises from the 'mixing' between the W_μ^3 of SU(2) and the B_μ of U(1).

We could certainly imagine more U(1)-gauge invariant terms than are in (3.44)—for example, the term

$$\frac{-ie\lambda}{M_W^2}(\partial_\nu W_\lambda - \partial_\lambda W_\nu)(\partial^\lambda W_\mu - \partial_\mu W^\lambda)F^{\mu\nu}. \tag{3.52}$$

Here the coefficient does indicate (mass)$^{-2}$ dimensions (we put in M_W^2 arbitrarily but conventionally), and (3.52) will be a 'non-renormalisable' interaction. Since (3.47) includes all the W-W-γ couplings in the SU(2)-symmetric model, λ is predicted to be zero. The term would add $+\lambda$ to g_W in (3.51), so that again any $\lambda \neq 0$ will spoil the cancellation of $\ln M$ in (2.105). It is also worth noting that in addition to a magnetic moment, a charged spin-1 particle can have an electric quadrupole moment Q_W: in general we have

$$g_W = 1 + \kappa + \lambda; \qquad Q_W = \frac{-e}{M_W^2}(\kappa - \lambda). \qquad (3.53)$$

It is evident that it is extremely important to test these W-W-γ couplings, which has not been done so far. To do so will be a challenge for LEP-II, where the e^+e^- energy will be above the W^+W^- creation threshold. In Born approximation, the contributing amplitudes are shown in Figure 17. In addition to the W-W-γ vertex predicted in (3.47), there is also the W-W-Z^0 one, which differs only by a factor $\cot\theta_W$ (c.f. (3.45)). As we shall see in Section 5.2, the neutrino coupling is also determined by the same parameter g (it is essentially the vertex (3.33)). The intricate relationships between the three graphs of Figure 17 are a direct consequence of the assumed gauge symmetry. They turn out to imply considerable cancellations among the diagrams, causing the cross sections to be smaller, and to rise less steeply with energy, than if these cancellations did not occur. The behaviour with energy, in particular, is related to the fact that the model as a whole is renormalisable. A recent discussion of experimental tests of the gauge structure in $e^+e^- \to W^+W^-$ is given by Hagiwara et al. (1987). Note that in the full theory a fourth graph involving the Higgs particle is also present (Figure 26 below). The 4-W vertex contained in (3.44) is given in Aitchison and Hey, 1989, for example.

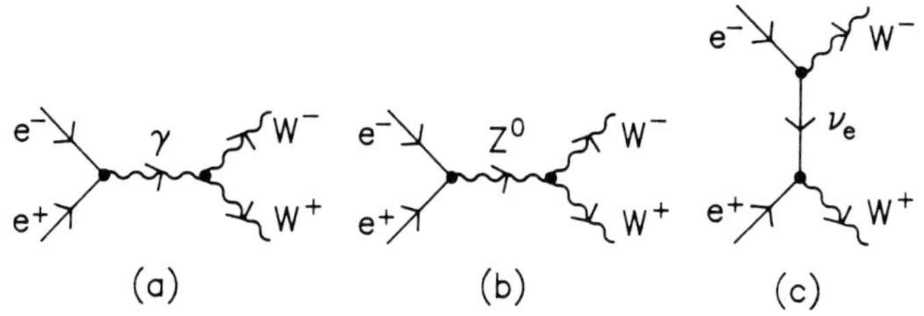

Figure 17.

Verification of this can be sought via a process such as that shown in Figure 18.

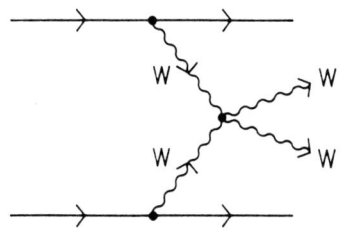

Figure 18.

3.3 Local SU(3) Symmetry: QCD and asymptotic freedom

Using what has been said about global SU(3) in Section 3.1, and about how to make a global SU(2) symmetry into a local one in Section 3.2, it is straightforward to discuss local SU(3). This is the gauge group of QCD, the labels 1, 2, 3 in (3.20) standing for colour, the ψ's being one flavour of quark. Under a local SU(3)$_c$ transformation, the triplet (3.20) transforms by

$$\delta\psi = (-ig_s\underline{\eta}(x)\cdot\underline{\lambda}/2)\psi \tag{3.54}$$

(c.f. (3.22) and (3.29)), where now there are eight field parameters $\eta_1(x), \eta_2(x)\ldots\eta_8(x)$ going with the eight λ's. To cancel off the unwanted $\slashed{\partial}\eta$ parts which occur when we try to make $\bar\psi\slashed{\partial}\psi$ invariant under (3.54), we now need eight vector gauge fields $A_a^\mu(x)$, $a = 1, 2, \ldots 8$. These A's transform according to

$$\delta A_a^\mu(x) = \partial^\mu \eta_a(x) + g f_{abc}\eta_b(x)A_c^\mu(x) \tag{3.55}$$

(c.f. (3.31) and (3.23)). The SU(3)$_c$ covariant derivative acting on a triplet is

$$D^\mu\psi = (\partial^\mu + ig_s\underline{\lambda}/2\cdot\underline{A}^\mu)\psi \tag{3.56}$$

giving the A-ψ-ψ vertex (c.f. (3.33))

$$-ig_s\frac{\lambda_a}{2}\gamma^\mu . \tag{3.57}$$

The quanta of the A_a^μ field are the (eight different) gluons.

As in local SU(2), there is an SU(3)$_c$ field strength tensor which is

$$F_a^{\mu\nu} = \partial^\mu A_a^\nu - \partial^\nu A_a^\mu - g_s f_{abc} A_b^\mu A_c^\nu . \tag{3.58}$$

The SU(3)$_c$ Yang-Mills term is then

$$-\frac{1}{4}F_a^{\mu\nu}F_{a\mu\nu} \tag{3.59}$$

and it contains triple and quadruple gluon couplings (see for example Aitchison and Hey, 1989), all involving the same 'strong' coupling g_s, and the constants f_{abc} determined from (3.23). Once again, there is no mass term allowed by (3.55), and the gluons are massless.

For one SU(3)$_c$ triplet ψ, then, our Lagrangian so far is

$$\mathcal{L} = \bar\psi(i\slashed{D} - m)\psi - \frac{1}{4}F_a^{\mu\nu}F_{a\mu\nu} \tag{3.60}$$

with $D\psi$ given by (3.56). For many different quark flavours f, the Dirac term is repeated for each, giving

$$\mathcal{L}_{\text{QCD}} = \sum_f \bar\psi_f(i\slashed{D} - m_f)\psi_f - \frac{1}{4}F_a^{\mu\nu}F_{a\mu\nu} . \tag{3.61}$$

Actually, however, matters are not quite that simple. As in QED, we need a gauge-fixing term to produce the propagator (2.25); in the non-Abelian case this turns out to be a more complicated affair, necessitating additional pieces in $\mathcal{L}_{\text{QCD}}$ called 'ghost terms'. We shall not give their form here: they are needed only for loop calculations, the details of which we shall not need. The Lagrangian of (3.61) is adequate at the tree level. Chapter 15 of Aitchison and Hey, 1989 gives an intuitive explanation of why ghosts are necessary; a full account is in Ryder, 1985.

We shall not have time in this course to look at any detailed QCD calculations. We must however introduce the fundamental result known as 'asymptotic freedom'. This is a property of the 'running strong fine structure constant', $\alpha_s(p^2) = g_s^2(p^2)/4\pi$ (c.f. (2.94)), which arises from loop corrections (note that QCD is renormalisable—see Ryder, 1985). First performed by Gross and Wilczek (1973), and by Politzer (1973), the calculation of $\alpha_s(p^2)$ is similar to that of $e^2(p^2)$ in (2.94), but rather more complicated. Apart from the extra internal ('colour') label in the couplings, and the attendant matrix structure in colour space (via the $\lambda^a/2$ in (3.57) for example), we have to remember that, in addition to the fermion loop insertion in the gauge boson (gluon) propagator which led to (2.94), we have to consider gluon self-coupling contributions as well. These, which were absent in QED, turn out to have a dramatic effect.

There is also a new point of principle. Recall that the 'making sense of loops' procedure described in Section 2.3 for QED involved taking from experiment the value of 'the' charge e. We saw that this had to be defined at some particular value of the photon momentum—in fact, for $p = 0$ in this case (see (2.89)). In QED two sources of charge can be physically separated to large distances, and it makes sense to define the 'physical' charge as $e^2(p^2 \to 0)$. But such a separation of two quarks, for example, seems *not* to be possible in QCD: we have no long-distance classical limit of gluon-quark scattering, corresponding to the Thompson limit in γe^-. Actually, though, one is not forced to define the coupling at $p^2 = 0$; any p^2 would do just as well. In QCD we introduce an arbitrary mass scale into the theory, so that the 'reference' value of α_s is $\alpha_s(p^2 = \mu^2)$. The effect of this is that the finite loop contribution to the gluon polarisation function $\Pi^g(p^2)$ will involve terms of the form $\ln(p^2/\mu^2)$—c.f. (2.93).

The contribution of the quark loop, shown in Figure 19 (a), to Π^g is exactly the same as the fermion loop contribution to Π^γ, except for a factor coming from the λ-matrix couplings, which is

$$\frac{1}{4}\text{Tr}(\lambda^a \lambda^b). \tag{3.62}$$

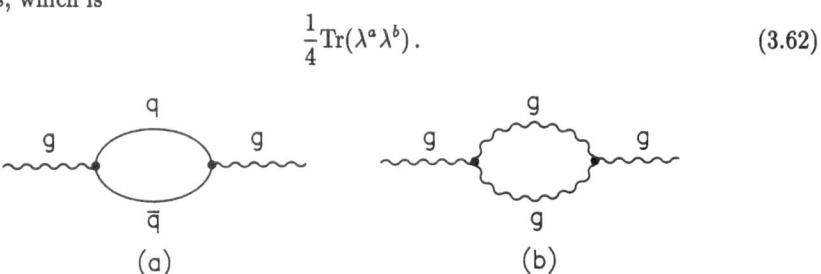

Figure 19.

(The trace arises in the same way as the γ-trace for a fermion loop.) The λ matrices are normalised to $\text{Tr}\lambda^a\lambda^b = 2\delta^{ab}$, so that (3.62) takes the value 1/2. Actually, this is a particular case of a more general group-theoretical result: for any matrices M^a providing a representation R of the algebra (*i.e.* obeying the commutation relations (3.14) or (3.26)) we have

$$\text{Tr}(M^a M^b) = T(R)\delta^{ab} \tag{3.63}$$

where $T(R)$ is a number characteristic of R and of the group G. In the case of an SU(N) group, if 'R' is the representation of dimension N (i.e. the doublet for SU(2), triplet for SU(3), ...), $T(R)$ has the universal value of $\frac{1}{2}$; the case of SU(2) is easily checked from (3.63) using the $\underline{\tau}/2$ matrices. Thus the quark loop contributes a term of the form (2.93) to Π^g, with $\alpha \to \alpha_s$ and Q_f^2 replaced by $T(R)$.

The gluon trilinear self-coupling leads to a purely gluonic analogue of Figure 19 (a), shown in Figure 19 (b). This will involve the factor

$$\sum_{cd} f_{acd} f_{bcd} \tag{3.64}$$

(c.f. (3.58) and (3.59): we take an 'fAA' piece in one 'F' and a '∂A' piece in the other). We saw in Section 3.1 that the f's can be regarded as the entries in the 8×8 matrices F_a via $(F_a)_{bc} = -i f_{abc}$. The factor can be rewritten as

$$\sum_{cd} f_{cda} f_{cdb} = -\sum_{cd} f_{cad} f_{cdb} = \sum_{cd} (F_c)_{ad} (F_c)_{db} = (\underline{F}^2)_{ab} \tag{3.65}$$

using the cyclic symmetry of the f's and their antisymmetry under interchange of a pair of indices. So the factor (3.64) is the (ab)-element of the sum of the squares of the eight matrices $\underline{F}$. In the more familiar case of SU(2), such a sum of squares is just the 'total angular momentum squared', which is just $j(j+1)$ times the unit matrix, where j is the spin of the representation in question. A similar result is true in general: the sum of the squares of the representation matrices is proportional to the unit matrix, the factor being called $C_2(R)$—the 'quadratic Casimir operator' characteristic of the group G and the representation R (e.g. C_2 for SU(2) is $j(j+1)$). For SU(N), the value of C_2 in the adjoint representation, as required for (3.64), is

$$C_2(\text{adj}) = N; \tag{3.66}$$

thus for SU(2), C_2 for the triplet is $2 = j(j+1)$ with $j = 1$, and $C_2 = 3$ for the octet of SU(3).

In addition to the two gluon self-energy insertions of Figure 19, one has also to include vertex corrections and ghost contributions. The final result is

$$\alpha_s(p^2) = \alpha_s(\mu^2) \Big/ \left[1 - A \frac{\alpha_s(\mu^2)}{12\pi} \ln\left(\frac{p^2}{\mu^2}\right)\right] \tag{3.67}$$

where

$$A = 4T(f) \cdot f - 11 C_2(\text{adj}) \tag{3.68}$$

and f is the number of quark flavours. The remarkable property of (3.67) is that it exhibits '*antiscreening*'—compare the discussion of (2.95). That is, the gluon contribution to the coefficient A is opposite in sign to the quark contribution, and tends to cause the effective α's to *decrease* at large p^2. Indeed, using $T(f) = 1/2$ and $C(\text{adj}) = 3$, (3.68) predicts that A will be *negative*, and (3.67) then implies that α_s will to tend to zero as $p^2 \to \infty$ (provided the number of flavours is less than 16). This is the famous property of *asymptotic freedom*, which is of central importance in understanding the parton model, and perturbative corrections to it (see the lectures by Bethke, these Proceedings).

Despite appearances, (3.67) does not depend on two parameters $\alpha_s(\mu^2)$ and μ^2, but only one. This can be brought out by writing

$$\ln \Lambda^2 = \ln \mu^2 + 12\pi/A\alpha_s(\mu^2), \qquad (3.69)$$

in terms of which (3.67) becomes

$$\alpha_s(p^2) = \frac{12\pi}{(33 - 2f) \ln(p^2/\Lambda^2)} \qquad (3.70)$$

for the case at hand. The parameter Λ is 'the QCD Λ', and is—roughly speaking—a measure of the energy scale at which, as p^2 is reduced, the strong interactions become non-perturbative.

At the end of Section 2.3 we noted that Equation (2.92) summed up the large '$\ln p^2$' terms in $\alpha(p^2)$, and the same is true of (3.67) or (3.70). In QCD it is particularly important to go further than this, and extend (3.70) to include terms of order $\ln(\ln p^2)$. The way to do this is via the 'renormalisation group equation' (Ryder 1985). In brief, this proceeds as follows. One introduces the variable $t = \ln p^2$, in terms of which (3.70) can be written in differential form as

$$\frac{d\alpha_s(t)}{dt} = -\frac{\beta_0}{4\pi}\alpha_s^2(t) \qquad (3.71)$$

where

$$\beta_0 = 11 - \frac{2}{3}f. \qquad (3.72)$$

The right hand side of (3.71) represents, in fact, the lowest order contribution to the 'β function', which controls the way in which the coupling constant runs. (Note that in QED the corresponding lowest order term would be positive, which signals a non-asymptotically free theory.) At the next (two-loop) order, a term $-\beta_1 \alpha_s^3/16\pi^2$ is added to the RHS of (3.71), where

$$\beta_1 = \left(102 - \frac{38}{3}f\right). \qquad (3.73)$$

Integration of the resulting equation then yields corrections to (3.70) involving (up to second order) $\ln(\ln p^2)$ terms.

It is important to determine experimentally an accurate value of α_s at one p^2, and to see whether it 'runs' in the predicted way. This is not at all easy. First, the observed strongly interacting states involve hadrons rather than quarks and gluons, and to interpret data in terms of underlying perturbative QCD interactions one has to use models of the hadronisation process, which introduces uncertainties. Also, the experimental meaning of the energy scale 'p^2' is ambiguous, unless an 'all orders' calculation is done (although where $O(\alpha_s^3)$ calculations are available they already indicate insensitivity to changes in the scale). These matters will be discussed in detail for e^+e^- annihilation in the lectures by Bethke.

4 Spontaneous symmetry breaking

4.1 Introduction, and chiral symmetries

Particle physics is almost as much about symmetry *breaking* as it is about symmetry. Any attempt to see the weak interaction as in some way the same as ('unified with') the electromagnetic one has to confront the gross fact that their ranges are completely different—or, equivalently, that the mediating quanta have extremely different masses ($M_\gamma < 3 \times 10^{-27}$ eV, $M_{W,Z} \sim 90$ GeV). Whatever symmetry may exist between the photon and the W/Z particles must be a broken symmetry. And there are other important examples of broken symmetries, as we shall see. Two kinds of symmetry breaking are commonly distinguished—'explicit' and 'spontaneous' (sometimes also called 'dynamical'). In the former, the observed breaking is associated with a specific term in the Lagrangian, in the absence of which the theory would lead to some exact observed symmetries. For example, the strong interaction Lagrangian, $\mathcal{L}_{\text{QCD}}$ of (3.61), would be invariant under a global (infinitesimal) $SU(2)_f$ 'flavour' symmetry of the type (3.10), namely

$$\psi = \begin{pmatrix} \psi_u \\ \psi_d \end{pmatrix} \to \psi + \delta\psi, \qquad \delta\psi = -i\underline{\epsilon}\cdot\underline{\tau}/2\,\psi, \qquad (4.1)$$

if the up and down quark masses were equal, $m_u = m_d$. This follows from the fact that the gluon interaction is independent of quark flavour. To the extent to which $m_u \neq m_d$, this $SU(2)_f$ symmetry is explicitly broken by the mass difference $(m_u - m_d)$. It is this approximate $SU(2)_f$ symmetry of $\mathcal{L}_{\text{QCD}}$ which leads to the observed approximate isospin symmetry of hadronic multiplets. Similarly, to the extent that $m_u = m_d = m_s$, $\mathcal{L}_{\text{QCD}}$ would have an approximate global $SU(3)_f$ symmetry, which is evident in the hadronic multiplets when extended to include strange hadrons. These flavour symmetries, discovered long before QCD, are now understood as following from $\mathcal{L}_{\text{QCD}}$.

But it is also possible to have a symmetrical Lagrangian, while the particle states and other physical observables seem to show no obvious (even approximate) sign of the symmetry. This is the 'spontaneously broken' case. This language is borrowed from condensed matter physics, where the ferromagnet is the frequently quoted example. The (Heisenberg) Hamiltonian is certainly rotationally invariant, yet below the transition temperature the spins are thought of as lining up in some particular direction, breaking the rotational symmetry 'spontaneously'. In the case of a field theory, there are striking differences in the physical consequences depending on whether the symmetry that is spontaneously broken is a global or a local one. In the global case, a general result due to Goldstone (1961) and others states that spontaneous breaking of a continuous symmetry is always associated with the appearance of a massless particle, or particles, called 'Goldstone bosons'. In the local case, these Goldstone bosons become the longitudinal components of the gauge field(s)—which, before symmetry breaking, always had only the two transverse components (*c.f.* (2.14)). The total of three 'spin' components in all is exactly what is required for a *massive* vector field. This is the essence of the theoretical loophole which allows gauge bosons to be massive even though the Lagrangian is locally (gauge-) invariant (*c.f.* the end of Section 2.2), and which is invoked to give masses to the W and the Z in the Standard Model.

In this section we shall be mostly concerned with aspects of spontaneous breaking of (local) gauge symmetries. However, there is one important spontaneously broken global symmetry which we should mention first, namely the chiral flavour symmetry of $\mathcal{L}_{\rm QCD}$. A massless Dirac Lagrangian

$$\bar{\psi} i \slashed{\partial} \psi \tag{4.2}$$

is invariant not only under the ordinary (infinitesimal) global U(1) symmetry of (2.32), but also under the 'γ_5-version' of (2.32), namely

$$\psi \to \psi' = (1 - i\eta\gamma_5)\psi. \tag{4.3}$$

This can be easily verified directly, using

$$\gamma^0 \gamma^5 = -\gamma^5 \gamma^0, \quad \gamma^i \gamma^5 = -\gamma^5 \gamma^i, \tag{4.4}$$

but it will be useful later to expand the discussion now to cover this type of symmetry, not considered in Section 3. We may write

$$\psi = \frac{(1-\gamma_5)}{2}\psi + \frac{(1+\gamma_5)}{2}\psi \equiv \psi_{\rm L} + \psi_{\rm R} \tag{4.5}$$

where the 'Left' and 'Right' pieces of ψ are called that because they correspond to states of left- and right-handed helicity respectively, which are projected out by the operators $\frac{1}{2}(1 \pm \gamma_5)$ (see Section 10.4 of Aitchison and Hey, 1989, for example). The ordinary U(1) symmetry (2.32) is then

$$\delta\psi_{\rm R} = -i\epsilon\psi_{\rm R}, \quad \delta\psi_{\rm L} = -i\epsilon\psi_{\rm L} \tag{4.6}$$

while (4.3) is

$$\delta\psi_{\rm R} = -i\eta\psi_{\rm R}, \quad \delta\psi_{\rm L} = +i\eta\psi_{\rm L}. \tag{4.7}$$

Transformations such as (4.7), which act differently on the L and R components are called 'chiral'. Using (4.4), (4.2) can be written as

$$\bar{\psi}i\slashed{\partial}\psi = \bar{\psi}_{\rm L}i\slashed{\partial}\psi_{\rm L} + \bar{\psi}_{\rm R}i\slashed{\partial}\psi_{\rm R}, \tag{4.8}$$

which clearly exhibits both the symmetries (4.6) and (4.7). It is also manifestly L $\leftrightarrow$ R symmetric, which means it conserves parity. On the other hand, a mass term $m\bar{\psi}\psi$ becomes

$$m\bar{\psi}\psi = m(\bar{\psi}_{\rm L}\psi_{\rm R} + \bar{\psi}_{\rm R}\psi_{\rm L}) \tag{4.9}$$

which is invariant under (4.6) but not under (4.7), while still preserving parity.

Consider then $\mathcal{L}_{\rm QCD}$ of (3.61), in the limit in which some quark masses—in particular the lightest, m_u and m_d—are regarded as negligible. The fact that $\slashed{\partial}$ in (4.8) is replaced by $\slashed{D}$ clearly makes no difference to the preceding discussion, which depended only on (4.4). Thus in this limit $\mathcal{L}_{\rm QCD}$ will be invariant under the γ_5-version of (4.3), namely

$$\delta\psi = -i\underline{\eta}\cdot\underline{\tau}/2\,\gamma_5\,\psi, \tag{4.10}$$

which is a chiral 'SU(2)$_{f5}$' transformation. Now this cannot be realised as an exact symmetry in nature, or else for every non-strange baryon made of u and d quarks there would have to exist another one, degenerate in mass, but with the opposite parity—and this is not seen. Instead, this chiral flavour symmetry seems to be *dynamically* broken—the u and d quarks acquire a constituent mass of 300–400 MeV, while the attendant Goldstone bosons are interpreted as the pions. Still useful, though more 'explicitly' broken than this chiral $SU(2)_{f5}$, is the corresponding chiral $SU(3)_{f5}$ in which we suppose $m_s \approx 0$; the kaons are the corresponding Goldstone bosons. This interpretation of the light mesons, which is phenomenologically successful, provides an important example of nature using the 'dynamical symmetry breaking' option.

Remarkably enough, these ideas are also relevant to the weak interactions. In this case, as we shall see, the interaction is most definitely not left-right symmetric (it violates parity)—indeed the 'V–A' structure means that the weak gauge fields couple only to the ψ_L components of the fermions, and not to the ψ_R components at all. This means that the corresponding local gauge symmetry is of the form

$$\delta\psi_L = -i\underline{\epsilon}\cdot\underline{\tau}/2\,\psi_L \qquad (4.11)$$
$$\delta\psi_R = 0, \qquad (4.12)$$

for a 'weak doublet' such as

$$\begin{pmatrix} \nu_e \\ e^- \end{pmatrix}. \qquad (4.13)$$

But this implies that any mass term of the form (4.9), which treats ψ_L and ψ_R the same, will break this 'left-handed' gauge symmetry. Although the neutrinos may be massless, the other leptons are definitely not, nor are the quarks. Thus, curiously enough, there is another 'mass problem' with the weak interactions: they would like not only the W and Z bosons but also the fermions to be massless. Once again, we shall have to suppose that the fermion masses arise 'spontaneously', if we want to save the (weak) gauge symmetry. In the Standard Model, one appeals to the same mechanism (the Higgs field) to give mass to the gauge bosons and to the fermions, which is an economical but not necessary step.

We now proceed with the discussion of spontaneous breaking of gauge symmetries. We shall not attempt a general treatment, concentrating instead on some particular examples which are relevant to the Standard Model.

4.2 Spontaneously broken U(1) symmetry

We begin by considering a simple classical field theory which shows the effect we want to study. Let ϕ be a complex scalar field, described by the Lagrangian

$$\mathcal{L}_\phi = \partial_\mu \phi^* \partial^\mu \phi - V(\phi) \qquad (4.14)$$

where the potential is taken to have the form

$$V(\phi) = -\mu^2 \phi^* \phi + \frac{\lambda}{4}(\phi^*\phi)^2. \qquad (4.15)$$

Clearly $\mathcal{L}_\phi$ is invariant under the global U(1) symmetry

$$\phi \to \phi' = e^{-i\alpha}\phi. \qquad (4.16)$$

(Note that a term like $(\phi^*\phi)^3$ would also be invariant under (4.16), but this would be a non-renormalisable interaction in the quantum theory of $\mathcal{L}_\phi$, so we exclude it.)

Application of the Euler-Lagrange equation (2.3) yields the equation of motion

$$(\Box - \mu^2)\phi = -\frac{\lambda}{2}|\phi|^2\phi. \tag{4.17}$$

This is nearly the standard Klein-Gordon equation for ϕ (with an interaction term on the right-hand side)—except for the fact that '$-\mu^2$' has the wrong sign for a mass term! This prevents us from making any quantum interpretation of (4.14) as yet; we therefore concentrate on $V(\phi)$ regarded simply as the potential energy of the classical field.

As a first step to understanding (4.14), we try to identify the configuration(s) of minimum energy, about which the system might be expected to oscillate. Generally, the energy will be a minimum when ϕ is a constant, which reduces the kinetic terms to zero. The minimum energy is then reached at the minimum of $V(\phi)$. This occurs at

$$|\phi| = v/\sqrt{2}, \quad v = 2\mu/\lambda^{1/2}, \tag{4.18}$$

where v is referred to as the 'symmetry breaking parameter'. To have a clearer picture, it is helpful to introduce two real fields ϕ_1 and ϕ_2 by

$$\phi = (\phi_1 - i\phi_2)/\sqrt{2} \tag{4.19}$$

and also the 'polar' variables θ and ρ defined by

$$\phi = (\rho/\sqrt{2})e^{i\theta/v}, \tag{4.20}$$

where the v is inserted so that θ has the same dimensions as ρ. Figure 20 shows $V(\phi)$ versus ϕ_1 and ϕ_2, from which it is obvious that the minimum of V is not at $\phi_1 = \phi_2 = 0$. In fact, there is *no* unique minimum point—rather, any value on the circle $\phi_1^2 + \phi_2^2 = v^2$ or equivalently $\rho = v$ will do. Before proceeding further, we briefly outline the condensed matter analogue of (4.14) and (4.15) which we mentioned earlier—namely the ferromagnet. In this case, one considers the free energy as a function of the magnetism $\underline{M}$ at a given temperature T, and makes an expansion of the form

$$F \approx F_0(T) + \frac{1}{2}\mu^2(T)\underline{M}^2 + \frac{1}{4}\lambda(T)(\underline{M}^2)^2 + \cdots, \tag{4.21}$$

valid for small magnetization. If the parameter μ^2 is positive, it is easy to see that F has a simple 'bowl' shape as a function of $|\underline{M}|$, with a minimum at $|\underline{M}| = 0$. This is the case for T greater than the ferromagnetic transition temperature T_C. However, if one assumes that $\mu^2(T)$ becomes negative for $T < T_C$ (so that $\mu^2(T_C) = 0$), then F will now look like Figure 20 and the minimum free energy will occur for $|\underline{M}| \neq 0$. The interpretation is that in this case the ground state will be magnetized. Any direction of $\underline{M}$ is possible (only $|\underline{M}|$ is specified); but when the system does settle into one actual configuration with $\underline{M} \neq \underline{0}$ the original full rotational invariance of (4.21) is lost—the magnetization, and the breaking of the symmetry, has occurred 'spontaneously'.

In the same way, any particular minimum on the circle $\rho = v$ will select out a particular θ in (4.20), breaking 'spontaneously' the invariance (4.16).

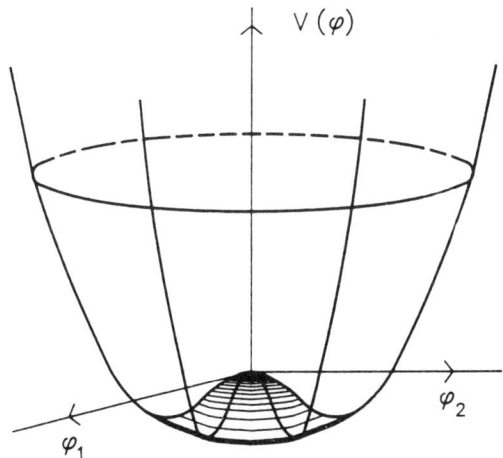

Figure 20.

In quantum field theory, particles are thought of as excitations from a ground state, which we call 'the vacuum'. Figure 20 strongly suggests that if we want a decent quantum interpretation of (4.14), we should consider expanding the fields about a point on the circle of minima, about which stable oscillations are likely. Any such point represents a possible vacuum state in which

$$\langle 0|\phi_1^2 + \phi_2^2|0\rangle = v, \quad \text{or} \quad \langle 0|\rho|0\rangle = v. \tag{4.22}$$

Bearing in mind (*c.f.* (4.15)) that for a field with a conventional (positive) mass² parameter the potential would be U-shaped, we might guess that 'radial' oscillations in Figure 20 would correspond to a conventional massive field, while 'angle' oscillations—which pass through all the degenerate minima (vacua)—have no 'restoring force' and are massless. Accordingly, we set (*c.f.* (4.20))

$$\phi(x) = \frac{1}{\sqrt{2}}(v + h(x))e^{-i\theta(x)/v} \tag{4.23}$$

and find that $\mathcal{L}_\phi$ becomes

$$\mathcal{L}_\phi = \frac{1}{2}\partial_\mu h \partial^\mu h - \mu^2 h^2 + \frac{1}{2}\partial_\mu \theta \partial^\mu \theta + \frac{\mu^4}{\lambda} + \text{terms cubic and quartic in } \theta, h. \tag{4.24}$$

Equation (4.24) exhibits the desired form of a conventional scalar field h with mass $\sqrt{2}\mu$ and a massless field θ, together with interaction terms. In particular, the quantum version of (4.24) will have $\langle 0|h(x)|0\rangle = \langle 0|\theta(x)|0\rangle = 0$, consistent with (4.22), so that h and θ will have the usual mode expansions (of the form (2.5) and (2.13) for example), allowing the usual particle interpretation. (The constant term in (4.24), which does not affect equations of motion, reflects the fact that $V(\min) = -\mu^2/\lambda$). Note that the symmetry (4.16), which is evident in (4.14), is well and truly hidden in (4.24)!

This model (due originally to Goldstone (1961)) contains the essence of spontaneous symmetry breaking in field theory: a non-zero value of a field in the ground state (vacuum), a zero mass mode or modes (the Goldstone bosons), and a massive excitation or excitations in the directions 'perpendicular' to the degenerate ground states.

It is interesting to find out what happens to the symmetry current corresponding to the invariance (4.16). Following the usual procedure, this current is

$$j_\phi^\mu = i\{\phi^* \partial^\mu \phi - (\partial^\mu \phi)^* \phi\} = v \partial^\mu \theta + 2h \partial^\mu \theta + h^2 \partial^\mu \theta / v. \tag{4.25}$$

The presence of the term involving just the *single* field θ is very remarkable: it tells us that (in the quantum theory) there is a non-zero matrix element of the form

$$\langle 0 | j_\phi^\mu(0) | \theta \rangle = -i p^\mu v, \tag{4.26}$$

where $|\theta\rangle$ stands for a state with one Goldstone boson θ, with momentum p^μ. That is, the symmetry current connects the Goldstone boson to the vacuum, with an amplitude proportional to the symmetry breaking parameter. In the case of the chiral $SU(2)_{f5}$ symmetry mentioned earlier, the analogue of j_ϕ^μ is the current of the global axial $SU(2)$ symmetry A_i^μ, and there are three θ modes which are identified with the physical pions. The parameter v in the corresponding Eq. (4.26) is then f_π (~ 94 MeV), the constant which enters into the pion decay $\pi \to \ell\nu$.

Although by the ansatz (4.23) we seem to have arrived at a viable particle interpretation of (4.14), we might well ask: how would such a negative (mass)^2 term arise in quantum field theory? One possible answer is that, as with the ferromagnetic analogy, the coefficient μ^2 in (4.15) could be temperature dependent: perhaps at extremely high temperatures, such as prevailed in the early universe, μ^2 had the opposite sign, suitable to a conventional mass term. In that case the potential would have a simple minimum at the origin, and the symmetry would not be spontaneously broken until T dropped below some T_C, where $\mu^2(T_C) = 0$. This simple picture is indeed popular in models of the early universe, where such phase transitions are proposed. On the other hand, it may be that some theory might predict the coefficient μ^2 in (4.15) to be negative, in a particular case. Or, one might simply postulate a $V(\phi)$ of the form (4.15), so as to 'trigger' the desired breakdown. The last alternative is essentially what is done in the Higgs sector of the Standard Model—and we now turn to the original Higgs model (Higgs (1964)).

The $U(1)$ Higgs model is just $\mathcal{L}_\phi$ of (4.14) extended so as to be locally $U(1)$ invariant; it provides a beautifully simple model for investigating what happens when a *gauge* symmetry is spontaneously broken. To make (4.14) locally $U(1)$ invariant, we need only replace ∂'s by D's as in (2.50), and add the Maxwell piece, giving

$$\mathcal{L}_h = [(\partial_\mu + ieA_\mu)\phi]^* [(\partial^\mu + ieA^\mu)\phi] - \frac{1}{4} F_{\mu\nu} F^{\mu\nu} - V(\phi) \tag{4.27}$$

where V is still (4.15), and of course $F^{\mu\nu} = \partial^\mu A^\nu - \partial^\nu A^\mu$. (4.27) is invariant under the local version of (4.16), namely

$$\phi \to \phi'(x) = e^{-i\alpha(x)} \phi(x) \tag{4.28}$$

when accompanied by a gauge transformation on A^μ

$$A^\mu \to A^{\mu\prime} = A^\mu + \frac{1}{e} \partial^\mu \alpha \tag{4.29}$$

as in Section 2.2. Before proceeding further, we note at this stage that we have four field degrees of freedom—two in ϕ and two in the massless A^μ ($F^{\mu\nu} = \partial^\mu A^\nu - \partial^\nu A^\mu$).

Now we have learned that the form of V in (4.15) does not lend itself to a natural particle interpretation, which only appears after making the 'shift to the minimum', as in (4.23). But there is a remarkable difference between the local and global cases. In the local case, the phase of ϕ is completely arbitrary, since any change in $\theta(x)$ in (4.23) can be compensated by an appropriate transformation (4.29) on A^μ, leaving $\mathcal{L}_h$ the same as before. Thus in fact the 'θ' field in (4.23) can be 'gauged away' altogether, if we like! This must mean that the massless Goldstone boson, described precisely by θ in the quantum theory, somehow no longer appears. This is the first unexpected result in the local case.

However, we cannot simply 'lose' degrees of freedom. Somehow the system must keep track of the fact that we started with four. To see what has happened, we substitute (4.23) into (4.27) with $\theta = 0$; i.e. set

$$\phi = \frac{1}{\sqrt{2}}(v + h(x)) \tag{4.30}$$

in $\mathcal{L}_h$. We find then

$$\mathcal{L}_h = \frac{1}{2}\partial_\mu h\, \partial^\mu h - \mu^2 h^2 + \frac{\mu^4}{\lambda} - \frac{1}{4}F_{\mu\nu}F^{\mu\nu} + \frac{1}{2}e^2 v^2 A_\mu A^\mu + \text{ interaction terms}, \tag{4.31}$$

where A^μ has to be understood as the gauge field after the transformation needed to reduce ϕ to (4.30). Equation (4.31) shows the second 'Higgs miracle': comparing it with (2.53) we see that the A^μ field now has a *mass*, equal to ev where v is the symmetry breaking parameter. The missing degree of freedom has reappeared as the third (longitudinal) polarisation state of the massive field A^μ. The fourth degree of freedom is still there, the massive h field as in (4.24).

Can such miracles ever occur? The answer is undoubtedly yes, at least in the non-relativistic case. The low-energy version of $\mathcal{L}_h$ is just the Ginzburg-Landau (GL) approximation for (again) the free energy in a superconductor. In this case (see Chapter 13 of Aitchison and Hey, 1989, for example) ϕ represents a composite (rather than elementary) field, such that $|\phi|^2$ is the density of bound Cooper pairs (of e^-e^-). Also, the mass for the A-field implies that the field is exponentially attenuated inside the superconductor, with a penetration length of order $1/ev$; this is the Meissner effect. It is worth noting that the GL free energy is not to be regarded as a fundamental theory, which must of course be derived from the physical electron-electron and electron-lattice interactions; this is what the BCS theory is all about, and the GL free energy is a phenomenological expression embodying much of the important physics of the BCS theory. In particle physics the question of whether the ϕ field in the Standard Model (see Section 4.3) is elementary or composite is completely unknown. However, whatever the truth of that may be, it seems pretty well inevitable that some such field, or effective field, is required to give mass to the W and Z (see Section 4.3, and Section 5)—and in that case it should have its own excitation quantum, the *Higgs boson*: hence the intense interest in hunting for it! (See Stirling's lectures, these Proceedings.)

Before proceeding further we can at this stage read off from (4.31) the propagator for the massive vector A-field. As in the discussion following (2.20), we need to invert the quantity $P_{\mu\nu}(M_A) = [(-k^2 + M_A^2)g_{\mu\nu} + k_\mu k_\nu]$, where $M_A = ev$ here. Unlike the quantity $P_{\mu\nu}(M_A{=}0)$ we met after (2.20), $P_{\mu\nu}(M_A)$ does have a straightforward inverse, which leads to the propagator

$$\left(-g_{\mu\nu} + \frac{k_\mu k_\nu}{M_A^2}\right) \frac{i}{k^2 - M_A^2} \,. \tag{4.32}$$

We see that (4.32) makes no sense as $M_A \to 0$, reflecting the difficulty with $P_{\mu\nu}(M_A{=}0)$. A more technical point concerns the fact that (4.32) follows only when the special choice of gauge, $\theta = 0$, is made in (4.30). In general, the vector propagator will contain a gauge parameter ξ (see equation (5.5) below) like the massless propagator of (2.25): this is after all a gauge theory. However, a scalar field will also be present with a ξ-dependent propagator (it is called the 'unphysical Higgs field', since it is associated with the degree of freedom suppressed in (4.30)), and the complete theory is ξ-independent. Further discussion of this is contained in Chapter 13 of Aitchison and Hey, 1989, for example.

Returning to (4.27), we can again look at the electromagnetic current in this 'spontaneously broken local U(1)' model. The gauge invariant form of (4.25) is

$$\begin{aligned} j^\mu_{\text{e.m.}} &= ie\,[\phi^*(\partial^\mu + ieA^\mu)\phi - \text{complex conjugate}] \\ &= ie(\phi^*\partial^\mu\phi - (\partial^\mu\phi^*)\phi) - 2e^2 A^\mu \phi^*\phi \,. \end{aligned} \tag{4.33}$$

Inserting (4.23) into (4.33) (this time in a gauge such that $\theta \neq 0$) we find (c.f. (4.25))

$$j^\mu_{\text{e.m.}} = -e^2 v^2 A^\mu + ev\partial^\mu\theta + \text{interaction terms} \,. \tag{4.34}$$

(4.34) tells us that there is a 'screening current' (the first term on the RHS) which leads to a mass ev of the A-field, once again; the second term shows that—as in (4.26)—the vacuum couples to the 'would-be Goldstone boson' (which has become the longitudinal part of the A-field) via the electromagnetic current.

This is an important observation as it leads to a somewhat different way of understanding the 'mechanism' whereby a gauge particle can become massive. In Section 2.3 we introduced the photon self-energy $\Sigma^\gamma_{\rho\sigma}$ which had the general form

$$\Sigma^\gamma_{\rho\sigma} = (g_{\rho\sigma}p^2 - p_\rho p_\sigma)\Pi^\gamma(p^2) \,. \tag{4.35}$$

When all the self-energy insertions are summed up, and after renormalisation, the photon propagator has the form (c.f. (2.92))

$$-ig^{\mu\nu}/p^2 \left(1 + \hat{\Pi}^\gamma(p^2)\right) \,. \tag{4.36}$$

in the Feynman gauge. The existence of the matrix element

$$\langle 0|j^\mu_{\text{e.m.}}(0)|\theta\rangle = -ip^\mu ev \tag{4.37}$$

means that $\Sigma^\gamma_{\rho\sigma}$ will now receive a contribution from the diagram of Figure 21, where the dotted line represents the massless θ quantum. This is now a tree diagram, not a loop as in the e^+e^- contribution of Figure 6(b), and so the contribution to $\Sigma^\gamma_{\rho\sigma}$ will involve simply the (massless) θ-propagator, with no momentum integration. The γ-θ vertex is given by (4.37), with the result that the contribution to $\hat{\Pi}^\gamma(p^2)$ in (4.35) is

$$\hat{\Pi}^\gamma_\theta(p^2) = -e^2 v^2/p^2 \,, \tag{4.38}$$

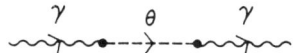

Figure 21.

so that the pole in the photon propagator (4.36) is now at $p^2 = e^2v^2$, and the photon has a mass ev, as before. We have been casual about questions of gauge choice in this argument, but the essential point is valid: a gauge quantum can acquire mass if (for some reason) its vacuum polarisation function has a zero mass pole. This pole can be associated with the 'elementary' massless quantum in a Higgs potential of the form (4.15), but it does not have to be. The massless quantum could equally well be a bound state in some strongly-interacting fermion-antifermion channel—in particular, a Goldstone boson arising from the spontaneous breaking of some global symmetry in a purely fermionic theory, for instance. All that is necessary is that it has a coupling of the form (4.37). The point of this latter interpretation is that only the product 'ev' has significance—there is no sign of Figure 20, or of 'v' alone as the vacuum value of a scalar field. Theories of this latter type do seem to produce a natural 'dynamical' mechanism for gauge boson mass generation. Both the '$t\bar{t}$' models (Nambu 1989; Miransky et al. 1989a, 1989b; Bardeen et al. 1990), and technicolour (Farhi and Susskind 1981), are of this type, but neither seem to be favoured by experiment at present. In the electroweak theory it is of course the W and Z particles that we want to be massive (while still being gauge bosons), not the photon. We therefore need to extend the above to the (non-Abelian) SU(2) case.

4.3 Spontaneously broken SU(2)×U(1) symmetry

We shall confine ourselves to the particular case which we need for the electroweak theory. We consider a complex scalar (spin-0) SU(2) doublet (called the 'Higgs doublet')

$$\phi = \begin{pmatrix} \phi^+ \\ \phi^0 \end{pmatrix} \tag{4.39}$$

where the complex ϕ^+ field destroys positively charged particles and creates negatively charged ones, and the complex ϕ^0 field creates neutral particles and antiparticles (a hadronic analogy would be the K^+ and K^0 fields under hadronic SU(2)$_f$). The Lagrangian

$$\mathcal{L}_\Phi = (\partial_\mu \phi)^\dagger (\partial^\mu \phi) + \mu^2 \phi^\dagger \phi - \frac{\lambda}{4}(\phi^\dagger \phi)^2 \tag{4.40}$$

then exhibits a global SU(2) invariance of the form (c.f. (3.10))

$$\phi \to \phi' = \exp(-i\underline{\alpha}\cdot\underline{\tau}/2)\phi, \tag{4.41}$$

but this is spontaneously broken, the minimum of the potential in (4.40) occurring at (c.f. (4.18))

$$(\phi^\dagger \phi)_{\min} = 2\mu^2/\lambda \equiv v^2/2. \tag{4.42}$$

As in the U(1) case, we interpret (4.42) in the quantum theory as (*c.f.* (4.22))

$$\langle 0| \phi^\dagger \phi |0\rangle = v^2/2, \tag{4.43}$$

so that the ϕ-field has a non-zero value in the vacuum. Once again, we exclude higher powers of $\phi^\dagger \phi$ in (4.40) on grounds of renormalisability.

As before, in order to get a sensible particle spectrum we must 'shift' the fields so as to deal with stable oscillations about the minimum (vacuum) given by (4.43). So we need to define '$\langle 0|\phi|0\rangle$' and expand about it, as in (4.22) and (4.23). In the present case, however, the situation is more complicated than (4.23), since the complex doublet (4.39) contains four real fields, parametrised for example as

$$\phi^+ = \frac{1}{\sqrt{2}}(\phi_1 - i\phi_2), \quad \phi^0 = \frac{1}{\sqrt{2}}(\phi_3 - i\phi_4); \tag{4.44}$$

(4.43) then becomes

$$\langle 0| \phi_1^2 + \phi_2^2 + \phi_3^2 + \phi_4^2 |0\rangle = v^2. \tag{4.45}$$

It is evident that we have a lot of freedom in choosing the $\langle 0|\phi_i|0\rangle$ so that (4.45) holds, and it is not at first obvious what an appropriate generalisation of (4.22) and (4.23) might be.

Furthermore, in this more complicated (non-Abelian) situation a qualititatively new feature can arise: it may happen that the chosen condition $\langle 0|\phi_i|0\rangle \neq 0$ is *invariant* under some subset of the allowed symmetry transformations. This would effectively mean that this particular choice of the vacuum state respected that subset of symmetries, which would therefore not be 'spontaneously broken' after all. Since each broken symmetry is associated with a massless Goldstone boson, we would then get fewer of these bosons than expected.

Just this happens (by design!) in the present case. To understand how it works, we must first recognize that, in addition to the global SU(2) symmetry of (4.41), $\mathcal{L}_\Phi$ of (4.40) is also invariant under a completely independent global U(1) symmetry of the form

$$\phi \rightarrow \phi' = e^{-i\beta} \phi \tag{4.46}$$

which just means that the phases of the upper and lower components of ϕ in (4.39) change simultaneously by the same amount. Thus the full symmetry of (4.40) is global SU(2)×U(1) (which will be made local in a moment, as is required in the Standard Model).

Suppose then that we could choose the $\langle 0|\phi_i|0\rangle$ so as to break this SU(2)×U(1) symmetry completely: we would then expect four massless fields. Actually, however, it is not possible to make such a choice. An analogy may make this point clearer. Suppose we were considering just SU(2), and the field $\underline{\phi}$ was an SU(2)-triplet. Then we could always write $\langle 0|\underline{\phi}|0\rangle = v\underline{n}$ where $\underline{n}$ is a unit vector; but this form is invariant under rotations about the $\underline{n}$-axis, irrespective of where that points. In the present case, by using the freedom of global SU(2)×U(1) phase changes, an arbitrary $\langle 0|\phi|0\rangle$ can be brought to the form

$$\langle 0|\phi|0\rangle = \begin{pmatrix} 0 \\ v/\sqrt{2} \end{pmatrix}. \tag{4.47}$$

In considering what symmetries are respected or broken by (4.47), it is easiest to look at infinitesimal transformations. It is then clear that the particular transformation

$$\delta\phi = -i\epsilon(1 + \tau_3)\phi \tag{4.48}$$

(which is a combination of (4.46) and the 'third component' of (4.41)) is still a symmetry of (4.47) since

$$(1 + \tau_3)\begin{pmatrix} 0 \\ v/\sqrt{2} \end{pmatrix} = \begin{pmatrix} 0 \\ 0 \end{pmatrix}, \tag{4.49}$$

so that

$$\langle 0|\phi|0\rangle = \langle 0|\phi + \delta\phi|0\rangle ; \tag{4.50}$$

we say that 'the vacuum is invariant under (4.48)', and when we look at the spectrum of oscillations about that vacuum we expect to find only three massless bosons, not four.

Oscillations about (4.47) are conveniently parametrized by

$$\phi = \exp(-i\underline{\theta}(x)\cdot\underline{\tau}/2v)\begin{pmatrix} 0 \\ (v + H(x))/\sqrt{2} \end{pmatrix}, \tag{4.51}$$

which is to be compared with (4.23). Inserting (4.51) into (4.40), we easily find that no mass term is generated for the $\underline{\theta}$ fields, while the H field piece is

$$\mathcal{L}_H = \frac{1}{2}\partial_\mu H \partial^\mu H - \mu^2 H^2 + \text{interactions} \tag{4.52}$$

just as in (4.24), showing that $m_H = \sqrt{2}\mu$.

As noted in Section 4.1, there is an interesting physical example of a spontaneously broken global SU(2) symmetry, the SU(2)$_{f5}$ symmetry of $\mathcal{L}_{\text{QCD}}$ (Equation (4.10)), in which the three massless modes are identified with the pions. We cannot consider this in any more detail here, however, being concerned rather to proceed to the local version of the SU(2)×U(1) model of (4.40). Such an extension is easily written down, just by using the SU(2) covariant form (3.28) and the U(1) covariant derivative of the form (2.50). In the notation we shall use in the next section, this means replacing (4.40) by

$$\mathcal{L}_{G\Phi} = (D_\mu\phi)^\dagger(D^\mu\phi) + \mu^2\phi^\dagger\phi - \frac{\lambda}{4}(\phi^\dagger\phi)^2 - \frac{1}{4}\underline{F}_{\mu\nu}\cdot\underline{F}^{\mu\nu} - \frac{1}{4}G_{\mu\nu}G^{\mu\nu} \tag{4.53}$$

where

$$D_\mu\phi = (\partial_\mu + ig\underline{\tau}\cdot\underline{W}_\mu/2 + ig'B_\mu/2)\phi, \tag{4.54}$$

$\underline{F}_{\mu\nu}$ is as in (3.42), and $G_{\mu\nu} = \partial_\mu B_\nu - \partial_\nu B_\mu$. Thus the $\underline{W}$'s are the SU(2) gauge fields, and the B is the U(1) gauge field. (4.53) is, in fact, the gauge and Higgs field sector of the Standard Model. As in the local U(1) case, the particle spectrum is most easily found by exploiting the local gauge freedom to choose the $\underline{\theta}$ fields in (4.51) to vanish, as in the ansatz (4.30): that is, we set

$$\phi = \begin{pmatrix} 0 \\ (v + H(x))/\sqrt{2} \end{pmatrix}. \tag{4.55}$$

Substituting (4.55) into (4.53) and retaining only terms which are of second order in the fields (*i.e.* kinetic energies or mass terms) we find

$$\begin{aligned}
\mathcal{L}_{G\Phi} &= \frac{1}{2}\partial_\mu H \partial^\mu H - \mu^2 H^2 \\
&\quad - \frac{1}{4}F_{1\mu\nu}F_1^{\mu\nu} + \frac{1}{8}g^2 v^2 W_{1\mu}W_1^\mu \\
&\quad - \frac{1}{4}F_{2\mu\nu}F_2^{\mu\nu} + \frac{1}{8}g^2 v^2 W_{2\mu}W_2^\mu \\
&\quad - \frac{1}{4}F_{3\mu\nu}F_3^{\mu\nu} - \frac{1}{4}G_{\mu\nu}G^{\mu\nu} + \frac{1}{8}v^2(gW_{3\mu} - g'B_\mu)(gW_3^\mu - g'B^\mu).
\end{aligned} \quad (4.56)$$

The first line of (4.56) tells us that we have a scalar field of mass $\sqrt{2}\mu$. The next two lines tell us that the components W_1 and W_2 of the triplet (W_1, W_2, W_3) acquire a mass

$$M_1 = M_2 = gv/2 \equiv M_W. \quad (4.57)$$

The last line shows us that the fields W_3 and B are mixed. But they can easily be unmixed by noting that the last term in (4.56) involves only the combination $gW_3 - g'B$, which evidently acquires a mass. This suggests introducing the linear combinations

$$Z^\mu = \cos\theta_W W_3^\mu - \sin\theta_W B^\mu \quad (4.58)$$
$$A^\mu = \sin\theta_W W_3^\mu + \cos\theta_W B^\mu \quad (4.59)$$

where

$$\cos\theta_W = g/(g^2 + g'^2)^{1/2}, \quad \sin\theta_W = g'/(g^2 + g'^2)^{1/2}. \quad (4.60)$$

We then find that the last line of (4.56) becomes

$$-\frac{1}{4}F_{Z\mu\nu}F_Z^{\mu\nu} + \frac{1}{8}v^2(g^2 + g'^2)Z_\mu Z^\mu - \frac{1}{4}F_{\mu\nu}F^{\mu\nu} \quad (4.61)$$

where

$$F_{Z\mu\nu} = \partial_\mu Z_\nu - \partial_\nu Z_\mu \quad \text{and} \quad F_{\mu\nu} = \partial_\mu A_\nu - \partial_\nu A_\mu.$$

Thus

$$M_Z = \frac{1}{2}v(g^2 + g'^2)^{1/2} = M_W/\cos\theta_W \quad (4.62)$$

and

$$M_A = 0. \quad (4.63)$$

Counting degrees of freedom as in the local U(1) case, we originally had 12 in (4.53)—three massless W's and one massless B, which is 8 in all, together with 4 ϕ-fields. After symmetry breaking, we have three massive vector fields W_1, W_2 and Z making 9 degrees of freedom, one massless vector field A with 2, and one massive scalar H. Of course, the physical application will be to identify the W and Z fields with those physical particles, and the A field with the massless photon. In the gauge (4.55), the W and Z particles have propagators of the form (4.32).

The identification of A^μ with the photon field is made clearer if we look at the form of $D_\mu \psi$ written in terms of A_μ and Z_μ, discarding the W_1, W_2 pieces:-

$$D_\mu \phi = \left\{\partial_\mu + ig\sin\theta_W \left(\frac{1+\tau_3}{2}\right)A_\mu + \frac{ig}{\cos\theta_W}\left[\frac{\tau_3}{2} - \sin^2\theta_W \left(\frac{1+\tau_3}{2}\right)\right]Z_\mu\right\}\phi. \quad (4.64)$$

Now the operator $(1 + \tau_3)$ acting on $\langle 0|\phi|0\rangle$ gives zero, as observed in (4.49), and this is why A_μ does not acquire a mass when $\langle 0|\phi|0\rangle \neq 0$ (gauge fields coupled to unbroken symmetries of $\langle 0|\phi|0\rangle$ do not become massive). Although certainly not unique, this choice of ϕ and $\langle 0|\phi|0\rangle$ (due to Weinberg (1967)) is undoubtedly very economical and natural. The zero eigenvalue of $(1+\tau_3)$ can be interpreted as the electromagnetic charge of the vacuum, which we would not wish to be non-zero. We would then tentatively expect the identification

$$e = g \sin \theta_W \tag{4.65}$$

in order to get the right 'electromagnetic D_μ' in (4.64).

We have at last assembled all the conceptual ingredients we need for the electroweak theory, to which we now turn.

5 The electroweak theory

5.1 Fermi theory, and non-gauge vector bosons

As originally conceived by Fermi in 1934, the interaction responsible for β-decay was of the 'four-fermion' type $\psi_p^\dagger \psi_n \psi_e^\dagger \psi_{\nu_e}$ with all fields interacting at the same point, as in Figure 22. The idea was extended to μ decay, via the purely leptonic interaction $\psi_{\nu_\mu}^\dagger \psi_\mu \psi_e^\dagger \psi_{\nu_e}$. After decades of experimentation, it was established that these simple 'scalar' (and parity conserving) interactions were incorrect, the right low-energy interaction having the form

$$(G_F/\sqrt{2})\overline{\psi}_4 \gamma_\mu (1-\gamma_5)\psi_3 \overline{\psi}_2 \gamma^\mu (1-\gamma_5)\psi_1 \,. \tag{5.1}$$

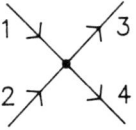

Figure 22.

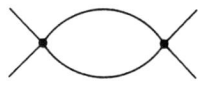

Figure 23.

The Fermi coupling constant G_F in (5.1) must have dimensions (mass)$^{-2}$, as was noted in the case of the similar effective $e^+e^- \to e^+e^-$ vertex considered in (2.83) and Figure 9. Its value, as extracted from μ-decay after inclusion of QED radiative corrections, is (PDP 1992)

$$G_F = 1.16639(2) \times 10^{-5} \text{ GeV}^{-2} \,. \tag{5.2}$$

Following the discussion in Section 2.4, we expect this fact to imply that the interaction (5.1) is non-renormalisable, in the sense that as we approach the natural energy scale $G_F^{-1/2}$ we shall need to take increasingly more parameters from experiment in order to control the cut-off dependences of loops such as Figure 23. This scale is $G_F^{-1/2} \approx 300$ GeV. Partly in connection with the search for a renormalisable theory of weak interactions, and also because it was felt desirable to see all forces as arising from some kind of 'Yukawa quantum exchange' mechanism, attempts were made very early on to regard (5.1) as an effective low energy theory, replacing Figure 22 by Figure 24 for which the amplitude would be (dropping i's) of the form

$$\hat{g}^2 \bar{u}_4 \gamma_\mu (1-\gamma_5) u_3 \frac{[-g^{\mu\nu} + q^\mu q^\nu / M_V^2]}{q^2 - M_V^2 + i\epsilon} \bar{u}_2 \gamma_\nu (1-\gamma_5) u_1 \quad (5.3)$$

where (c.f. (2.29)) $\hat{g}$ is the coupling of the 'V-particle' to the fermions. The propagator in (5.3) (c.f. (4.32)) is associated with the exchange of a massive V-quantum which has to be a vector (spin 1) particle because of the couplings in (5.1). For momentum transfer $q^2 \ll M_V^2$, (5.3) evidently reduces to (5.1) with $G_F/\sqrt{2} = \hat{g}^2/M_V^2$ (we shall shortly see the the correct version of this in equation (5.16) below).

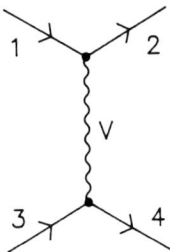

Figure 24.

Prior to the formulation of the electroweak theory (Glashow 1961, Salam 1968, Weinberg 1967), and before the parameter θ_W had been measured in neutral current processes (see below), the mass M_V of the vector boson was unknown. Nevertheless, it was attractive to suppose that perhaps it coupled with an intrinsic strength of the same order of magnitude as the photon does, so that $\hat{g} \sim e$. Knowing the value of G_F then gave an estimate for M_V, namely $M_V \sim e/G_F^{1/2} \sim 100\,\text{GeV}$—as is, indeed, roughly correct for the W and Z particles (see (5.16)). Is a theory of massive charged vector bosons (suitably extended to include neutral ones mediating weak processes which do not change quark or lepton charge) perfectly satisfactory? The answer is: not entirely, because it is not renormalisable either. This may come as a surprise, since the vertex factor for the fermion-V coupling in (5.3) corresponds to the interaction term

$$\hat{g} \bar{\psi} \gamma_\mu (1 - \gamma_5) \psi V^\mu \quad (5.4)$$

where $\hat{g}$ is dimensionless. The discussion of Sections 2.3 and 2.4 suggested that a coupling with negative mass dimension signalled a non-renormalisable theory, but that theories with dimensionless couplings were all right. Or at least QED was.... Indeed, there is a problem with an ordinary massive charged vector boson theory.

The nature of the trouble appears, for example, if we consider a weak interaction analogue of one of the e^+e^- box graphs shown in Figure 5, namely the process in Figure 25. Noting that the V propagators behave no better than a constant for large k^2 (from (4.32) or (5.3)), we seem to have a quadratic divergence, a sure signal of trouble in a two fermion → two fermion process. It is clear that the 'scale' of the trouble is set by the vector boson mass M_V, since for loop momenta much less than M_V^2 the propagator behaves like $1/k^2$, which would make the integral rapidly insensitive to the cut-off. The discussion of Section 2.4 therefore implies, once again, that a simple massive charged V theory will only make sense up to an energy scale of order the V mass. Beyond that point we begin to be sensitive to a higher energy cut-off.

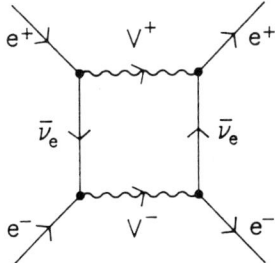

Figure 25.

In the case of QED, the analogous box graph is convergent, essentially because of the better large-k behaviour of the γ-propagator, (2.25). Looking at (2.25) more closely, we see that the $k^\mu k^\nu$ part is actually gauge dependent—and this raises the thought that perhaps, if our V's could somehow be regarded as *gauge* quanta (even though massive), then we could get rid of the offending $k^\mu k^\nu$ part of the massive vector propagator. Indeed this is the case. As we remarked after (4.32), the form given there corresponds to one particular choice of gauge in the U(1) Higgs model. There is a general class of gauges in which the propagator is (see Chapter 13 of Aitchison and Hey, 1989)

$$\left[-g_{\mu\nu} + \frac{(1-\xi)k_\mu k_\nu}{k^2 - \xi M_A^2} \right] \frac{1}{k^2 - M_A^2}. \tag{5.5}$$

We see that (4.32) is a special case of (5.5) with the gauge parameter $\xi \to \infty$: this is called 'unitary gauge'. With $\xi = 1$, on the other hand, we have a propagator which behaves like $1/k^2$ for large k^2, which if used in Figure 25 will make the loop integral converge. The problem with any finite ξ is that (5.5) contains an unphysical pole at $k^2 = \xi M_A^2$, but this is actually cancelled by a contribution from an 'unphysical Higgs field'. Thus one gauge is suitable for exhibiting the physical particle spectrum, while another is best for considering loops and renormalisation. The hope would be that the underlying gauge invariance of the theory (even after spontaneous symmetry breaking) would ensure that all results were independent of ξ, so that the theory was *both* renormalisable *and* possessed a sensible spectrum. To prove that this was indeed so was the achievement of 't Hooft (1971a, b). The discovery of the renormalisability of spontaneously broken gauge theories in the early 1970's dramatically revived interest in the Glashow (1961)-Salam (1968)-Weinberg (1967) model, which until then had not been much pursued. In principle, the property of renormalisability does mean that by introducing a few uncalculable but measured parameters into the theory, higher order effects are insensitive to the cut-off scale, and are actually finite even as the latter goes to ∞. In a real sense, the Higgs mass m_H (which is an arbitrary parameter) essentially plays the role of a cut-off. We shall see, however, that there are serious reasons for doubting that the simple Higgs sector of Section 4.3 can really make sense for m_H much beyond 500–1000 GeV.

5.2 The electroweak theory for one fermion family

Electroweak interactions are described by a gauge theory based on a spontaneously broken local $SU(2)_L \times U(1)$ invariance. The 'L' means that the SU(2) part (with the gauge fields $\underline{W}^\mu$ of Section 4.3) acts only on the left-handed parts of fermion fields (see (4.5) and (5.1)); it is therefore 'maximally' parity violating. The U(1) part (with the gauge field B^μ) acts on both right-(if any) and left-handed components, in such a way that the particular combination (4.59) conserves parity, as is required for the electromagnetic interaction; the other combination (4.58), which mediates neutral weak interactions, will turn out not to couple in the 'pure V−A' form, as is indeed observed. One could imagine different schemes: for example (Bludman 1958), a simple $SU(2)_L$ in which the vector boson masses were put in by hand, the neutral currents were pure V−A and electromagnetism was quite separate; or an $SU(2)_L$ in which (by suitable symmetry breaking) the neutral boson remained massless and was identified with the photon—but then electromagnetism would have to violate parity though a V−A interaction. The simplest group structure allowing connection between the weak and parity conserving electromagnetic forces is the $SU(2)_L \times U(1)$ one, originally proposed by Glashow (1961). The $SU(2)_L$ is often called 'weak isospin' and the U(1) 'weak hypercharge'.

In this theory, the basic fields are fermions (leptons and quarks), gauge bosons, and Higgs fields. The left-handed parts of the fermion fields form (weak isospin) doublets under $SU(2)_L$

$$`\psi_L' = \begin{pmatrix} \nu_e \\ e^- \end{pmatrix}_L, \begin{pmatrix} \nu_\mu \\ \mu^- \end{pmatrix}_L, \begin{pmatrix} \nu_\tau \\ \tau^- \end{pmatrix}_L, \begin{pmatrix} u \\ \tilde{d} \end{pmatrix}_L, \begin{pmatrix} c \\ \tilde{s} \end{pmatrix}_L, \begin{pmatrix} t \\ \tilde{b} \end{pmatrix}_L, \quad (5.6)$$

where the ˜ denotes states which are mixed with respect to the strong interaction states d, s and b (see the following section, and note that the colour labels will be suppressed throughout), while the right-handed components are $SU(2)_L$ singlets

$$`\psi_R' = e_R^-, \mu_R^-, \text{ etc.}, \quad (5.7)$$

where for simplicity we shall generally assume in this section that the neutrinos are massless (see also Section 5.3). We shall confine the discussion in the present section to just one 'family', comprising ν_e, e^-, u and d (which should really be $\tilde{d}$ but we are ignoring mixing for the moment).

The Lagrangian can be looked at in many ways, but we shall write it as

$$\mathcal{L} = \mathcal{L}_S + \mathcal{L}_{SB} \quad (5.8)$$

where S stands for 'symmetrical' under $SU(2) \times U(1)$ and SB stands for 'symmetry breaking'. In $\mathcal{L}_S$ we have a gauge invariant Lagrangian $\mathcal{L}_f$ describing the interactions of the fermions with the $\underline{W}$ and B fields, together with the SU(2) Yang-Mills Lagrangian $\mathcal{L}_W$ (3.44) for the $\underline{W}$ fields and the U(1) Lagrangian $\mathcal{L}_B$ for the B field as in (4.53);

in $\mathcal{L}_{SB}$ we will have everything involving the Higgs fields. In Section 3.2 we learned how to construct a locally SU(2) invariant gauge theory with a fermion doublet (see (3.28)). The difference now is that we want the SU(2)$_L$ to act only on the L-component of the doublet. However, there is no problem with this for *massless* fields: (4.8) shows us that the 'kinetic' operator $\not{\partial}$ does not mix L and R components, and hence there is no objection to 'gauging' each of them differently (*i.e.* using a different $\not{D}$ on ψ_L and on ψ_R). On the other hand, (4.9) shows that this is *not* true for the mass terms—a difficulty we will deal with shortly by getting the mass terms from $\mathcal{L}_{SB}$. First, we simply state that the appropriate D's are in fact

$$D_\mu = \partial_\mu + ig\boldsymbol{\tau}\cdot\boldsymbol{W}_\mu/2 + ig'yB_\mu/2 \quad \text{on } \psi_L\text{'s} \tag{5.9}$$

and

$$D_\mu = \partial_\mu + ig'yB_\mu/2 \quad \text{on } \psi_R\text{'s}, \tag{5.10}$$

where the condition

$$Q = \tau_3/2 + y/2 \tag{5.11}$$

is imposed, Q being the electric charge in units of e (the positron charge). The factor of $1/2$ in the B-term of (5.9) is conventional, but (5.11) fixes the normalisation of the coupling g'. The eigenvalues of the $\tau_3/2$ operator in (5.9) are as indicated by the placings in (5.6): namely $+1/2$ for $(\nu_e, \nu_\mu, \nu_\tau, u, c, s)_L$ and $-1/2$ for e_L^-, etc. For the (lepton)$_L$ fields the y eigenvalue is -1, while for the (quark)$_L$ fields it is $+1/3$; for the R-fields y is just $2Q$ since the $\tau_3/2$ eigenvalue is zero.

The gauge invariant Lagrangian $\mathcal{L}_f$ (for massless fermions) is therefore

$$\mathcal{L}_f = \bar{\ell}_{eL}i\not{D}\ell_{eL} + \bar{q}_L i\not{D} q_L + \bar{e}_R i\not{D} e_R + \bar{u}_R i\not{D} u_R + \bar{d}_R i\not{D} d_R \tag{5.12}$$

where

$$\ell_{eL} = \begin{pmatrix}\nu_e \\ e^-\end{pmatrix}_L, \quad q_L = \begin{pmatrix}u \\ d\end{pmatrix}_L \tag{5.13}$$

and a ν_{eR} term can be added to (5.12) if desired. From (5.12) we can already read off the couplings of the charged W's to the fermions (the W_3 and B will mix, as we saw in Section 4.3). The correct normalisation for charged fields is that $W^\mu = (W_1 - iW_2)/\sqrt{2}$ destroys the W^+ or creates W^-, so that the $\boldsymbol{\tau}\cdot\boldsymbol{W}/2$ terms are

$$\frac{1}{\sqrt{2}}\left\{\tau_+\frac{(W_1 - iW_2)}{\sqrt{2}} + \tau_-\frac{(W_1 + iW_2)}{\sqrt{2}}\right\} + \tau_3\frac{W_3}{2} \tag{5.14}$$

where $\tau_\pm = (\tau_1 \pm i\tau_2)/2$ are the raising and lowering operators for the doublet. Thus the first term in (5.14) picks out the process $e^- \to \nu_e W^-$ for example, with the result that the corresponding vertex is

$$-\frac{ig}{\sqrt{2}}\gamma_\mu\frac{(1-\gamma_5)}{2}, \tag{5.15}$$

and similarly for the quarks (if unmixed), and other families. Hence (*c.f.* (5.3)) we can immediately make a connection with the original Fermi theory of these charged current processes, namely

$$G_F/\sqrt{2} = g^2/8M_W^2. \tag{5.16}$$

Although the quark couplings can also be read off from (5.12), they are unphysical at this stage since mixing has not yet been introduced.

We must now consider how to bring fermion masses into this theory. We begin by noting, again, that a typical mass term has the form (4.9), which is clearly not invariant under transformations which treat ψ_L and ψ_R differently. Would it matter if we just added in such a mass term? The answer is that if we did this the theory would (once again) not be renormalisable. An indication of the trouble can once more be obtained by considering the box graph of Figure 25, with $V^\pm$ replaced by $W^\pm$. It seems intuitively plausible (and is essentially correct) that the box graph of Figure 25 can be obtained by somehow multiplying two ν_e-exchange tree graphs together, of the kind shown in Figure 17, and integrating over the internal momentum. From this point of view, the bad high-energy behaviour of the box integral can be related to that of the tree graph. Now we remarked in Section 3.2 that the sum of all three tree graphs in Figure 17, with coefficients determined by the gauge symmetry, had a much less rapid rise with energy than did any one of them alone—for example the ν_e-exchange one. This is part of the reason why the 'disease' in Figure 25 is tamed by the gauge symmetry: other graphs have to be added to it, in the same order of perturbation theory. But more is needed. If one calculates the amplitude from *all* of Figure 17 in a helicity non-conserving channel, one finds (Appelquist and Chanowitz 1987) that the amplitude is proportional to $G_F m_e \sqrt{s}$ where s is the square of the c.m. momentum. This behaviour is bad enough to cause a 'dangerous' divergence in the box graphs built out of Figure 17. We therefore require a further cancellation—which can be provided, for example, by a new scalar particle in the s-channel, as shown in Figure 26. In this case the scalar particle will have to couple to the $e^+ e^-$ with a strength proportional to m_e, for the cancellation to occur. Of course, as a contributor to the physical process $e^+ e^- \to W^+ W^-$ at tree level, Figure 26 will be negligible until very high energies, $\sqrt{s} \sim 10^5$ TeV, because of the small value of m_e. Note, however, that a similar contribution must also be present in $t\bar{t} \to W^+ W^-$, and in this case it will begin to be significant at much lower energies (of order 10 TeV for $m_t \sim 150$ GeV).

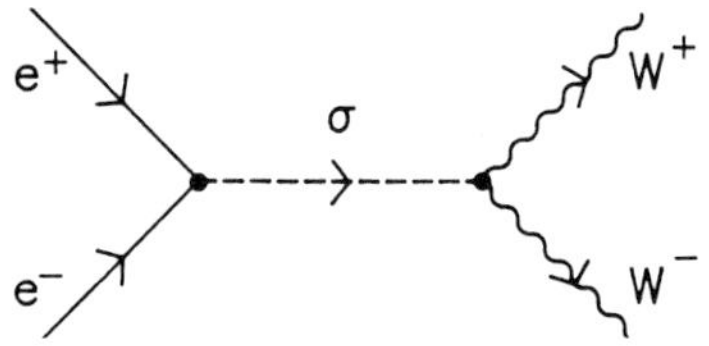

Figure 26.

The study of the high energy behaviour of tree graphs, and the constraints which the requirement of 'good' high energy behaviour imposes on the couplings and/or the particle spectrum, is a long-established and fascinating approach to gauge theories (see for example Cornwall *et al.* (1974) and Llewellyn Smith (1974))—but we have not been following it in the present course. However, it may come as no surprise that the 'scalar particle' of the preceding paragraph is (or can be) none other than the Higgs boson H. In other words it is possible to 'blame' the fermion masses on the Higgs sector, just as the W and Z masses are. Once again, 'something like this' seems to be necessary for the theory to be consistent much beyond the W–Z mass range.

The idea for the fermion masses is easily illustrated for the ν_e and e^- fields, in the simple case that the ν_e is indeed massless. Consider a 'Yukawa' interaction of the following type between these electron-type fields and the Higgs field ϕ (which, as in Section 4.3, is also an SU(2) doublet):

$$-g_e(\bar{l}_{eL}\phi e_R + \bar{e}_R\phi^\dagger l_{eL}). \tag{5.17}$$

Remembering that e_R is an SU(2) scalar, we see that (5.17) is Lorentz invariant, and invariant under global SU(2) transformations (because $\bar{l}\phi$ and $\phi^\dagger l$ are invariant); it is also invariant under U(1)$_y$ transformations, with the y assignments made after (5.11), if $y(\phi) = 1$ (which is what we actually assumed in (4.54)). In fact, since no derivatives are involved in (5.17), it is also invariant under local SU(2)×U(1) transformations. But the Higgs sector contains the potential $V(\phi)$ of (4.40), which 'triggers' spontaneous symmetry breaking. The vacuum value (4.47) for ϕ when inserted into (5.17), yields

$$-(g_e v/\sqrt{2})(\bar{e}_L e_R + \bar{e}_R e_L) \tag{5.18}$$

which is precisely a mass term for the electron if we identify

$$g_e = m_e \sqrt{2}/v. \tag{5.19}$$

When oscillations about this vacuum are considered, in the simple gauge of (4.55), one easily finds that the H-field couples to the electron with a vertex

$$-im_e/v. \tag{5.20}$$

Sure enough, the coupling is proportional to the electron mass—and on dimensional grounds to v^{-1}.

It might seem from the following that only a mass for the $t_3 = -\frac{1}{2}$ component of the fermion doublets could be generated this way, because of the form of $\langle 0|\phi|0\rangle$. Remarkably enough, however, the same Higgs field can also provide a mass for the $t_3 = +\frac{1}{2}$ component (and this is of course necessary for the quarks, if not for the neutrinos). It can be shown that the field ϕ_c defined by

$$\phi_c = i\tau_2\phi^* = \begin{pmatrix} (\phi_3 + i\phi_4)/\sqrt{2} \\ -(\phi_1 + i\phi_2)/\sqrt{2} \end{pmatrix} = \begin{pmatrix} \bar{\phi}^0 \\ -\phi^- \end{pmatrix}, \tag{5.21}$$

where (4.44) has been used, is also an isodoublet. (The notation in (5.21) is reminiscent of the K-meson doublet $(\bar{K}^0, K^-)$; alternatively, we may think of a quark isospin doublet like $\begin{pmatrix} u \\ d \end{pmatrix}$ and its conjugate doublet $\begin{pmatrix} \bar{d} \\ -\bar{u} \end{pmatrix}$, with the $I = 0$ combination being $(\bar{d}d - \bar{u}u)$). With the help of ϕ_c we can write down another gauge invariant coupling in the ν_e-e sector, namely

$$-g_{\nu_e}\left(\bar{l}_{eL}\phi_c\nu_{eR} + \bar{\nu}_{eR}\phi_c^\dagger l_{eL}\right) \tag{5.22}$$

which produces

$$-\left(g_{\nu_e}v/\sqrt{2}\right)\left(\bar{\nu}_{eL}\nu_{eR} + \bar{\nu}_{eR}\nu_{eL}\right) \tag{5.23}$$

in the Higgs vacuum (4.47), which is a neutrino mass term (if required) provided $g_{\nu_e} = \sqrt{2} m_{\nu_e}/v$. Once again, the H-field will couple with an amplitude of the form (5.20), with $m_e \to m_{\nu_e}$. The procedure can obviously be repeated for the u and d quarks.

It is clearly possible to go on like this, and arrange for as many fermion families to have a mass as is required—and we will look at this a little more closely in the next section. However, one must note that the theory does no more than accommodate itself to the mass difficulty: in no sense do the fermion masses 'come out' of the theory, since each has simply to be inserted by hand via a new Yukawa coupling. In essence, these Yukawa couplings are *not* gauge interactions, and hence not universal.

By contrast, this is not the case—or at least not to quite the same extent—for the vector boson masses. They also arise through symmetry breakdown via the Higgs sector (in the Standard Model)—see section 4.3. But in this case their couplings to the Higgs field are (largely) determined by the gauge symmetry. Let us complete this section by looking again at the Higgs sector Lagrangian $\mathcal{L}_\Psi$ of (4.40), which together with (5.17)—and terms like (5.22) if needed—makes up $\mathcal{L}_{SB}$ for this one family case.

As shown in Section 4.3, after spontaneous breaking we have

$$M_W = gv/2 = \cos\theta_W M_Z \tag{5.24}$$
$$\cos\theta_W = g/(g^2 + g'^2)^{1/2} \tag{5.25}$$
$$e = g\sin\theta_W \tag{5.26}$$
$$m_H = \sqrt{2}\mu \tag{5.27}$$

in terms of the fundamental coupling parameters g, g' of the $SU(2) \times U(1)$ gauge group, and the parameters v and μ of the Higgs potential. There is also the low-energy connection (5.16), which we can write as

$$\frac{v}{\sqrt{2}} = 2^{-3/4} G_F^{-1/2} = 174.1\,\text{GeV}, \tag{5.28}$$

using the value (5.2). This gives us the scale of $\langle 0|\phi|0\rangle$, for which as yet there is no theoretical explanation. We may also write (5.16) as

$$M_W = (\pi\alpha/\sqrt{2}\,G_F)^{1/2}/\sin\theta_W \tag{5.29}$$
$$= 37.2802\,\text{GeV}/\sin\theta_W \tag{5.30}$$

using the conventional low-energy value of α. Note that all the above relations are between parameters in the Lagrangian, and hold at the tree level only; they can be changed by loop corrections (see Section 5.5).

The Higgs coupling to fermions can now be written generally as

$$-iem_f/2\sin\theta_W M_W. \tag{5.31}$$

There are also trilinear and quadrilinear Higgs self-couplings arising from the $\lambda(\phi^\dagger\phi)^2$ term in (4.53). Recalling that $\lambda = 4\mu^2/v^2$ and that $m_H = \sqrt{2}\mu$, we can write the trilinear coupling as

$$-i3m_H^2 e/8 M_W \sin\theta_W \tag{5.32}$$

and the quadrilinear as

$$-i3m_H^2 e^2/16 M_W^2 \sin^2\theta_W. \tag{5.33}$$

There are also the trilinear H-W^+-W^-

$$ieM_W g_{\lambda\mu}/\sin\theta_W \tag{5.34}$$

and H-Z-Z

$$i2eM_Z g_{\lambda\mu}/\sin 2\theta_W \tag{5.35}$$

couplings, together with quadrilinear $\phi^2 W^2$, $\phi^2 Z^2$ couplings which we shall not give here. Note that all these couplings are determined by the existing set of parameters—and, in particular, that the Higgs couples most strongly to the heaviest particles, so that decays to heavy channels offer the largest rates.

There are also couplings of the Higgs to the Z^0, and of the Z^0 to fermions. To find these, we need to rewrite the neutral part of the D's in (5.0) and (5.10) in terms of the Z and A fields defined in (4.58) and (4.59) (c.f. (4.64)). We find

$$D_\mu(\text{neutral}) = \partial_\mu + ieQA_\mu + \frac{igZ_\mu}{2\cos\theta_W}(v_f - a_f\gamma_5) \tag{5.36}$$

where

$$v_f = \frac{\tau_3}{2} - 2Q\sin^2\theta_W \tag{5.37}$$

and

$$a_f = \frac{\tau_3}{2}. \tag{5.38}$$

We see that, as remarked earlier, the Z (or 'neutral-current') coupling is not pure V−A. The Z-couplings analogous to (5.15) are therefore

$$\frac{-ig}{2\cos\theta_W}\gamma_\mu(v_f - a_f\gamma_5). \tag{5.39}$$

(5.39) is the coupling observed around the Z^0 peak.

Finally, we may write effective four-fermion interactions (valid for energies much less than M_W, M_Z) as

$$\frac{G_F}{\sqrt{2}} j^C_{\mu+} j^{C\mu}_- \tag{5.40}$$

for the charged current processes, with (c.f. (5.1))

$$j^C_{\mu\pm} = (\bar\psi_2 \gamma_\mu (1-\gamma_5)\tau_\pm \psi_1), \tag{5.41}$$

and as

$$\sqrt{2} G_F \rho j^N_\mu j^{N\mu} \tag{5.42}$$

for the neutral current processes, where

$$j^N_\mu = \bar\psi_f \gamma_\mu (v_f - a_f\gamma_5)\psi_f \tag{5.43}$$

and the quantity

$$\rho = M_W^2 / M_Z^2 \cos^2\theta_W \tag{5.44}$$

has the value 1 in the Standard Model, at tree level.

5.3 The three-family model

We now extend the preceding discussion to the three family case, which will involve the important subjects of quark flavour mixing in charged current processes (and of no mixing—the GIM mechanism (Glashow et al. 1970)—in neutral current processes), and CP violation. We shall here assume that there are just three families (see for example the lectures by Blondel for a recent review of the evidence for that).

We cannot do better than follow the excellent treatment given by Jarlskog (1985) in an earlier School. We introduce three doublets of left handed fields

$$\begin{pmatrix} f_{1L} \\ f'_{1L} \end{pmatrix}, \quad \begin{pmatrix} f_{2L} \\ f'_{2L} \end{pmatrix}, \quad \begin{pmatrix} f_{3L} \\ f'_{3L} \end{pmatrix} \tag{5.45}$$

and the corresponding six singlets

$$f_{1R}, \quad f'_{1R}, \quad f_{2R}, \quad f'_{2R}, \quad f_{3R}, \quad f'_{3R}, \tag{5.46}$$

which transform in the now familiar way under $SU(2)_L \times U(1)$. The unprimed fields correspond to the $t_3 = +\frac{1}{2}$ components of $SU(2)_L$, the primed ones to the $t_3 = -\frac{1}{2}$ components, and to their 'R' partners. The labels 1, 2, and 3 refer to the family number; for example, with no mixing at all, $f_{1L} = u_L$, $f'_{1L} = d_L$, etc. (We are thinking of (5.45) and (5.46) as quark fields, but the discussion will be quite general and could just as well apply to leptons if they should need mixing too—we return to leptons later). We have to consider what is the most general $SU(2)_L \times U(1)$-invariant interaction between the Higgs field (assuming we can still get by with only one) and these various fields. Apart from the symmetry, the only other theoretical requirement is renormalisability—for, after all, if we drop this we might as well abandon the whole motivation for the 'gauge' concept. This implies (as in the discussion of the Higgs potential V) that we cannot have terms like $(\bar{\psi}\psi\phi)^2$ appearing—which would have a coupling with dimensions $(\text{mass})^{-4}$ and would be non-renormalisable. In fact the only renormalisable Yukawa coupling is of the form '$\bar{\psi}\psi\phi$', which has a dimensionless coupling (as in the g_e and g_{ν_e} of (5.17) and (5.22)). However, there is no a priori requirement for it to be 'diagonal' in the family index. The allowed generalisation of (5.17) and (5.22) is therefore an interaction of the form

$$\mathcal{L}_{\psi\phi} = c_{ij}\bar{\psi}_{iL}\phi^c f_{jR} + c'_{ij}\bar{\psi}_{iL}\phi f'_{jR} + \text{h.c.} \tag{5.47}$$

where

$$\psi_{iL} = \begin{pmatrix} f_{iL} \\ f'_{iL} \end{pmatrix} \tag{5.48}$$

and a sum on the family indices i and j (from 1 to 3) in (5.47) is assumed. After symmetry breaking, using the gauge (4.55), we find

$$\mathcal{L}_{f\phi} = -\left(1 + \frac{H}{v}\right)\left[\bar{f}_{iL}m_{ij}f_{jR} + \bar{f}'_{iL}m'_{ij}f'_{jR} + \text{h.c.}\right] \tag{5.49}$$

where the 'mass matrices' are

$$m_{ij} = -\frac{v}{\sqrt{2}}c_{ij}, \quad m'_{ij} = -\frac{v}{\sqrt{2}}c'_{ij}. \tag{5.50}$$

Although we have not indicated it, the m and m' matrices could involve a 'γ_5' part as well as a '1' part in Dirac space. It can be shown (Weinberg 1973, Feinberg *et al.* 1959) that m and m' can both be made Hermitean, γ_5-free, and diagonal by making four separate unitary transformations on the 'family triplets'

$$f_L = \begin{pmatrix} f_{1L} \\ f_{2L} \\ f_{3L} \end{pmatrix}, \quad f'_L = \begin{pmatrix} f'_{1L} \\ f'_{2L} \\ f'_{3L} \end{pmatrix}, \quad etc. \tag{5.51}$$

via

$$q_L = U_L f_L, \quad q'_L = U'_L f'_L, \quad q_R = U_R f_R, \quad q'_R = U'_R f'_R \tag{5.52}$$

with

$$q_L = \begin{pmatrix} q_{1L} \\ q_{2L} \\ q_{3L} \end{pmatrix} \quad etc. \tag{5.53}$$

That is to say, if we insert (5.52) into the first term of (5.49)

$$\overline{f}_{iL} m_{ij} f_{jR} = \overline{f}_L m f_R = \overline{q}_L U_L m U_R^\dagger q_R, \tag{5.54}$$

the matrices U_L and U_R can be chosen so that

$$U_L m U_R^\dagger = \begin{pmatrix} m_u & & \\ & m_c & \\ & & m_t \end{pmatrix}, \tag{5.55}$$

and the q's can be identified with the indicated strong interaction (mass) eigenstates: $q_1 = u$, $q_2 = c$, $q_3 = t$. Similarly, the other term in (5.49) yields

$$U'_L m' U'^{\dagger}_R = \begin{pmatrix} m_d & & \\ & m_s & \\ & & m_b \end{pmatrix} \tag{5.56}$$

and the q''s are again the indicated physical states: $q'_1 = d$, $q'_2 = s$, $q'_3 = b$. Indeed, (5.49) becomes

$$\mathcal{L}_{q\phi} = -\left(1 + \frac{H}{v}\right) \left[m_u \overline{u} u + \cdots + m_b \overline{b} b \right]. \tag{5.57}$$

The first thing we can conclude from this is that, rather remarkably, we can still manage with only the one Higgs doublet. It couples to each fermion with a strength proportional to the mass of that fermion divided by M_W. Next, we must note that the $SU(2)_L \times U(1)$ gauge invariant interaction part of the Lagrangian is (by definition) written in terms of the f and f' fields as (*c.f.* (5.9) and (5.10))

$$\mathcal{L}_{fW,B} = i \left(\overline{f}_{jL}, \overline{f}'_{jL} \right) \gamma^\mu (\partial_\mu + ig\underline{\tau} \cdot \underline{W}_\mu/2 + ig'y B_\mu/2) \begin{pmatrix} f_{jL} \\ f'_{jL} \end{pmatrix}$$
$$+ i \overline{f}_{jR} \gamma^\mu (\partial_\mu + ig'y B_\mu/2) f_{jR} + i \overline{f}'_{jR} \gamma^\mu (\partial_\mu + ig'y B_\mu/2) f'_{jR} \tag{5.58}$$

where a sum on j is understood. This now has to be rewritten in terms of the mass-eigenstates q and q'.

Consider first the neutral current part of (5.58). This involves the t_3 and the y part of the first line of (5.58), and all of the second line. A typical term in the latter can be written in terms of the q fields as

$$i\overline{f}_{jR}[\ ]f_{jR} = i\overline{q}_{kR}(U_R)_{kj}[\ ](U_R^\dagger)_{j\ell}\, q_{\ell R}. \tag{5.59}$$

But the content of [ ] is a diagonal matrix in family space. Hence the matrix part of (5.59) is just

$$(U_R)_{kj}(U_R^\dagger)_{j\ell} = \delta_{k\ell} \tag{5.60}$$

since U_R is unitary, and so (5.59) is diagonal in the q's. A similar argument clearly applies to the last term in (5.58), with the matrix U_R', and to the y term in the first line of (5.58) using the matrices U_L, U_L'. Finally, the t_3 part of (5.58) causes no trouble either, since it leads to

$$i\overline{f}_{jL}[\gamma^\mu(\partial_\mu + igW_{3\mu}/2)]f_{jL} + i\overline{f}'_{jL}[\gamma^\mu(\partial_\mu - igW_{3\mu}/2)]f'_{jL} \tag{5.61}$$

and the first piece is diagonal in the q's using $U_L U_L^\dagger = 1$, while the second is also using $U_L' U_L'^\dagger = 1$. Thus the neutral current interactions, in this standard version of the Standard Model, do not change the flavour of the physical (mass eigenstate) quarks (Glashow *et al.* 1970). Avoiding such flavour-changing neutral currents, at the experimentally required level, imposes severe restrictions on any extension to the Standard Model.

The charged current processes, however, involve the *non*-diagonal matrices τ_1 and τ_2 in (5.58), and this stops the previous argument holding. Indeed, using (5.14) we find that the charged current piece is

$$\begin{aligned}
\mathcal{L}_{cc} &= -\frac{g}{\sqrt{2}}(\overline{f}_{jL}, \overline{f}'_{jL})\gamma_\mu \tau_+ W_\mu \begin{pmatrix} f_{jL} \\ f'_{jL} \end{pmatrix} + \text{h.c.} \\
&= -\frac{g}{\sqrt{2}}\overline{f}_{jL}\gamma^\mu f'_{jL} W_\mu + \text{h.c.} \\
&= -\frac{g}{\sqrt{2}}\overline{q}_L U_L U_L'^\dagger \gamma^\mu q'_L W_\mu + \text{h.c.}
\end{aligned} \tag{5.62}$$

where the matrix

$$V \equiv U_L U_L'^\dagger \tag{5.63}$$

is not diagonal, though it is unitary. V therefore has 9 real parameters, which can be reduced to 4—three 'rotational angles' and one phase—by redefinitions of the quark fields (Jarlskog 1985). This is the famous CKM matrix, (Cabibbo 1963, Kobayashi and Maskawa 1973) the interaction (5.62) having the form

$$-\frac{g}{\sqrt{2}}W_\mu (\overline{u}_L\ \overline{c}_L\ \overline{t}_L) \begin{pmatrix} V_{ud} & V_{us} & V_{ub} \\ V_{cd} & V_{cs} & V_{cb} \\ V_{td} & V_{ts} & V_{tb} \end{pmatrix} \begin{pmatrix} d_L \\ s_L \\ b_L \end{pmatrix} + \text{h.c.} \tag{5.64}$$

The entries in the V-matrix modify the vertex (5.15) in an obvious way. The single phase δ in the V-matrix accommodates CP-violation. In the case of only two flavours, V has only 1 real parameter, which is the Cabibbo angle, and there is no freedom to have a CP violation phase in the family mixing matrix. It is an important challenge to experiment to find out whether all CP-violating phenomena can be described with just this one parameter δ in the CKM matrix.

Returning finally to the leptons, all of the above will apply (with three more mixing angles and one more phase) if the neutrinos do in fact have a mass. We would then have leptonic flavour mixing in c.c. processes, involving a term of the form $[\bar{\ell}_L V_\ell \gamma^\mu \ell'_L W_\mu + \text{h.c.}]$ (c.f. (5.62)), and lepton mass terms $[\bar{\ell}_L m_\ell \ell_R + \text{h.c.}]$ and $[\bar{\ell}'_L m'_\ell \ell'_R + \text{h.c.}]$, where V_ℓ is the leptonic analogue of (5.63), and m_ℓ, m'_ℓ are the analogues of the diagonal matrices (5.55) and (5.56). However, if the diagonal matrix m_ℓ is actually zero (i.e. the neutrinos are massless) or if (improbably, no doubt) the masses in m_ℓ are all equal, so that m_ℓ is proportional to the unit matrix, then we are free to redefine ℓ_L and ℓ_R by

$$\tilde{\ell}_L = V_\ell^\dagger \ell_L, \qquad \tilde{\ell}_R = V_\ell^\dagger \ell_R \qquad (5.65)$$

so as to reduce the c.c. term to family-diagonal form $[\bar{\tilde{\ell}}_L \gamma^\mu \ell'_L W_\mu + \text{h.c.}]$. Thus if the neutrinos are indeed massless, it is these 'tilde' states that we can identify with ν_e, ν_μ and ν_τ. Again, experimental information on neutrino masses and mixing is of great importance. There is nothing in the Standard Model which requires the neutrinos to be massless, and indeed they generally do have (small) masses in GUTs.

5.4 Some aspects of the Higgs sector

This course was intended to present the elements of the Standard Model. It seems fair to say that 'the' SM is usually taken to mean just what has been discussed so far—in particular, it is a model with only one Higgs doublet, which is responsible for all symmetry-breaking via the potential

$$V(\phi) = -\mu^2 \phi^\dagger \phi + \frac{\lambda}{4}(\phi^\dagger \phi)^2 \qquad (5.66)$$

of (4.53). Although it may be somewhat beyond the remit of this course, I want to include here a brief discussion of some of the reasons why theorists are not disposed to regard this 'minimal Higgs' version of electroweak theory as valid all the way up to (say) the Planck or GUT scales, preferring instead to think of it as either a part of, or an effective approximation to, a more elaborate theory with an energy scale which is likely to be around 1 TeV (± 0.5 TeV). These remarks may provide a bridge to Peccei's lectures (these Proceedings).

First, it is worth stressing again that, while the coupling of the Higgs field to the gauge fields is determined by the gauge symmetry, the Yukawa couplings to the fermions are completely unconstrained. True, they are renormalisable couplings, but this basically means that their values are not calculable and have to be taken from experiment. Also, we used to think of Yukawa couplings as having to do with forces—do we really believe there are so many independent 'forces'! Furthermore, the magnitudes of these couplings, being proportional to the fermion masses, vary very considerably, over a range of some five orders of magnitude between the (probable) top quark mass and the e^- (even assuming the neutrinos were for some reason exactly massless). This seems all very arbitrary and suggests that these couplings are phenomenological rather than fundamental; but, in fact, no really plausible solution to the fermion mass spectrum is in sight.

Secondly, there are two *fine tuning* problems with the minimal Higgs model. The first has to do with the vacuum energy density implied by (4.45). At its minimum, V of (5.66) takes the value

$$V_{\min} = -\mu^4/\lambda = -m_H^2 M_W^2/2g^2 \qquad (5.67)$$

in terms of the known parameters M_W and g, and the unknown m_H. $V_{\min}$ can be regarded as a vacuum energy density $\langle \rho \rangle$, which from (5.67) has a value

$$\langle \rho \rangle \gtrsim -(10^5 \text{ to } 10^6)\,\text{GeV}^4 \qquad (5.68)$$

if we take $m_H \geq 100\,\text{GeV}$. However, Weinberg (1989) gives a rough bound, from the present expansion rate of the universe, of

$$|\langle \rho \rangle| \leq 10^{-47}\,\text{GeV}^4, \qquad (5.69)$$

which is an enormous discrepancy. Of course, we could always add a constant V_0 to (5.66) so as to cancel (5.68). Alternatively, or in addition, we could appeal to a cancellation from the contribution of a cosmological constant term, λ, which adds $\lambda/8\pi G_N$ to the total effective vacuum energy (and defines the 'curvature of the vacuum'). Here G_N is Newton's gravitational constant. Obviously such a cancellation requires a very *unnatural* degree of 'fine-tuning'. On the other hand, it is commonly thought likely that in the early stages of the universe's evolution the sign of the μ^2 term in (5.66) was positive (*c.f.* the discussion of the ferromagnet in Section 4.2), so that the minimum of $V(\phi)$ was at the symmetric point $\phi = 0$, where $V(\phi) = 0$. Thus to get a very small value of $\langle \rho \rangle$ now, we need a very large effective λ before the symmetry-breaking transition. This is currently regarded as not at all unwelcome, as it would cause inflation which is favoured on other grounds (see Kolb's lectures, these Proceedings). The problem is the unnatural cancellation required to get the very small value now. It is fair to say that there is no solution to this problem yet, either.

The second fine-tuning problem presents itself when we begin to think about the place of the SM, and especially the Higgs sector, in a grander scheme of things. As we stressed in Section 2.3, we do not really expect its predictions to be true at arbitrarily high energies, even though (due to its renormalisability) it is certainly capable of making finite predictions up to any energy, with no more free parameters, as the cut-off M goes to ∞. There is the question of quantum gravity at the Planck scale ($M_P \sim 10^{19}\,\text{GeV}$), and of the possible Grand Unified scale somewhat lower. So we should take quite seriously the idea that the fields and the parameters of the SM represent a kind of 'low-energy physics' on the scale of $M \geq 10^{16}\,\text{GeV}$, with masses and couplings depending on the physics at the high scale M (*c.f.* the discussion of 'levels' in Section 2.3). In particular, the parameters of the Higgs Lagrangian should depend on M.

This raises a fine-tuning problem when we consider the Higgs mass. So far, we have considered no loop effects in the Higgs sector, but when we do we meet Figure 27, which is a self-energy correction to the Higgs propagator, Π_H, and it therefore contributes a shift in the Higgs mass—or in fact to the (mass)^2, since that is how the mass enters the Lagrangian (and the propagator) for a scalar particle. Counting powers of k as usual, the loop in Figure 27 is quadratically divergent, so that it contributes a shift of order

$$\Pi_H(M) \sim \lambda M^2. \qquad (5.70)$$

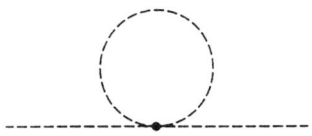

Figure 27.

Now we could of course follow the same route that we did in Section 2.3 for QED, and absorb (5.70) into the 'original' $m_H^2(M)$ so as to arrive at a 'real' m_H^2 (at our present energy scale) given by

$$m_H^2 = m_H^2(M) + \Pi_H(M), \tag{5.71}$$

which is insensitive to changes in M. Indeed, this is exactly what *is* done in a conventional renormalisation of the Higgs sector, and it can be carried through consistently even as $M \to \infty$. But if we do not take $M \to \infty$, and instead—as seems reasonable—regard it as having a physical value in the region 10^{16} to 10^{19} GeV, then (5.71) looks very different. As we shall see in a moment, there are theoretical arguments which quite strongly imply $m_H \lesssim 1$ TeV; also, present precision tests of the SM (see Blondel's lectures, and the following Section), suggest that m_H may be less than 1 TeV. Consequently, if M in $\Pi(M)$ of (5.71) is of order in 10^{16} to 10^{19} GeV, we shall need a very finely tuned $m_H(M)$ to give us an m_H of order 1 TeV.

Why did we not get similarly upset in the case of QED? Looking back at (2.60) we see that in that case the shift in the electron mass $\Delta m(M)$ was proportional to the original mass $m(M)$, while the M-dependence arising from the loop in $\Sigma(M)$ was only logarithmic. So the cancellation required in $m(M) + \Sigma(M)$ was not nearly so delicate, even if $M \sim 10^{16}$ GeV. The fact that Δm is proportional to $m(M)$ (which already implies on dimensional grounds that the loop integral rises only logarithmically) means that if $m(M)$ were actually set to zero initially, it would (at least at the one loop level) remain zero even after quantum corrections. Indeed, a very similar thing happened with the photon self-energy $\Sigma^\gamma(p^2)$, which (provided Π^γ has no pole!) contained a p^2-factor, showing that the on-shell photon received no quantum correction to its mass. Both these results are actually true to all orders in perturbation theory, and are due to symmetry principles—in the first, to the chiral symmetry (see Section 4.1) of all 'covariantised' Dirac Lagrangians as the fermion masses go to zero, and in the second to gauge invariance. These symmetries ensure that it is natural (t'Hooft 1980) to have light fermions and gauge quanta, in the sense that no fine-tuning will be needed, even for $M \sim M_P$. In the case of fundamental scalars, no useful symmetry restricts the mass to be zero. Of course, the actual fermions and the massive gauge bosons do acquire masses—but via the spontaneous symmetry breaking mechanism, which preserves the symmetry in the Lagrangian, and hence in the loop corrections. This mechanism has, itself, an associated mass scale—namely ~ 200 GeV (as in (5.28)) for the electroweak theory. The relationship of this value to an $M \sim 10^{16}$ GeV is *not* explained by any of the above reasoning.

Thus there is a fine-tuning problem with an 'elementary' Higgs, from this point of view. One way out ('technicolour') is to suppose that there is a new, strongly interacting and confined sector in the TeV region, say, with certain global symmetries which are spontaneously broken, the resulting Goldstone fields being precisely the massless Higgs modes required to give masses to the W and Z particles. Thus the Higgs fields are, on this view, composites. This has problems, so far, in accounting for the fermion masses in a convincing way. A second possibility is supersymmetry, which arranges cancellations among the divergences of boson and fermion loops, since it is a symmetry which relates fermions and bosons. This symmetry cannot be exact, or else there would be a mass-dependent bosonic counterpart of every known fermion, and fermionic counterparts of all bosons, and these states are not yet observed. The net (finite) parts of loop corrections will then be proportional to the mass differences between the known particles and their superpartners. To avoid the fine-tuning problem once again, this difference—which is a measure of the scale of supersymmetry breaking—cannot be much above 1 TeV, and this presents a new and unexplained energy scale on its own account.

A final interesting aspect of the 'minimal Higgs' model, which results from a closer study (see for example, Maiani 1991) of the dynamics of the ϕ^4 theory itself, suggests quite firm conclusions regarding the value of m_H. We first note that, for a given fixed value of v as in (5.28), the Higgs mass is

$$m_H = v\lambda^{1/2}/\sqrt{2} \sim \lambda^{1/2} \times 174\,\text{GeV}. \tag{5.72}$$

Now λ is a dimensionless coupling constant: if it is $O(\alpha)$ we would say the theory is perturbative, while if it is $O(1)$ we would say it was strongly coupled. It is clear from (5.72), and the present lower bounds on m_H, that we are already not far from the strongly coupled region. But we can ask: can λ (the renormalised coupling) take *any* value at all? That is, can m_H (for fixed v) be arbitrarily large?

To answer this we must recall that, in a renormalisable theory, 'the' value of λ has to be defined at a certain scale, and that at another scale λ is different (*i.e.* it 'runs'). For the interaction (5.66), calculation shows that the analogue of (2.94) is

$$\lambda(E) = \lambda(M) \Big/ \left[1 - \frac{3}{8\pi^2}\lambda(M)\ln\left(\frac{E}{M}\right)\right]. \tag{5.73}$$

(We note that this theory, like QED, is not asymptotically free). In (5.73), M is the cut-off scale. We see from (5.73) that, whatever the value of $\lambda(M)$, $\lambda(E)$ can be no larger than

$$\lambda_{\text{max}}(E) = 8\pi^2/3\ln(M/E). \tag{5.74}$$

In particular, at the scale v where (presumably) m_H is defined, we have from (5.74) and (5.72)

$$m_{H\,\text{max}}/v = 2\pi \Big/ \left[\frac{3}{2}\ln(M/v)\right]^{1/2} \tag{5.75}$$

with $v \approx 174\,\text{GeV}$. Now from (5.75) we see an interesting and curious connection between the cut-off scale and $m_{H\text{max}}$: the lighter $m_{H\text{max}}$ is, the higher the cut-off can be, while for heavier $m_{H\text{max}}$ the cut-off is lower. For example, if $M \sim 10^{19}\,\text{GeV}$, we have $m_{H\text{max}} = 144\,\text{GeV}$, while for $M \sim 1\,\text{TeV}$ we have $m_{H\text{max}} = 675\,\text{GeV}$. Thus in principle a 'light' Higgs particle would allow us consistently to employ a cut-off of order M_P, while as we go up in m_H the cut-off scale drops very rapidly. Indeed, it would seem absurd to have $m_H(v)$ larger than some sizable fraction of the cut-off scale itself—and this (if believed) seems to imply $m_H \leq$ few times $100\,\text{GeV}$.

Of course, this is a crude argument. It is based on (renormalisation group improved) perturbation theory, which led to (5.73) and which will presumably fail as m_H rises and thus (at fixed v) so does λ, entering the non-perturbative regime. Also, the evolution of λ should include contributions from fermions (especially the top) and gauge fields. But in fact a more refined calculation, including these effects, gives much the same conclusion (Lindner 1986). True, this is still subject to the criticism that it is based on perturbation theory—but non-perturbative lattice calculations appear to confirm the picture, and imply $m_H \leq 600\text{--}800\,\text{GeV}$ (see, for example, Chapter 9 of Einhorn (1991)).

Thus from all the preceding arguments we seem to reach the conclusion that *some* kind of new physics lies below or around the TeV region. This is a most exciting prospect for the next generation of accelerators.

5.5 Higher order corrections and the symmetry of the symmetry-breaking sector

Anything approaching a review of higher order corrections in the electroweak theory would be far outside the scope of this course. However, after all the stress on the renormalisability of the electroweak theory, and the introduction to one loop calculations in QED in Section 2, it seems a pity not to touch on just a few of the simpler aspects of radiative corrections, especially as the accuracy of much current data is sensitive to these effects, and very interesting phenomenological conclusions (about the top quark mass, for instance) can be drawn from the comparison of theory and experiment. The literature on radiative corrections to the electroweak theory is vast and growing: I refer here simply to a standard volume (Altarelli *et al.* 1989), especially the clear and pedagogical account by Consoli *et al.* (1989); see also the equally approachable lectures by Hollik (1991).

As we have seen, one obtains cut-off independent results from loop corrections in a renormalisable theory by taking from experiment certain parameters, which have to be defined in some suitable way at a convenient energy scale. For the electroweak theory, the parameters in the Lagrangian are

$$g,\ g',\ \lambda,\ \mu^2,\ g_f \tag{5.76}$$

where the first two are the gauge coupling constants, the next two are the parameters of the Higgs potential, and the g_f's are the Higgs-fermion Yukawa couplings. The set (5.76) can be replaced by the more physical-looking one

$$e\,(\text{or } \alpha),\ M_W,\ M_Z,\ m_H,\ m_f, \tag{5.77}$$

in which each can be directly measured. In the conventional presentation which we followed, however, another quantity frequently appeared, namely $\sin\theta_W$. This is, of course, not an independent parameter, being defined in terms of the Lagrangian parameters (5.76) by $\sin\theta_W = g'/(g'^2 + g^2)^{\frac{1}{2}}$. Note that the renormalised g and g' will 'run' and hence so will $\sin\theta_W$ defined this way. It is also possible to define $\sin\theta_W$ in terms of the set (5.77) (or a related one) via (5.24), or via (5.16) and (5.26). Alternatively one can define it via the effective neutral current coupling (5.43). These latter definitions are, however, only valid at *tree* level, and so will also in general be modified by loop corrections. Hence it is important to be precise in stating what definition of this quantity is used. For LEP physics it is now usual to choose the set

$$\alpha, \quad G_F, \quad M_Z, \quad m_H, \quad m_f \qquad (5.78)$$

since G_F is known to better accuracy (c.f.(5.2)) than m_W or $\sin^2\theta_W$. (The QCD fine-structure constant α_s should also be included in the list, in the full $SU(3)\times SU(2)\times U(1)$ Standard Model, together with the CKM parameters and the QCD θ-parameter.)

After renormalisation, one can derive radiatively-corrected values for physical quantities in terms of the input parameter set, e.g. (5.78). For example, the tree-level relation (5.29) takes the following form at one loop:

$$M_W^2 = \left[(\pi\alpha/\sqrt{2}\,G_F)/\sin^2\theta_W\right](1+\Delta r) \qquad (5.79)$$

where $\sin\theta_W$ has been defined by

$$\sin^2\theta_W \equiv 1 - M_W^2/M_Z^2, \qquad (5.80)$$

and Δr is the one loop correction, calculated in terms of the parameters (5.78). Note that (5.79) can be regarded as an equation either for m_W, or for $\sin^2\theta_W$ as defined by (5.80).

We cannot go into detail about all the contributions to Δr, but we do want to focus on two features of the result—which are surprising, important phenomenologically, and related to a symmetry we have not so far discussed. It turns out (Consoli et al. (1989), Hollik (1991)) that the leading terms in Δr have the form

$$\Delta r = \Delta\alpha - \cot^2\theta_W\Delta\rho + (\Delta r)_{\text{rem}}. \qquad (5.81)$$

In (5.81), $\Delta\alpha$ is precisely the quantity $-\text{Re}\hat{\Pi}^\gamma(M_Z^2)$ which entered into the running QED constant α discussed in Section 2.3, and which has the value (2.95). $\Delta\rho$ is given by

$$\Delta\rho = \frac{3G_F(m_t^2 - m_b^2)}{8\pi^2\sqrt{2}}, \qquad (5.82)$$

while the 'remainder' $(\Delta r)_{\text{rem}}$ contains a non-negligible term proportional to $\ln(m_t/M_Z)$, and a contribution from the Higgs boson which is (for $m_H \gg M_W$)

$$(\Delta r)_{\text{rem},H} \approx \frac{\sqrt{2}\,G_F M_W^2}{16\pi^2}\frac{11}{3}\left[\ln\left(\frac{m_H^2}{M_W^2}\right) - \frac{5}{6}\right]. \qquad (5.83)$$

The running of α from $p^2 = 0$ to $p^2 = M_Z^2$ is no surprise. But (5.82) and (5.83) are not what we might have expected. As regards (5.82), it is associated with top quark loops in vacuum polarisation amplitudes, of the kind discussed for $\hat{\Pi}^\gamma$. But in the latter case we saw in (2.90) that the contribution of heavy fermions ('$|p^2| \ll m_f^2$') was 'decoupled', in the sense that the large fermion mass appeared in the denominator. Why is $m_t^2 - m_b^2$ in the numerator in (5.82)? Clearly this fact makes a very big difference to the impact of the unknown parameter m_t^2 in (5.82). Secondly, as regards the Higgs contribution, that too is surprising, as we might well have expected it to involve m_H^2

in the numerator, as appropriate for a scalar particle in a loop (*c.f.* (5.70)). Δr would then have been very sensitive to m_H^2—but instead the sensitivity is only logarithmic.

First of all, we can understand the appearance of the fermion masses (squared) in the numerator as follows. The shift $\Delta\rho$ is associated with vector boson vacuum polarisation contributions, for example the one shown in Figure 28. Consider in particular the contribution from the longitudinal polarisation components of the W's. As we have seen, these components are nothing but three of the four Higgs components which the $W^\pm$ and Z^0 'swallowed' to become massive. But the couplings of these 'swallowed' Higgs fields to fermions are determined by just the same Higgs-fermion Yukawa couplings as we introduced to generate the fermion masses via spontaneous symmetry breaking. Hence we expect the fermion loops to contribute (to these longitudinal W states) something of order $g_f^2/4\pi$ where g_f is the Yukawa coupling. Since $g_f \sim m_f/v$ (see (5.19)) we arrive at an estimate $\sim m_f^2/4\pi v^2 \sim G_F m_f^2/4\pi$ as in (5.82). An important message is that particles which acquire their mass spontaneously do not 'decouple'.

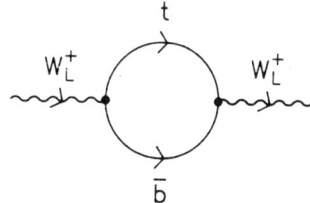

Figure 28.

But we now have to explain why $\Delta\rho$ in (5.82) would vanish if $m_t^2 = m_b^2$, and why only $\ln m_H^2$ appears. Both these facts are related to a symmetry of the assumed minimal Higgs sector which we have not yet discussed. As the notation suggests, $\Delta\rho$ is a leading contribution to the parameter ρ introduced in (5.42) and (5.44). At tree level, ρ has the value 1, and if $\sin^2\theta_W$ is defined by (5.80)—as in our discussion it is—then by definition the ratio (5.44) remains 1. However, the ratio of the effective neutral to charged current couplings will received radiative corrections, so the ρ in (5.42) will change—and this is what $\Delta\rho$ contributes to. But why does ρ have the value 1 at tree level? First of all, it may be shown (Ross and Veltman 1975) that $\rho = 1$ is a natural consequence of having the symmetry broken by an SU(2)$_L$ doublet Higgs field (rather than a triplet, say)—or indeed by any number of doublets. The nearness of the measured ρ parameter to 1 is good support for the hypothesis that there are only doublet Higgs fields.

At tree level, it is simplest to think of ρ in connection with the mass ratio (5.44). To see the significance of this, let us go back to the Higgs-gauge field Lagrangian $\mathcal{L}_{G\Phi}$ of (4.53) which produced the masses. With the doublet Higgs of the form (4.44), it is a striking fact that the Higgs potential only involves the highly symmetrical combination of fields

$$\phi_1^2 + \phi_2^2 + \phi_3^2 + \phi_4^2, \qquad (5.84)$$

as does the vacuum condition (4.45). This suggests that there may be some extra symmetry in (4.53) which is special to the doublet structure. But of course, to be of any interest, this symmetry has to be present in the $(D_\mu\phi)^\dagger(D^\mu\phi)$ term as well.

The nature of this symmetry is best brought out by introducing a change of notation for the Higgs doublet $\phi^\dagger$ and ϕ^0: instead of (4.44), we now write

$$\phi = \begin{pmatrix} (\pi_2 + i\pi_1)/\sqrt{2} \\ (\sigma - i\pi_3)/\sqrt{2} \end{pmatrix} \tag{5.85}$$

while the ϕ_c field of (5.21) is

$$\phi_c = \begin{pmatrix} (\sigma + i\pi_3)/\sqrt{2} \\ -(\pi_2 - i\pi_1)/\sqrt{2} \end{pmatrix}. \tag{5.86}$$

We then find that these can be written as

$$\phi = \frac{1}{\sqrt{2}}(\sigma + i\underline{\tau}\cdot\underline{\pi})\begin{pmatrix} 0 \\ 1 \end{pmatrix}, \quad \phi_c = \frac{1}{\sqrt{2}}(\sigma + i\underline{\tau}\cdot\underline{\pi})\begin{pmatrix} 1 \\ 0 \end{pmatrix}. \tag{5.87}$$

Consider now the covariant $SU(2)_L \times U(1)$ derivative acting on ϕ, as in (4.54), and suppose to begin with that $g' = 0$. Then

$$\begin{aligned}
D_\mu \phi &= \frac{1}{\sqrt{2}}(\partial_\mu + ig\underline{\tau}\cdot\underline{W}_\mu/2)(\sigma + i\underline{\tau}\cdot\underline{\pi})\begin{pmatrix} 0 \\ 1 \end{pmatrix} \\
&= \frac{1}{\sqrt{2}}\left\{\partial_\mu \sigma + i\underline{\tau}\cdot\partial_\mu\underline{\pi} + i\frac{g}{2}\sigma\underline{\tau}\cdot\underline{W}_\mu - \frac{g}{2}\left[\underline{\pi}\cdot\underline{W}_\mu + i\underline{\tau}\cdot(\underline{W}_\mu \times \underline{\pi})\right]\right\}\begin{pmatrix} 0 \\ 1 \end{pmatrix}
\end{aligned} \tag{5.88}$$

using $\tau_i\tau_j = \delta_{ij} + i\epsilon_{ijk}\tau_k$. Now the vacuum choice (4.47) corresponds to $\sigma = v$, $\underline{\pi} = 0$, so that when we form $(D_\mu\phi)^\dagger(D^\mu\phi)$ from (5.88) we will get just

$$\frac{1}{2}(0\ 1)\left\{\frac{g^2}{4}v^2(\underline{\tau}\cdot\underline{W}_\mu)(\underline{\tau}\cdot\underline{W}^\mu)\right\}\begin{pmatrix} 0 \\ 1 \end{pmatrix} = \frac{1}{2}M_W^2\underline{W}_\mu\cdot\underline{W}^\mu \tag{5.89}$$

with $M_W = gv/2$ as usual. This of course corresponds to (c.f. (4.60)) $\theta_W = 0$, and thus to $W_{3\mu} = Z_\mu$, and so (5.89) says that in the limit of $g' \to 0$, $M_W = M_Z$ as expected if $\cos\theta_W = 1$. It is clear from (5.88) that the three components $\underline{W}_\mu$ are treated on a precisely equal footing by the Higgs field (5.85), and indeed the notation suggests that $\underline{W}_\mu$ and $\underline{\pi}$ should perhaps be regarded as some kind of *new* triplets.

It is straightforward to calculate $(D_\mu\phi)(D^\mu\phi)$ from (5.88); one finds

$$\begin{aligned}
(D_\mu\phi)^\dagger D^\mu\phi &= \frac{1}{2}(\partial_\mu\sigma)^2 + \frac{1}{2}(\partial_\mu\underline{\pi})^2 - \frac{g}{2}\partial_\mu\sigma\underline{\pi}\cdot\underline{W}^\mu \\
&+ \frac{g}{2}g\sigma\partial_\mu\underline{\pi}\cdot\underline{W}^\mu + \frac{g}{2}\partial_\mu\underline{\pi}\cdot(\underline{\pi} \times \underline{W}^\mu) \\
&+ \frac{g^2}{4}\underline{W}_\mu^2(\sigma^2 + \underline{\pi}^2) + \frac{g^2}{4}\underline{\pi}^2\underline{W}_\mu^2.
\end{aligned} \tag{5.90}$$

This expression now reveals what the symmetry is: (5.90) is invariant under global SU(2) transformations under which $\underline{W}_\mu$ and $\underline{\pi}$ are vectors—i.e. (c.f.(3.9))

$$\left.\begin{aligned} \underline{W}_\mu &\to \underline{W}_\mu + \underline{\epsilon} \times \underline{W}_\mu \\ \underline{\pi} &\to \underline{\pi} + \underline{\epsilon} \times \underline{\pi} \\ \sigma &\to \sigma \end{aligned}\right\}. \tag{5.91}$$

Standard Model Foundations

This is why, from the term $\underline{W}^2_\mu \sigma^2$, all three $\underline{W}$ fields have the same mass in this $g' \to 0$ limit.

If we now reinstate g', and use (4.58) and (4.59) to write $W_{3\mu}$ and B_μ in terms of the physical fields Z_μ and A_μ as in (4.64), (5.88) becomes

$$\frac{1}{\sqrt{2}}\left\{\partial_\mu + ig\frac{\tau_1}{2}W_{1\mu} + ig\frac{\tau_2}{2}W_{2\mu} + ig\frac{\tau_3}{2}\frac{Z_\mu}{\cos\theta_W} + ig\sin\theta_W\left(\frac{1+\tau_3}{2}\right)A_\mu\right.$$
$$\left. - \frac{ig}{\cos\theta_W}\sin^2\theta_W\left(\frac{1+\tau_3}{2}\right)Z_\mu\right\}(\sigma + i\underline{\tau}\cdot\underline{\pi})\begin{pmatrix}0\\1\end{pmatrix}. \qquad (5.92)$$

We see from (5.92) that $g' \neq 0$ has two effects. First, there is a '$\underline{\tau}\cdot\underline{W}$'-like term, as in (5.88), except that the 'W_3' part of it is now $Z/\cos\theta_W$. In the vacuum $\sigma = v$, $\underline{\pi} = \underline{0}$ this simply means that the mass of the Z is $M_Z = M_W/\cos\theta_W$ i.e. $\rho = 1$; and this relation is preserved under 'rotations' of the form (5.91), since they do not mix $\underline{\pi}$ and σ. Hence this mass relation (and $\rho = 1$) is a consequence of the global SU(2) symmetry of the interactions and the vacuum under (5.91), and of the relations (4.58) and (4.59) which embody the requirement of a massless photon.

On the other hand, there are additional terms in (5.92) which single out the 'τ_3' component, and therefore break this global SU(2). These terms vanish as $g' \to 0$, and do not contribute at tree level, but we expect that they will cause $O(g'^2)$ corrections to $\rho = 1$ at the one loop level.

None of the above, however, yet involves the quark masses, and the question of why m_t^2 appears in the numerator in (5.82). We can now answer this question. Consider a typical mass term, of the form discussed in Section 5.3, for a quark doublet of the ith family

$$\mathcal{L}_m = -g_+ (\bar{q}_{iL}\ \bar{q}'_{iL})\,\phi_c q_{iR} - g_- (\bar{q}_{iL}\ \bar{q}'_{iL})\,\phi q'_{iR} \qquad (5.93)$$

(recall that the primed and unprimed fields refer to the $t_3 = \pm\frac{1}{2}$ components). Using (5.85) and (5.86), this can be written as

$$\mathcal{L}_m = \frac{-g_+}{\sqrt{2}}(\bar{q}_{iL}\ \bar{q}'_{iL})(\sigma + i\underline{\tau}\cdot\underline{\pi})\begin{pmatrix}q_{iR}\\0\end{pmatrix} - \frac{g_-}{\sqrt{2}}(\bar{q}_{iL}\ \bar{q}'_{iL})(\sigma + i\underline{\tau}\cdot\underline{\pi})\begin{pmatrix}0\\q'_{iR}\end{pmatrix}$$
$$= -\frac{(g_+ + g_-)}{2\sqrt{2}}(\bar{q}_{iL}\ \bar{q}'_{iL})(\sigma + i\underline{\tau}\cdot\underline{\pi})\begin{pmatrix}q_{iR}\\q'_{iR}\end{pmatrix}$$
$$- \frac{(g_+ - g_-)}{2\sqrt{2}}(\bar{q}_{iL}\ \bar{q}'_{iL})(\sigma + i\underline{\tau}\cdot\underline{\pi})\tau_3\begin{pmatrix}q_{iR}\\q'_{iR}\end{pmatrix}. \qquad (5.94)$$

Consider now a simultaneous (infinitesimal) global SU(2) transformation on the two doublets $(q_{iL}, q'_{iL})^T$ and $(q_{iR}, q'_{iR})^T$:

$$\begin{pmatrix}q_{iL}\\q'_{iL}\end{pmatrix} \to (1 - i\underline{\epsilon}\cdot\underline{\tau}/2)\begin{pmatrix}q_{iL}\\q'_{iL}\end{pmatrix}, \quad \begin{pmatrix}q_{iR}\\q'_{iR}\end{pmatrix} \to (1 - i\underline{\epsilon}\cdot\underline{\tau}/2)\begin{pmatrix}q_{iR}\\q'_{iR}\end{pmatrix}. \qquad (5.95)$$

Under (5.95), the first term of (5.94) becomes (to first order in $\underline{\epsilon}$)

$$-\frac{(g_+ + g_-)}{2\sqrt{2}}(\bar{q}_{iL}\ \bar{q}'_{iL})[\sigma + i\underline{\tau}\cdot(\underline{\pi} + \underline{\pi}\times\underline{\epsilon})]\begin{pmatrix}q_{iR}\\q'_{iR}\end{pmatrix}. \qquad (5.96)$$

From (5.96) we see that if, at the same time as (5.95), we *also* make the transformation of $\underline{\pi}$ given in (5.91), then this first term in $\mathcal{L}_m$ will be invariant under these combined transformations. The second term in (5.94), on the other hand, will not be invariant under (5.95), but only under transformations with $\epsilon_1 = \epsilon_2 = 0$, $\epsilon_3 \neq 0$. We conclude that the global SU(2) symmetry of (5.91), which was responsible for $\rho = 1$ at the tree level, can be extended also to the quark sector; but—because the $g_\pm$ in (5.93) are proportional to the masses of the quark doublet—this symmetry is explicitly broken by the quark mass difference. This is why a t-b loop in a W vacuum polarisation correction can produce the 'non-decoupled' contribution (5.82) to ρ, which grows as $m_t^2 - m_b^2$.

Returning to (5.95), the transformation on the L-components is just the same as a standard SU(2)$_L$ transformation, except that it is global; so the gauge interactions of the quarks obey this symmetry also. As far as the R-components are concerned, they are totally decoupled in the gauge dynamics, and we are free to make the transformation (5.95) if we wish. The resulting complete transformation, which does the same to both the L and R components, is a non-chiral one—in fact it is precisely an ordinary 'isospin' transformation of the type

$$\begin{pmatrix} q_i \\ q'_i \end{pmatrix} \to (1 - i\underline{\epsilon}\cdot\underline{\tau}/2) \begin{pmatrix} q_i \\ q'_i \end{pmatrix}. \tag{5.97}$$

This analysis of the symmetry of the Higgs (or a more general symmetry breaking sector) was first given by Sikivie *et al.* (1980). The isospin-SU(2) is frequently called 'custodial SU(2)' since it 'protects' $\rho = 1$.

What about the *absence* of m_H^2 corrections? Here the position is rather more subtle. Without the Higgs particle H the theory is non-renormalisable, and hence one might expect to see some radiative correction becoming very large $(0(m_H^2))$ as one tried to 'banish' H from the theory by sending $m_H \to \infty$. However, even without a Higgs contribution the theory is renormalisable at the one-loop level for zero fermion masses (Veltman (1968), (1970)). Thus one suspects that the large m_H^2 effects will not be so dramatic after all. In fact, calculation shows (Veltman 1977; Chanowitz *et al.* 1978, 1979) that one-loop radiative corrections grow at most like $\ln m_H^2$ for large m_H. While there are finite corrections which are approximately $O(m_H^2)$ for $m_H^2 \ll M_{W,Z}^2$, for $m_H^2 \gg M_{W,Z}^2$ the $O(m_H^2)$ pieces cancel out from all observable quantities, leaving only $\ln m_H^2$ terms. Similarly, at the two-loop level, the expected $O(m_H^4)$ behaviour becomes $O(m_H^2)$ instead (van der Bij and Veltman 1984, van der Bij 1984)—and of course appears (relative to the one-loop contributions) with an additional factor of $O(\alpha)$. This relative insensitivity of the radiative corrections to m_H, in the limit of large m_H, was discovered by Veltman (1977) and called a 'screening' phenomenon by him: for large m_H (which also means, as we have seen, large λ) we have an effectively strongly interacting theory whose principal effects are screened off from observables at lower energy. It was shown by Einhorn and Wudka (1989) that this screening is also a consequence of the (approximate) isospin-SU(2) symmetry we have just discussed in connection with (5.82).

As mentioned above, $\Delta \rho$ of (5.82) is a contribution to the radiative corrections to the effective neutral current coupling (5.42). It turns out that, to a good approximation, these corrections in $e^+e^- \to Z^0 \to f\bar{f}$ can be parametrised by replacing the '$\sqrt{\rho} j_\mu^N$'

factors of (5.42) by

$$\sqrt{\rho_f}\,\bar{\psi}_f\left[\gamma_\mu\left(\frac{T_3}{2}-2Q(\sin^2\theta_W)\kappa_f\right)-\frac{T_3}{2}\gamma_\mu\gamma_5\right]\psi_f \quad (5.98)$$

together with corrections in the Z^0 propagator. The combination $\sin^2\theta_W\kappa_f$ is often called $\sin^2\theta^f_{\text{eff}}$. Then, the form factors ρ and κ have 'universal' parts which are the same for all fermion species f (and which are due to boson self-energies), and 'non-universal' parts which do depend on f. One finds (Consoli et al. 1989, Hollik 1991)

$$\rho = 1 + (\Delta\rho)_{\text{un}} + (\Delta\rho)_{\text{non-un}} \quad (5.99)$$

and similarly for κ, with $(\Delta\rho)_{\text{un}}$ as in (5.82) and

$$(\Delta\kappa)_{\text{un}} = \cot^2\theta_W \times (\Delta\rho \text{ in } (5.82)). \quad (5.100)$$

For the light fermions ($f \neq b, t$) the non-universal contributions are small, and almost independent of the unknown parameters m_t and m_H, which enter only the universal part. For the b-quark, however, there is a strong non-universal dependence on m_t (Akhundov et al. 1986, Beenakker and Hollik 1988) coming from the virtual top quark in the vertex correction (Figure 29); indeed the ratio $\Delta\Gamma(Z^0 \to b\bar{b})/\Gamma(Z^0 \to b\bar{b})$, where $\Delta\Gamma$ is the radiative correction, contains the term

$$-\frac{20}{13}\frac{\alpha}{\pi}\frac{m_t^2}{M_W^2} \quad (5.101)$$

which is a sizable shift in percentage terms (about -1.8% at $m_t \sim 150\,\text{GeV}$).

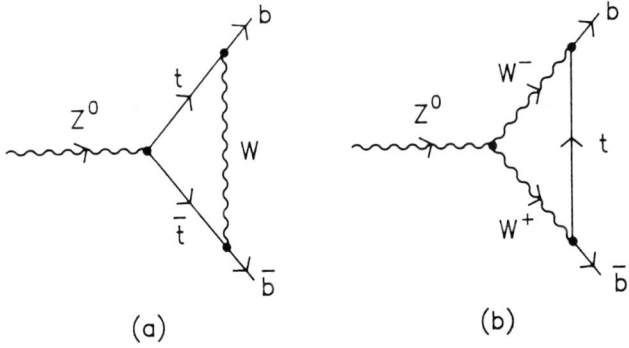

Figure 29.

Finally we mention that other precisely measurable quantities at LEP are (Altarelli et al. 1989) the various forward-backward asymmetries, the τ-polarisation asymmetry and the left-right polarised beam asymmetry, all of which test (5.98). Precision tests of the electroweak theory were reviewed by Rolandi (1993) at the Dallas Conference in 1992, and are covered by Blondel (these Proceedings). The predictions agree extremely well with the data (whose accuracy is often better than 0.5%), using only the two parameters m_t and m_H. The sensitivity to m_t of the radiative corrections determined its mass to be (Rolandi 1993) $145 \pm 25\,\text{GeV}$, but this figure is revised by Blondel. As discussed above, the radiative corrections are much less sensitive to m_H: however, m_t and m_H are correlated in the fits, a light/heavy top (within the above limits) going with a light/heavy Higgs. For example (Rolandi 1993), if $m_t \approx 120\,\text{GeV}$ the fit could exclude $m_H > 200\text{--}300\,\text{GeV}$ at the $1\,\sigma$ level. If $m_t \approx 160\,\text{GeV}$ then m_H is larger and so is its error, so that no secure upper limit could be placed on m_H. Although cautious commentators are reluctant to quote bounds on m_H at present, the fits tend to look (to the non-statistical eye) as if they preferred a relatively light m_H.

Discovery of the top and measurement of its mass would further constrain the fit and thus improve the sensitivity to m_H.

Acknowledgements

I am most grateful to David Saxon and the SUSSP Committee for the invitation to participate in what was a most stimulating, timely and successful School. It is a pleasure to thank all the organisers, and especially Alan Walker, for creating such an excellent atmosphere. I am particularly indebted to Lance Vick for all of the care and time he put into the preparation of the final typescript of these lectures; I am also very grateful to Nikki Fathers for her hard work in producing the original typescript. Finally I have benefitted very much from many conversations about renormalisation with Richard Ball and Robert Thorne (even though they may not think so, from the evidence herein).

References

Aitchison I J R and Hey A J G, 1989, *Gauge Theories in Particle Physics, 2nd edition* (Adam Hilger, Bristol and Philadelphia).
Akhundov A A, Bardin D and Riemann T, 1986, *Nucl Phys* **B276** 1.
Altarelli G, Kleiss R and Verzegnassi C, 1989, (eds) *Z Physics at LEP-I* (CERN 89-08, Geneva).
Appelquist T and Chanowitz M S, 1987, *Phys Rev Lett* **59** 2405.
Beenakker W and Hollik W, 1988, *Z Phys* **C40** 569.
Bludman S A, 1958, *Nuov Cim* **9** 443.
Cabibbo N, 1963, *Phys Rev Lett* **10** 531.
Chanowitz M, Furman E and Hinchliffe I, 1978, *Phys Lett* **B78** 285.
Chanowitz M, Furman E and Hinchliffe I, 1979, *Nucl Phys* **B153** 402.
Consoli M, Hollik W and Jegerlehner F, 1989, Electroweak Radiative Corrections for Z Physics, in *Z Physics at LEP-I*, eds Altarelli G, Kleiss R and Verzegnassi C (CERN 89-08, Geneva).
Cornwall J M, Levin D N and Tiktopoulos, 1974, *Phys Rev* **D10** 1145.
Einhorn M B, 1991, (ed) *The Standard Model Higgs Boson* (North-Holland, Amsterdam).
Einhorn M B and Wudka J, 1989, *Phys Rev* **D39** 2758.
Farhi E and Susskind L, *Phys Rep* **74C** 277.
Feinberg G, Kabir P and Weinberg S, 1959, *Phys Rev Lett* **3** 527, especially footnote 9.
Glashow S L, 1961, *Nucl Phys* **22** 579.
Glashow S L, Iliopoulos J and Maiani L, 1970, *Phys Rev* **D2** 1285.
Goldstone J, 1961, *Nuov Cim* **19** 154.
Gross D J and Wilczek F, 1973, *Phys Rev Lett* **30** 1343.
Hagiwara K, Peccei R D and Zeppenfeld D, 1987, *Nucl Phys* **B281** 253.
Higgs P W, 1964, *Phys Rev Lett* **13** 508.

Hollik W, 1991, in *Proc 1989 CERN-JINR School of Physics* (CERN 91 - 07) 50.
't Hooft G, 1971a, *Nucl Phys* **B33** 173.
't Hooft G, 1971b, *Nucl Phys* **B35** 167.
't Hooft G, 1980, in *Recent Developments in Gauge Theories*, Proc 1979 NATO ASI, Cargèse, France, eds 't Hooft G *et al.* (Plenum, New York) Lecture III.
Hughes V W, 1989, *AIP Conference Proceedings Number* **187** 326.
Jackiw R and Weinberg S, 1972, *Phys Rev* **5** 2396.
Jauch J M and Rohrlich F, 1976, *The Theory of Photons and Electrons*, 2nd expanded edition (Springer-Verlag, New York).
Jarlskog C, 1985, in *Fundamental Forces*, Proc. XXVII Scott.Univ.Summer School, St Andrews 1984 (Scott.Univ.Sum.Sch.in Phys. 1985) 1.
Kobayashi M and Maskawa K, 1973, *Prog Theor Phys* **49** 652.
Lee T D and Yang C N, 1962, *Phys Rev* **128** 885.
Lindner M, 1986, *Z Phys* **C31** 295.
Llewellyn Smith C H, 1974, in *Phenomenology of Particles at High Energies*, ed Crawford R L and Jennings R (Academic Press, London) 459.
Maiani L, 1991, in Z^0 *Physics*, Proc 1990 NATO ASI, Cargèse, France, eds Lèvy M *et al.* (Plenum, New York) 237.
May M, 1988, *AIP Conference Proceedings Number* **176** 168.
Méry P, Moubarik S E, Perrotet M and Renard F M, *Z Phys* **C46** 229.
Miransky V A, Tanabashi M and Yamawaki K, 1989a, *Mod Phys Lett* **A4** 1043.
Miransky V A, Tanabashi M and Yamawaki K, 1989b, *Phys Lett* **B221** 177.
Nambu Y, 1989, in *New Theories in Physics*, proc XI Int.Symp.on Elem.Part.Phys., Kazimierz, Poland, eds Ajduk Z, Pokorski S and Trautman A (World Scientific, Singapore).
PDP 1992, *Review of Particle Data Properties, Phys Rev* **D45** Part 2.
Politzer H D, 1973, *Phys Rev Lett* **30** 1346.
Rolandi L, 1993, in *Proc XXVI Int Conf on High Energy Physics, Dallas 1992*, ed Sanford J R (*AIP Conference Proceedings Number* **272**) 56.
Ross D A and Veltman M, 1975, *Nucl Phys* **B95** 135.
Ryder L H, 1985, *Quantum Field Theory* (Cambridge University Press, Cambridge).
Salam A, 1968, in *Elementary Particle Theory*, ed Svartholm N (Almqvist, Stockholm) 367.
Schwinger J, 1948, *Phys Rev* **73** 416.
Sikivie P, Susskind L, Voloshin M and Zakharov V, (1980), *Nucl Phys* **B173** 189.
van der Bij J J and Veltman M, 1984, *Nucl Phys* **B231** 205.
van der Bij J J, 1984, *Nucl Phys* **B248** 141.
Veltman M, 1977, *Acta Phys Polon* **B8** 475 (reprinted in Einhorn (1991)).
Veltman M, (1968), *Nucl Phys* **B7** 637.
Veltman M, (1970), *Nucl Phys* **B21** 288.
Veltman M, (1977), *Nucl Phys* **B123** 89.
Weinberg S, 1967, *Phys Rev Lett* **19** 1264.
Weinberg S, (1973), *Phys Rev* **D8** 605, especially footnote 8.
Weinberg S, 1979, *Physica* **A96** 327.
Weinberg S, 1980, *Rev Mod Phys* **52** 515.
Weinberg S, 1989, *Rev Mod Phys* **61** 1.
Weinberg S, 1993, in *Proc XXVI Int Conf on High Energy Physics, Dallas 1992*, ed Sanford J R (*AIP Conference Proceedings Number* **272**) 346.
Yang C N and Mills R L, 1954, *Phys Rev* **96** 191.

Hadronic Physics in Electron-Positron Annihilation

Siegfried Bethke

University of Heidelberg
Heidelberg, Germany

1 Introduction

Hadronic physics, which is the physics of the strong interaction, has made remarkable progress during the past 20 to 25 years. This is essentially due to the success of Quantum Chromodynamics (QCD), which is commonly accepted as the fundamental theory of the strong interactions. QCD consistently predicts and describes many (if not all or, at least, most) aspects of hadronic interactions up to the highest available energies, while it depends only on very few free parameters which are not determined by theory: the coupling 'constant' α_s and the quark masses. Due to the nature of this theory, however, which does not directly describe the dynamics of observable hadrons but of quarks and gluons instead, experimental tests of this theory are not nearly as precise as tests of Quantum Electrodynamics (QED) or of the Standard Model (SM) of electro-weak interactions.

Hadronic final states of highly energetic e^+e^- annihilations have proved to be an ideal laboratory for performing precise tests of the strong interaction and of QCD in general. According to our current understanding, these processes can be characterised by Feynman diagrams of the following type,

where the electron and positron annihilate into a highly virtual photon or a Z^0 boson, which in turn decays into a quark-antiquark pair. These quarks may then emit one or more hard gluons.

The basic advantages of e^+e^- annihilations for performing precise tests of QCD, if compared to other deep inelastic scattering processes like hadron-hadron collisions or lepton-nucleon scattering, are as follows:-

- Due to the point-like coupling of electrons, positrons and quarks to the gauge bosons, the quantum numbers and the energy of the hard scattering process are well known. There is no underlying event of target remnants.

- Hadronic final states can be easily identified due to the clear signatures of such events (large multiplicity of particles; momentum and energy balance of the event, i.e. $\sum_i \vec{p}_i = 0$ and $\sum_i E_i = E_{cm}$). Backgrounds due to τ-lepton pair production or 2-photon exchange processes can be efficiently suppressed.

- Data are available in a large range of well defined centre-of-mass energies, starting at the energies of heavy quark resonances up to the mass of the Z^0 boson, $M_{Z^0} = 91.19$ GeV. These data thus allow us to perform tests of asymptotic freedom as well as studies in the region of quark confinement.

Especially during the last four years, a wealth of rather precise studies of hadronic final states became available at the large electron-positron collider LEP, which operates at the highest energies which have been achieved in e^+e^- annihilation so far, $E_{cm} \sim M_{Z^0}$. The advantage of LEP with respect to former e^+e^- machines is not only its high centre-of-mass energy, which provides a closer relationship between hadron jets and the underlying partons (quarks and gluons), but also the large cross-section at the peak of the Z^0 resonance. So far, each of the four LEP experiments has collected more than 1.5 million hadronic final states of Z^0 decays. The advantage of the Z^0 resonance is illustrated in Figure 1, where the measured cross-sections for the annihilation processes $e^+e^- \to$ hadrons and $e^+e^- \to \mu^+\mu^-$, as well as for the purely electromagnetic process of $e^+e^- \to \gamma\gamma$, are displayed. Also shown, as lines, are the corresponding theoretical predictions which match the data well over the entire energy range.

This lecture will start with a short historical overview of the development of the theory of strong interactions and of QCD in general. Section 3 contains a list of the e^+e^- accelerators which are relevant for the topics discussed in this lecture, as well as a description of a typical, modern detector. A short discussion of the basics of QCD and of hadron production in e^+e^- annihilation is given in Section 4. The following Sections will then contain discussions of selected topics; due to the limited amount of time which is available for this lecture, and due to the enormous number of studies and experimental results which have become available in this field, this selection cannot be complete. The topics which will be covered here in detail are studies of hadronic event shapes (Section 5), of the physics of jets (Section 6), tests of the basic quantum numbers of quarks and gluons (Section 7), measurements of α_s (Section 8) and studies of the differences between quark and gluon jets (Section 9). Section 10 contains a short summary and an outlook to future studies of hadron production and tests of QCD in e^+e^- annihilations at higher centre-of-mass energies.

Since this article is a contribution to a summer school rather than a conference summary report, the selection of exemplary results which will be presented was not

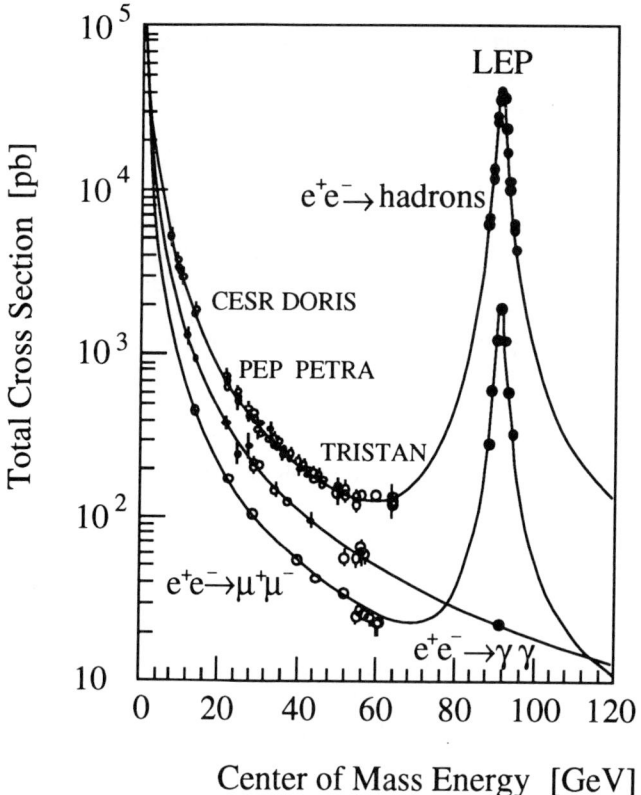

Figure 1. *Total cross-sections for various processes in e^+e^- annihilation and comparison with Standard Model Predictions. The data are from experiments at the e^+e^- storage rings* CESR *and* DORIS *($E_{cm} \sim 10\,\text{GeV}$),* PETRA *($14$–$46\,\text{GeV}$),* PEP *($29\,\text{GeV}$),* TRISTAN *($48$–$64\,\text{GeV}$) and* LEP *($\sim 91\,\text{GeV}$).*

made on the basis of a democratic treatment of all the different groups of authors and experiments, but was determined by pedagogic suitability or simply by optimal availability of data and illustrations. In addition, if similar experimental results are available from several groups, references will only be given to those studies which are discussed here in detail. Further descriptions of available studies and results can be found in several review articles, conference summary reports and books, to which the reader will be referred for further reading. In preparing the historic overview and the theoretical basics, I greatly benefited from the books of Halzen and Martin (1984) and Nachtmann (1986).

2 A short historical review

The necessity of introducing a new fundamental theory besides QED became apparent when in 1933 it was discovered that the magnetic moment of the proton is about

2.5 times as big as was expected by the Dirac theory: the proton apparently had to have a substructure, which was attributed to new types of interaction. The discovery of further baryons (like the neutron) and of meson states (like the pions $\pi^\pm$ and π^0, which were in fact predicted by Yukawa as carriers of the strong interaction in 1935), and the observation of a new class of long-lived particles (like the Kaon K and the Λ hyperons), which were also called 'V'-particles, opened a whole new field of theoretical investigation and experimental activity. Since 1953, it has been possible to produce V-particles at accelerators. The production cross-sections were found to correspond to the geometrical cross-sections of hadrons $\sim (10^{-13}\text{cm})^2$. The relatively long lifetime of such particles, like $\tau_\Lambda \sim 2.63 \times 10^{-10}$ s, was found to be much larger than the typically expected decay time of a strongly decaying particle, $\tau_{\text{strong}} \sim (\text{hadron radius})/c \sim 10^{-23}$ s. This led to the introduction of a new 'inner' quantum number, called strangeness. The strong and the electromagnetic interaction were supposed to conserve strangeness, and only the weak interaction should violate the conservation of this quantum number.

After these discoveries, it was possible to attribute several quantum numbers to all known particles: the charge Q, baryon number B, isospin I and I_3, strangeness S and hypercharge $Y = S + B$. Isospin and hypercharge were taken to form the generators of an invariant group of the strong interactions, namely of $SU(2) \times U(1)$, which however did not provide a connection between the properties of K and π mesons, or between Λ hyperons and the nucleons.

In 1961, Gell-Mann and Ne'eman introduced a scheme which could classify all known hadrons according to the representation of the SU(3) group $(SU(3)_{\text{flavour}})$. Mesons and baryons were believed to belong to the 8-dimensional representation of SU(3), which was also called the 'theory of the eightfold way'. One of the biggest successes of this theory was the prediction and the following observation of the Ω^- particle, which has a strangeness of $S = -3$.

In 1964, Gell-Mann and Zweig extended this model by introducing new particles of spin 1/2 which transform according to the fundamental representation of $SU(3)_{\text{flavour}}$. These particles were named *quarks* and should appear in 3 species, namely u, d, and s for 'up', 'down' and 'strange'. The famous 'quark model' was born! Within this theory, mesons and baryons are states which transform like states made of quark-antiquark and of three quarks, respectively.

Since the spin of the Ω is 3/2, it is naturally described by the *symmetric* wave function of three strange quarks with parallel spin. In order to enforce an *antisymmetric* wave function, which, according to the Pauli principle, is required for a state being composed of identical particles with spin 1/2, an additional degree of freedom, named 'colour', was ascribed to quarks. According to this picture, each quark exists in three versions or 'colours'. Hadrons then had to be composed of quarks such that they are invariant under colour transformations. These transformations again constitute an SU(3) group, $SU(3)_{\text{colour}}$, which should not be confused with $SU(3)_{\text{flavour}}$. $SU(3)_{\text{colour}}$ plays the fundamental rôle in the theory of the strong interactions.

After the success of the non-relativistic, static quark model in describing the spectrum of all presently known hadrons, the dynamic parton model was introduced by Feynman (1969) and others. The parton model was successfully used to describe scattering processes at high energies such as, for instance, electron-positron annihilation into hadrons ($e^+e^- \to X$), deep inelastic lepton-nucleon scattering ($\ell N \to \ell X$) and μ-pair production in highly energetic nucleon-nucleon collisions ($NN \to \mu^+\mu^- X$). For the process of e^+e^- annihilation, it is believed that the electron and positron annihilate into a highly virtual photon (γ^*), which primarily materialises into partons, *i.e.* into a quark-antiquark pair:

$$e^+e^- \to \gamma^* \to q\bar{q}.$$

These partons are then believed to 'hadronise' into the hadrons which are observed to originate from the high energetic scattering process. At high enough energies, the hadrons should emerge as 'jets' of collimated particles, which reflect the direction and also the quantum numbers of primary partons (quarks).

In 1975, the first evidence for jet structure in hadron production by e^+e^- annihilation was reported (Hanson 1975). The data, taken with the SLAC-LBL magnetic detector at the SPEAR storage ring, showed increasing evidence for the production of two-jet like events when the centre-of-mass energy, E_{cm}, was raised from 3 to 7.4 GeV. The jet structure manifested itself as a decrease of the mean sphericity, which is a measure of the global shape of hadronic events. The angular distribution of the jet axes from the same measurement provided evidence that the underlying partons must have spin 1/2. These observations were further corroborated by similar measurements at E_{cm} = 14 to 34 GeV at the e^+e^- storage ring PETRA (see Wu 1984).

The bosons, which were supposed to mediate the force field between quarks and antiquarks, were called 'gluons'. In analogy to the photon in QED, gluons were believed to be vector particles of spin 1.

Until 1972 there was no real theory of strong interactions; there were several models (*e.g.* the parton model), which were able to describe certain parts of the experimental data, but none of the models, of which several had many free parameters, was able to give a consistent and overall description of the experimental results.

The situation changed significantly with the discovery of asymptotic freedom (Gross and Wilczek 1973a, Politzer 1973a). The couplings between quarks and gluons were described with a non-Abelian theory, which in turn allowed quantitative calculations to be performed and certain features of strong interactions to be predicted (Gross and Wilczek 1973b, Politzer 1973b, Fritzsch 1973). The theory of the strong interaction, Quantum Chromodynamics (QCD), was born! Its only free parameters are the strong coupling strength, α_s, and the masses of the different quark flavours.

Experimentally, QCD witnessed its largest triumph when in 1979 a small fraction of planar, well separated 3-jet events were observed by the PETRA experiments around $E_{cm} \approx 30$ GeV. These events could be convincingly attributed to the emission of a third parton with zero electric charge and spin 1, exactly as expected for gluon bremsstrahlung predicted by QCD. The first evidence for 3-jet like events came from visual scans of the energy flow within hadronic events, followed by more detailed statistical analyses of global event shapes like oblateness and planarity. Not much later, at the end of 1979, the first determination of α_s was reported, in first-order perturbative QCD, from measurements of hadronic event shape distributions. A detailed description of these exciting, first years of experimentation at the PETRA e^+e^- storage ring at DESY in Hamburg is described in (Wu 1984). The further historical development of hadronic physics in e^+e^- annihilation, up until the recent experiments at LEP can be found *e.g.* in (Naroska 1987, Bethke 1989, Bethke 1992).

3 Accelerators and detectors

Those e^+e^- accelerators which have been, still are or which will be important for the subjects discussed in this lecture, are listed, together with some of their main parameters, in Table 1. The high energy frontier is currently covered by LEP and by the SLC. PETRA, PEP, TRISTAN and LEP have (had) 4 interaction regions which are equipped with multipurpose detectors, the others have one or two experiments. Note that most of the machines listed in Table 1 are storage rings; the only linear e^+e^- accelerator to date is the SLC. The currently planned next generation of higher energy e^+e^- accelerators is also a linear collider (NLC). More technical details about accelerators, and especially about LEP, are discussed in Blondel's lectures (these Proceedings).

Name (Laboratory)	Start/End date	Circumf. or Length [km]	E_{beam}^{max} [GeV]	Luminosity 10^{30}cm^{-2}s^{-1}	Beam ⌀ [μm]
SPEAR (SLAC)	1972/1990	0.234	4	10	700 × 50
DORIS (DESY)	1973/1992	0.289	5.6	33	540 × 30
CESR (Cornell)	1979/——	0.768	6	250	500 × 11
PETRA (DESY)	1978/1986	2.304	23.4	24	430 × 13
PEP (SLAC)	1980/1990	2.2	15	60	340 × 14
TRISTAN (KEK)	1987/——	3.02	32	35	310 × 8
SLC (SLAC)	1989/——	1.45 + 1.47	50	0.5	3 × 3
LEP (CERN)	1989/——	26.66	55	13	200 × 8
LEP-II (CERN)	(1995)	26.66	90–110	11–40	
NLC (n.n.)	???	~ O(40)	250	> 1000	0.3 × 0.003

Table 1. *Past, present and future e^+e^- colliders, the results of which are discussed in these lectures, and some of their main parameters. The numbers were collected in 1991; many of them changed over time, in which case the latest values are given (from RPP 1992).*

At least 24 different experiments have taken data at the existing e^+e^- accelerators listed in Table 1. One of these, Mark-II, operated—with different configurations and after several upgrades—at SPEAR, PEP and at the SLC. Instead of discussing the features of all of these detectors, only one of the most recent and most modern ones will be described here briefly.

Figure 2 shows a schematic view of OPAL, one of the four multipurpose detectors at LEP. The OPAL detector, when going from the interaction region in the centre of the detector to the outside, consists of the following main components.

- A small, high resolution silicon vertex detector, consisting of 2 layers of double-

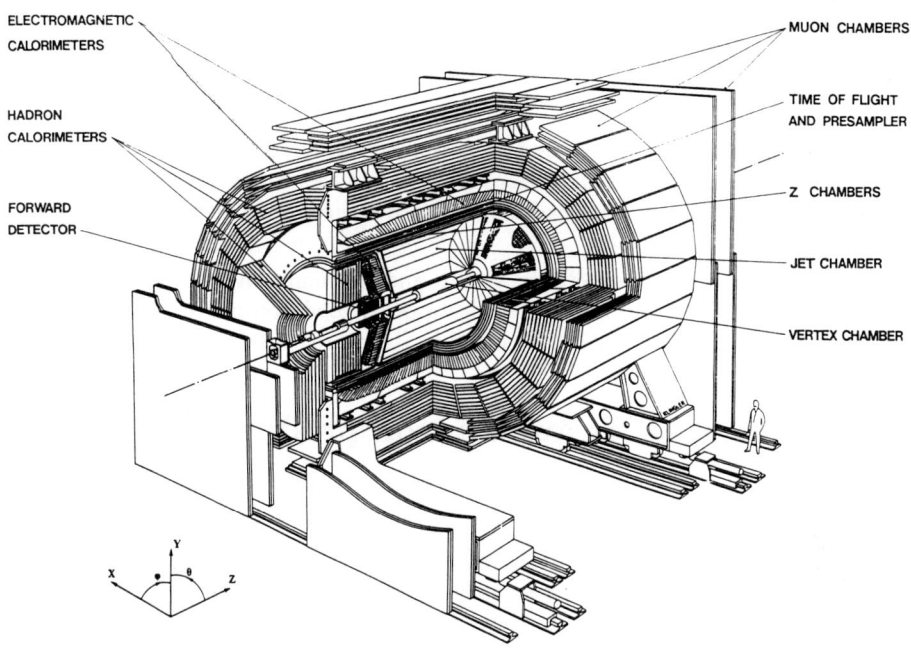

Figure 2. *The* OPAL *detector at* LEP.

sided silicon strips, providing a three-dimensional read-out of the coordinates of penetrating tracks of charged particles.

- The central detector consisting of a small vertex drift chamber, the big central jet chamber and an outer barrel of z-chambers to precisely reconstruct tracks of charged particles in a volume roughly 4m in length and 4m in diameter.

- A warm conductor solenoid magnet providing a uniform magnetic field of 0.44T inside the central detector volume.

- A time of flight scintillator barrel mounted around the solenoid.

- An electromagnetic calorimeter which covers 98% of the solid angle and which consists of about 9400 lead glass blocks in the barrel and about 2300 lead glass blocks in the end-caps. Inside this calorimeter and surrounding the pressure vessel of the central detector are thin gas detectors ('presamplers') which provide measurements of the position and the energy of electromagnetic showers which start in front of the main calorimeter.

- A hadron calorimeter instrumented by streamer tubes and thin gap wire chambers which are mounted inside the gaps of the magnet iron return yoke. The calorimeter has three main parts, the barrel, the outer end-caps and the pole tips.

- An external muon detector consisting of four layers of drift chambers in the barrel part and of two layers of streamer tubes in the end-caps.
- A forward detector including a silicon-tungsten luminosity monitor and an electromagnetic calorimeter system.

The general structure of the other LEP detectors, ALEPH, DELPHI and L3, is similar, but they differ in the overall dimensions and types of the various detector components. A detector view of a 'typical' hadronic event measured with OPAL can be found in Figure 3.

Some subdetectors, especially the vertex- and the luminosity-detectors at LEP, will be discussed in more detail in Blondel's lectures. A good overview of the LEP machine and of the four LEP detectors can be found in Burkhardt (1991).

4 Basics of QCD and hadron production

4.1 Some basics of QCD

The basic ideas and features of QCD are best reviewed by comparing this non-Abelian gauge theory with the (Abelian) gauge theory of QED. Such a comparison is schematically given in Table 2. While this table is rather simplistic, and far from exhaustive, it serves its purpose as an elementary introduction to the field. For more details and a deeper theoretical background, see Aitchinson's lectures (these Proceedings) or the books of Halzen and Martin (1984) and Nachtmann (1986).

QCD describes the interaction of quarks with gluons, which are the exchange quanta of the strong interaction. The force couples to the 'colour' charge of quarks and gluons. Since gluons carry colour charge by themselves, they exhibit the feature of gluon self coupling, in contrast to photons of QED which cannot couple to themselves. As a direct consequence, the effective coupling strength, α_s, has a strong dependence on the momentum scale of the interaction. The renormalisation scale dependence of α_s is controlled by the so-called β-function ::

$$\mu \frac{\partial \alpha_s}{\partial \mu} = -\frac{\beta_0}{2\pi}\alpha_s^2 - \frac{\beta_1}{8\pi^2}\alpha_s^3 - \cdots$$

$$\beta_0 = 11 - \frac{2}{3}N_f,$$

$$\beta_1 = 102 - \frac{38}{3}N_f, \qquad (1)$$

where N_f is the number of quark flavours with mass less than the energy scale μ. In solving this equation for α_s, a constant of integration is introduced. This constant is not predicted by QCD; it is the one fundamental parameter of QCD which must be deduced from experiment. The most common procedure is to introduce a dimensional parameter Λ, and to write as a solution of Equation 1, in second-order expansion and in the so-called $\overline{MS}$ (modified minimal subtraction) scheme (see Bardeen 1978),

$$\alpha_s(\mu) = \frac{12\pi}{(33-2N_f)\ln(\mu/\Lambda_{\overline{MS}})^2}\left(1 - 6\frac{153-19N_f}{(33-2N_f)^2}\frac{\ln\left(\ln(\mu/\Lambda_{\overline{MS}})^2\right)}{\ln(\mu/\Lambda_{\overline{MS}})^2} + \mathcal{O}(\alpha_s^3)\right). \qquad (2)$$

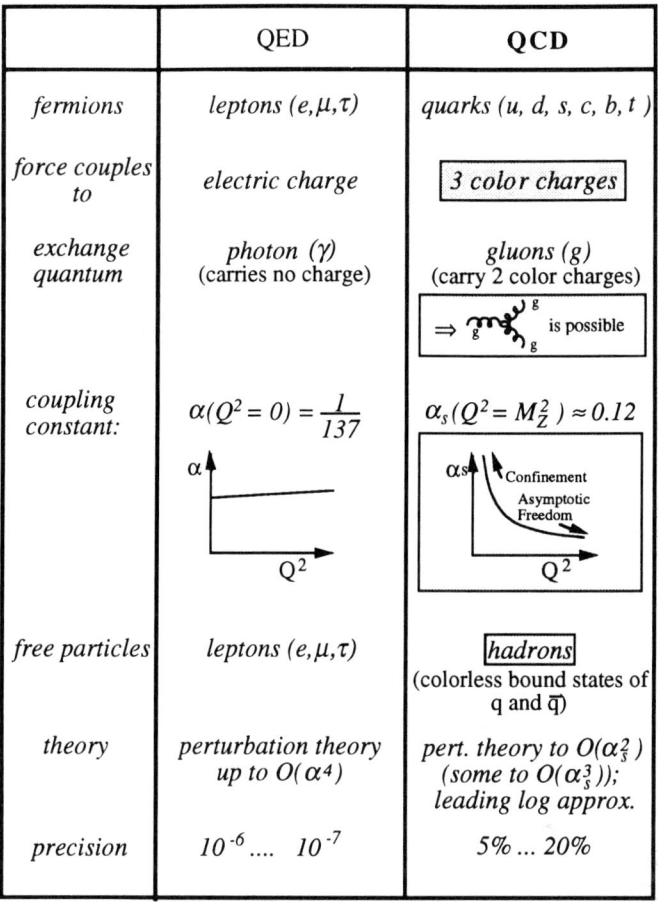

Table 2. *Comparison of the basic properties and features of Quantum Electrodynamics (QED) and Quantum Chromodynamics (QCD).*

At large energy scales μ, or equivalently at small distances, α_s vanishes logarithmically, an effect which is also called 'asymptotic freedom'.

In regions where α_s is sufficiently small, QCD may be solved by perturbation theory. At large distances or small momentum scales, however, α_s becomes very large, so that perturbation theory no longer applies. In this region of phase space, non-perturbative methods such as calculations on a lattice must be used to describe the strong interaction between quarks and gluons. Such methods, however, have not yet reached the same level of predictive power as perturbation theory has in the regime of large μ, so that the dynamics of QCD at large distances may be regarded as currently unsolved. Some phenomenological models were developed to describe the 'confinement' at small scales of μ, for instance by modelling the process of 'hadronisation', *i.e.* the transformation of quarks and gluons into hadrons, to an extent which allows the connection to be made between measurements of hadrons and the dynamics of quarks and gluons predicted by perturbative QCD.

Thus hadronisation is a non-perturbative process which typically happens at energy scales smaller than or of the order of a few GeV, at the scale of hadron masses. The average transverse momentum p_t of hadrons with respect to the direction of the quark or gluon motion should be independent of the energy of the parton. It was therefore expected that at large enough energies, hadrons appear as jets of particles, more or less collimated around the direction of the quark or gluon. These hadron jets were indeed observed in e$^+$e$^-$ annihilations, for the first time, around centre-of-mass energies E_{cm} of about 7 GeV, as mentioned in the historical review above. Currently, at E_{cm} as large as 95 GeV, the jet structure of hadronic events is rather distinct and is easily visible by eye, as can be seen in Figure 3. The physics of hadron jets is thus an important tool for precise tests of QCD, such as for instance the determination of the quantum numbers of quarks and gluons, and of the coupling strength $\alpha_s(\mu)$.

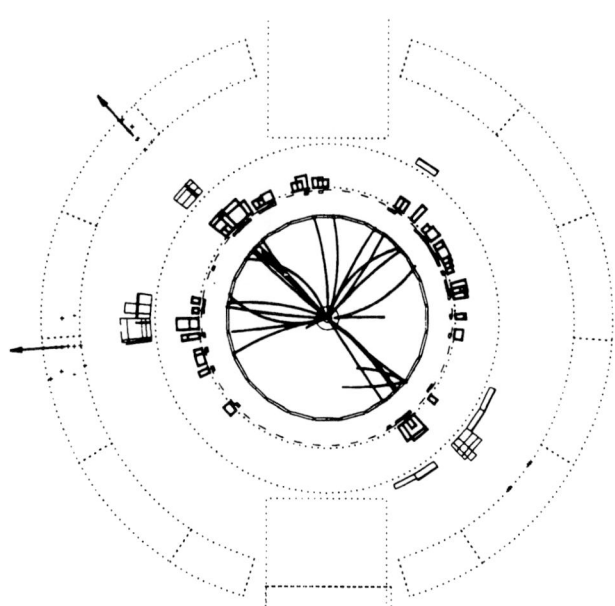

Figure 3. *A typical hadronic event of the type* e$^+$e$^- \to 4$ *jets, observed with the* OPAL *detector at* LEP *(c.f. Figure 2), viewed along the electron beam axis. The observation of two energetic muons, penetrating the calorimeters and the outer muon chambers, in one high and one of the two lower energy jets, makes it possible to tag this events as* e$^+$e$^- \to$ q$\bar{\text{q}}$q$\bar{\text{q}}$ *with a high level of confidence.*

4.2 Some predictions for $e^+e^- \to$ hadrons.

In the e^+e^- continuum, *i.e.* away from any resonance, the electron and positron annihilate into a virtual photon γ^*, which primarily materialises into a quark-antiquark pair; the partons then fragment into hadrons. The total cross-section of this reaction can be derived from the point-like cross-section of muon pair production, $e^+e^- \to \mu^+\mu^-$, which is

$$\sigma(e^+e^- \to \mu^+\mu^-) = \frac{4\pi\alpha^2}{3}\frac{1}{s}, \qquad (3)$$

where α is the electromagnetic coupling strength and $s \equiv E_{cm}^2$. For partons of spin 1/2, with masses which can be neglected for a given s, the same cross-section occurs, however multiplied by the sum of the squares of the parton charges, $\sum Q_q^2$, and with the number of colour degrees of freedom which applies to quarks. For the ratio of the hadronic to the leptonic cross-section, R, we therefore expect, in lowest order,

$$R_0 = \frac{\sigma(e^+e^- \to \text{hadrons})}{\sigma(e^+e^- \to \mu^+\mu^-)} = 3\sum_q Q_q^2. \qquad (4)$$

Experimental measurements of R give typical values of around 11/3 for $\sqrt{s} \geq 10\,\text{GeV}$, which is the predicted value for the production of 2 quark flavours with charge 2/3 and three flavours with charge 1/3, see Figure 4. Such measurements are an impressive confirmation of the hypothesis of three colour degrees of freedom for quarks.

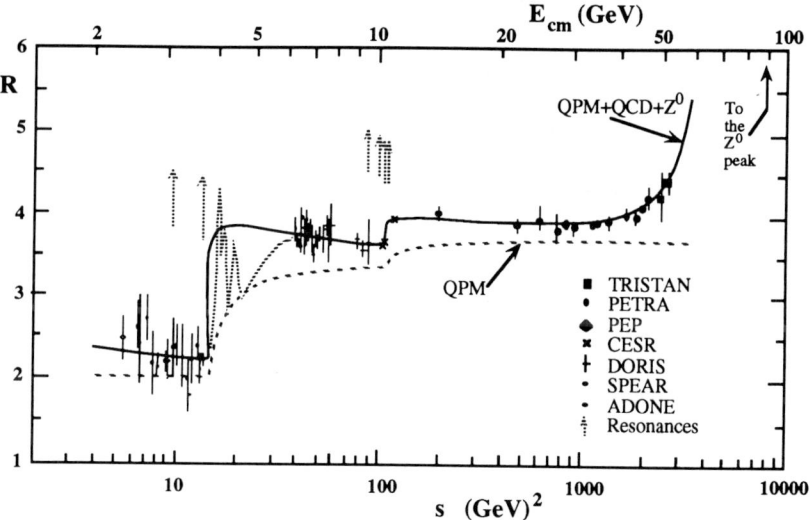

Figure 4. *Data for $R = \sigma(e^+e^- \to \text{hadrons})/\sigma(e^+e^- \to \mu^+\mu^-)$, together with the predictions of the pure quark-parton model and with the best fit to the data, including $\mathcal{O}(\alpha_s^3)$ QCD corrections and the effects of the Z^0 pole (Marshall 1988).*

On a closer look at Figure 4, the data in the e^+e^- continuum ($\sqrt{s} \geq 10\,\text{GeV}$) lie somewhat above the naïve value of 11/3; this discrepancy can be explained by higher-order QCD corrections to R, which are calculated, in complete third-order perturbation theory ($\mathcal{O}(\alpha_s^3)$), to be (Gorishny 1991)

$$R = R_0 \left(1 + \left(\frac{\alpha_s}{\pi}\right) + 1.4 \left(\frac{\alpha_s}{\pi}\right)^2 - 12.8 \left(\frac{\alpha_s}{\pi}\right)^3 + \mathcal{O}(\alpha_s^4)\right), \tag{5}$$

and by the existence of the Z^0 pole at $\sqrt{s} = 91.19$ GeV. This deviation of the data on R from the prediction of the naive quark-parton model, R_0 (see Equation 4), can thus be used to determine the value of α_s, as will be discussed in more detail later.

Another basic observable is the production cross-section for events which have one energetic gluon radiated off the quark or the antiquark, i.e. for the process $e^+e^- \rightarrow q\bar{q}g$, which, at large enough energies, may be observed as events with 3 clearly separated hadron jets. If the laboratory frame of the e^+e^- annihilation is identical to the centre-of-mass frame of the reaction, the 3 jets must be, for an ideal detector and without electromagnetic radiation effects in the initial e^+e^- state, momentum balanced (i.e. $\sum_i \vec{p}_i = 0$), and the normalised jet energies, $x_i = 2E_i/E_{cm}$, must add up to $\sum_i x_i = 2$.

In leading order ($\mathcal{O}(\alpha_s)$), and for gluons assumed to be vector particles, one obtains for the differential cross-section of 3-jet events

$$\frac{1}{\sigma}\frac{d\sigma}{dx_q dx_{\bar{q}}} = \frac{2\alpha_s}{3\pi} \frac{x_q^2 + x_{\bar{q}}^2}{(1-x_q)(1-x_{\bar{q}})}, \tag{6}$$

while for scalar gluons (i.e. spin 0), the result (assuming vector coupling) would be

$$\frac{d\sigma_{\text{scalar}}}{dx_q dx_{\bar{q}}} \sim \frac{[(1-x_q) + (1-x_{\bar{q}})]^2}{(1-x_q)(1-x_{\bar{q}})}. \tag{7}$$

First measurements of the 3-jet cross-section at PETRA gave evidence for gluons being vector particles, as expected from QCD (Wu 1984).

4.3 QCD and hadronisation models in e^+e^- annihilation

For experimental studies of hadron production and of QCD, model calculations which generate complete final states of hadrons are an essential tool. Such models are used to study the signature, experimental feasibility and backgrounds of the process under investigation. They are also used to study and correct for the limited acceptance and resolution of the detector, and—if appropriate—to analyse and correct for effects of hadronisation. A good overall description of QCD plus hadronisation models is given by Sjöstrand in the *Yellow Book* 'Z Physics at LEP-I' (LEP 1989). As sketched in Figure 5, such models consist of several different stages.

Electroweak phase. In this phase, the generation of $q\bar{q}$ pairs from the annihilation of $e^+e^- \rightarrow \gamma^*$ or of $e^+e^- \rightarrow Z^0$ is performed, whereby the angular orientation as well as the relative flavour composition of the overall event sample is modelled according to the predictions of the electroweak theory. In addition, most models also allow for the emission of a photon from the initial e^+e^- state.

Perturbative QCD phase. Gluon radiation off the initial $q\bar{q}$ state is generated according the predictions of perturbative QCD. In general, the different models use either of the two basic classes of QCD calculations which are available:

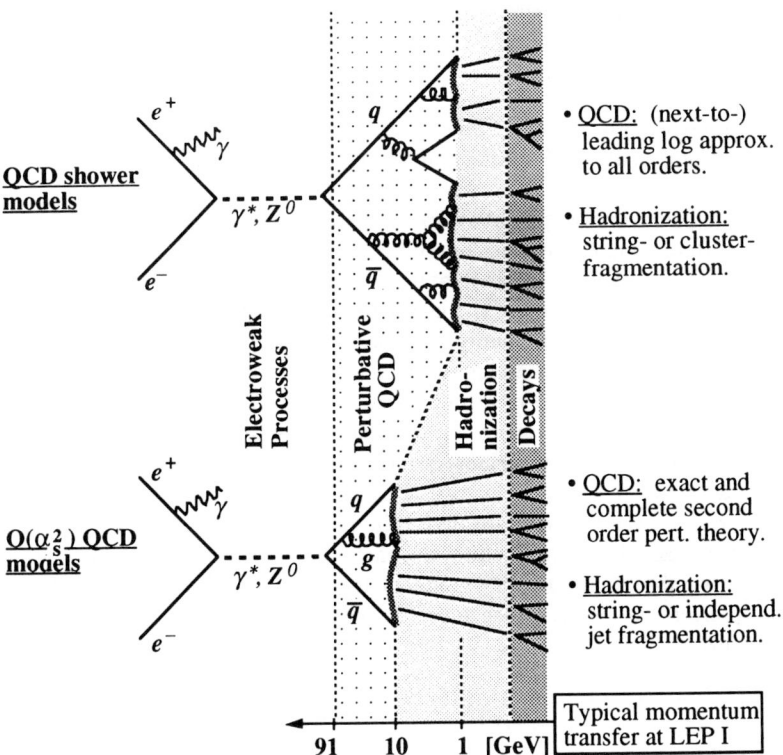

Figure 5. QCD *plus hadronisation models which are typically used to model hadronic final states of* e^+e^- *annihilation. The typical momentum scale where hadronisation sets in is given, taking* LEP *energies as an example, at the bottom axis.*

QCD shower models are based on calculations which derive probabilities for the processes q→qg, g→gg and g→q$\bar{\text{q}}$ by summing the leading (and sometimes also the next-to-leading) logarithms of the perturbative series to all orders. Depending on the total centre-of-mass energy of the reaction and on the energy scale Q_0 where perturbative gluon emission stops (typically of $\mathcal{O}(1\,\text{GeV})$), such models can generate large numbers of secondary quarks and gluons in the final state: typical numbers at LEP energies ($\sim 91\,\text{GeV}$) vary between 5 to 10 quarks and gluons per event, on average. These models, however, do not include the full fixed order matrix element for parton splitting; some of the QCD shower models may correct for this deficiency by a probabilistic reweighting method (according to, for example, the first-order matrix element) at the first q→qg vertex.

Fixed order QCD models are based on the complete QCD matrix element for q$\bar{\text{q}}$, q$\bar{\text{q}}$g (in 1st-order) and q$\bar{\text{q}}$gg and q$\bar{\text{q}}$q$\bar{\text{q}}$ (in 2nd-order or $\mathcal{O}(\alpha_s^2)$, which is the highest order available to date) final states. In order to render the cross-sections for 2, 3 and 4-parton final states positive, a cutoff against collinear and soft gluon emission, usually in terms of a minimum required scaled parton-parton invariant mass, $y_{\min} = M_{ij}^2/E_{\text{cm}}^2$, is applied. Thus, in practice, this minimum pair mass cannot be smaller than about 10% of the total centre-of-mass energy (*i.e.* $y_{\min} \geq 0.01$).

Hadronisation phase. Only phenomenological hadronisation models are available to describe the transition of quarks and gluons into final hadrons.

> **Independent fragmentation model.** In this simplest, but nowadays outdated, hadronisation model, each outgoing quark generates a chain of secondary $q'\bar{q}'$ pairs out of the vacuum. At each step of this chain, a meson is formed which contains the initiating and one of the newly generated, secondary quark or antiquark. The meson takes over an energy fraction z of the initiating quark, while the chain is carried on by the remaining secondary antiquark or quark, with a fractional energy of $(1-z)$. A special treatment is applied to the quark which remains at the end of the chain, when there is not enough energy left to create further $q\bar{q}$ pairs. In addition, since invariant (jet-) mass is generated out of a massless primary quark, momentum and energy sums are not conserved, which at the end must be corrected for. Gluons are first split into $q\bar{q}$ pairs which then in turn hadronise as described above. In this way, each of the partons created in the perturbative QCD phase hadronises independent from the others; this procedure thus neglects interference effects which must be present from a quantum-mechanical point of view.

> **String hadronisation model.** In this, the most commonly used hadronisation model, colour strings are formed between the colour charges of quarks and gluons, and each string then fragments into several $q\bar{q}$ pairs, out of which colourless hadrons are formed. Dynamical effects which are predicted by the string model were first observed in 1980 by the JADE collaboration at the PETRA e^+e^- storage ring (JADE 1981). String models are utilised both within QCD shower and fixed order QCD models.

> **Cluster model.** This is another hadronisation model which is basically used in QCD shower models. Here, each final state gluon is split into a $q\bar{q}$ pair; colourless clusters of $q\bar{q}$ pairs are then formed, which subsequently decay into hadrons according to simple phase-space and spin conservation rules.

Decays of unstable hadrons. The last step in generating hadronic final states is to let unstable hadrons decay. This is done invoking the decay tables which are available, for example, from the particle data book (RPP 1992).

A compilation of the most widely used QCD plus hadronisation models is given in Table 3. Each of these models usually has a (large) number of free parameters, which must be tuned to describe the experimental data:

- **QCD parameters:** for $\mathcal{O}(\alpha_s^2)$ models, these are the QCD scale parameter $\Lambda_{\overline{MS}}$ and the cutoff for soft and collinear gluons, $y_{\min}$, which is usually chosen to have the smallest possible value. For QCD shower models, these are either $\Lambda_{\overline{MS}}$ (in next-to-leading log approximation, NLLA) or a more general Λ (in LLA), plus a parameter

Model	QCD	Hadronisation	Reference
JETSET	$\mathcal{O}(\alpha_s^2)$ matrix element	string	Sjöstrand 1986,
	LLA QCD shower + $\mathcal{O}(\alpha_s)$	indep.	Sjöstrand 1987
HERWIG	NLLA QCD shower	cluster	Marchesini 1988
ARIADNE	colour-dipole QCD shower	string	Lönnblad 1988
COJETS	non-coherent QCD shower	indep.	Odorico 1988
NLLJET	NLLA QCD shower	string	Kamae 1988

Table 3. *The most commonly used* QCD + *hadronisation models in* e^+e^- *annihilation.*

Q_0 specifying the minimal parton mass where the shower stops.

In addition, the assignment of physical parton (quark and gluon) masses can be regarded as free parameters. Some models also allow switching between coherent and non-coherent parton showers.

- **Hadronisation parameters:** the most important are, in the case of string models and independent-fragmentation models, the parametrisations of the fragmentation functions $f(z)$, which govern the longitudinal momentum distribution (with respect to the string direction or to the primary quark) of hadrons, and the width of the (usually gaussian) distribution of transverse momenta, σ_q.

  In addition, a large number of parameters which determine the relative production, for example, of vector and scalar mesons and other details must be considered.

Depending on the specific model, there are in general about 5 'main' parameters which must be tuned to the data, plus a similar number of 'less important' (hadronisation) parameters which do not significantly influence the overall event structures but which may be important in describing special features and exclusive hadron production.

5 Hadronic event shapes

Hadronic event shape variables are tools for studing both the amount of gluon radiation and details of the hadronisation process. Since the laboratory frame and the centre-of-mass system of the annihilation are usually identical, events from $q\bar{q}$ final states without hard gluon radiation result in two collimated, back-to-back jets of hadrons. The emission of one hard gluon leads to planar 3-jet events, while the emission of two or more energetic gluons can cause nonplanar multi-jet event structures.

A typical event shape observable is the *thrust*, T, which is defined as the normalised sum of the momentum components of all particles of a given event along a specific axis; this axis is chosen such that T is maximised. Ideal 2-jet events result in $T = 1$, while $2/3 \leq T < 1$ for planar 3-jet events and $T = 1/2$ for completely spherical events.

Name of Observable	Definition	Typical Value for: (pencil)	Typical Value for: (3-jet)	Typical Value for: (spherical)	QCD calculation
Thrust	$T = \max_{\vec{n}} \left(\frac{\sum_i |\vec{p}_i \vec{n}|}{\sum_i |\vec{p}_i|} \right)$	1	$\geq 2/3$	$\geq 1/2$	(resummed) $O(\alpha_s^2)$
Thrust major	Like T, however T_{maj} and $\vec{n}_{maj}$ in plane $\perp \vec{n}_T$	0	$\leq 1/3$	$\leq 1/\sqrt{2}$	$O(\alpha_s^2)$
Thrust minor	Like T, however T_{min} and $\vec{n}_{min}$ in direction $\perp$ to $\vec{n}_T$ and $\vec{n}_{maj}$	0	0	$\leq 1/2$	$O(\alpha_s^2)$
Oblateness	$O = T_{maj} - T_{min}$	0	$\leq 1/3$	0	$O(\alpha_s^2)$
Sphericity	$S = 1.5(Q_1 + Q_2);\ Q_1 \leq ... \leq Q_3$ are Eigenvalues of $S^{\alpha\beta} = \frac{\sum_i p_i^\alpha p_i^\beta}{\sum_i p_i^2}$	0	$\leq 3/4$	≤ 1	none (not infrared safe)
Aplanarity	$A = 1.5\, Q_1$	0	0	$\leq 1/2$	none (not infrared safe)
Jet (Hemisphere) masses	$M_\pm^2 = \left(\sum_i E_i^2 - \sum_i \vec{p}_i^2 \right)_{i \in S_\pm}$ ($S_\pm$: Hemispheres $\perp$ to $\vec{n}_T$) $M_H^2 = \max(M_+^2, M_-^2)$ $M_D^2 = |M_+^2 - M_-^2|$	0 0	$\leq 1/3$ $\leq 1/3$	$\leq 1/2$ 0	(resummed) $O(\alpha_s^2)$
Jet broadening	$B_\pm = \frac{\sum_{i \in S_\pm} |\vec{p}_i \times \vec{n}_T|}{2 \sum_i |\vec{p}_i|};\ B_T = B_+ + B_-$ $B_w = \max(B_+, B_-)$	0 0	$\leq 1/(2\sqrt{3})$ $\leq 1/(2\sqrt{3})$	$\leq 1/(2\sqrt{2})$ $\leq 1/(2\sqrt{3})$	(resummed) $O(\alpha_s^2)$
Energy-Energy Correlations	$EEC(\chi) = \sum_{events} \sum_{i,j} \frac{E_i E_j}{E_{vis}^2} \int_{\chi - \frac{\Delta\chi}{2}}^{\chi + \frac{\Delta\chi}{2}} \delta(\chi - \chi_{ij})$	(histogram)	(histogram)	(histogram)	(resummed) $O(\alpha_s^2)$
Asymmetry of EEC	$AEEC(\chi) = EEC(\pi - \chi) - EEC(\chi)$	(histogram)	(histogram)	(histogram)	$O(\alpha_s^2)$
Differential 2-jet rate	$D_2(y) = \frac{R_2(y - \Delta y) - R_2(y)}{\Delta y}$				$O(\alpha_s^2)$

Table 4. *Definition and properties of observables which are typically applied to e^+e^- hadronic final states in various* QCD *studies.*

The definitions of thrust and other observables which are applied to hadronic final states of e^+e^- annihilations are summarised in Table 4. The *thrust, thrust major, thrust minor, oblateness, sphericity, aplanarity, jet masses* and the *jet broadening* measures are the so-called event shape observables, which are sampled in histograms where each hadronic event contributes with one entry. The *energy-energy correlation* (EEC) is different in this respect: basically, it is a histogram of the angles between all pairs of particles, where each entry is weighted by the product of the energies of the particles. The *asymmetry* of EEC (AEEC) is computed from the EEC histogram, and thus is also based on particle pairs and not on one characteristic number per event. The *differential change* of the 2-jet event rate, as a function of the jet resolution parameter y, will be discussed in detail in Section 6.

The list of observables shown in Table 4 is not entirely complete; other event shape parameters have been used in the past, however less frequently than those given in the table. Many of these observables are, to some degree, correlated with each other; after all, they are all sensitive to the relative amount of gluon radiation from the primary $q\bar{q}$ pair. In leading order ($\mathcal{O}(\alpha_s)$) QCD, some are even identical, as for instance $(1 - T)$ and M_H/E_{cm} as well as B_T and B_w; however the higher order QCD corrections are different for all observables listed in Table 4. With the exception of S and A, which depend *quadratically* on particle momenta and are thus not infra-red safe, all the observables given in the list are calculated to complete $\mathcal{O}(\alpha_s^2)$ perturbation theory, for massless quarks and gluons (Kunszt and Nason 1989). In addition, for some (but not all) observables the leading and next-to-leading logarithms of all orders have been resummed; see Section 8.2.2 for more details.

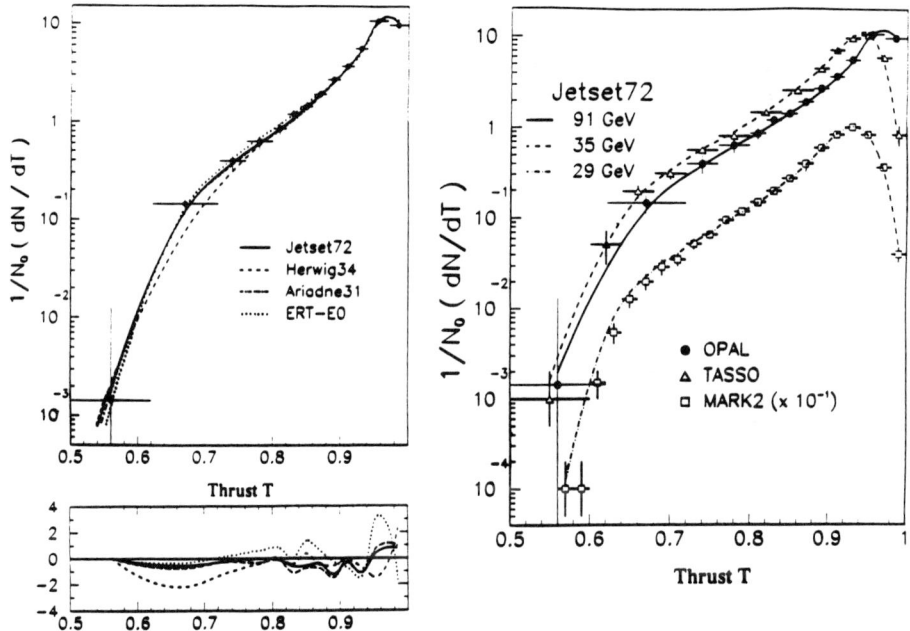

Figure 6. Left: *The thrust distribution as measured by* OPAL *at* E_{cm}=91 GeV, *compared to various QCD plus hadronisation models tuned to describe hadronic* Z^0 *decays. The insert below shows the difference between the model predictions and the data in units of standard deviations.* Right: *The thrust distribution measured at various centre-of-mass energies, compared to model predictions with parameters tuned to the* 91 GeV *data.*

In Figure 6 the thrust distribution of hadronic Z^0 decays, measured by the OPAL collaboration (OPAL 1990) at centre-of-mass energies close to 91 GeV, is compared to the predictions of several QCD plus hadronisation models. The data are corrected for detector acceptance, resolution, and for the effects of initial-state photon radiation; the latter are almost negligible at the Z^0 pole. The model parameters, like the QCD scale parameter Λ and several parameters specifying the hadronisation process, were adjusted to provide an overall good description of hadronic Z^0 decays. In general, such studies show that most global event shape distributions can be well described by the QCD plus hadronisation models studied so far. QCD shower models are somewhat superior compared to $\mathcal{O}(\alpha_s^2)$ QCD models, presumably because the latter cannot generate events with more than four parton jets. At present, the overall best description of hadronic events is provided by the JETSET parton shower model.

QCD predicts scaling violations for observables which do not depend on absolute energies or momenta, as for example the T-distribution. These scaling violations are caused by the energy dependence of α_s, which determines the amount of gluon radiation. In leading order the probability of gluon radiation is proportional to α_s. It is thus expected that fewer 3-jet and multi-jet events should be observed at higher centre-of-mass energies, and that event shape distributions evolve towards the 2-jet limit. The OPAL collaboration compared their measured event shape distributions with similar measurements done at lower centre-of-mass energies. As seen in Figure 6 for the T-distribution, small but significant differences are observed. To distinguish scaling violations from effects of a possible energy dependence of the hadronisation process, the data at 29 and 35 GeV are also compared with the predictions of the JETSET QCD shower model. The model predictions, with no parameters changed except the centre-of-mass energy, describe the data well at all energies. Within this model, the energy dependence of these distributions is thus explained by QCD scaling violations plus an energy independent parametrisation of the hadronisation process. Thus effects of asymptotic freedom are clearly supported by the data. More specific studies of scaling violations and of the energy dependence of α_s will be discussed in Section 8.2.5. Explicit determinations of $\alpha_s(M_{Z^0})$ will be presented in Section 8.

6 Jet physics

Studies of jet production rates provide the most intuitive tests of the underlying parton structure of hadronic events at high energies. Hadron jets and the corresponding jet multiplicity of individual events can often be inferred from graphical displays of hadronic events, *c.f.* Figure 3. Quantitative studies of jet production, however, require an exact definition of *resolvable* jets, which is usually given in terms of one or more resolution parameters, and a detailed description of how to combine jets (or clusters of particles) which do not fulfill the resolution criteria. Ideally, such a definition should be infra-red and collinear safe, *i.e.* the result should not change abruptly if a low energetic particle is added to the final state or if one particle is split into two. The definition must be applicable to experimental analyses and also to theoretical calculations in order to make a comparison between the two meaningful.

6.1 Jet recombination schemes

The most commonly used algorithm which fulfills these requirements was introduced by the JADE collaboration (JADE 1988): the scaled pair mass of any two resolvable jets i and j in a hadronic event, $y_{ij} = M_{ij}^2/E_{\text{vis}}^2$, is required to exceed a threshold value y_{cut}, where E_{vis} is the total visible (measured) energy of the event. In a recursive process, the pair of particles or clusters of particles which has the smallest value of y_{ij} is replaced by (or 'recombined' into) a single jet k with four-momentum $p_k = p_i + p_j$, as long as $y_{ij} < y_{\text{cut}}$. The procedure is repeated until all y_{ij} are larger than the jet resolution parameter y_{cut}, and the remaining clusters of particles are called jets.

For the JADE-type jet algorithms, the detailed partition of particles into jets and the jet multiplicity for a given value of y_{cut} not only depend on the definition of M_{ij}, but also on the prescription for combining two unresolvable jets. The reasons for introducing several recombination schemes are basically of a theoretical nature: higher-order QCD calculations are carried out for massless partons only; however, when the four-vectors of two jets (or partons) are added together, the resulting jet acquires, in general, a non-zero invariant mass. Several jet definition and recombination schemes exist, therefore, which either render jets massless or which neglect explicit mass terms in the definition of the jet resolution parameter (Bethke 1992a). Table 5 summarises those jet algorithms that have been adopted in theoretical calculations and experimental analyses.

Algorithm	Resolution	Combination	Remarks
E	$\dfrac{(p_i+p_j)^2}{E_{\text{vis}}^2}$	$p_k = p_i + p_j$	Lorentz invariant
p	$\dfrac{(p_i+p_j)^2}{E_{\text{vis}}^2}$	$\vec{p}_k = \vec{p}_i + \vec{p}_j$; $E_k = |\vec{p}_k|$	violates $\sum E$
E0	$\dfrac{(p_i+p_j)^2}{E_{\text{vis}}^2}$	$E_k = E_i + E_j$; $\vec{p}_k = \dfrac{E_k(\vec{p}_i + \vec{p}_j)}{|\vec{p}_i + \vec{p}_j|}$	conserves $\sum E$, but violates $\sum \vec{p}$
JADE	$\dfrac{2E_i E_j(1-\cos\theta_{ij})}{E_{\text{vis}}^2}$	$p_k = p_i + p_j$	conserves $\sum E$, $\sum \vec{p}$
D (k_T)	$\dfrac{2\min(E_i^2, E_j^2)(1-\cos\theta_{ij})}{E_{\text{vis}}^2}$	$p_k = p_i + p_j$	resummable

Table 5. *Resolution criteria and recombination schemes of various jet algorithms; $\vec{p}_i$ denotes a 3-vector and $p_i \equiv (E_i, \vec{p}_i)$ is the corresponding 4-vector.*

The algorithm which is entirely Lorentz invariant is the so-called E-scheme, where unresolved jets are combined by simply adding their four-vectors and for which the exact, Lorentz invariant definition of the pair mass, namely $M_{ij}^2 = (p_i + p_j)^2$, is used in the definition of y_{cut}. Within the E0 and the p-scheme, the momentum or the energy components of the resulting, recombined four-vector are scaled such that it appears to have zero mass. The original JADE scheme is based on the exact four-vector recombination, but neglects explicit mass terms in the definition of M_{ij}. Up to second-order perturbation theory, the E0 and the JADE schemed are equivalent.

The latest development is the 'Durham' (D) jet algorithm (Brown and Stirling 1992, Bethke 1992a), where the invariant jet pair mass M_{ij} is replaced by the minimum transverse energy of one jet with respect to the other; see Table 5. This algorithm exhibits certain features which make it very attractive both from an experimental and a theoretical point of view: the algorithm largely avoids 'unintuitive' recombinations of particles with large angles between them, and—as a consequence—leading and next-to-leading logarithms can be resummed, in addition to the usual $\mathcal{O}(\alpha_s^2)$ calculations, in theoretical predictions of D-scheme jet rates (see Brown and Stirling 1992 and Section 8.2.2).

6.2 QCD predictions in $\mathcal{O}(\alpha_s^2)$

QCD predictions in complete $\mathcal{O}(\alpha_s^2)$ perturbation theory, based on the QCD matrix elements of Ellis, Ross and Terrano (Ellis 1981), exist for all the jet algorithms listed in Table 5 (LEP 1989, Bethke 1992a). In $\mathcal{O}(\alpha_s^2)$, jet production rates are quadratic functions of the running coupling constant $\alpha_s(\mu)$:

$$R_2(y_{\text{cut}}, \mu) \equiv \frac{\sigma_2}{\sigma_{\text{tot}}} = 1 - A(y_{\text{cut}})\frac{\alpha_s(\mu)}{2\pi} - \{B(y_{\text{cut}}, x_\mu) + C(y_{\text{cut}})\}\left(\frac{\alpha_s(\mu)}{2\pi}\right)^2$$

$$R_3(y_{\text{cut}}, \mu) \equiv \frac{\sigma_3}{\sigma_{\text{tot}}} = A(y_{\text{cut}})\frac{\alpha_s(\mu)}{2\pi} + B(y_{\text{cut}}, x_\mu)\left(\frac{\alpha_s(\mu)}{2\pi}\right)^2$$

$$R_4(y_{\text{cut}}, \mu) \equiv \frac{\sigma_4}{\sigma_{\text{tot}}} = C(y_{\text{cut}})\left(\frac{\alpha_s(\mu)}{2\pi}\right)^2, \qquad (8)$$

where σ_{tot} is the total hadronic cross-section, σ_n are the cross-sections for n-parton event production, $\mu = x_\mu E_{\text{cm}}$ is the renormalisation scale at which α_s is evaluated and x_μ is the renormalisation scale factor. The leading-order QCD coefficients A and C depend only on the jet resolution parameter y_{cut}; the next-to-leading order coefficient B is recombination-scheme dependent and exhibits an explicit dependence on the renormalisation scale factor x_μ:

$$B(y_{\text{cut}}, x_\mu) = B(y_{\text{cut}}, 1) + A(y_{\text{cut}})2\pi\beta_0 \ln(x_\mu^2), \qquad (9)$$

where $\beta_0 = 11 - 2N_f/3$ and N_f is the number of active quark flavours ($N_f = 5$ at LEP energies).

6.3 Quality of jet algorithms

The usefulness of a jet algorithm depends on several features such as the size of the higher-order perturbative corrections, the stability against variations of the renormalisation scale and the magnitude of non-perturbative hadronisation effects. These and other characteristics were extensively studied in previous analyses (Bethke 1992a) which

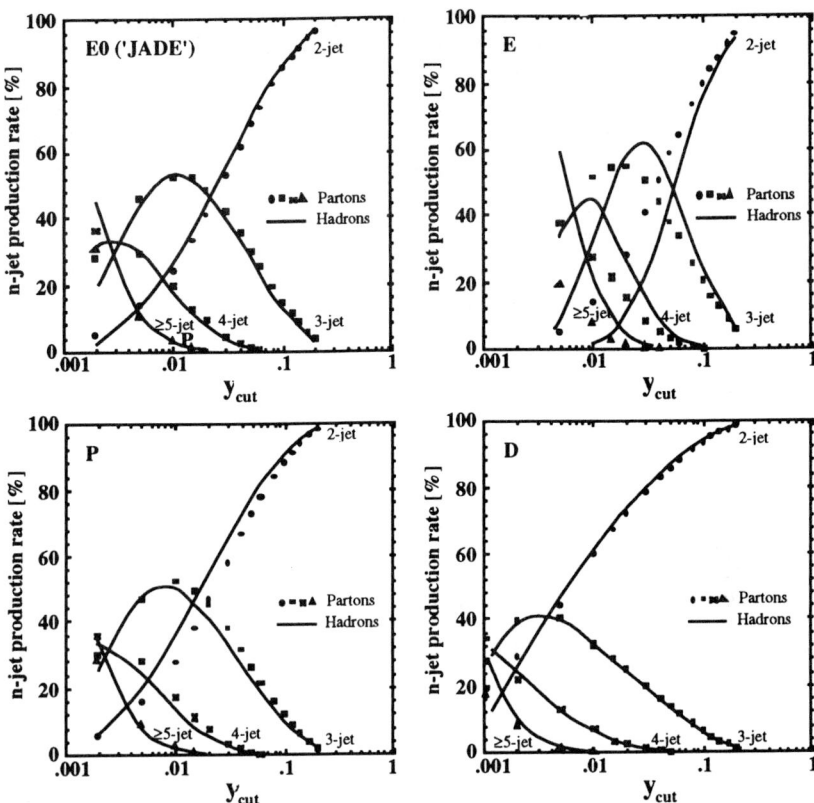

Figure 7. *Jet rates as a function of y_{cut}, as predicted by the* JETSET QCD *shower model before (partons) and after (hadrons) the hadronisation process ($E_{cm} = 91.2$ GeV).*

demonstrated that there are sizable differences between the various jet algorithms. For example, the average hadronisation corrections for n-jet event rates ($n = 2, 3, 4$ and ≥ 5) are demonstrated in Figure 7, where the relative n-jet production rates are shown as a function of the jet resolution parameter y_{cut}. The calculations are done with the JETSET QCD shower plus string hadronisation program, both for the quarks and gluons at the end of the parton shower ('parton level') and for the particles after hadronisation ('hadron level').

From Figure 7 it can be seen that the absolute numbers of n-jet events for a given value of y_{cut} are quite different for the various algorithms. More importantly, the size of the hadronisation correction, *i.e.* the difference between the hadron and the parton levels, varies significantly. The hadronisation corrections are smallest (about 5%) for the E0 and the D jet algorithms, while they are largest (about 30%) for the E-scheme.

Due to its overall small hadronisation corrections, the JADE (E0) algorithm was always preferred for experimental studies in the past, while the E-scheme was, at best, utilised for systematic checks. The D-scheme, whose hadronisation corrections are even smaller (at LEP energies) than those of the JADE scheme and which has a superior jet resolution power, is now taking over the rôle of the standard jet algorithm in e^+e^- annihilation. An additional reason for this development is the fact that within the JADE scheme, two soft (*i.e.* low-energy) gluons or jets, which may be even back-to-back, may be combined, in some rare cases, instead of each being assigned to its closest high-energy jet. This nonintuitive behaviour, which also spoils the exponentiation of logarithms of y_{cut} for $y_{\text{cut}} \to 0$ and which thus prevents the resummation of these logs in recent resummation calculations, is not present in the D-scheme.

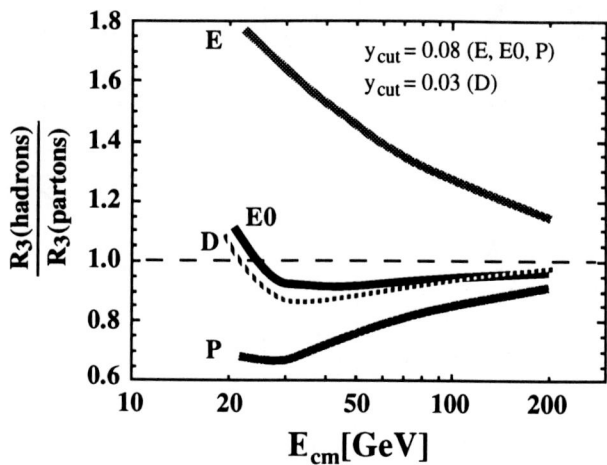

Figure 8. *The ratio r of 3-jet event rates, calculated from JETSET QCD shower model events before and after hadronisation, as a function of E_{cm}, for 4 different jet algorithms.*

The JADE scheme, on the other hand, shows smaller and more stable hadronisation corrections at lower centre-of-mass energies. This is demonstrated in Figure 8, where the ratio $r = R_3(\text{hadrons})/R_3(\text{partons})$ predicted by QCD shower model calculations is plotted as a function of E_{cm}, for constant values of y_{cut}. In order to consider the different definitions of y_{cut} in the different schemes, y_{cut} was chosen to be 0.08 in the case of the E, E0 and p-scheme and 0.03 for the D-scheme, such that the 3-jet rates at $E_{\text{cm}} = 91.2\,\text{GeV}$ are around 20% in each case. The QCD and hadronisation parameters of the parton shower model were kept at the values which were optimised at LEP energies since these proved to be a good description of the lower energy data; see Figure 6. For all jet algorithms, the quantity $(1 - r)$ shows an approximate $1/E_{\text{cm}}^2$ behaviour at large E_{cm}, as expected for non-perturbative hadronisation effects. At smaller energies, usually for $\sqrt{y_{\text{cut}}}E_{\text{cm}} < 7\,\text{GeV}$ ($< 4\,\text{GeV}$ for the D-scheme), r increases with decreasing E_{cm} because of misassignments of jets, caused by hadronisation fluctuations and heavy quark decays. For the JADE (E0) algorithm, the energy dependence of r is rather flat for E_{cm} between 25 and 100 GeV, and can in fact be approximated by a constant within a systematic uncertainty of ±2%. This remarkable stability in the overall size of hadronisation effects is not provided by any of the other jet algorithms (although the D-scheme is coming close), which will make an important impact on the experimental evidence for asymptotic freedom.

6.4 Evidence for asymptotic freedom.

According to Equation 2, α_s is predicted to decrease with increasing momentum transfer Q like $\alpha_s \sim 1/\ln Q$; this behaviour is also called 'asymptotic freedom'. The multitude of measurements of α_s from many different processes and for energy scales which range between the mass of the τ lepton, $M_\tau = 1.78\,\text{GeV}$, and the Z^0 mass, $M_{Z^0} = 91.19\,\text{GeV}$, provides a consistent picture of the running α_s, as will be demonstrated in Section 8.3.

An alternative way of verifying the energy dependence of α_s, without determining actual values of α_s, is to study the energy dependence of observables which should, ideally, fulfill the following requirements: the observable must depend distinctively on α_s, and hadronisation corrections must be small and energy independent. For practical analyses, the latter requirement can be further specified: within the range of energies under study, hadronisation corrections must be significantly smaller than the size of the effect which is expected from the running α_s; a remaining energy dependence of the hadronisation correction can then be safely corrected for, or—more conservatively—can even be absorbed into a systematic point-to-point uncertainty.

One observable which fulfills these demands is the 3-jet event production rate R_3 using the JADE jet algorithm, in regions of y_{cut} where $\sqrt{y_{\text{cut}}}E_{\text{cm}} > 7\,\text{GeV}$ (see Figure 8). We recall that for constant values of the jet resolution parameter y_{cut}, the energy dependence of R_3 is only determined by α_s (see Equation 8), and hadronisation corrections are rather small and vary little with energy ($6\% \pm 2\%$ for $y_{\text{cut}} = 0.08$ and $25 < E_{\text{cm}} < 100\,\text{GeV}$; see Figure 8). It has been verified that similar results hold for y_{cut} values between 0.04 and 0.20, as long as $\sqrt{y_{\text{cut}}}E_{\text{cm}} > 7\,\text{GeV}$, and that other models, like HERWIG and the $\mathcal{O}(\alpha_s^2)$ QCD model of JETSET, predict similar numbers. Since α_s is expected to change by 20–30% in the PETRA/PEP, TRISTAN and LEP energy range, measurements of R_3—without explicit corrections for hadronisation effects—are thus expected to provide a rather clean test of the running α_s and of asymptotic freedom.

A compilation of measurements of R_3, analysed with the JADE (E0) jet finder using $y_{\text{cut}} = 0.08$, from PETRA, PEP, TRISTAN and LEP is presented in Figure 9. The data, which are not corrected for hadronisation effects, are compared with analytic $\mathcal{O}(\alpha_s^2)$ QCD calculations (Kunszt and Nason 1989), using two different renormalisation scales, $\mu = E_{\text{cm}}$ and $\mu = 0.07 E_{\text{cm}}$, and with the hypothesis of an energy *independent* coupling constant (which is however not predicted by any gauge theory). The free parameters $\Lambda_{\overline{MS}}$ or a constant average 3-jet rate $\langle R_3 \rangle$ were determined by minimising χ^2 for the data from $E_{\text{cm}} = 29$ to 91 GeV. The data points at $E_{\text{cm}} = 22\,\text{GeV}$ were not included in the fit since hadronisation effects may already bias the measurements at this energy; see Figure 8. The results for these parameters and the corresponding values of χ^2 are listed in Table 6. In order to account for the small, energy-dependent hadronisation effects as predicted by the model calculations shown in Figure 8, a relative systematic point-to-point uncertainty of $\pm 2\%$ was included in the fits. Also shown in Figure 9 are the predictions of the Abelian vector theory in $\mathcal{O}(\alpha_A^2)$, where α_A was adjusted such that the jet rates at $E_{\text{cm}} = 44\,\text{GeV}$ are reproduced (see Section 7.2 for a further discussion of the Abelian vector theory).

The measured jet production rates decrease significantly with increasing energy, excluding the possibility of an energy-independent coupling with a significance of more than 7 standard deviations (3.1 standard deviations without the LEP data). The Abelian

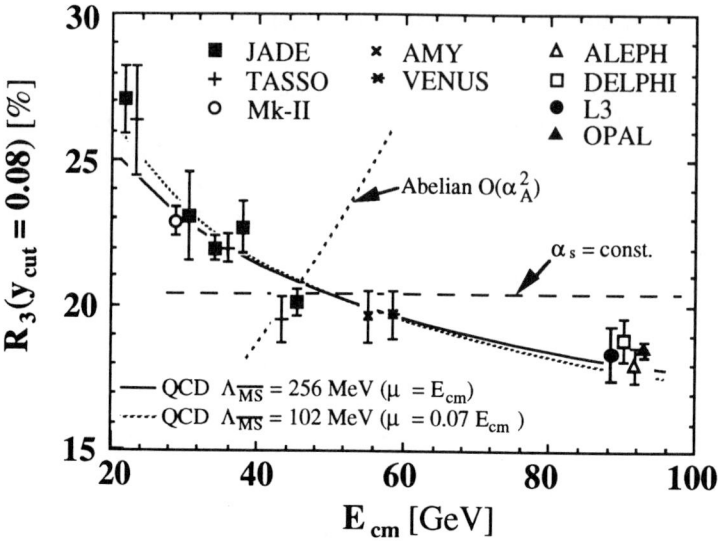

Figure 9. *Energy dependence of three-jet event production rates R_3, using the JADE scheme with $y_{cut} = 0.08$. The measurements are compared with predictions of analytic $\mathcal{O}(\alpha_s^2)$ QCD calculations, with the hypothesis of an energy independent α_s and with the Abelian vector theory in $\mathcal{O}(\alpha_A^2)$.*

Theory	fit result	χ^2/d.o.f.
QCD, $x_\mu = 0.07$	$\Lambda_{\overline{MS}} = (120 \pm 4)\,\text{MeV}$	14.2 / 12
QCD, $x_\mu = 1.0$	$\Lambda_{\overline{MS}} = (256\,{}^{+\,12}_{-\,11})\,\text{MeV}$	8.9 / 12
$\alpha_s = \text{const.}$	$\langle R_3 \rangle = 20.3 \pm 0.2$	76.2 / 12
Abelian theory	$\alpha_A(44\,\text{GeV}) = 0.26$	∞

Table 6. *Fit results for various assumptions about the energy dependence of 3-jet production rates, obtained for the data between $E_{cm} = 29\,\text{GeV}$ and $91\,\text{GeV}$ shown in Figure 9 (an additional relative systematic point-to-point error of $\pm 2\%$ was included in the fit).*

theory with an increasing coupling strength is completely ruled out; the differences between data and the Abelian theory are too large to be explained by any reasonable assumption about a hypothetical Abelian hadronisation model. The $\mathcal{O}(\alpha_s^2)$ QCD calculations describe both the absolute values of R_3 *and* their energy dependence well, with only one free parameter, namely $\Lambda_{\overline{MS}}$. The predicted degree of energy dependence is largely insensitive to the renormalisation scale chosen; however, the actual values of $\Lambda_{\overline{MS}}$ and of α_s do depend strongly on the scale (see also Section 8). Thus the data provide convincing evidence for the 'running' of α_s.

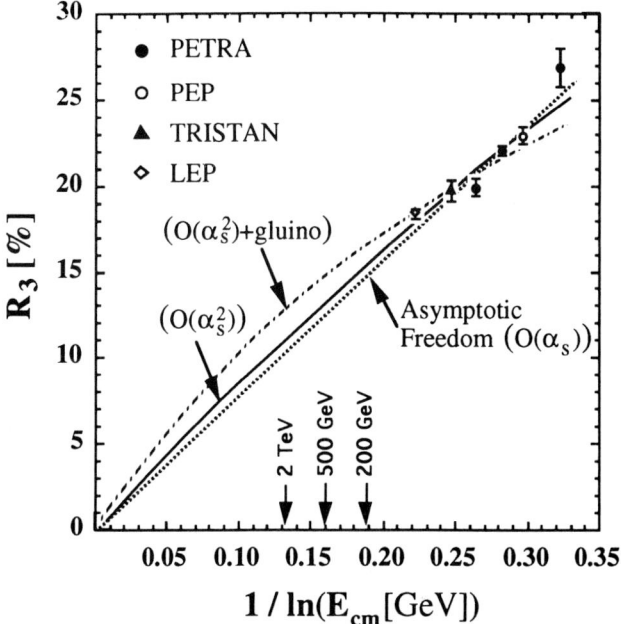

Figure 10. *The same data as in Figure 9, as a function of $1/\ln(E_{cm})$, compared to the prediction of asymptotic freedom [the gluino curve is discussed at the end of Section 8.3].*

An obvious way to demonstrate the evidence for asymptotic freedom is to plot R_3 as a function of $1/\ln(E_{cm})$, as shown in Figure 10. The data, again for $y_{cut} = 0.08$, are now combined at similar centre-of-mass energies. The dashed line is a fit to the leading order QCD prediction, namely $R_3 \propto \alpha_s \propto 1/\ln E_{cm}$. The corresponding prediction in $\mathcal{O}(\alpha_s^2)$ is also shown, indicating that higher-order terms affect the energy dependence of R_3 only slightly. At infinite energy $(1/\ln(E_{cm}) \to 0)$, R_3 and α_s are expected to vanish; an assumption which is in good agreement with the data. Note that a new data point from a future e^+e^- linear collider with $E_{cm} \approx 500$ GeV would add tremendous significance to the evidence for asymptotic freedom, even with modest data statistics of a few thousand events (Bethke 1992b). From the point of view of PETRA/PEP energies and through the eyes of QCD, *i.e.* on a logarithmic energy scale, a 500 GeV collider is already halfway towards infinite energies. Sooner, we can expect a data point at $E_{cm} \sim 200$ GeV, where the second phase of LEP operation will provide sufficient data for this type of analysis, probably by the end of 1995.

7 Tests of basic quantum numbers of quarks and gluons

As was mentioned in Section 2, evidence for quarks having spin 1/2 was already found in 1975 (Hanson 1975), where it could be shown that the orientation of the sphericity axis (*i.e.* the axis corresponding to the largest eigenvalue Q_3 of the sphericity tensor $S^{\alpha\beta}$; see Table 4) follows the prediction for spin 1/2 particles, to be proportional to $(1 + \cos\theta)^2$, where θ is the angle between the e^+e^- beam axis and the sphericity axis.

Evidence for the vector (*i.e.* spin 1) nature of the gluon was reported in earlier studies of 3-jet events at PETRA (Wu 1984), and first evidence for the gluon self-coupling, *i.e.* for the gluon carrying colour charge, was obtained from 4-jet angular correlations measured at TRISTAN (AMY 1989). The current state-of-the-art of these latter two studies will be discussed further, especially since the LEP data, due to the higher centre-of-mass energy, provide reduced hadronisation corrections, higher data statistics and a much more reliable jet definition at low values of y_{cut}, where the characteristic features of the gluons are most pronounced.

7.1 The gluon spin.

Two variables, which are defined in 3-jet events, are especially sensitive to the difference between vector and (hypothetical) scalar, *i.e.* spin 0, gluons. The differential 3-jet cross-section for vector gluon production, in leading order perturbation theory, was given in Equation 6. For a practical analysis in which, in general, the gluon jet cannot be identified, the jets are labelled such that $x_1 > x_2 > x_3$. Due to the bremsstrahlung character of gluon emission, x_3 is most likely to correspond to the gluon jet and we may replace x_q and $x_{\bar{q}}$ by x_1 and x_2 (or vice versa). The cross-section is singular for $x_2 \to 1$ (and hence $x_1 \to 1$), while this is not the case for spin 0 gluons (see Equation 7). It thus appears that the shape of the x_2 distribution is rather sensitive to the gluon spin. Another observable is the Ellis-Karliner angle λ, defined as the angle between jets 1 and 3 in the rest frame of jets 2 and 3 (Ellis and Karliner 1979). The leading-order QCD prediction has a pole for $\cos\lambda \to 1$, while the scalar gluon hypothesis does not.

The experimental distributions of x_2 and λ, as measured by the L3 collaboration (L3 1991), are shown in Figure 11. The figure also shows the predictions of vector and scalar gluon models. The vector gluon hypothesis is clearly favoured, and the data also demonstrate the pole structure for x_2, $\cos\lambda \to 1$. As usual, similar results are also available from other LEP experiments, (see *e.g.* Bethke and Pilcher 1992).

7.2 Gluon self coupling and QCD group constants.

The non-Abelian nature of QCD manifests itself in the process of gluon self coupling. In e^+e^- annihilation, the dominant contribution to 4-jet final states comes from the triple gluon vertex (TGV): about 95% of all 4-jet events are expected to be due to $q\bar{q}gg$ (*i.e.* double gluon bremsstrahlung and TGV) final states and only 5% are $q\bar{q}q\bar{q}$ (from $g \to q\bar{q}$ splitting) events (Bethke, Ricker and Zerwas 1991). An Abelian model, in which the TGV does not exist and which can be constructed by simply replacing the group constants of SU(3), C_f, N_c and T_f, with those of U(1) (Kramer 1984), predicts significantly different numbers: about 68% $q\bar{q}gg$ and 32% $q\bar{q}q\bar{q}$ events; see Table 7.

The different spin structure of the processes $g \to gg$ and $g \to q\bar{q}$ thus gives rise to different predictions of angular correlations within 4-jet events, which may be used to discriminate between QCD and the Abelian scapegoat model. OPAL and L3 (see Bethke

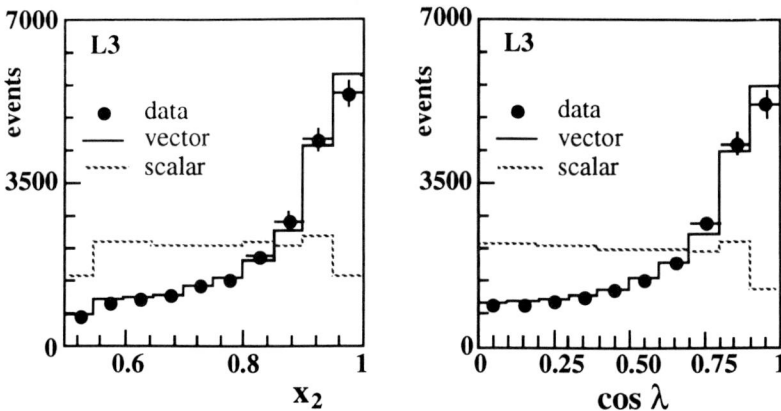

Figure 11. *Distributions sensitive to the gluon spin in 3-jet events: reduced momentum of the second-most energetic jet, x_2, and the Ellis Karliner angle λ.*

	QCD	Abelian
C_f	4/3	1
N_c	3	0
T_f	1/2	3
q q̄ gg	≈ 95%	≈ 68%
q q̄ q q̄	≈ 5%	≈ 32%

Table 7. *Group constants of QCD and of the Abelian vector theory, as well as the expected relative fractions of 4-jet events from qq̄gg and qq̄qq̄ final states.*

and Pilcher 1992) have studied the distributions of two such angles defined by the axes of reconstructed 4-jet events, which are especially sensitive to the relative contribution of qq̄qq̄ final states (see Bethke, Ricker and Zerwas 1991). The distribution of the angle between the two planes spanned by the two most-energetic and the two least-energetic jets, $\chi_{\rm BZ}$ (Bengtsson and Zerwas 1988), as measured by the L3 collaboration (L3 1990), is shown in Figure 12. The data are well described by the predictions of QCD, while Abelian vector models fail to reproduce the shape of the data distribution.

DELPHI and ALEPH have followed another strategy of analysis, which provides more direct evidence for the existence of the TGV. DELPHI (DELPHI 1993) studies the two-dimensional distribution of the modified Nachtmann-Reiter angle, $\cos \theta^*_{\rm NR}$ (Nachtmann and Reiter 1982), which is the cosine of the angle between the 4-jet event momentum vectors $(\vec{p}_1 - \vec{p}_2)$ and $(\vec{p}_3 - \vec{p}_4)$, and α_{34}, which is the angle between the two lowest momentum jets. While $\theta^*_{\rm NR}$ gives good discrimination between four-quark and qq̄gg final states but cannot distinguish double gluon bremsstrahlung from TGV events, the distribution of α_{34} is expected to discriminate between the latter. ALEPH (ALEPH 1992) analyses a five-dimensional distribution of invariant jet pair masses, which contains

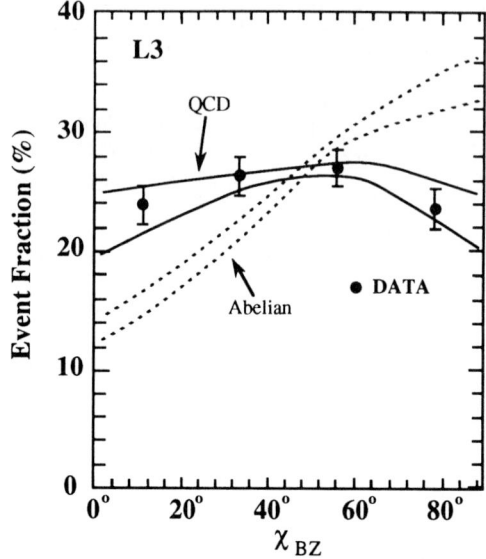

Figure 12. *Distribution of the Bengtsson-Zerwas angle χ_{BZ} for 4-jet events. Solid lines show the range of predictions for* QCD, *while dashed lines show the predictions for Abelian vector models.*

the full information of the 4-jet kinematic phase-space, and thus of the 4-jet QCD matrix element. In a fit of the $\mathcal{O}(\alpha_s^2)$ (*i.e.* leading-order) 4-jet matrix element to their experimental distributions, both DELPHI and ALEPH determine the group constants N_c/C_f and T_f/C_f, by separating the events into five classes whose partial cross-sections are proportional to different combinations of the group constants C_f, N_c and T_f. Note that the TGV contribution is proportional to N_c, the numbers of colours, while the process of g → q$\bar{\text{q}}$ is basically proportional to $T_f N_f$, where N_f is the number of quark flavours.

The results of both DELPHI and ALEPH are summarised in Table 8. They are in good agreement with the predictions of QCD, while Abelian models are significantly ruled out (*c.f.* Table 7). In particular, the non-zero result for N_c/C_f is interpreted as direct evidence for the TGV. In Figure 13, the fit result of DELPHI is displayed in the two-dimensional $N_c/C_f - T_f/C_f$ plane, together with the predicted values for several different gauge theories.

8 Measurements of α_s

8.1 Determination of α_s in the e$^+$e$^-$ continuum

Since gluon jets have been seen explicitly in experiments at PETRA and PEP, the determination of α_s was always one of the key analysis in e$^+$e$^-$ annihilations at high energies. Typically, α_s was determined from comparisons of hadronic event shape distributions,

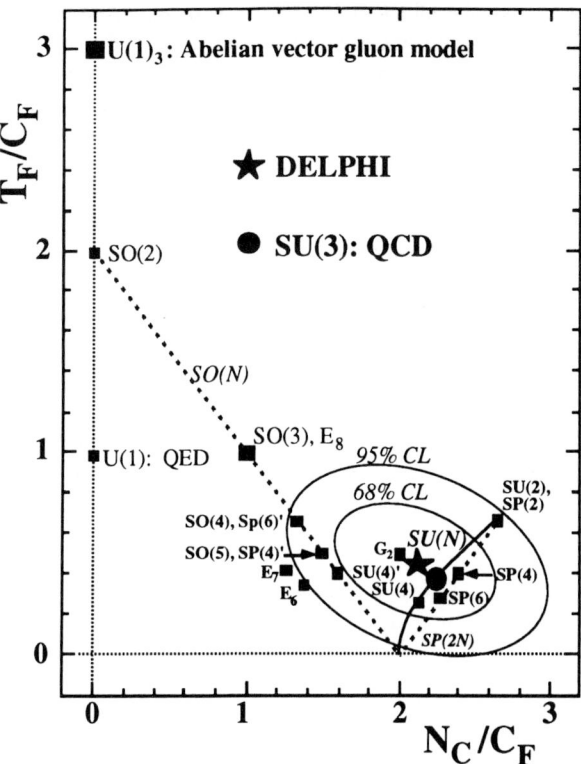

Figure 13. *Results of fitting the group constants N_c/C_f and T_f/C_f from a study of 4-jet events by* DELPHI, *compared with the predictions for several gauge theories.*

Experiment	N_c/C_f	T_f/C_f
DELPHI	2.12 ± 0.35	0.46 ± 0.19
ALEPH	2.24 ± 0.40	0.58 ± 0.29

Table 8. *Results of fitting the group constants N_c/C_f and T_f/C_f from studies of 4-jet events.*

energy correlations and jet production rates (like those defined in Sections 5 and 6) with the predictions of $\mathcal{O}(\alpha_s^2)$ QCD plus hadronisation models. Precise measurements of the total hadronic cross-section R, normalised by the leptonic cross-section, which according to Equation 5 exhibits a QCD correction which is even known to complete $\mathcal{O}(\alpha_s^3)$, were also employed to determine α_s.

The result of one of the most comprehensive compilations of measurements of R between $E_{cm} = 2.65$ GeV and 52 GeV are shown in Figure 4 (Marshall 1988), together with the predictions of the quark-parton model and those including the full

$\mathcal{O}(\alpha_s^3)$ QCD correction plus the effect of the Z^0 resonance. From a fit to these data, $\alpha_s(35\,\text{GeV}) = 0.157 \pm 0.018$ was obtained (this value is already corrected for an initially faulty 3rd-order correction, on which the original analysis was based). In a more recent re-analysis of the e^+e^- continuum data (Haidt 1993), including the latest updates on R from the TRISTAN experiments, the electroweak corrections of all these different data were treated in a consistent and more 'modern' way (i.e. similar to that currently done at LEP). The result, based on the data in the energy range of 20–65 GeV, is now $\alpha_s(35\,\text{GeV}) = 0.146^{+0.031}_{-0.026}$ in $\mathcal{O}(\alpha_s^3)$ QCD. This is, as we shall see later, in very good agreement with α_s determinations at LEP.

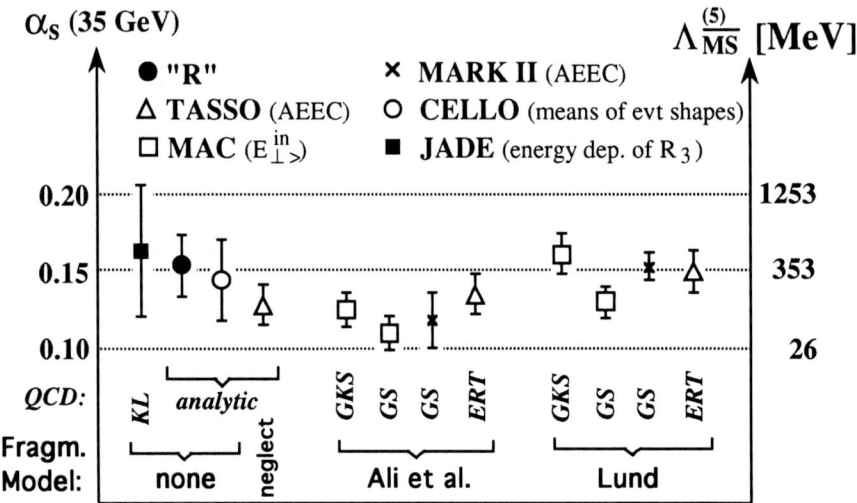

Figure 14. A summary of α_s results obtained from hadronic final states in the e^+e^- continuum between 1987 and 1989; from (Bethke 1989).

Further summaries of early (i.e. pre-LEP) determinations of α_s can be found in (Wu 1984, Naroska 1987); a compilation of the more 'recent' status in 1989 is shown in Figure 14. Here, the results are grouped according to the detailed assumption about the hadronisation on which the analyses were based: those labelled 'Lund' and 'Ali et al.' were based on studies of event shapes and energy correlations, using different $\mathcal{O}(\alpha_s^2)$ QCD models with string or with independent hadronisation, respectively, to extract the value of $\Lambda_{\overline{MS}}$ or of α_s. It was a common knowledge that independent hadronisation models yielded systematically smaller values of α_s than those using string hadronisation. A number of analyses which made no assumption on the hadronisation process were available, either through direct fits of analytic calculations to measured observables (like R) or to their energy dependence. Those results, however, usually had larger experimental or statistical uncertainties, as can be seen from Figure 14.

The authors of the respective $\mathcal{O}(\alpha_s^2)$ QCD calculations which were used in those studies are also given in Figure 14, indicating that there were differences between different sets of calculations. For instance, Gutbrod, Kramer and Schierholz (GKS; Gutbrod 1984) had neglected certain terms in $\mathcal{O}(\alpha_s^2)$, resulting in slightly larger values of α_s than the complete calculations of Ellis, Ross and Terrano (ERT; Ellis 1981) or of Gottshalk and Shatz (GS; Gottschalk and Shatz 1985). Currently, the correct ERT matrix elements, and further calculations based on those, are used almost exclusively. Furthermore, independent fragmentation models are no longer considered in the determination of systematic uncertainties of α_s, since the string effect and related coherence effects, which are well reproduced by string models, are experimentally well established.

Since a more or less 'complete' determination of the overall systematic uncertainty of α_s determinations, including estimates of those imposed by the unknown higher orders, were not available for measurements before the era of LEP, general interpretations of data like those shown in Figure 14 concluded that α_s, expressed for and measured around $E_{cm} = 35\,\text{GeV}$, is

$$\alpha_s(35\,\text{GeV}) = 0.14 \pm 0.02.$$

8.2 Determination of α_s at LEP and SLC

While measurements of α_s at the Z^0 resonance at LEP and SLC benefit from much higher event statistics and smaller hadronisation corrections, a number of new calculations and experimental techniques have further improved the possibilities of performing precise determinations of $\alpha_s(M_{Z^0})$.

8.2.1 Hadronic event shapes in $\mathcal{O}(\alpha_s^2)$ QCD

A recent new development in α_s determinations from e^+e^- annihilations is to study many different observables in one analysis, using the same data set and identical experimental methods. This procedure provides the possibility of verifying the size of the theoretical uncertainty which must be taken into account in order to describe all data distributions with one universal value of α_s. Nowadays, next-to-leading order (NLO) QCD predictions (*i.e.* in $\mathcal{O}(\alpha_s^2)$) are available for more than 15 different hadronic event shape, jet production and energy correlation observables, some of which are listed in Table 4. Most of these observables are more or less correlated with each other, but they all have different NLO coefficients and thus, presumably, different (unknown) higher-order contributions.

The general procedure of determining $\alpha_s(M_{Z^0})$ will be demonstrated for a measurement of jet production rates, which is especially instructive due to the simple and direct relationship between jet rates and α_s (see Equation 8). The results of one of the recent studies which are based on a combination of many different observables and a detailed analysis of the theoretical uncertainties will be discussed thereafter.

Whereas previously α_s was usually determined from jet production rates defined at selected values of the jet resolution, y_{cut} ('integral' jet production rates), it is now determined from the 'differential' two-jet rate $D_2(y)$, which is defined by ($y \equiv y_{cut}$)

$$D_2(y) = \frac{R_2(y) - R_2(y - \Delta y)}{\Delta y}. \tag{10}$$

This function measures the distribution of y_{cut} values for which the classification of events changes from a 3-jet to a 2-jet configuration. Therefore, each hadronic event contributes only once to this distribution, similar to the case of event shape observables like the thrust T.

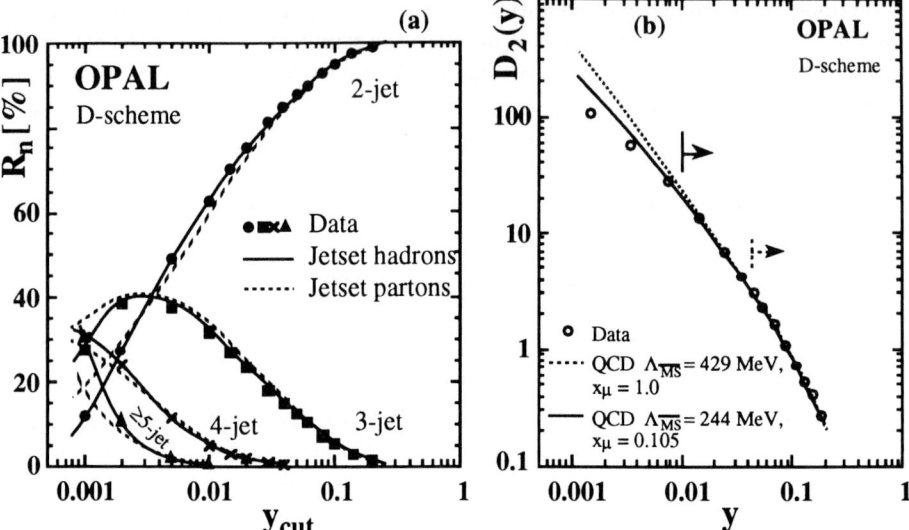

Figure 15. *Jet production rates as a function of y_{cut}, corrected for detector acceptance and resolution (a); differential two-jet distribution after correction for hadronisation effects (b).*

Figure 15a shows the D-scheme jet rates measured by OPAL (OPAL 1992), corrected for acceptance and resolution of the detector, as a function of y_{cut}. The JETSET QCD shower model describes these data in detail (also other QCD models like HERWIG, in general, may provide a good description of the data); the model can therefore be used to further correct the data for hadronisation effects such that the analytic $\mathcal{O}(\alpha_s^2)$ QCD calculations can be fitted to the data. The differential two-jet distribution, $D_2(y)$, obtained from the data shown in Figure 15a, after correction for hadronisation effects, is given in Figure 15b, together with the fit results of QCD calculations for two choices of the renormalisation scale factor x_μ (c.f. Equation 9). The arrows indicate the regions of fits, which are determined by the requirement that $R_5 < 1\%$ if both $\Lambda_{\overline{MS}}$ and x_μ are fitted (5-jet events are only predicted in higher than second-order QCD), and by demanding that $R_4 < 1\%$ if $\Lambda_{\overline{MS}}$ is fitted for $x_\mu = 1$ (with this scale, data cannot be described in regions where a sizable fraction of 4-jet events is observed).

In agreement with previous observations (Bethke 1989a, Bethke 1992), it is found that $\Lambda_{\overline{MS}}$ largely depends on the choice of μ. This is also true if data at small values of y_{cut} are not considered in the fit: even at the largest values of y_{cut}, $\Lambda_{\overline{MS}}$ largely depends on x_μ, as can be inferred from the good agreement of both theoretical curves and data in this region.

The dependence of $\Lambda_{\overline{MS}}$ on x_μ, transformed into a variation of $\alpha_s(M_{Z^0})$ using Equation 2, is currently included in the systematic uncertainty of α_s and is believed to provide an estimate of the theoretical uncertainties due to the unknown higher-order contributions. This may be a meaningful assumption, since to infinite order in QCD, there is strictly no dependence on the renormalisation scale. Unfortunately, no commonly accepted method exists to define the range of renormalisation scales which must be included in the systematic uncertainty of α_s. At LEP, the range of x_μ is typically taken between 1 and values as low as 0.008, depending on the experiment and on the observable under study. The scale ambiguity is usually the largest source of error on the final values of $\alpha_s(M_{Z^0})$.

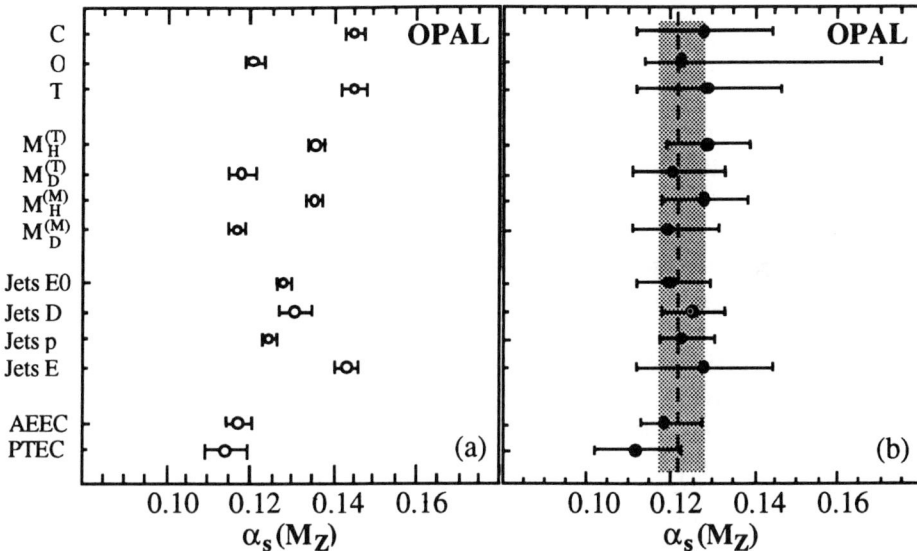

Figure 16. *Measurements of $\alpha_s(M_{Z^0})$ from different observables in $\mathcal{O}(\alpha_s^2)$, including experimental uncertainties and performing the QCD fits for a fixed renormalisation scale $\mu = E_{\rm cm}$; no theoretical uncertainties are taken into account (a). The final values of $\alpha_s(M_{Z^0})$ from the same analysis, now including also theoretical uncertainties (b). The vertical line and the shaded band indicate the final, combined value of $\alpha_s(M_{Z^0})$ and its overall uncertainty, respectively.*

As already mentioned, the size of the higher-order uncertainties can be verified by analysing several different observables in a consistent study of one event sample, from one experiment. The results of a recent OPAL study of 13 different observables (OPAL 1992) are shown in Figure 16. The data are corrected for detector acceptance and for hadronisation effects, using a number of different QCD plus hadronisation models. The coupling constant $\alpha_s(M_{Z^0})$ is then extracted from fits of the analytic $\mathcal{O}(\alpha_s^2)$ QCD predictions to the data, using, initially, a renormalisation scale of $\mu = E_{\rm cm}$. This leads to the values of $\alpha_s(M_{Z^0})$ plotted in Figure 16a, where only statistical and experimental but no theoretical uncertainties are considered. Within the experimental errors of typically 1% to 3% in α_s, the results from different observables are not compatible with each other; they disagree by as much as eight standard deviations. The need to include theoretical uncertainties is thus obvious, if QCD is required to give a consistent picture.

In the next step of the analysis, the renormalisation scale dependence of $\alpha_s(M_{Z^0})$ and of the quality of fit, in terms of $\chi^2/\text{d.o.f.}$, is analysed for each observable as shown in Figure 17. The functional form of this dependence is different for each observable, typically leading to a decrease of $\alpha_s(M_{Z^0})$ with decreasing renormalisation scale, and a minimum at rather small scales ($x_\mu \sim 0.01$ to 0.1, which corresponds to energy scales of a few GeV at LEP).

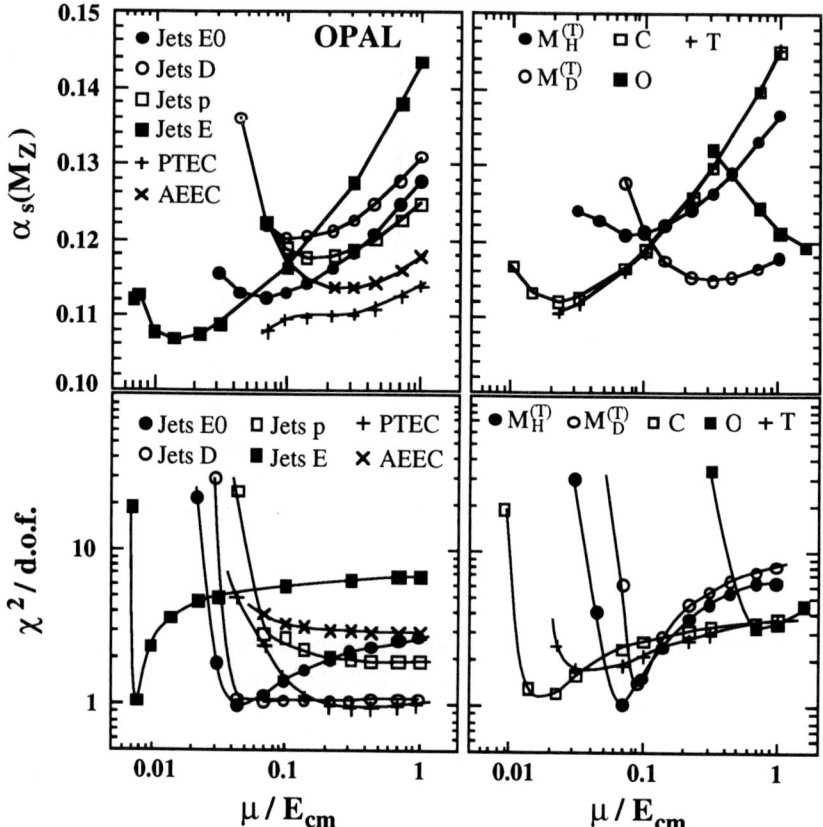

Figure 17. *Dependence of $\alpha_s(M_{Z^0})$ and the $\chi^2/\text{d.o.f.}$ of the fits on the renormalisation scale factor $x_\mu^2 = \mu^2/E_{cm}^2$, for different observables in $\mathcal{O}(\alpha_s^2)$.*

The methods for defining the central values of $\alpha_s(M_{Z^0})$ and their theoretical (scale) uncertainty are different, in general, for different experiments. In the OPAL analysis, the central value of $\alpha_s(M_{Z^0})$ for each observable is defined as the arithmetic mean of the result for $x_\mu \equiv \mu/E_{cm} = 1$ and for x_μ from the best fit; the difference between these two choices is then assigned to be the scale uncertainty. As can be seen from Figure 17, the best fit result for x_μ is typically close to the point at which $\alpha_s(M_{Z^0})$ reaches the minimum of its functional dependence on x_μ, and thus the uncertainty for each observable usually includes the results from theoretical estimates of scale optimisation (Stevenson

1981, Grunberg 1980, Brodsky 1983). Including also hadronisation uncertainties and uncertainties due to the minimum parton virtuality to which the data are corrected, leads to the results of $\alpha_s(M_{Z^0})$ shown in Figure 16b. Comparing this with Figure 16a, where theoretical uncertainties are not included, one realises that the overall uncertainties have increased by about an order of magnitude, but that the results from different observables are now in good agreement with each other. Within the theoretical uncertainties assigned in this analysis, the different observables thus lead to consistent results. Averaging over all observables, taking correlations between them into account, finally gives $\alpha_s(M_{Z^0}) = 0.122^{+0.006}_{-0.005}$, where the error is almost entirely theoretical.

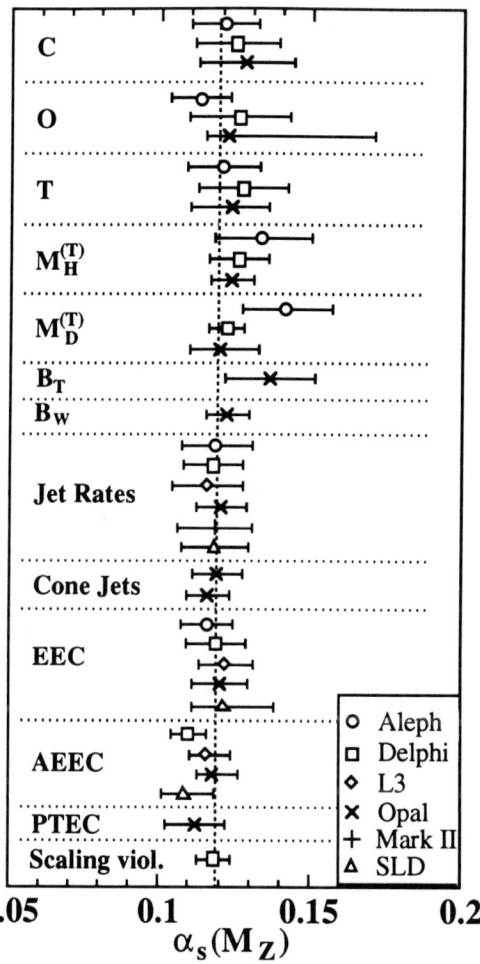

Figure 18. *Compilation of measurements of $\alpha_s(M_{Z^0})$ from event shapes, jet rates and energy correlations, in $\mathcal{O}(\alpha_s^2)$, at LEP and SLC. The errors contain the experimental and theoretical uncertainties, added in quadrature.*

Since each experiment has chosen its own method to determine and quote the central value of $\alpha_s(M_{Z^0})$ and its systematic uncertainty, it is mandatory to apply a consistent definition of these quantities before an overall average value of $\alpha_s(M_{Z^0})$ can be quoted. This was done in a comprehensive analysis of the published $\alpha_s(M_{Z^0})$ results (Bethke and Pilcher 1992), basically applying the method which was used by OPAL, as briefly described above. Following this method, an update of the results of $\alpha_s(M_{Z^0})$, from all experiments at LEP and SLC, is presented in Figure 18. Good agreement is found between the experiments, but note that the errors are largely dominated by theoretical uncertainties, which are common to all the experiments. The overall combined result is

$$\alpha_s(M_{Z^0}) = 0.120 \pm 0.006$$

from hadronic event shapes, energy correlations and jet rates in $\mathcal{O}(\alpha_s^2)$ QCD, whereby the final error was taken from estimates of the overall remaining, uncorrelated uncertainties (see *e.g.* OPAL 1992).

8.2.2 Measurements of $\alpha_s(M_{Z^0})$ using resummed $\mathcal{O}(\alpha_s^2)$ QCD calculations

In general, calculations which are based on the $\mathcal{O}(\alpha_s^2)$ matrix elements are unsuccessful in describing the back-to-back two-jet region of phase space. In this region multiple emissions of soft gluons may be expected to be important. An alternative approach may be taken to the QCD calculations of hadronic final states in e^+e^- annihilations, based on the resummation of leading logarithms which arise from soft and collinear singularities in gluon emission. The consequence is that the effective expansion parameter is not simply α_s, but $\alpha_s L^2$ (to leading order in L), where $L = \ln(1/y)$ and y is some generic observable which tends to zero in the two-jet region. At small y the value of $\alpha_s L^2$ is not small, and therefore these terms must be summed to all orders in α_s in order to provide a satisfactory calculation. For certain observables it has proved possible to sum both the leading and next-to-leading logarithms, which is referred to as the 'Next-to-Leading Log Approximation' or NLLA. Such calculations are available for thrust, heavy jet mass, two measures of jet broadening, energy-energy correlations, two-jet rates and average jet multiplicities (see *e.g.* Catani 1992), though in the latter two cases the next-to-leading terms have been only partially resummed. For all the variables for which resummation is possible, with the exception of the average jet multiplicity, the *cumulative* cross-section may be written in the general form:

$$R(y) \equiv \int_0^y \frac{1}{\sigma}\frac{d\sigma}{dy} dy = C(\alpha_s)\exp\{G(\alpha_s, L)\} + D(\alpha_s, y), \qquad (11)$$

where y is $(1-T)$, M_H^2/s, B_T or B_W in the case of the event shapes, $\cos^2(\chi/2)$ in the case of EEC (an additional factor $\frac{1}{2}$ precedes the exponential for EEC, in this and subsequent equations), and y_{cut} for the jet rates, with $L = \ln(1/y)$. The function $D(\alpha_s, y)$ is a remainder function which should vanish as $y \to 0$. The functions C and G may be written:

$$C(\alpha_s) = 1 + \sum_{n=1}^{\infty} C_n \overline{\alpha}_s^n ; \qquad (12)$$

$$G(\alpha_s, L) = \sum_{n=1}^{\infty}\sum_{m=1}^{n+1} G_{nm}\overline{\alpha}_s^n L^m$$

$$\equiv L g_1(\alpha_s L) + g_2(\alpha_s L) + \alpha_s g_3(\alpha_s L) + \alpha_s^2 g_4(\alpha_s L)\cdots, \qquad (13)$$

where for brevity we write $\bar{\alpha}_s$ for $(\alpha_s/2\pi)$. The functions $Lg_1(\alpha_s L)$ and $g_2(\alpha_s L)$ represent the sums of the leading and next-to-leading logarithms respectively, to all orders in α_s.

The general structure of the cross-section in powers of α_s and of large logarithms is indicated in Table 9. Here, the NLLA calculations provide the terms in the first two columns, while the $\mathcal{O}(\alpha_s^2)$ calculations yield the sums of the terms in the first two rows.

	Leading logs	Next-to-Leading logs	Subleading logs	Non-log. terms	
$\ln R(y) =$	$G_{12}\bar{\alpha}_s L^2$	$+G_{11}\bar{\alpha}_s L$		$+\alpha_s\mathcal{O}(1)$	$\mathcal{O}(\alpha_s)$
	$+G_{23}\bar{\alpha}_s^2 L^3$	$+G_{22}\bar{\alpha}_s^2 L^2$	$+G_{21}\bar{\alpha}_s^2 L$	$+\alpha_s^2\mathcal{O}(1)$	$\mathcal{O}(\alpha_s^2)$
	$+G_{34}\bar{\alpha}_s^3 L^4$	$+G_{33}\bar{\alpha}_s^3 L^3$	$+G_{32}\bar{\alpha}_s^3 L^2$	$+\cdots$	$\mathcal{O}(\alpha_s^3)$
	$+\cdots$	$+\cdots$	$+\cdots$	$+\cdots$	$\vdots$
$=$	$Lg_1(\alpha_s L)$	$+g_2(\alpha_s L)$	$+\cdots$	$+\cdots$	

Table 9. *Decomposition of the cumulative cross-section, $\ln R(y)$, in powers of $\bar{\alpha}_s = (\alpha_s/2\pi)$ and $L = \ln(1/y)$. The NLLA calculations provide the terms in the first two columns, while the $\mathcal{O}(\alpha_s^2)$ calculations yield the sums of the terms in the first two rows. The matching procedures involve combining these without double-counting the terms in common.*

If the maximum available information from both NLLA and $\mathcal{O}(\alpha_s^2)$ calculations are to be utilised, they must be added such that double counting of the four terms which are common to both these calculations (terms G_{12}, G_{11}, G_{23} and G_{22} in Table 9) is avoided. This can be done involving several (approximate), so-called 'matching schemes'. For instance, terms to $\mathcal{O}(\alpha_s^2)$ in the *logarithm* of the above NLLA expression for $R(y)$ can be removed and replaced by the full expression of the logarithm of $R(y)$ in exact $\mathcal{O}(\alpha_s^2)$, which yields the so-called $\ln R$ matching, which is most widely used. A similar procedure may be carried out for the functions $R(y)$ instead of $\ln R(y)$, yielding the so-called R matching schemes. A comprehensive overview of these and other matching schemes can be found, for example, in OPAL 1993a. Note that different matching schemes can yield different values of α_s in experimental analyses, and thus may contribute as a new source of systematic, theoretical uncertainties.

Resummation is expected to provide more accurate predictions of the distributions, especially at high thrust or low jet masses, and they should also have a much reduced dependence on the renormalisation scale if compared to the $\mathcal{O}(\alpha_s^2)$ calculations alone. Both these predictions were confirmed in several recent experimental studies. For instance, the experimental distribution of the heavy jet mass M_H, corrected for detector acceptance and for hadronisation effects, is shown in Figure 19, together with the best fits of the pure $\mathcal{O}(\alpha_s^2)$ (dashed) and the resummed $\mathcal{O}(\alpha_s^2)$ calculation (full line), both for a renormalisation scale factor of $x_\mu = 1$. The $\mathcal{O}(\alpha_s^2)$ prediction fails to describe the shape of the distribution at small jet masses (*i.e.* the two-jet region) for $x_\mu = 1$; only rather small values of x_μ provide a significantly better description of the data in the two-jet as well as in 3 and 4-jet region of the distribution, which is also true for most hadronic event shapes and jet rate observables. The resummed $\mathcal{O}(\alpha_s^2)$ calculations are able to give a much better account of the data, even at $x_\mu \approx 1$, as was theoretically expected.

Note, however, that resummed $\mathcal{O}(\alpha_s^2)$ + NLLA calculations often give a poor description in regions of hard 3-jet and multi-jet events, where the higher-order NLLA terms do not vanish sufficiently rapidly towards the kinematic limit of the observable. Essentially the NLLA calculation is expected to be most applicable in the low mass or high thrust region, where the description of data is greatly improved, while the $\mathcal{O}(\alpha_s^2)$ calculation is likely to be reliable at large masses.

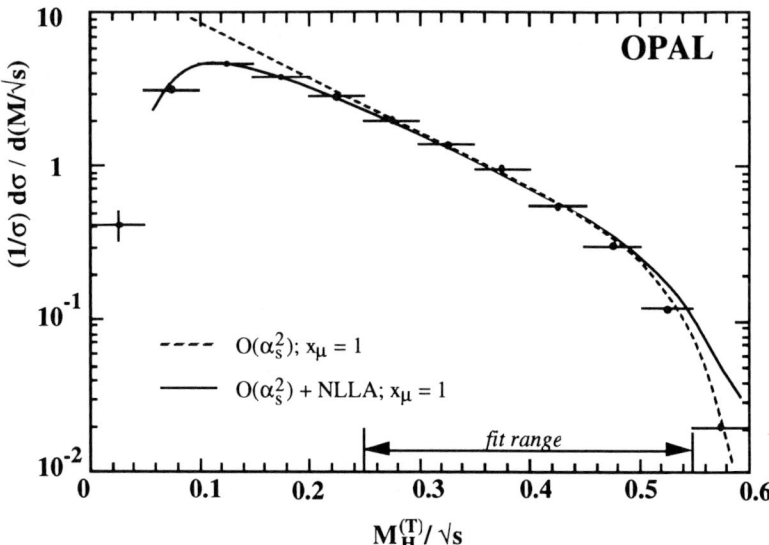

Figure 19. *Measured distribution of $M_H^{(T)}$, compared with fits to the $\mathcal{O}(\alpha_s^2)$ and to the resummed $\mathcal{O}(\alpha_s^2)$+NLLA calculations, with a renormalisation scale factor $x_\mu=1$ in both cases. The range over which the fits were performed is indicated by the arrow.*

A summary of $\alpha_s(M_{Z^0})$ from resummed calculations is given in Figure 20. In contrast to the results in $\mathcal{O}(\alpha_s^2)$ QCD (c.f. Figure 18), the central values of $\alpha_s(M_{Z^0})$ are always given for $\mu = E_{cm}$, since small renormalisation scales are no longer preferred in resummed calculations. In fact, best fit results are obtained for renormalisation scales much closer to $\mu = E_{cm}$ than in the case of $\mathcal{O}(\alpha_s^2)$ QCD. Nevertheless, an uncertainty in α_s due to the detailed choice of the renormalisation scale, usually taking $0.5 \leq x_\mu \leq 2$, individually for each observable, is still present in resummed $\mathcal{O}(\alpha_s^2)$, which is however smaller than in $\mathcal{O}(\alpha_s^2)$ alone.

The errors presented in Figure 20 include experimental and theoretical (*i.e.* mainly scale) uncertainties, added in quadrature. Combining these results provides an average value of

$$\alpha_s(M_{Z^0}) = 0.123 \pm 0.006,$$

which is in good agreement with the final value from analyses in $\mathcal{O}(\alpha_s^2)$ alone. The error of this combined result was taken from detailed studies (OPAL 1993a, DELPHI 1993a) of

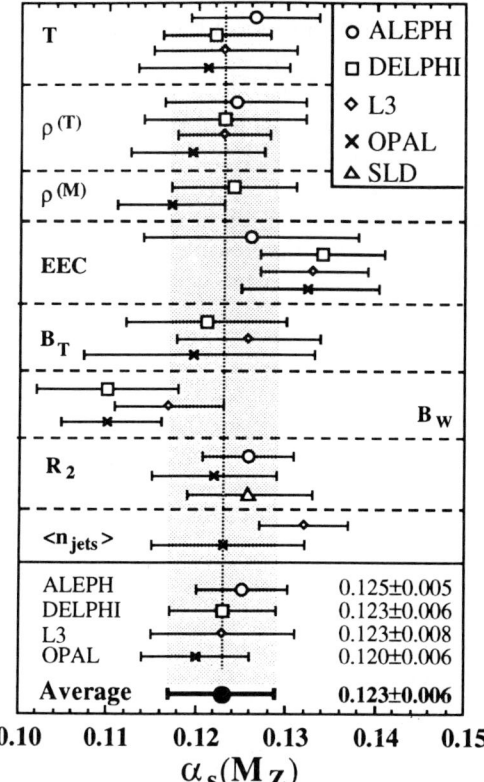

Figure 20. *Summary of measurements of $\alpha_s(M_{Z^0})$ from LEP and SLC, using resummed $\mathcal{O}(\alpha_s^2)$ QCD calculations.*

several observables and their overall remaining, uncorrelated uncertainties, similar to the case of $\mathcal{O}(\alpha_s^2)$ studies described in the previous subsection.

8.2.3 Measurements of $\alpha_s(M_{Z^0})$ from the hadronic width of the Z^0

An attractive way of determining $\alpha_s(M_{Z^0})$ is to make a precise measurement of the ratio R_Z of the hadronic and leptonic partial widths of the Z^0,

$$R_Z \equiv \left(\frac{\Gamma_{\text{had}}}{\Gamma_{\text{lept}}}\right)_{\text{exp}} = \left(\frac{\Gamma_{\text{had}}}{\Gamma_{\text{lept}}}\right)_0 (1 + \delta_{\text{QCD}}), \tag{14}$$

since R is not affected by hadronisation effects and because the QCD correction δ_{QCD}, which is exactly the same as for the total hadronic cross-section R in the e^+e^- continuum, as given in Equation 5, has been calculated to complete third-order ($\mathcal{O}(\alpha_s^3)$) perturbation theory (Gorishny 1991). Including quark mass corrections, δ_{QCD} is of the form (Hebbeker 1991)

$$\delta_{\text{QCD}} = 1.05 \left(\frac{\alpha_s}{\pi}\right) + 0.9 \left(\frac{\alpha_s}{\pi}\right)^2 - 13 \left(\frac{\alpha_s}{\pi}\right)^3. \tag{15}$$

The prediction for $(\Gamma_{\text{had}}/\Gamma_{\text{lept}})_0$, without QCD corrections, is 19.943 with an uncertainty of ± 0.02 from varying the unknown masses of the top quark and of the Higgs particle, in the ranges $M_t = 90\text{–}250$ GeV and $M_H = 50\text{–}1000$ GeV, respectively. The value of 19.943 corresponds to $M_t = 300$ GeV and $M_H = 300$ GeV.

The average value of R_Z, summarized from the measurements of the four LEP experiments, is $R = 20.766 \pm 0.049$ (Blondel 1993). This is based on a total of about 4.5×10^6 hadronic and 495,000 leptonic Z^0 decays. From this result one infers that $\alpha_s(M_{Z^0}) = 0.122 \pm 0.007 \pm 0.003 \pm 0.002$, where the first error is statistical, the second comes from uncertainties in M_t and M_H, and the third error is due to uncertainties in the mass of the b-quark. Adding these errors in quadrature results in

$$\alpha_s(M_{Z^0}) = 0.122 \pm 0.008,$$

which is in perfect agreement with the measurements from hadronic event shapes, jet rates and energy correlations. Note, however, that there exist different calculations of the mass corrected QCD coefficients given in Equation 15, which, at the extreme, can result in central values of $\alpha_s(M_{Z^0})$ which vary between 0.120 and 0.124. Whether this is an additional theoretical uncertainty or can be explained by inadequate calculations remains to be clarified.

During recent years of LEP running, the value of $\alpha_s(M_{Z^0})$ from R_Z fell from previously rather high values (starting at 0.16), however always within the statistical error—which also gradually decreased. Such an effect can be explained by a statistical fluctuation of the initial measurement, which then, by adding more and more data and by increasing the statistical precision, leads to a monotonic decrease towards the 'true' value.

From a combined fit of M_t and $\alpha_s(M_{Z^0})$, using all the LEP data on the hadronic and leptonic Z^0 line shape and measurements of lepton asymmetries, one obtains $M_t = 164^{+18+19}_{-20-22}$ GeV and $\alpha_s(M_{Z^0}) = 0.120 \pm 0.006 \pm 0.002$, where the first error is experimental and the second is due to the unknown mass of the Higgs boson (Blondel 1993). The renormalisation scale dependence of α_s in these fits is about ± 0.003, for $x_\mu = 0.5$ to 2.

8.2.4 Measurements of α_s from τ decays

The ratio R_τ of the hadronic and electronic branching fractions of the τ lepton,

$$R_\tau = \frac{B(\tau \to \text{hadrons} + \nu_\tau)}{B(\tau \to e\bar{\nu}_e\nu_\tau)} \equiv \frac{1 - B_e - B_\mu}{B_e},$$

which can be reliably determined by measurements of the electronic and muonic branching fractions B_e and B_μ, is theoretically expected to be given by (Braaten 1992)

$$R_\tau = 3.058\,(1.001 + \delta_{\text{pert}} + \delta_{\text{nonpert}}).$$

Here, δ_{pert} and δ_{nonpert} are perturbative and non-perturbative QCD corrections; δ_{pert} was calculated to complete $\mathcal{O}(\alpha_s^3)$ and is of a similar structure to that for R_Z (Braaten 1992). Recently, a revised $\mathcal{O}(\alpha_s^3)$ prediction of δ_{pert} became available, which also includes the resummation of leading logarithmic terms (Le Diberder and Pich 1992). The non-perturbative correction was estimated to be small, $\delta_{\text{nonpert}} = -0.007 \pm 0.004$ (Braaten 1992).

The average value, combined from the four LEP experiments (Fernandez 1993), is $R_\tau = 3.617 \pm 0.034$, which, according to the predictions given above, leads to $\alpha_s(M_\tau) = 0.360 \pm 0.04$, in $\mathcal{O}(\alpha_s^3)$ and for three quark flavours, $N_f = 3$. This result, which also contains a renormalisation scale uncertainty of ± 0.03, added in quadrature, is significantly larger than the value of $\alpha_s(M_{Z^0})$ from event shapes and jet rates measured in Z^0 decay, as is expected for an energy dependent α_s.

A new method of analysis was proposed, in which δ_{nonpert} can be simultaneously determined from the data (Le Diberder and Pich 1992a), in addition to $\alpha_s(M_\tau)$, instead of relying on the estimates of δ_{nonpert} mentioned above. This method requires the measurement of weighted integrals of the hadronic invariant mass spectrum of τ decays. The ALEPH collaboration has contributed an analysis (ALEPH 1993), which is based on this new method and on the improved QCD predictions. The result is $\alpha_s(M_\tau) = 0.330 \pm 0.046$, which is in good agreement with the result (0.360) given above and therefore indicates that non-perturbative corrections to R_τ are indeed very small: the ALEPH fit gives $(0.3 \pm 0.5)\%$.

When extrapolating this value of α_s from $\mu = M_\tau$ to $\mu = M_{Z^0}$, $\Lambda_{\overline{MS}}$ is adjusted when crossing a quark threshold so that α_s is a continuous function of μ (Marciano 1984). This results in

$$\alpha_s(M_{Z^0}) = 0.122 \pm 0.005$$

from τ decays at LEP, where the relative size of the error is decreased because of the logarithmic dependence of α_s on μ. The agreement with $\alpha_s(M_{Z^0})$ obtained from hadronic event shapes is remarkable, especially if one considers the large difference in the effective energy scale. This agreement may be one of the most important QCD tests performed at LEP so far!

8.2.5 Other studies of α_s at LEP

Flavour dependence of α_s

There is only one universal coupling constant, α_s, in QCD. This means that the amount of gluon radiation off quarks should be independent of the quark flavour. Experimental tests of the flavour independence of of α_s typically involve tagging of one or several quark flavours. For instance, events of the type $e^+e^- \to b\bar{b}$ can be selected by tagging a lepton with high momentum, p, and with a high transverse momentum component with respect to the jet axis, p_T, making use of the leptonic decay as well as of the relatively large mass and the hard fragmentation function of the b-quark. A more modern and more efficient method is to tag b-quarks by their relatively long lifetime and thus, by secondary vertices which can be reconstructed using high precision vertex detectors.

The most significant studies of the flavour dependence of α_s are available from LEP and SLC. Typically, the ratio of 3-jet event rates, defined using the JADE or the DURHAM jet algorithm, for the tagged and for the total event samples are employed:

$$R_q = \frac{R_3(\text{tagged sample})}{R_3(\text{all events})},$$

as a function of the jet resolution y_{cut}.

Several LEP and SLC experiments have analysed R_b and extracted the relative size of α_s for b-quarks. In leading-order QCD and for massless quarks, one has

$$\frac{R_3(\text{tagged sample})}{R_3(\text{all events})} = \frac{\alpha_s(b)}{\alpha_s(udsc)}.$$

Due to the large b-quark mass, however, a phase-space suppression of gluon radiation of about 10% is expected, which should lead to a reduction in R_b of approximately 10%, even if α_s is flavour independent. This can be corrected for by using QCD plus hadronisation model calculations, and/or by the predictions of a QCD calculation for *massive* quarks, which however is available in leading order (*i.e.* $\mathcal{O}(\alpha_s)$) only. The measured R_q must also be corrected for the limited quark tagging purity, which is again obtained from model calculations.

More recently, a study became available in which the flavour dependence of α_s is tested for *all* five quark flavours (OPAL 1993b). In addition to the usual tagging of b-quarks, c-quarks are enriched by analysing events with a highly energetic $D^{*\pm}$ meson, s-quark events are selected by requiring a reconstructed, highly energetic K_s^0 meson, and a sample of enriched light quark (*i.e.* u, d, s) events is obtained by tagging highly energetic charged particles (*i.e.* pions, protons and kaons). The idea behind tagging highly energetic particles to select certain primary quark flavours is that such particles are most likely to be produced in the earliest stages of the hadronisation chain, and they should thus preferentially contain the primary quark. Naturally, rather large kinematic biases are introduced when analysing event samples with a highly energetic particle. These biases must again be corrected for by using model calculations, leading to increased systematic uncertainties of the results. Note that from the different event samples described above, it is possible to determine R_q for *each* quark flavour separately, due to the different amount of backgrounds to the latter two samples.

A summary of studies on the flavour dependence of α_s, from LEP and SLC experiments, is given in Figure 21. All results are compatible with the strict flavour independence of α_s, as predicted by QCD. It was also verified that there is no dependence of α_s on weak isospin or on the generation of quarks (OPAL 1993b). Combining different experiments, the most significant result is

$$\frac{\alpha_s(b)}{\alpha_s(udsc)} = 1.01 \pm 0.03,$$

which tests the flavour independence of α_s for b-quarks with an overall precision of 3%.

Measurement of α_s from scaling violations

An elegant way of determining α_s is by analysing the degree of scaling violations which is observed in observables for which data at different center of mass energies are available. According to the QCD β-function discussed in Section 4.1 (Equation 1), the degree of energy dependence, *i.e.* the slope $\partial \alpha_s / \partial \mu$, is larger for larger values of α_s. One can thus determine α_s from precise measurements of suitable observables at two (or more) different energies. In Figure 6 it was already demonstrated that significant scaling violations exist in, for example, hadronic event shape distributions like the thrust T.

The first analysis of this type was probably performed by the JADE collaboration in 1988 (JADE 1988), who studied the energy dependence of 3-jet event production rates,

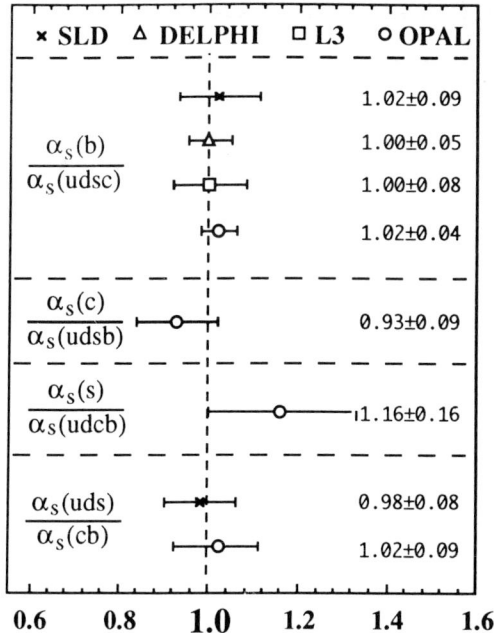

Figure 21. *Summary of studies on the flavour independence of α_s.*

R_3; see also Section 6.4. They determined α_s from the measured ratio r of 3-jet rates R_3 at 44 and 34 GeV centre-of-mass energy, for identical values of $y_{\rm cut}$. This ratio was found, for values of $y_{\rm cut}$ that were not too small, to be consistently less than unity, as expected from QCD and the running α_s.

The data on r are shown in Figure 22, together with the predictions of $\mathcal{O}(\alpha_s^2)$ QCD calculations of jet production, for different values of $\Lambda_{\overline{\rm MS}}$. Note that these predictions are almost independent of $y_{\rm cut}$, in good agreement with the data. The data show a strong increase of r for small values of $y_{\rm cut}$, which was attributed to the effects of hadronisation or of heavy quark decays ($y_{\rm cut} = 0.02$ at $E_{\rm cm} = 34$ GeV corresponds to a mass of 4.8 GeV). Later it was found that by using a small renormalisation scale of $\mu/E_{\rm cm} \sim 0.05$, the QCD prediction exactly follows the data, also at low values of $y_{\rm cut}$. In 1988, however, it was not yet customary to change the renormalisation scale. For the measurement of $r = 0.935 \pm 0.023$, JADE determined (for $\mu = E_{\rm cm}$) $\alpha_s = 0.154 \pm 0.038$ at $E_{\rm cm} = 44$ GeV.

Naturally, one can determine α_s from the ratio of R_3 measured at LEP (91 GeV) and at PETRA (35 GeV), which is a much larger lever arm and thus promises a more accurate result than in the case of the JADE analysis. This requires, however, use of data from different experiments, with unknown point-to-point systematic uncertainties. We therefore just mention here that the energy dependence of R_3 is indeed well described by QCD, over a wide range of energies, as was seen in Figure 9 and in Table 6.

Recently, α_s has also been determined from scaling violations in the fragmentation functions (DELPHI 93), using the energy dependence of the inclusive momentum cross-

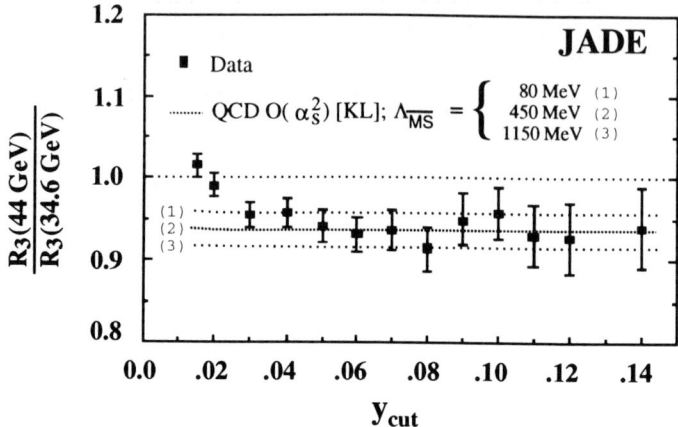

Figure 22. *The ratio of 3-jet events, observed at $E_{cm} = 34$ GeV and 44 GeV, compared with $\mathcal{O}(\alpha_s^2)$ QCD predictions for different values of $\Lambda_{\overline{MS}}$ and a renormalisation scale $\mu = E_{cm}$.*

sections of charged particles, $\frac{1}{\sigma}\frac{d\sigma}{dx}$, where x is the fraction of beam energy carried by a hadron ($x = 2E_h/E_{cm}$). DELPHI measured the momentum spectrum of charged hadrons at LEP and compared it with similar measurements made at lower energies. They find significant scaling violations. This is demonstrated in Figure 23, where the ratio between the momentum spectra from DELPHI and from TASSO ($E_{cm} = 35$ GeV) is plotted as a function of x. If there were no scaling violations present, the data should scatter around unity.

DELPHI compared the measured scaling violations with the predictions of $\mathcal{O}(\alpha_s^2)$ QCD plus hadronisation models. Simultaneous fits of the QCD scale $\Lambda_{\overline{MS}}$ and of the parameters for the light and the heavy quark fragmentation functions, a and ϵ_b, were made in the range $0.18 < x < 0.80$. For these fits, the cut-off parameter y_{min} against soft and collinear gluons (see Section 4.3) was kept constant at all energies, in order to maintain the same phase space for hadronisation. In detail, $y_{min} = \min(M_{ij}^2/E_{cm}^2)$ was chosen to be $(9.1\,\text{GeV})^2/E_{cm}^2$, which is $y_{min} = 0.01$ and 0.068 at $E_{cm} = 91$ and 35 GeV, respectively. For smaller values of M_{ij}, the 2-jet cross-section at $E_{cm} = 91$ GeV would become unphysical.

DELPHI found that with this choice of model parameters, the measured momentum spectra can be well described at all energies analysed in this analysis. Note, however, that other important distributions of hadronic final states, such as the energy-energy-correlations at the low energies, are known to be poorly described by the $\mathcal{O}(\alpha_s^2)$ model if y_{min} is chosen to be as large as 0.068; the reliability of this model at the lower energies thus seems at least to be doubtful, even if the distribution under consideration is well described. Nevertheless, if one ascribes the observed scaling violation to the running of α_s alone, DELPHI obtains $\alpha_s(M_{Z^0}) = 0.118 \pm 0.005$ in $\mathcal{O}(\alpha_s^2)$, where the error contains the experimental as well as some estimates of theoretical uncertainties. This result is included in the summary of α_s determinations in $\mathcal{O}(\alpha_s^2)$ QCD (see Figure 18).

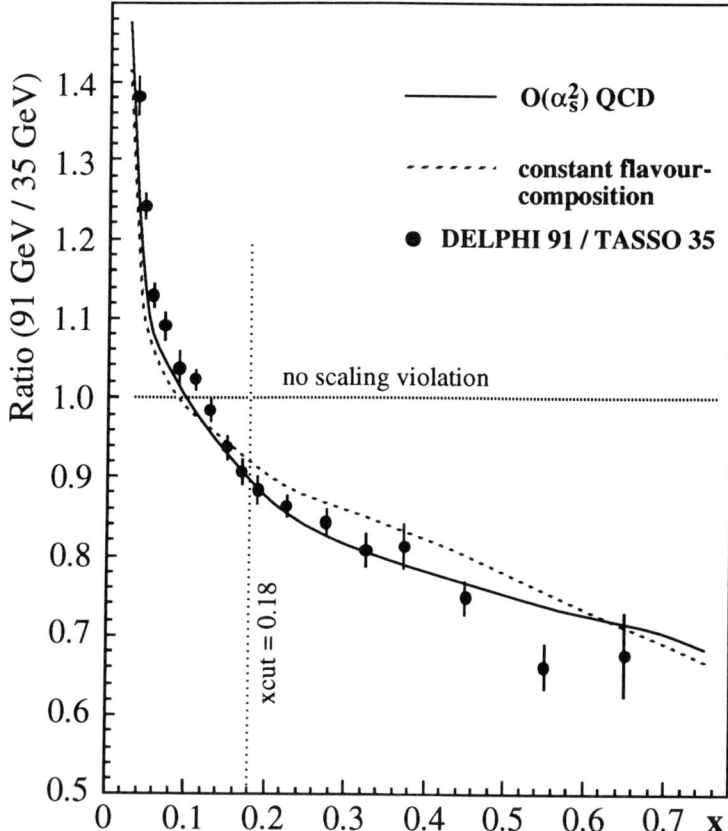

Figure 23. *The ratio of the inclusive momentum spectra $\frac{1}{\sigma}\frac{d\sigma}{dx}$ measured at $E_{cm} = 91\,\text{GeV}$ and $35\,\text{GeV}$, together with the prediction of $\mathcal{O}(\alpha_s^2)$ QCD model calculations (solid line). The dashed line is based on an artificial model that assumes equal quark flavour composition at both energies, showing that the increase of b-quark production at the Z^0 resonance does not influence this analysis strongly.*

8.3 World summary of α_s measurements

A world summary of α_s measurements, updated from a previous compilation (Bethke 1992c), is presented in Table 10. The values of α_s are given at typical energy scales Q where actual measurements were done, and are also converted, using Equation 2 to $\mathcal{O}(\alpha_s^3)$, to the energy scale M_{Z^0} by analytically solving the condition that α_s be continuous at quark thresholds (Marciano 1984). The errors given in columns 3 and 4 contain experimental and theoretical errors, added in quadrature; column 5 presents a breakdown of the total error into these two classes of uncertainties. The last column indicates the degree of QCD perturbation theory used to determine α_s, where 'LGT' means lattice gauge theory, 'resum.' stands for resummed $\mathcal{O}(\alpha_s^2)$ calculations, and (N)NLO stands for (next-)next-to-leading order.

Process	Q [GeV]	$\alpha_s(Q)$	$\alpha_s(M_{Z^0})$	$\Delta\alpha_s(M_{Z^0})$ exp.	theor.	Theory
GLS [ν-DIS]	1.73	0.320 ± 0.05	0.115 ± 0.006	0.005	0.003	NNLO
R_τ [LEP]	1.78	0.360 ± 0.040	0.122 ± 0.005	0.002	0.004	NNLO
DIS [ν]	5.0	$0.193 ^{+0.019}_{-0.018}$	0.111 ± 0.006	0.004	0.004	NLO
DIS [μ]	7.1	0.180 ± 0.014	0.113 ± 0.005	0.003	0.004	NLO
$c\bar{c}$ mass splitting	5.0	0.174 ± 0.012	0.105 ± 0.004	0.000	0.004	LGT
$J/\Psi + \Upsilon$ decays	10.0	$0.167 ^{+0.015}_{-0.011}$	$0.113 ^{+0.007}_{-0.005}$	0.001	$^{+0.007}_{-0.005}$	NLO
e^+e^- [σ_{had}]	34.0	$0.146 ^{+0.031}_{-0.026}$	$0.124 ^{+0.021}_{-0.019}$	$^{+0.021}_{-0.019}$	–	NLO
e^+e^- [ev. shapes]	35.0	0.140 ± 0.02	0.119 ± 0.014	–	–	NLO
e^+e^- [ev. shapes]	58.0	0.130 ± 0.008	0.122 ± 0.007	0.003	0.007	NLO
$p\bar{p} \to b\bar{b}X$	20.0	$0.138 ^{+0.028}_{-0.019}$	$0.109 ^{+0.016}_{-0.012}$	$^{+0.012}_{-0.007}$	$^{+0.011}_{-0.010}$	NLO
$p\bar{p} \to W$ jets	80.6	0.123 ± 0.025	0.121 ± 0.024	0.017	0.016	NLO
$\Gamma(Z^0 \to$ had.)	91.2	0.122 ± 0.008	0.122 ± 0.008	0.007	0.004	NNLO
Z^0 [ev. shapes]	91.2	0.119 ± 0.006	0.119 ± 0.006	0.001	0.006	NLO
Z^0 [ev. shapes]	91.2	0.123 ± 0.006	0.123 ± 0.006	0.001	0.006	resum.

Table 10. *A summary of measurements of α_s.*

There is significant evidence for the running of α_s, as can be seen in a plot of α_s as a function of Q, from the numbers summarized in Table 10, as shown in Figure 24. The data are compared with the QCD predictions of a running coupling constant, calculated in $\mathcal{O}(\alpha_s^3)$ for four different values of $\Lambda_{\overline{MS}}$ which are given for $N_f = 5$ quark flavours. The energy dependence of α_s is distinct, and is in very good agreement with the QCD prediction. However, small systematic trends appear to be visible, the lower energy results from deep inelastic scattering (DIS) and from quarkonia decays, but *not* those from R_τ, seem to prefer smaller values of $\Lambda_{\overline{MS}}$ than those from LEP.

This systematic trend becomes more visible in Figure 25 where the corresponding values of $\alpha_s(M_{Z^0})$ are displayed. In general, the measurements agree quite well with each other, within their overall errors which mainly are of theoretical nature in most cases. One can also see that the results from e^+e^- annihilation and from deep inelastic scattering just touch each other within their overall assigned uncertainties. Since these are estimates, *i.e. lower limits* at best, of the theoretical uncertainties, the difference between these results is not regarded to be significant, and is in fact most likely to be due to the contributions and treatments of the (probably different but unknown) higher orders of perturbation theory.

Forming a weighted average of the results listed in Table 10 and shown in Figures 24 and 25, leaving out the result from lattice gauge theory (see Bethke 1992c), one obtains $\alpha_s(M_{Z^0}) = 0.118$ as the central value. Since the errors of most results are estimates of the theoretical uncertainties, the final error of the overall combined value of $\alpha_s(M_{Z^0})$ cannot be obtained by using standard techniques of error calculation.

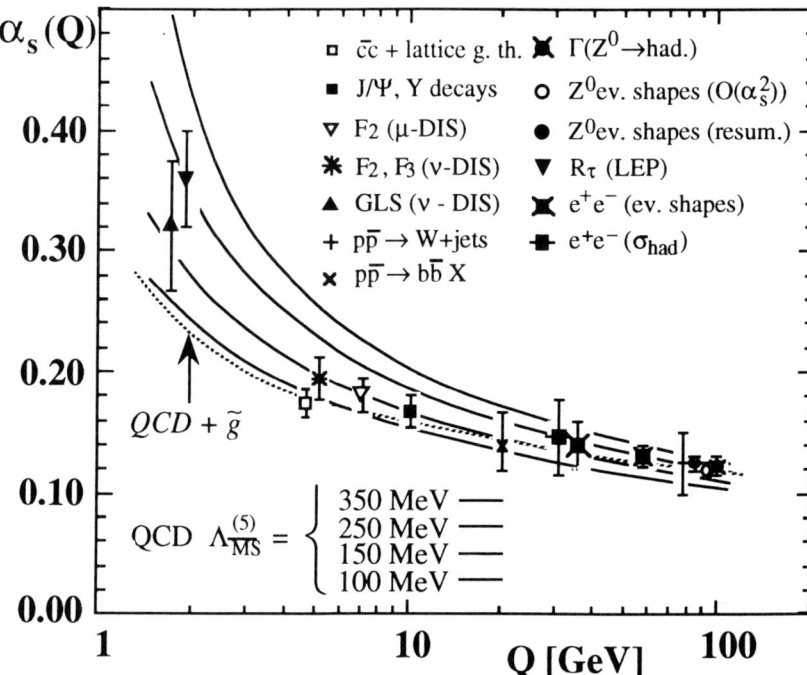

Figure 24. *A summary of measurements of $\alpha_s(Q)$, compared with the QCD predictions of the running α_s, for 4 different values of $\Lambda_{\overline{MS}}$ which are given for the region where $N_f = 5$ (i.e. $Q \geq 10$ GeV). The dashed line indicates the expected energy dependence of α_s in the presence of a light, neutral, coloured fermion, with a production threshold at $Q = 4$ GeV; see the last paragraphs of this section for further information.*

In order to make an educated estimate of the overall error, we look again at Figure 24. The data, taking an optimistic point of view, can all be well described by QCD with 150 MeV $< \Lambda_{\overline{MS}}^{(5)} <$ 250 MeV, which corresponds to an error in $\alpha_s(M_{Z^0})$ of ± 0.004. With a more pessimistic attitude, however, one can also argue that the range in $\Lambda_{\overline{MS}}$ which is necessary to describe the data is 100 MeV $< \Lambda_{\overline{MS}}^{(5)} <$ 350 MeV, which then corresponds to an error of ± 0.011. In the absence of a more scientific approach to estimate the overall uncertainty on $\alpha_s(M_{Z^0})$, the final world average is thus quoted to be

$$\boxed{\alpha_s(M_{Z^0}) = 0.118 \pm 0.007},$$

which corresponds to

$$\Lambda_{\overline{MS}}^{(5)} = 210^{+90}_{-70} \text{ MeV}, \quad \text{or} \quad \Lambda_{\overline{MS}}^{(4)} = 315^{+125}_{-100} \text{ MeV},$$

in $\mathcal{O}(\alpha_s^3)$ and for $N_f = 5$ or 4 quark flavours, respectively. The error corresponds to the average between the optimistic and the pessimistic view given above. Due to the remarkably good agreement between the results quoted from many different processes and observables, there is no apparent reason to argue for an uncertainty which should be much larger than the one given above.

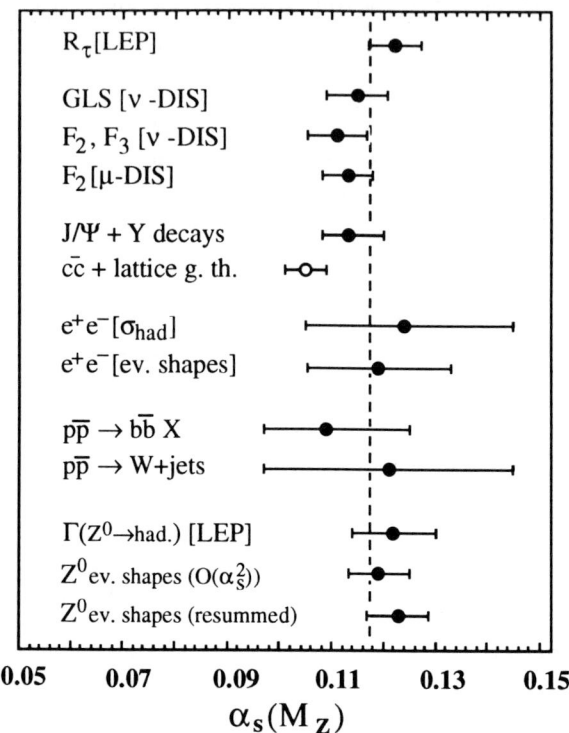

Figure 25. *A summary of measurements of $\alpha_s(M_{Z^0})$.*

Possibility of the existence of a light gluino?

In the past, the possibility for the existence of a new, electrically neutral, coloured fermion of low mass (here called $\tilde{g}$) has been discussed (see, for example, Jezabek and Kühn (1993) and Ellis (1993)). Such an object would add to loop diagrams involving gluon exchange, and would thus add to the effective number of quark flavours in the QCD β-function, such that $N_f \to N_f + 3N_{\tilde{g}}$. A larger effective number of quark flavours results in a slower energy dependence of α_s, which nourished speculations that this might be a possible explanation of the systematic difference of α_s results from LEP and from deep inelastic scattering.

Therefore in Figure 24, the prediction of the running of α_s, assuming a production threshold for a $\tilde{g}$ at 4 GeV, is also shown. In this case, a value of $\Lambda_{\overline{MS}} = 12$ MeV is necessary to obtain $\alpha_s(M_{Z^0}) = 0.123$. This curve would imply a better agreement between the LEP and the deep inelastic scattering data than in the case of standard QCD; however, the difference between e^+e^- and deep inelastic scattering data was not regarded to be significant, and the results from τ decays, at yet smaller energies, become inconsistent with the $\tilde{g}$ curve. Also note that for comparison with the $\tilde{g}$ case, all those results which were obtained for $Q > M_{\tilde{g}}$ should be re-analysed with the QCD coefficients of the respective theoretical calculations changed according to the increased effective number of quark flavours, instead of only changing the running α_s (Ellis 1993). Indeed, if calculated properly, α_s from R_Z would increase such that the difference between this result and α_s from DIS would still remain.

A fit of the QCD predictions to the 3-jet production rates measured at $E_{cm} = 29$ GeV to 91 GeV, taking one light $\tilde{g}$ into account when calculating both the running α_s and the $\mathcal{O}(\alpha_s^2)$ QCD coefficients, is displayed in Figure 10 (dashed-dotted line), which is the figure from which the evidence for asymptotic freedom was obtained. It can be seen that this prediction describes the current data about equally as well as does the fit of (standard) $\mathcal{O}(\alpha_s^2)$ QCD; indeed the slightly high but very precise value from LEP seems to indicate some preference for the $\tilde{g}$ case. While there is certainly no evidence for or against a light $\tilde{g}$ as yet, future measurements of 3-jet production rates at higher energies (e.g. at LEP-II or at a 500 GeV linear collider) are likely to add significant information to this topic.

9 Differences between quark jets and gluon jets

The question as to whether jets which are initiated by a highly virtual quark or gluon posses different properties has been the object of considerable theoretical and experimental study. In QCD, the gluon is associated with a colour charge $N_c = 3$ and the quark with a charge $C_f = 4/3$. At asymptotic jet energies, one thus expects, in leading order QCD, a ratio of $(N_c/C_f) = 9/4$ for the multiplicity of soft gluons produced from the two jet types. For equal quark and gluon jet energies, this implies that the particle energy spectrum of the gluon jet should be softer (i.e. the particles from gluon jets have lower energies, on average). Also, since the mean transverse energy of particles w.r.t. the jet axis is expected to be about the same, the angles of particles relative to their jet axes are expected to be larger in gluon jets, i.e. gluon jets are expected to be broader than quark jets.

Until recently, studies of differences between quark and gluon jets were performed, in the absence of any possibility of identifying quark and gluon jets on an event-by-event basis, by either using quark and gluon jets of different energies (as, for example, simply by energy ordering jets in 3-jet events, which determines that on average the lowest energy jet is from a gluon), or by comparing jets from 2-jet events at lower E_{cm} with the low-energy jet from 3-jet events at higher E_{cm}. In either case, the different jet species are embedded in different event environments, and so a direct comparison is likely to be biased. The experimental results were thus often contradictory or inconclusive, and they at best had to rely on comparisons with model studies, in order to try to correct for these biases.

At LEP, the OPAL collaboration invented a new method of tagging quark and gluon jets in symmetric 3-jet events, such that (quark and) gluon jets of roughly equal energy are identified with a relatively high purity (OPAL 93). This method is explained in Figure 26. Symmetric 3-jet events are selected using the 'Durham' jet finder, whereby the angles between the two lower-energy jets and the highest energy jet (called 'jet 1') are about equal and within 150 ± 10°. Starting from a total sample of 1,003,140 hadronic

Z^0 decays, and applying a few further quality cuts, 22,637 symmetric 3-jet events were thus obtained.

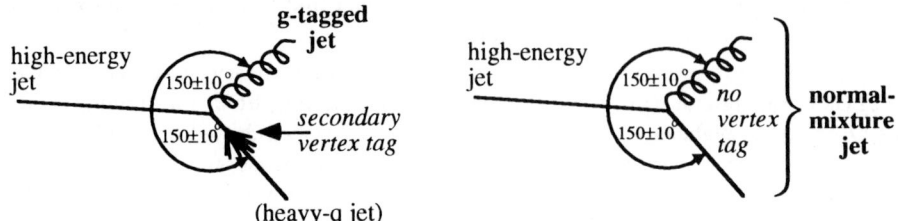

Figure 26. *Principle of selecting 'gluon-tagged' jets and 'normal mixture' jets in symmetric 3-jet events.*

Due to the bremsstrahlung nature of gluon radiation, jet 1 can then be identified as being a quark (or antiquark) jet, with a very high purity of about 97%. The two lower energy jets, having about equal energy, are thus known to be an equal mixture of quark and gluon jets; the averaged properties of these jets are thus defined to be those of a 'normal mixture' jet. A sample in which gluon jets are enriched is obtained by selecting events, as a subsample of the symmetric 3-jet events, in which a secondary vertex is reconstructed in one of the two lower-energy jets, using the silicon vertex detector. Such secondary vertices are due to the relatively long lifetime of bottom quarks ($\tau \sim 10^{-12}$s). The other low energy jet is then taken to be the gluon jet. With this method 1175 anti-tagged gluon jets are selected; they are named 'gluon-tagged jets'.

The purity of the gluon-tagged jet sample is estimated to be about 80%, *i.e.* there is a contamination of about 20% quark jets in this sample. These numbers are obtained from a wide range of model studies, but also from the data themselves, using 3-jet events with two identified secondary vertex tags.

OPAL then compares properties of the 'gluon-tagged jets' with those of the 'normal mixture jets'. In other words, one compares a sample of jets which consists of 80% gluons and 20% quarks, with a set which consists of 50% quark and 50% gluon jets. Note that both jet samples originate from identical jet and event environments, and that the average jet energies of both samples are about equal, namely about 21.5 GeV. No comparison with model calculations is therefore necessary in this analysis. Also note that the tagged (bottom) quark jet is *not* used in this analysis, since b-quark jets might have distinct properties, due to the heavy mass of this quark, which could bias the results.

The particle multiplicities of these two jet samples are compared in Figure 27. It can be seen that the gluon-tagged jet sample has a significantly larger particle multiplicity than the normal-mixture jets, leading to a ratio of average particle multiplicities $\langle n \rangle$ of

$$\frac{\langle n \rangle_{\text{gluon-tagged}}}{\langle n \rangle_{\text{normal-mixture}}} = 1.081 \pm 0.011.$$

In addition, distributions of particle energies and angles w.r.t. the jet axes show that the gluon-tagged jets are generally broader, and their particles are less energetic, on average.

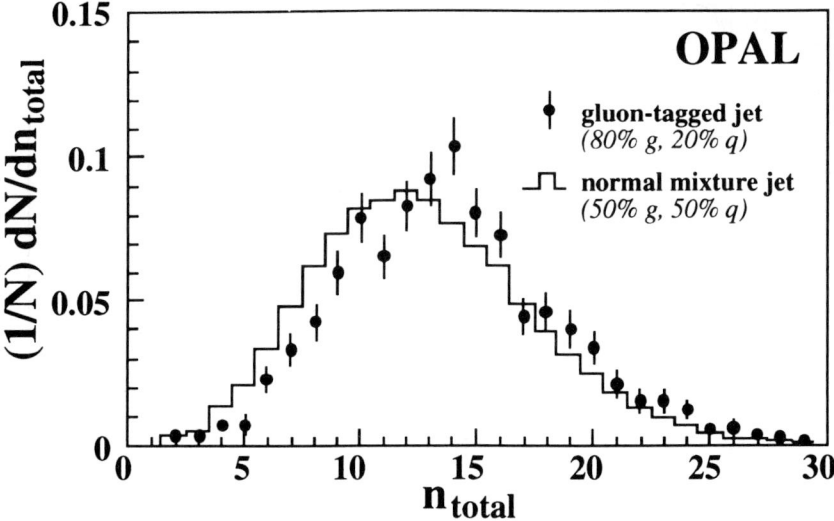

Figure 27. *Particle multiplicity distributions of the (uncorrected) gluon-tagged and normal-mixture jet samples.*

While the analysis so far has not depended on any model calculation (it was sufficient to assume that the gluon content in the gluon-tagged sample is larger than in the normal-mixture sample), it is interesting to further extract numbers and distributions for *pure* gluon and quark jet samples, by unfolding the measured distributions according to the quark contents and gluon contents of the two samples given above. For instance, the ratio of particle multiplicities in gluon jets and in quark jets is then determined to be

$$\frac{\langle n \rangle_{\text{gluon}}}{\langle n \rangle_{\text{quark}}} = 1.27 \pm 0.04 (\text{stat.}) \pm 0.06 (\text{syst.}),$$

which is significantly larger than unity, but also lower than the naïve expectation of 9/4 (which is strictly valid in leading order and for asymptotic jet energies only). A lower value than 9/4 is indeed expected from higher-order QCD calculations; however a direct prediction for the experimental number given above still does not exist.

A further example of the results obtained after unfolding for pure quark and gluon jets is shown in Figure 28. Here, the corrected distributions of particle energies in gluon and in quark jets is shown, which significantly demonstrates that gluon jets have a softer energy spectrum than quark jets, as expected (but not further specified) by QCD. These results also demonstrate the many new possibilities for QCD studies at LEP, due to the increased data statistics, the improved detectors and the reduced hadronisation effects if compared to lower energy machines.

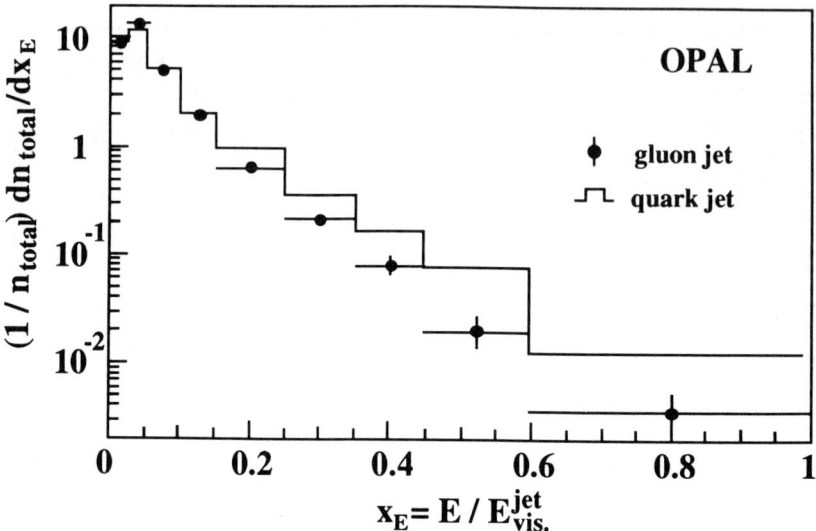

Figure 28. *Corrected, scaled inclusive energy distributions of particles in gluon and in quark jets.*

10 Summary and outlook

Studies of hadronic final states of e^+e^- annihilations have provided deep insights into the kinematics of quarks and gluons, and thus allowed significant tests of the predictions of perturbative QCD. Pioneering results, like the first evidence for jet production and the discovery of the gluon, were obtained at the lower energy colliders SPEAR and PETRA, and further studies at PETRA, PEP and TRISTAN provided the basis for our current knowledge of the physics of hadron production and hadronic jets. The most precise tests of QCD in e^+e^- annihilations are obtained at LEP, where the resonant cross-section at the Z^0 mass provides large data samples, and where non-perturbative hadronisation effects are significantly reduced due to the higher centre-of-mass energy.

LEP provided many significant results such as tests of the basic quantum numbers of quarks and gluons, the experimental determination of the group constants of QCD, N_c and T_f, and evidence for the existence of the gluon self coupling and for asymptotic freedom. An improved understanding of the virtues and the limits of the experimental methods and of the corresponding QCD predictions, as well as the availability of new calculations which, in some cases, include the next-to-next-to-leading order or the resummation of leading logarithms and next-to-leading logarithms to all orders, have led to a new level of precision at LEP. In particular, the value of $\alpha_s(M_{Z^0})$ has been determined, from many different event shape, jet rate and energy correlation observables, to an overall precision of about 5%. The remaining uncertainty is almost entirely of a theoretical nature: it can be attributed to our limited knowledge of the contributions of higher-order perturbative terms. The typical theoretical uncertainty for individual observables appears to be much larger, about 10–20% in $\alpha_s(M_{Z^0})$, and only combined analyses of many observables, whose higher-order corrections are supposedly different, provide the means of decreasing the overall uncertainty to the level of 5%, and to test the reliability of the methods chosen to determine these systematic uncertainties.

The final result from such studies, using resummed $\mathcal{O}(\alpha_s^2)$ calculations which contain the leading logarithms and next-to-leading logarithms of the perturbation series to all orders, is $\alpha_s(M_{Z^0}) = 0.123 \pm 0.006$. This value matches well with the result of $\alpha_s(M_{Z^0})$ from the hadronic width of the Z^0, normalised by the leptonic width, which gives $\alpha_s(M_{Z^0}) = 0.122 \pm 0.008$, where the error is still dominated by statistics. This is also consistent with the result obtained from the ratio of branching fractions of the τ lepton, which is $\alpha_s(M_\tau) = 0.36 \pm 0.04$ or, equivalently, $\alpha_s(M_{Z^0}) = 0.122 \pm 0.005$. The agreement between these different results, which all have different sources of uncertainties and which are determined at largely different energy scales, may constitute the most significant and precise test of QCD which is available so far.

In addition to these tests of 'hard' QCD, there are many studies which are related to the emission of 'soft' gluons, hadronisation and other non-perturbative effects, like gluon coherence, string hadronisation, differences between quark and gluon jets, particle and multiplicity correlations, inclusive particle distributions, Bose-Einstein correlations and studies of intermittency. Of these subjects, only the measurement of differences between quark and gluon jets, from a model independent study involving gluon jet tagging, has been discussed in these lectures. For general references to those studies which were not covered in this lecture, see, for example, (Drees 1992, Bethke and Pilcher 1992, Bethke 1992c) or other recent review articles.

The future of QCD studies in e^+e^- annihilation will largely depend on the further development of theoretical calculations, such as on jet production and hadronic event shapes to $\mathcal{O}(\alpha_s^3)$, and the further development of resummation techniques and their application to more and—maybe—new observables. The future will also depend on refinements of experimental techniques, on the interplay between theorists and experimentalists, on a significant increase of data statistics at LEP (for some studies which require, for example, tagging of special event classes), and on future data at higher centre-of-mass energies. The latter will be available from LEP-II and—hopefully—from a linear collider at 500 GeV centre-of-mass energy.

The general prospects of a linear collider are currently discussed in a series of workshops (LinColl 1991, 1993), where—among others—the idea of running such a machine in a large range of centre-of-mass energies has been put forward. With a design luminosity of $\mathcal{L} \geq 10^{33} \text{s}^{-1} \text{cm}^{-2}$ and assuming that this does not significantly change with energy, one gets the following number of hadronic events *per day* of efficient running:

E_{cm}	# of hadronic events/day
500 GeV	~ 350
91 GeV	$\sim 4{,}000{,}000$
60 GeV	$\sim 15{,}000$
35 GeV	$\sim 30{,}000$

These numbers do not contain radiative corrections nor experimental selection cuts; however the value of running a linear collider, with its high luminosity, at smaller centre-of-mass energies is obvious. Within a rather short time (of about a few months at most), one can perform an energy scan down to LEP/TRISTAN/PETRA energies, and get much more event statistics than was ever collected with the previous or existing experiments. This can be achieved for one and the same experimental setup, thereby minimising experimental point-to-point uncertainties. Such a program should allow extremely interesting and precise testing of all aspects of the standard model which in general depend on energy, such as QCD scaling violations, muon-asymmetries and many other topics. This will provide the possibility, for instance, of finding evidence for or excluding the existence of light, neutral, coloured objects such as a gluino ($\tilde{g}$), by a precise measurement of the degree of energy dependence of α_s, which may be the only practical way to search for such hypothetical particles.

A linear collider can also be a used as a very efficient factory for Z^0 bosons, when running on the Z^0 pole, so that studies of CP-violation in bottom hadrons are within reach. The proposal for a large energy scan implies strong demands on the construction and operation of the detector(s), of the electronics and data acquisition, and on the accelerator itself. However, the rich physics potential for precise tests of the standard model of the strong and electroweak interactions which is offered by such a program, in addition to the large potential for top-quark physics and for discovering new physics at the highest energies, will undoubtedly be worth the effort.

Acknowledgements

It has been a real pleasure to present these lectures at the 1993 Scottish Universities Summer School in Physics, in the pleasant surrounding of the old city of St. Andrews. I thus wish to thank David Saxon for his kind invitation. Many thanks go to Tony Doyle, Peter Negus, Ken Peach and Alan Walker for creating such a nice working and living atmosphere at the school, and to Lance Vick for his help in editing this manuscript.

References

ALEPH collaboration, 1992, Decamp D *et al. Phys Lett* **B284** 151.
ALEPH collaboration, 1993, Buskalic D *et al.* CERN-PPE/93-41.
AMY collaboration, 1989, Park I *et al. Phys Rev Lett* **62** 1713.
Bardeen W A *et al.* 1978, *Phys Rev* **D18** 3998.
Bengtsson M and Zerwas P, 1988, *Phys Lett* **B208** 306.
Bethke S, 1989, *Proceedings of the Workshop on the Standard Model at Present and Future Accelerator*, Budapest, and preprint LBL-28112 .
Bethke S, 1989a, *Z Phys* **C43** 331.
Bethke S, Ricker A and Zerwas P, 1991, *Z Phys* **C49** 59.
Bethke S, 1992, *Proceedings of the Aachen 'QCD – 20 years later' workshop* (World Scientific), and preprint HD-PY 92-12.
Bethke S, Kunszt Z, Soper D E and Stirling W J, 1992a, *Nucl Phys* **B370** 310.
Bethke S, 1992b, in: (LinnColl 1991, LinColl 1993), and preprint HD-PY 92-01.
Bethke S, 1992c, *Proceedings of the XXVI International Conference on High Energy Physics*, Dallas, Texas, August 1992; preprint HD-PY 92/13.
Bethke S and Pilcher J E, 1992, *Ann Rev Nucl Part Sci* **42** 251.
Blondel A, 1993, *these Proceedings*.

Braaten E, Narison S and Pich A, 1992, *Nucl Phys* **B373** 581.
Brodsky S J, Lepage G P and Mackenzie P B, 1983, Phys Rev **D28** 228.
Brown N and Stirling W J, 1992, *Z Phys* **C53** 629.
Burckhardt H and Steinberger J, 1991, *Annu Rev Nucl Part Sci* **41** 55.
Catani S *et al.* 1992, CERN-TH-6640-92.
DELPHI collaboration, 1993, Abreu P *et al.* CERN-PPE/93-69.
DELPHI collaboration, 1993a, Abreu P *et al.* CERN-PPE/93-43.
Drees J, 1992, *Proceedings of the Aachen 'QCD – 20 years later' workshop* (World Scientific).
Ellis J and Karliner I, 1979, *Nucl Phys* **B148** 141.
Ellis R K, Ross D A and Terrano A E, 1981, *Nucl Phys* **B178** 421.
Ellis J, Nanopoulos D V and Ross D A, 1993, *Phys Lett* **B305** 375.
Fernandez E, 1993, *Proceedings of the EPS Conference*, Marseille
Feynman R P, 1969, *Phys Rev Lett* **23** 1415.
Fritzsch H and Gell-Mann M, 1973, *Proceedings XVth International Conference on High Energy Physics*, Chicago-Batavia.
Gorishny SG, Kataev A L and Larin S A, 1991, *Phys Lett* **B259** 194.
Gottschalk T D and Shatz M P, 1985, *Phys Lett* **B150** 451.
Gross D J and Wilczek F, 1973a, *Phys Rev Lett* **30** 1343.
Gross D J and Wilczek F, 1973b, *Phys Rev* **D8** 3633.
Grunberg G, 1980, *Phys Lett* **B95** 70.
Gutbrod F *et al.* 1984, *Z Phys* **C21** 235.
Halzen F and Martin A D, 1984, Quarks and Leptons (John Wiley and Sons).
Haidt D, 1993, *Directions in High Energy Physics* **14**, Precision Tests of the Standard Electroweak Model, ed. Langacker P, (World Scientific).
Hanson G *et al.* 1975, *Phys Rev Lett* **35** 1609.
Hebbeker T, 1991, Aachen report PITHA 91/08 (revised version).
JADE collaboration, 1988, Bethke S *et al. Phys Lett* **B213** 235.
JADE collaboration, 1981, Bartel W *et al. Phys Lett* **B101** 129.
Jezabek M and Kühn H J, 1993, *Phys Lett* **B301** 121.
Kamae T *et al.* 1988, contribution to the *XXIVth International Conference on High Energy Physics*, Munich.
Kramer G, 1984, *Springer Tracts of Modern Physics* **102**.
Kunszt Z and Nason P, 1989, in LEP 1989.
LEP, 1989, *Z Physics at LEP-I*, eds G Altarelli, R Kleiss and C. Verzegnassi, CERN 89-08.
L3 collaboration,1990, Adeva B *et al. Phys Lett* **B248** 227.
L3 collaboration,1991, Adeva B *et al. Phys Lett* **B263** 551.
Le Diberder F and Pich A, 1992, *Phys Lett* **B286** 147.
Le Diberder F and Pich A, 1992a, *Phys Lett* **B289** 165.
LinColl 1991, *Proceedings of the Workshop on Physics and Experiments with Linear Colliders*, Saariselkä, Finland.
LinColl 1993, *Proceedings of the Workshop on Physics and Experiments with Linear Colliders*, Waikoloa, Hawaii.
Lönnblad L and Petterson U, 1988, Lund preprint LU TP 88-15.
Marchesini G and Webber B R, 1988, *Nucl Phys* **B310** 461.
Marciano W J, 1984, *Phys Rev* **D29** 580.
Marshall R, 1988, *Z Phys* **C43** 595.
Nachtmann O and Reiter A, 1982, *Z Phys* **C16** 45.
Nachtmann O, 1986, Elementarteilchenphysik, Phänomene und Konzepte (Vieweg).
Naroska B, 1987, *Phys Rep* **148** 67.
Odorico R, 1988, Bologna preprint DFUB 88-27.

OPAL collaboration, 1990, Akrawy M Z et al. *Z Phys* **C47** 505.
OPAL collaboration, 1992, Acton P D et al. *Z Phys* **C55** 1.
OPAL collaboration, 1993, Acton P D et al. CERN-PPE/93-02.
OPAL collaboration, 1993a, Acton P D et al. CERN-PPE/93-38.
OPAL collaboration, 1993b, Akers A et al. CERN-PPE/93-118.
Politzer H D, 1973a, *Phys Rev Lett* **30** 1346.
Politzer H D, 1973b, *Phys Rep* **14C** 129.
RPP 1992, *Review of Particle Properties*, *Phys Rev* **D45** Part2.
Sjöstrand T, 1986, *Comp Phys Commun* **39** 347.
Sjöstrand T and Bengtsson M, 1987, *Comp Phys Commun* **43** 367.
Stevenson P M, 1981, *Phys Rev* **D24** 1622.
Wu S L, 1984, *Phys Rev* **59** 107.

HERA Physics

Günter Wolf

Deutsches Elektronen Synchrotron (DESY)
Hamburg, Germany

1 Introduction

Scattering experiments have been marvelously successful in unravelling the structure of matter. The basic concept is quite simple, requiring a point-like and energetic test particle which is scattered on the probe and for which the angular and energy distribution is measured. An early example is the experiment by Rutherford and his coworkers, Geiger and Marsden (1909–1911), who scattered α-particles on a metal foil. The observation of occasional scatterings at large angles led Rutherford (1911) to the assumption of a hard core in the atom. The experimental verification of his scattering formula by Geiger and Marsden resulted in the conclusion that the atom has a positively charged core with a radius of less than 30 fm.

The structure of the nucleon has been explored mostly with lepton beams. Elastic scattering of electrons with beam energies of the order of 1 GeV showed that the proton is an extended object with a radius of 0.8 fm (*e.g.* Janssens *et al.* 1966). Inelastic scattering of 20 GeV electrons on nucleons (SLAC-MIT; *e.g.* Taylor 1969) revealed approximate scaling of the structure functions (Bjorken 1969) resulting from scattering on charged, point-like constituents in the nucleon as explained by the quark-parton model (Feynman 1972). The observation of logarithmic violations of scaling (Fox *et al.* 1974, Watanabe *et al.* 1975, Chang *et al.* 1975, SLAC-MIT 1974, 1975), which became particularly clear with beam energies of several hundred GeV, was instrumental in the formulation of Quantum Chromodynamics, QCD (Fritzsch and Gell-Mann 1972, Fritzsch *et al.* 1973, Weinberg 1973, Gross and Wilczek 1973).

The object size Δ that can be resolved in the scattering process is determined by the four momentum Q transferred to the probe particle. From the uncertainty relation follows $\Delta \approx 1/Q$. Better resolution requires larger momentum transfers and hence higher energies. This is best achieved in a storage ring where the test and the probe particles collide head-on. With the electron-proton collider HERA at DESY centre-of-mass energies of 300 GeV can be reached, compared to 30 GeV in current fixed target experiments. The result is a vast increase in phase space and physics potential for lepton-nucleon scattering, the maximum momentum transfer Q rising by a factor of ten and the energy transfer ν by a factor of hundred.

The physics that can be done at an electron-proton collider, such as HERA, has been developed in numerous theoretical and experimental studies. An introduction can be found in the proceedings of two recent workshops held in 1987 (Peccei) and 1991 (Buchmüller and Ingelman).

The HERA project (HERA 1981, Wiik 1982, 1992, Voss 1988) was approved in 1984. Operation of HERA for physics started in 1992 with the two large general purpose detectors H1 (1986) and ZEUS (1986) taking data. A third experiment, HERMES (1990), which has been approved recently for the measurement of the nucleon spin structure by colliding the polarised electron beam with a gas jet of polarised nucleons, is under construction. A fourth experiment, HERA-B (HERA-B 1992), which aims at measuring CP violation in the $b\bar{b}$ system by scattering beam protons on a fixed target, is under discussion.

These lectures are intended for a newcomer to the field and focus on the first results obtained by H1 and ZEUS.

2 The HERA collider

2.1 Layout

The layout of HERA is shown in Figure 1. Two separate magnet systems guide the e and p beams around the 6.3 km long ring. DESY and PETRA serve as injectors. There are four interaction regions, two of which are occupied by H1 and ZEUS. A third region has been allocated to HERMES.

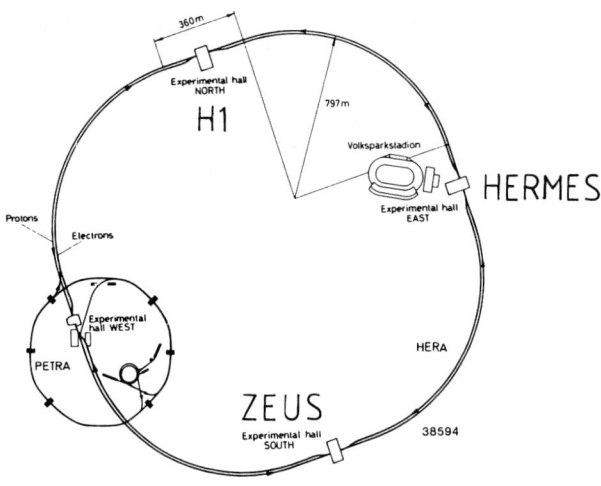

Figure 1. *Layout of* HERA.

Table 1 shows some of the parameters of the collider. In order to maximise the luminosity up to 210 bunches of particles can be stored for each beam. The distance between bunches is 96 ns.

	electron ring		proton ring
circumference		6336 m	
energy	30 GeV		820 GeV
e-p c.m. energy		314 GeV	
magnetic bending field	0.164 T		4.682 T
bending radius of dipoles	610 m		584 m
circulating current	60 mA		160 mA
number of particles/beam	0.8×10^{13}		2.1×10^{13}
number of bunch buckets	220		220
number of bunches	210		210
current/bunch	0.3 mA		0.8 mA
time between beam crossings		96 ns	
luminosity		1.5×10^{31} cm^{-2}s^{-1}	
specific luminosity		3.3×10^{29} cm^{-2}s^{-1}mA^{-2}	
polarisation time at $E_e = 30$ GeV	25 min		

Table 1. HERA *design parameters.*

2.2 Performance

For physics runs the beam energies chosen were 26.67 GeV for electrons and 820 GeV for protons. Since May 1992, when H1 and ZEUS observed the first ep collisions in their central detectors, data were taken in three running periods July 1992, September–October 1992 and July–October 1993. Over this time, beam currents and luminosity were gradually increased. In 1993 HERA was operated with 90 bunches per beam and typical (maximum) beam currents of 13 (25) mA for electrons and 13 (20) mA for protons yielding a typical (maximum) luminosity of 0.7×10^{30} cm^{-2}s^{-1} (1.5×10^{30} cm^{-2}s^{-1}). The maximum specific luminosity observed was 6×10^{29} cm^{-2}s^{-1}mA^{-2}, which is above the design value. The evolution of the luminosity during the three running periods is shown in Figure 2. The total integrated luminosity per experiment reached 33 nb^{-1} in 1992 and 600 nb^{-1} in 1993.

After coasting for some time the electrons become polarised with spins being antiparallel to the direction of the bending field as a result of the Sokholov-Ternov effect (Sokholov and Ternov 1964). The build-up time for the polarisation is determined by the synchrotron radiation and is given by

$$P(t) = P_0 \{1 - \exp(-t/t_p)\} ; \qquad t_p = 98 \, r^2 R E^{-5} \qquad (2.1)$$

with $P_0 = 92\%$, t_p in seconds, the bending radius r in metres, the average radius R in metres and the electron beam energy E in GeV. This prediction holds for a

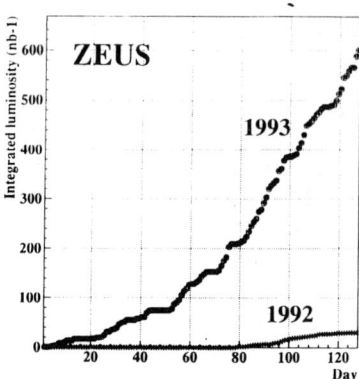

Figure 2. *Integrated luminosity collected by* ZEUS *during the 1992 and 1993 running periods.*

perfect machine. Already small imperfections, *e.g.* in the magnet lattice may produce depolarisation effects and make the depolarisation time shorter than the build-up time thereby reducing the polarisation.

Electron beam polarisation was observed for the first time in the autumn of 1991 at the level of $P = 5\text{-}9\%$. Studies of the polarisation (Barber *et al.* 1992, 1993) were performed at 26.67 GeV also in parallel to data taking of H1 and ZEUS. After realignment of magnets and empirical tuning of correction magnets, with guidance from a tracking program, polarisation values close to 70% could be attained (Figure 3).

For particle physics, electrons of definite helicity (left-handed or right-handed) rather than with transverse polarisation are needed. This can be achieved with the help a pair of spin rotators installed at the interaction regions. A pair of spin rotators was designed and built (Buon and Steffen 1985) and is being installed in hall East for HERMES. Given satisfactory performance, the interaction regions of H1 and ZEUS will also be equipped with spin rotators.

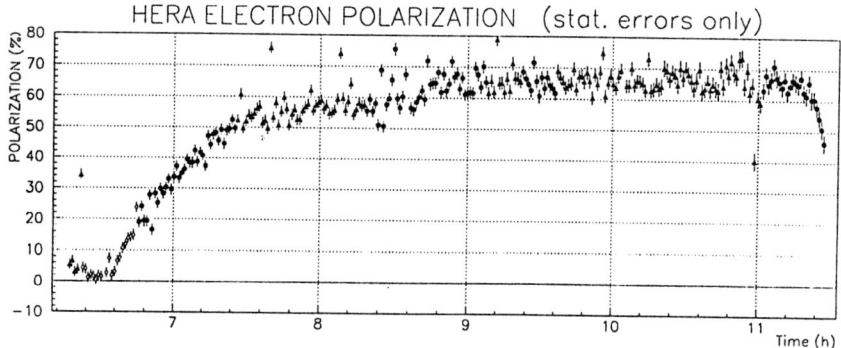

Figure 3. *Build-up of transverse electron polarisation at $E_e = 26.67$ GeV as a function of time. A change of the machine tune made at around 8:30 leads to an increase in the polarisation.*

3 The experiments

3.1 Detector challenges

HERA produces a large variety of reactions with widely differing energy flows. This feature together with the desire to detect and identify the constituents which participate in these reactions such as electron, neutrino, photon, quark and gluon places different requirements on the detector. The large momentum imbalance between incident electron and proton and the nature of space-like processes send most particles into a narrow cone around the proton direction. The observation of deep inelastic (DIS) neutral current (NC) scattering, $ep \to eX$ (Figure 4a), is fairly straightforward. It produces an energetic electron whose transverse momentum is balanced by the current jet. The remnants from the breakup of the proton escape mostly unseen down the beam pipe. The variables x and Q^2 which describe the process (see below) can be determined from the energy and angle of either the electron or the current jet. This requires a precise electromagnetic and hadronic calorimeter with the calibration known at the 1–2% level.

In charged current (CC) scattering, $ep \to \nu X$ (Figure 4b), only the current jet can be observed. The identification of such events is based on the observation of missing transverse momentum carried away by the neutrino. It requires a hermetic calorimeter which covers the full solid angle such that e.g. photons, neutrons or K^0's cannot escape undetected. The variables x and Q^2 are measured from the current jet.

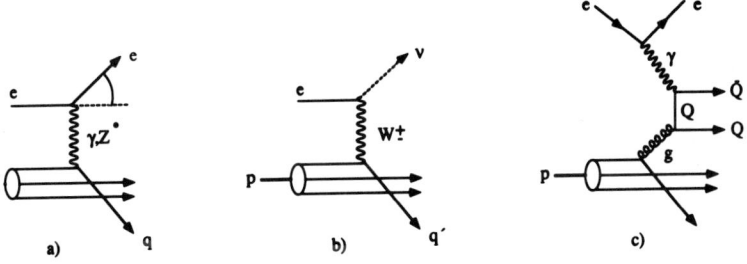

Figure 4. *Diagrams for NC and CC scattering and for photon-gluon fusion.*

The observation of processes at low Q^2, with scattering on soft partons (e.g. Figure 4c), is more difficult. The energy deposited in the calorimeter often is only a few GeV. Additional information from tracking detectors which surround the interaction point is necessary for their identification.

Background presents another challenge. The number of events from e-p interactions is tiny (10^{-3}–10^{-5}) compared to the background events produced for instance by beam protons on the beam pipe wall or in the residual gas. What is worse, this type of background deposits a large amount of energy in the detector. A typical background event is shown in Figure 5 where 225 GeV are observed in the calorimeter and many tracks in the tracking detector. At proton design current the background rate is expected to be around 10–100 kHz. The detector must be able to discriminate quickly—within a few microseconds—and dead-time free against background events although both beams cross each other every 96 ns. The high background rates combined with the short bunch crossing interval forced the HERA experiments to develop novel concepts of electronic readout and triggering, concepts which are suitable also for detectors at the next generation of pp colliders, such as the LHC.

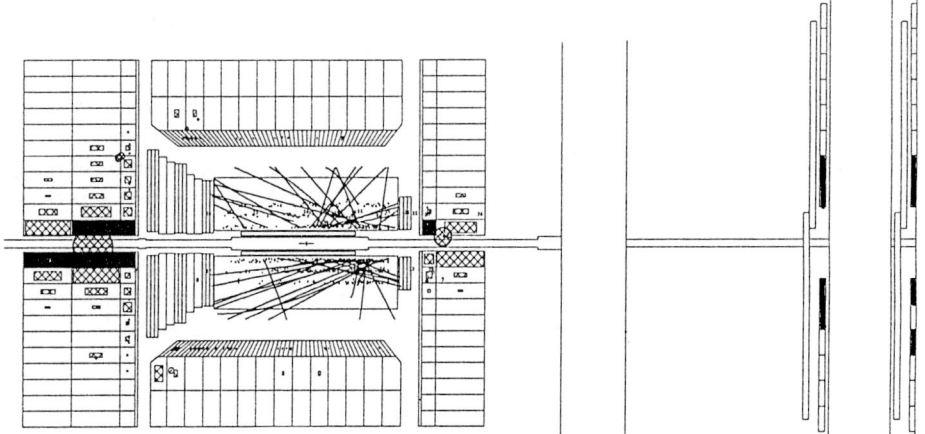

Figure 5. *A background event produced by a proton interaction upstream (to the right) of the ZEUS detector.*

The H1 and ZEUS detectors are driven by their choice of calorimeter. ZEUS uses a compensating uranium-scintillator calorimeter which provides the best possible energy resolution for hadrons. Compensation means that electromagnetic particles (electrons, photons) and hadrons of the same energy yield the same pulse height, $e/h = 1$. The radioactivity of the (depleted) uranium provides an extremely stable calibration signal, the mean life time of ^{238}U being 6.5×10^9 years. The H1 calorimeter uses liquid argon for readout which allows for a very stable and precise energy calibration and a high transverse and longitudinal segmentation. The calorimeter is noncompensating, $e/h = 1.1$–1.25. However, due to the high segmentation and using software weighting with the observed shower profile equal signals for electrons and hadrons can be obtained.

3.2 The H1 detector

The H1 collaboration at present consists of about 390 physicists from 36 institutes and 12 countries. The H1 detector (H1 1986, 1993) is displayed in Figure 6. The proton beam enters from the right, the electron beam from the left (the bottom as shown).

Charged particles are tracked in a magnetic field of 1.2 T which is produced by a superconducting solenoid that surrounds the calorimeter. The effect of the magnetic field on the beams is compensated by a small solenoid (in Figure 6 located on the right-hand side). The tracking system consists of two cylindrical jet-and z-drift chambers in

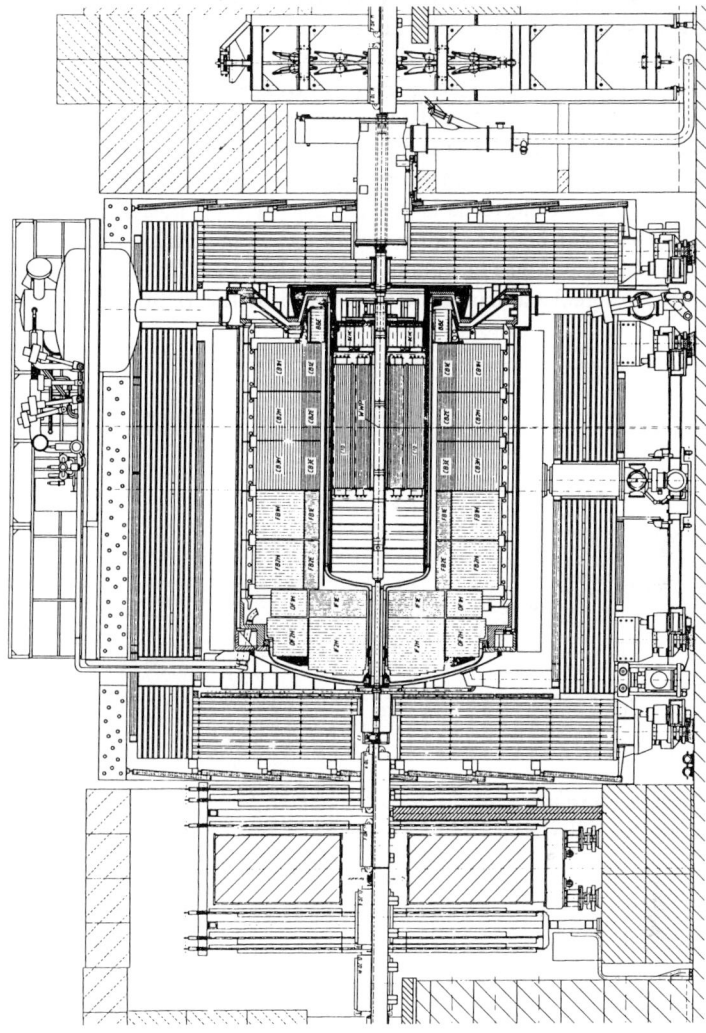

Figure 6. *Longitudinal cut through the H1 detector along the beam line.*

the central region, and of three radial and three planar drift chambers in the forward direction. The drift chambers are interleaved with proportional wire chambers for a fast trigger selection. The backward direction is covered by a four layer proportional wire chamber providing space points up to a scattering angle of 175°. In the forward direction a transition radiation detector (TRD) enhances the electron identification. The central drift chambers provide up to 56 space point with a resolution of 170 μm in $r\phi$ and 2.2 cm in z. The z-drift chambers measure the z-coordinate with an accuracy of 200–260 μm.

The liquid-argon calorimeter (LAC) covers the angular region $4° < \theta < 153°$ (the forward direction, $\theta = 0°$, is given by the direction of the proton beam). The calorimeter is longitudinally subdivided into an electromagnetic section with lead plates and a hadronic section with stainless steel plates as absorbers. The total depth varies between 8 and 4.5 absorption lengths. The calorimeter is finely segmented into a total of 45,000 channels. The calorimeter was optimised for a precise measurement and identification of electrons and for a stable energy calibration for electrons and hadrons. The energy resolution σ/E for electrons is $(12\%/\sqrt{E}) \oplus 1\%$ ($\oplus$ means quadratic addition) and $(50\%/\sqrt{E}) \oplus 2\%$ for hadrons (after weighting) as measured with test beams (Figures 7, 8). The calorimeter has been in operation since April 1991. With the same load of liquid argon the signal attenuation was found to be less than 0.5% per year.

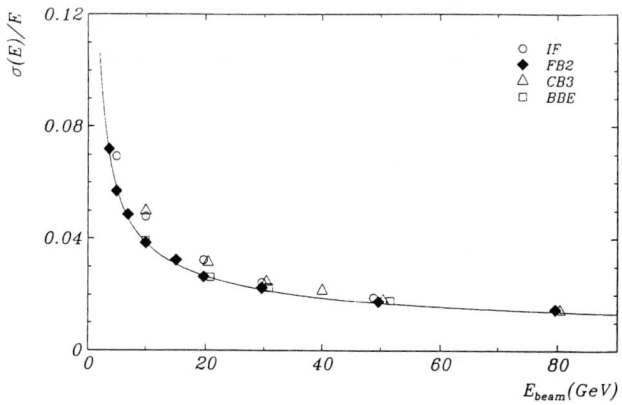

Figure 7. *Energy resolution of the H1 calorimeter for electrons for different parts of the calorimeter (H1 1993).*

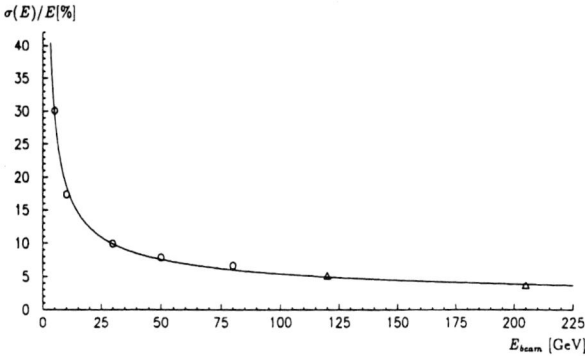

Figure 8. *Energy resolution of the H1 calorimeter for pions (H1 1993).*

In the backward direction the LAC is supplemented by an electromagnetic lead-scintillator calorimeter (BEMC), covering $151° < \theta < 177°$, followed by time-of-flight counters (TOF-VETO). In the forward direction the plug calorimeter with copper plates and silicon diode readout extends the energy measurement for hadrons down to angles of 0.7°.

The magnet yoke is made of 10 layers of 7.5 cm thick iron plates. The gaps are instrumented with limited streamer tube (LST) chambers for measuring energy which has not been fully absorbed in the liquid argon calorimeter and for tracking of muons. Large area LST chambers in front and behind the iron yoke and an iron toroid magnet plus 6 layers of drift chambers in the forward direction complete the muon detection system.

The luminosity is measured by observing the bremsstrahlung process $ep \to ep\gamma$ at very small angles to the electron beam direction. The final state electron and photon are detected in coincidence in electromagnetic calorimeters of the luminosity detector LUMI positioned at 33 m (electron tagger ET) and 100 m (photon tagger PD) upstream (in the proton direction) of the central detector (Figure 9) in the electron direction. At nominal luminosity the rate of luminosity events is between 50–100 kHz depending on the selection criteria.

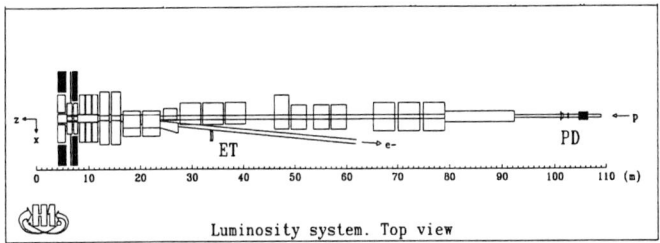

Figure 9. *Layout of the H1 luminosity detector.*

3.3 The ZEUS detector

The ZEUS experiment is performed by a joint effort of about 460 physicists from 12 countries and 50 institutes. A view of the central part of the ZEUS detector (ZEUS 1986, 1993a) is shown in Figure 10.

Charged particles are tracked by the inner tracking system, consisting of a vertex detector (VXD), a central tracking detector (CTD), planar drift chambers in the forward and rear directions (FTD, RTD). A transition radiation detector (TRD) helps with electron identification in the forward direction. The VXD consists of 12 layers of axial sense wires. The CTD has 9 superlayers (5 axial and 4 small angle stereo), each with 8 layers of sense wires. The three innermost axial layers (1, 3 and 5) are additionally instrumented with z-by-timing electronics mainly for triggering purposes. A thin (0.9 radiation length) superconducting solenoid surrounding the inner tracking system produces a magnetic field of 1.43 T. A compensator solenoid supresses the effects on the beams.

A uranium-scintillator calorimeter (CAL) encloses the solenoid. It is subdivided mechanically into the forward (FCAL), barrel (BCAL) and rear (RCAL) calorimeters. The CAL covers polar angles from 2.6° to 176.1°, which is 99.7% of the total solid angle.

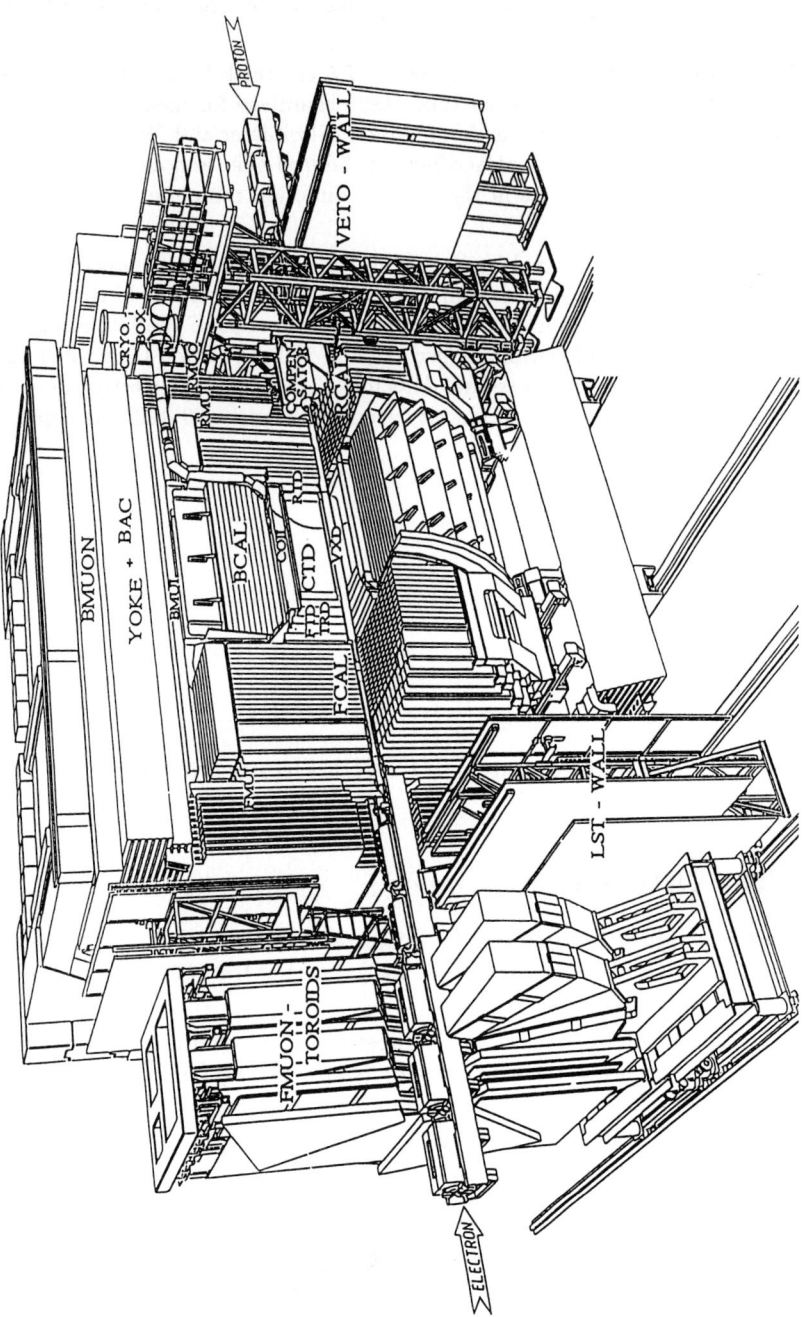

Figure 10. *Schematic view of the central part of the* ZEUS *detector.*

In the HERA reference system the coverage in pseudorapidity $\eta = -\ln\tan(\theta/2)$ is as follows: FCAL: $4.3>\eta>1.1$, BCAL: $1.1>\eta>-0.75$ and RCAL: $-0.75>\eta>-3.8$. The calorimeter consists of a total of 80 modules. Every module is made of up to 185 layers of 3.3 mm thick depleted uranium plates plus 2.6 mm thick scintillator plates. Wavelength shifter bars transport the light to photo-multipliers. The modules are subdivided longitudinally into an electromagnetic and two (one) hadronic sections in FCAL, BCAL (RCAL) presenting a total depth of 7 to 4 absorption lengths. The scintillator plates form $5\times 20\,\text{cm}^2$ ($10\times 20\,\text{cm}^2$) cells in the electromagnetic section and $20\times 20\,\text{cm}^2$ in the hadronic sections of FCAL, BCAL (RCAL). The total number of cells is 5918.

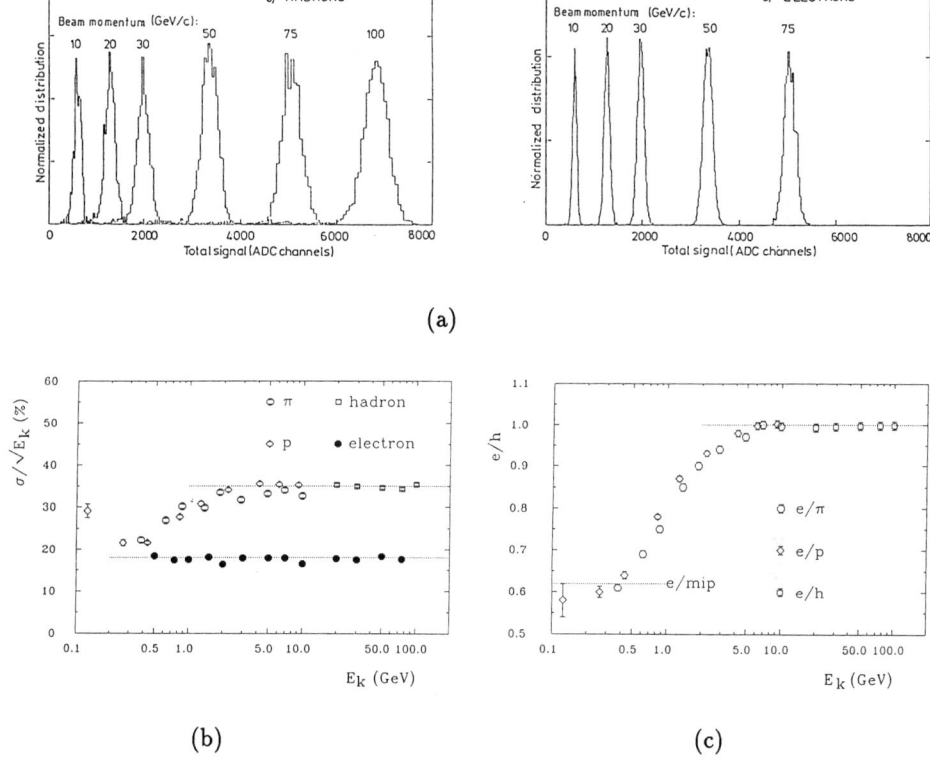

Figure 11. *Measurements with the ZEUS prototype calorimeter. (a) Pulse height distributions for electrons and hadrons; (b) Energy resolution for electrons and hadrons; (c) e/h ratio as a function of particle momentum.*

The calibration of the photomultipliers is being monitored with the signal from the radioactivity of the uranium (UNO) to a precision of less than 0.2%. The pulse heights of electrons and hadrons (Figure 11a) are equal to within 3%, i.e. $e/h = 1.0 \pm 0.03$ (Figure 11b), for momenta above 3 GeV/c. The energy resolution as measured in the test beam is for electrons $\sigma/E = 18\%/\sqrt{E}$ (E in GeV) and for hadrons $35\%/\sqrt{E}$ (Figure 12). The calorimeter noise, which is dominated by the uranium radioactivity, is typically

15 MeV in the EMC cells and 25 MeV in the HAC cells. The calorimeter yields also an accurate time measurement. The time resolution of a calorimeter cell is $\sigma = 1.5/\sqrt{E} \oplus 0.5\,\mathrm{ns}$ or less than 1 ns above 3 GeV.

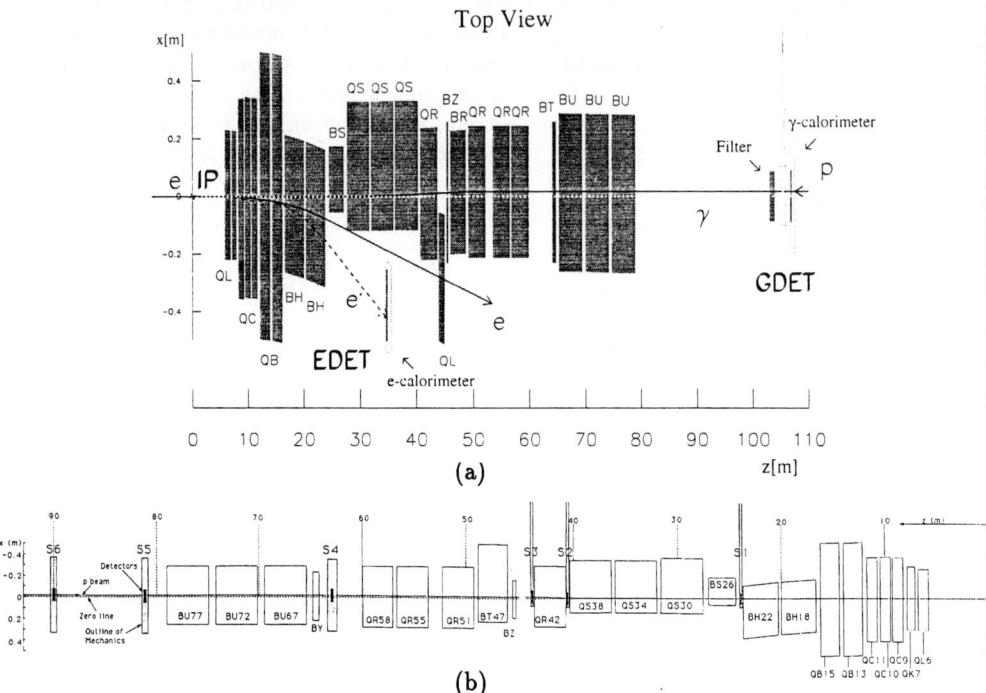

Figure 12. *(a) The ZEUS luminosity detector: shown are the detectors for plotting photons (GDET) and the scattered electron (EDET) together with the beam elements; (b) Layout of the leading photon spectrometer of ZEUS.*

In the course of an upgrade program, the transverse segmentation of the rear and forward parts of the calorimeter is being increased by inserting a plane of $3 \times 3\,\mathrm{cm}^2$ silicon diodes after the first 3 radiation lengths.

A small tungsten-silicon calorimeter (BPC) positioned at the beam pipe behind RCAL tags electrons scattered with Q^2 up to $0.5\,\mathrm{GeV}^2$.

The iron yoke serves as the absorber for the backing calorimeter and as a muon filter. It is made of 7.5 cm thick iron plates and is instrumented with proportional tube chambers for measuring the energy not absorbed in the uranium calorimeter. For identification and momentum measurement of muons the yoke is magnetised to 1.6 T with copper coils. Large area LST chambers measure the position and direction of muons in front and behind the iron yoke (BMUON, RMUON). In the forward direction a spectrometer of two iron toroids and drift chambers and LST chambers (FMUON) identifies muons and measures their momenta up to 100–150 GeV/c.

For luminosity measurement the same reaction and a setup similar to that of H1 is used (Figure 12a).

Very forward scattered protons (transverse momenta less than 1 GeV/c) are measured in the leading proton spectrometer (LPS) which uses the proton ring magnets for momentum analysis and detects the scattered protons in 6 stations with silicon strip detectors mounted very close to the beam at distances between 26 and 96 m from the interaction point (Figure 12b). First data were taken in 1993 with the stations S4–S6.

In 1993 a prototype calorimeter (FNC) was installed behind S6 for detecting very forward produced neutrons.

Particles produced by the proton beam upstream of the detector are detected in the VETOWALL. For monitoring the time structure and other properties of the two beams a ring counter C5 has proven to be invaluable It is made of two lead-scintillator layers and mounted on the beam pipe behind RCAL. C5 registers the halo particles accompanying both beams.

3.4 Trigger selection

The selection of interesting events during data acquisition proceeds in four (three) trigger steps for H1 (ZEUS). The selection at trigger level one is made after $2.4\mu s$ (H1) and $4.4\mu s$ (ZEUS). Up to this point information from 200,000 to 300,000 electronic channels are stored dead time free in analogue or digital pipelines for 25 (H1) and 46 (ZEUS) consecutive beam crossings, respectively. Global information from various components like the calorimeter energy sums obtained by summing over specific regions of the calorimeter are processed in trigger pipelines. In case an interesting event is detected, the signal pipelines are stopped and the data for the bunch crossing(s) in question are digitised. The digitised data are used on the next trigger level(s) for a more restrictive event selection.

The final event selection is done in computer farms. At this point the complete digitised information for the event is available and a first reconstruction of the event is performed. The filter farms consist of a large number of fast processors with computing power of about 1000 MIPS (million instructions/s), each processor processing one event at a time The rate of accepted events varies in both experiments between about 3 and 7 Hz; typical event sizes are 60 kByte (H1) to 140 kByte (ZEUS). The accepted events are reconstructed off-line in processor farms with sufficient computing power to have the reconstructed events available for analysis within a few hours of data taking.

4 Running conditions

The typical bunch configuration of HERA is sketched in Figure 13: consecutive proton bunches and electron bunches are filled which collide with each other. In addition there are unpaired proton bunches which have no electron partner, and vice versa. The unpaired bunches are used for the study of beam induced background and for the determination of the luminosity.

The luminosity was measured by detecting the process $ep \rightarrow e'p\gamma$, as mentioned previously. Figure 14 shows the scatter plot of the energies E'_e, E_γ for the final state electron and photon as determined by the luminosity detector. There is a well isolated band

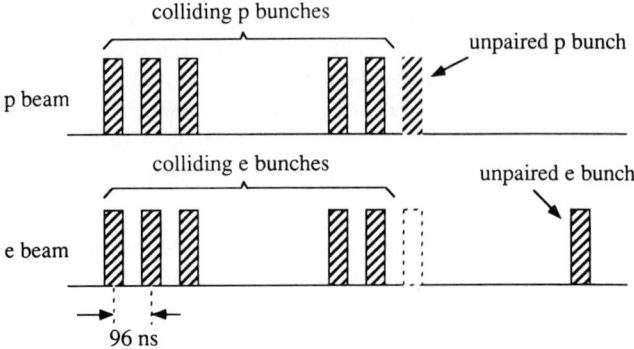

Figure 13. *Sketch of the HERA bunch configuration.*

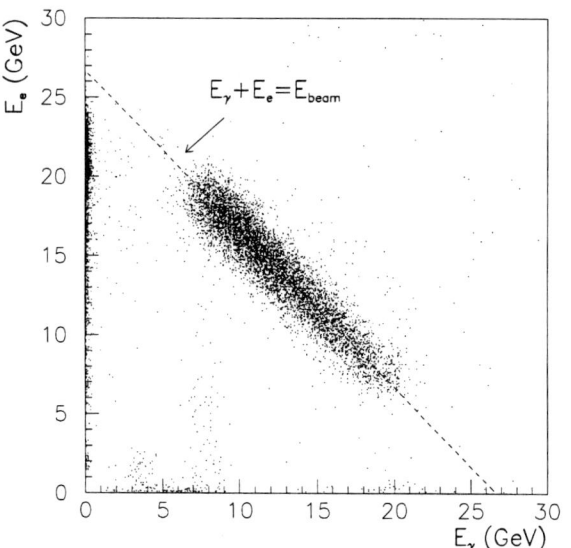

Figure 14. *Scatter plot of the electron energy versus the photon energy measured in the ZEUS luminosity detector.*

of events for which the sum of the two energies is equal to the energy of the electron beam, $E'_e + E_\gamma = E_e$, as expected for the luminosity reaction. However, bremsstrahlung of the electron beam on the residual gas in the beam pipe, $eA \to e'A'\gamma$, satisfies the same condition. The subtraction of this background was done by measuring the gas bremsstrahlung with the unpaired electron bunch and scaling with the currents of the unpaired electron bunch and of the total electron beam (see Figure 15). The achieved precision of the luminosity measurement is at present about 5%.

The high background in combination with the 96 ns bunch spacing makes triggering at HERA a challenging task. The major source of background, proton interactions

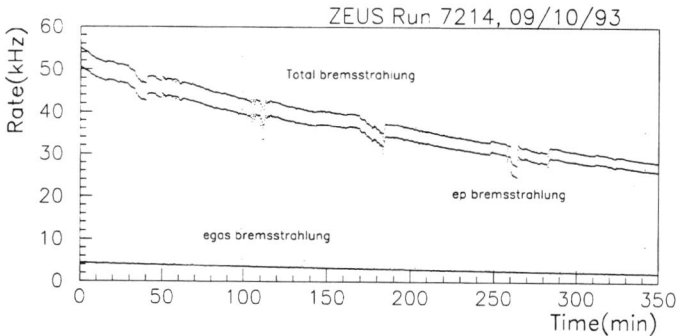

Figure 15. *The total bremsstrahlung rate (upper curve) and the luminosity after subtraction of beam gas bremsstrahlung (lower curve) as a function of time (from ZEUS).*

with the residual gas and the beam pipe, called beam gas (see Figure 5), must be eliminated without losing efficiency for deep inelastic e-p scattering and possible new physics processes of very low cross section. Not all of the large photoproduction cross section can be recorded, so the trigger acceptance is reduced for the standard processes and is maximised for specific sub-processes such as hard scattering and heavy quark production.

The strategies for suppressing unwanted background and selecting electron-proton collisions were different for the two experiments.

4.1 ZEUS data taking

In the ZEUS experiment, the trigger selection was made on the basis of the total transverse energy E_T and missing E_T measured by the calorimeter, veto signals from the veto wall and the C5 counter, electron tagging by the luminosity detector and combinations of tracking and calorimeter information at the first and second levels, making tight timing cuts at the second and third levels and running many physics filters at the third level.

The calorimeter time information has turned out to be a powerful handle for rejecting proton beam background. This is illustrated in Figure 16. In an e-p collision, particles are emitted from the interaction point, and arrive at the calorimeter cells at times $t = 0$, while a proton interaction upstream of the detector, such as shown in Figure 5, deposits energy in the RCAL about 10 ns earlier. The 10 ns difference corresponds to twice the distance between RCAL and the interaction point. Of course, in FCAL the proton induced background also arrives at $t = 0$. The measured distribution of FCAL (t_F) versus RCAL (t_R) times is shown in Figure 16 for events with more than 1 GeV deposited in a calorimeter cell in both FCAL and RCAL. The e-p events with $t_F \approx t_R \approx 0$ are well separated from the background which clusters around $t_R = -10$ ns, $t_F - t_R = 10$ ns. Note that there are about 1000 times more background than e-p events.

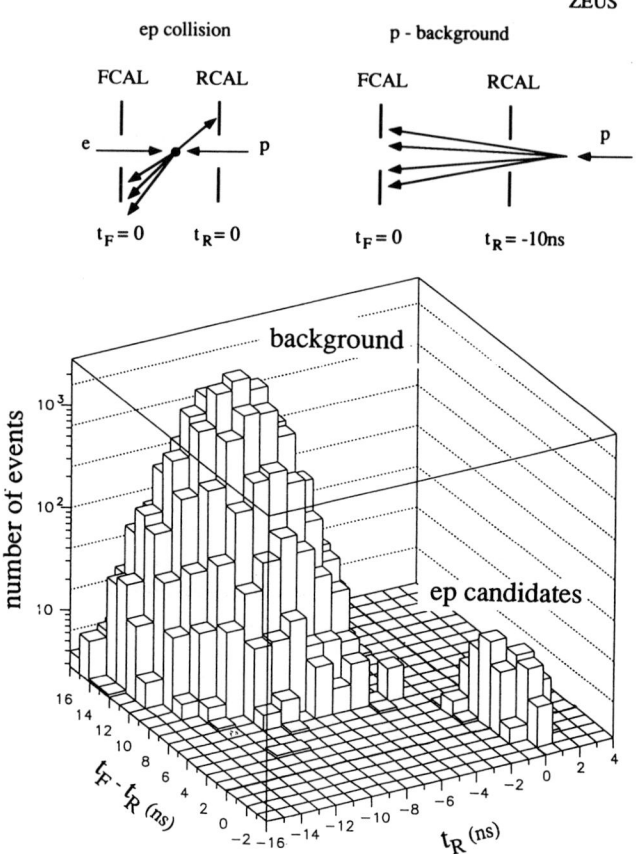

Figure 16. *Distribution of the signal time measured by* ZEUS *in the* RCAL (t_R) *versus the difference* $t_F - t_R$ *between signal times seen in* FCAL *and* RCAL.

Samples of event pictures are shown in Figure 17. The first event stems from neutral current scattering at $Q^2 \approx 5300\,\text{GeV}^2$, $x \approx 0.11$, with an electron seen in BCAL and a high-energy jet in FCAL (Figure 17a). The jet near the proton beam is presumably produced by the proton remnants. The high energy jet and the electron are back-to-back in the transverse plane and balance transverse momentum as expected for an NC event. The interaction point is marked by tracks detected in the cylindrical drift chamber (CTD). The second event (Figure 17b) shows a low-Q^2, low-x event ($Q^2 \approx 48\,\text{GeV}^2$, $x \approx 0.003$). The electron is produced very close to the beam and is only seen in RCAL. The third event (Figure 17c) is due to NC scattering ($Q^2 \approx 53\,\text{GeV}^2$) an shows a large rapidity gap between the proton direction and the first particle observed.

The event pictures were produced by including every calorimeter cell with an energy more than 60 (100) MeV in the electromagnetic (hadronic) section. The calorimeter is seen to be very clean.

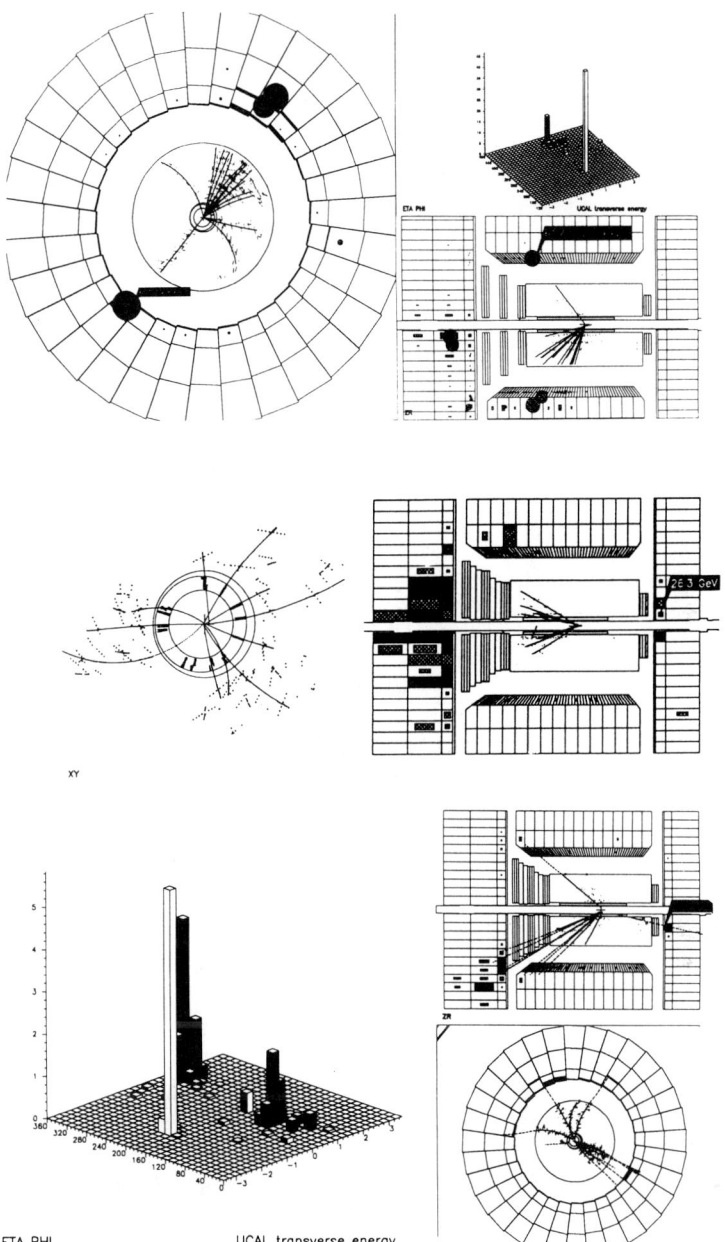

Figure 17. *Event pictures of deep inelastic scattering observed by ZEUS. Top: at $Q^2 \approx 5300\,\text{GeV}^2$, $x \approx 0.11$; the scattered electron is opposite a jet of particles. Middle: at $Q^2 \approx 48\,\text{GeV}^2$, $x \approx 0.003$; on the left a blow-up of the vertex and CTD region. The stereo hits (layers 2, 4, 6, 8) are plotted assuming $\theta = 90°$. Bottom: at $Q^2 \approx 53\,\text{GeV}^2$, $x \approx 0.004$ for an event with a large rapidity gap and two jets.*

4.2 H1 data taking

Events were accepted by H1 at the trigger stage using the information on E_T and missing E_T measured in the calorimeter, on electron tagging by the luminosity detector and on charged particles recorded by the tracking detectors. Background was rejected primarily with the time measurement provided by the time-of-flight counters and the z-coordinate of the event vertex determined by the proportional wire chambers.

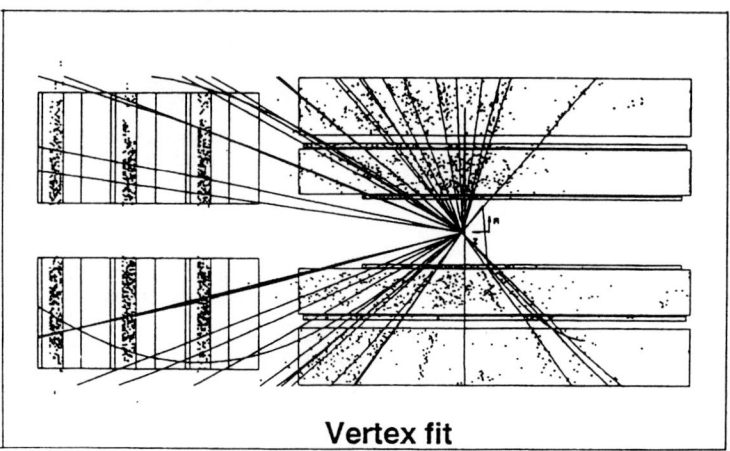

Figure 18. *Example of a proton-gas interaction in the H1 detector with 21 final state protons identified by dE/dx.*

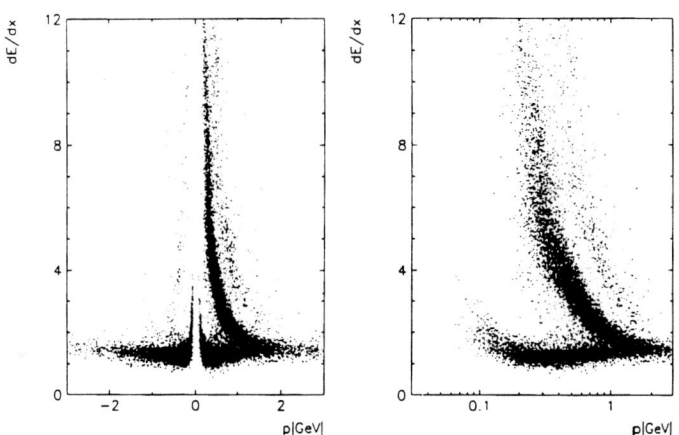

Figure 19. *Specific ionisation loss dE/dx as a function of momentum p observed by the central detector of H1 for negative $(p < 0)$ and positive $(p > 0)$ particles.*

A spectacular event produced by a proton interaction on the rest gas in the beam pipe is shown in Figure 18. It has protons identified via dE/dx in the central jet chamber. The dE/dx distribution for background events shows well isolated bands of π, K, p and d (Figure 19).

The overall response of the detector is illustrated in Figure 20 with events from NC scattering and photoproduction. The high longitudinal and transverse segmentation of the liquid argon calorimeter gives a detailed account of the energy deposition for single particles and jets.

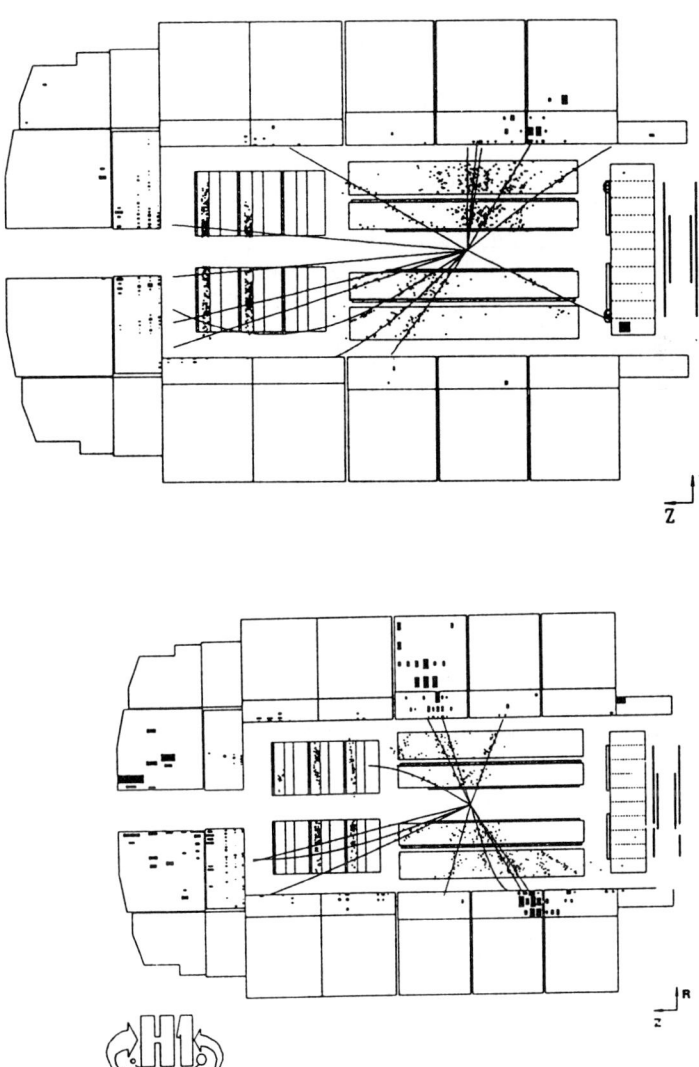

Figure 20. *Event pictures observed by* H1.

5 Introduction to deep inelastic lepton nucleon scattering

In deep inelastic scattering the incoming electron couples to the electroweak current J which probes the structure of the proton. The neutral (NC) and charged (CC) components of the current can be distinguished by the final state electron or neutrino. The basic deep inelastic scattering process (DIS) is illustrated in Figure 21. For Q^2 much larger than the mass squared of the proton the proton can be thought of as a group of quasi-free constituents—quarks, gluons, ...—one of which interacts with the current while the rest of the group (*i.e.* the proton remnant) moves on unperturbed.

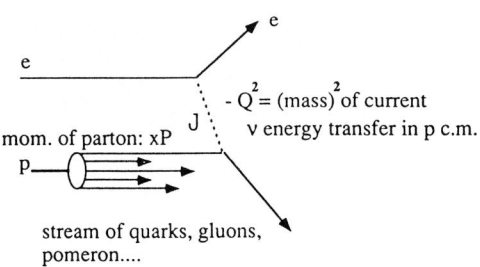

Figure 21. *Diagram for e-p scattering.*

5.1 Kinematics

The large imbalance between electron and proton beam momenta makes the kinematics at HERA quite different from that of other colliders where both beams have equal momenta or from fixed target experiments where the nucleon is stationary. Since a good grasp of the kinematical situation is useful for an understanding of the underlying processes, it will be discussed in some detail for DIS.

The relevant kinematic variables are:-

E_e, E_p — electron and proton beam energies
$e = (0, 0, -E_e, E_e) \quad p = (0, 0, E_p, E_p)$ — four-momenta of incoming e, p
$e' = (E'_e \sin\theta'_e, 0, E'_e \cos\theta'_e, E'_e)$ — four-momenta of scattered e'
$s = (e + p)^2 = 4E_e E_p$ — square of total c.m. energy
$q^2 = (e - e')^2$
$\quad = -2E_e E'_e (1 + \cos\theta'_e) = -Q^2$ — square of four momentum transfer
$Q^2_{max} = s$ — maximum possible Q^2 value
$\nu = q \cdot p / m_p$ — energy of current J in rest system of incoming proton
$\nu_{max} = s/(2m_p)$ — maximum energy transfer
$y = (q \cdot p)/(e \cdot p) = \nu/\nu_{max}$ — fraction of energy transfer
$x = Q^2/(2q \cdot p) = Q^2/(2m_p \nu) = Q^2/(ys)$ — Bjorken scaling variable

$$W^2 = (p+q)^2 = m_p^2 - Q^2 + 2m_p\nu \quad \text{mass-squared of the total hadronic}$$
$$= m_p^2 + Q^2(1/x - 1) \quad \text{system}$$
$$\Delta = \hbar/Q \quad \text{smallest object size that can be resolved in the proton}$$

where E'_e and θ'_e are the energy and angle (w.r.t. the incoming proton) of the scattered lepton and where the electron and proton masses, m_e, m_p, have been neglected.

For fixed c.m. energy, inclusive scattering, $ep \to eX$, is described by two variables which for example x and Q^2 can be used. The same variables describe the lowest order process where the electron scatters elastically on a free constituent of the proton (Figure 21). For NC events these variables can be determined either from the energy and angle of the scattered electron, or of the final state hadron system, or from a mixture of both. For CC events, only the hadron system is accessible for measurement. The electron side yields:

$$\begin{aligned} y &= 1 - (E'_e/2E_e)(1 - \cos\theta'_e) \\ Q^2 &= 2E_e E'_e(1 + \cos\theta'_e) \\ x &= E'_e(1 + \cos\theta'_e)/(2yE_p) \end{aligned} \qquad (5.1)$$

and the inverse relations

$$\begin{aligned} E'_e &= (1-y)E_e + xyE_p \\ \cos\theta'_e &= [xyE_p - (1-y)E_e]/[xyE_p + (1-y)E_e] \\ E'^2_e \sin^2\theta'_e &= 4xy(1-y)E_e E_p. \end{aligned} \qquad (5.2)$$

From the hadron system j (excluding the proton remnant) with energy E_j and production angle θ_j one finds:–

$$\begin{aligned} y &= (E_j/2E_e)(1 - \cos\theta_j) \\ Q^2 &= E_j^2 \sin^2\theta_j/(1-y) \\ x &= E_j(1 + \cos\theta_j)/[(1-y)(2E_p)] \end{aligned} \qquad (5.3)$$

and

$$\begin{aligned} E_j &= yE_e + x(1-y)E_p \\ \cos\theta_j &= [-yE_e + (1-y)xE_p]/[yE_e + (1-y)xE_p] \\ E_j^2 \sin^2\theta_j &= 4xy(1-y)E_e E_p = Q^2(1-y). \end{aligned} \qquad (5.4)$$

Using the method of Jacquet-Blondel (1979) the hadron variables can be determined approximately by summing the energies (E_h) and transverse (p_{Th}) and longitudinal momenta (p_{zh}) of all final states. The method rests on the assumption that the total transverse momentum carried by those hadrons which escape detection through the beam hole in the proton direction as well as the energy carried by particles escaping through the beam hole in the electron direction can be neglected. The result is

$$\begin{aligned} y_{JB} &= \sum_h (E_h - p_{zh})/(2E_e) \\ Q^2_{JB} &= \left[(\sum_h p_{xh})^2 + (\sum_h p_{yh})^2 \right]/[1 - y_{JB}] \\ x_{JB} &= Q^2_{JB}/(y_{JB}s). \end{aligned} \qquad (5.5)$$

The mixed or double-angle method (Bentvelsen et al. 1991) uses the electron scattering angle and the angle γ_h which characterises the longitudinal and transverse momentum flow of the hadronic system (in the naïve parton model γ_h is the scattering angle of the struck quark):–

$$\cos\gamma_h = \frac{(\sum_h p_{xh})^2 + (\sum_h p_{yh})^2 - (\sum_h (E_h - p_{zh}))^2}{(\sum_h p_{xh})^2 + (\sum_h p_{yh})^2 + (\sum_h (E_h - p_{zh}))^2}$$

$$Q_{DA}^2 = 4E_e^2 \sin\gamma_h (1 + \cos\theta_e')/[\sin\gamma_h + \sin\theta_e' - \sin(\gamma_h + \theta_e')]$$

$$x_{DA} = (E_e/E_p)[\sin\gamma_h + \sin\theta_e' + \sin(\gamma_h + \theta_e')]/[\sin\gamma_h + \sin\theta_e' - \sin(\gamma_h + \theta_e')]$$

$$y_{DA} = Q_{DA}^2/(x_{DA} s). \tag{5.6}$$

As the double-angle method relies on ratios of energies it is less sensitive to scale uncertainty in the energy measurement of the final state particles.

For NC events the precision of x and Q^2 can be improved by a simultaneous fit to all measured variables (Chaves et al. 1991).

The Bjorken variable x, defined above from the electron side, for the lowest order process (Figure 22a) is equal to the fraction η of the proton momentum carried by the struck quark. This can be seen as follows:–

ηp incident momentum of struck quark

$j = \eta p + q$ outgoing momentum of struck quark (system)   (5.7)

$j^2 = \eta^2 p^2 + 2\eta p \cdot q - Q^2$ mass squared of the outgoing struck quark (system)

If the mass of the outgoing quark is zero, $M_j^2 \equiv j^2 = 0$ and for $Q^2 \gg \eta^2 p^2 = \eta^2 m_p^2$:

$$Q^2 \approx 2\eta p \cdot q \tag{5.8}$$

and therefore $\eta = x$.

If the mass of the outgoing struck quark is not equal to zero, for instance due to gluon radiation (e.g. Figure 22b), then

$$\eta = x + M_j^2/2(q \cdot p) = x + M_j^2/(2m_p \nu). \tag{5.9}$$

In this case $\eta > x$. In order to produce a *massive* hadronic system for the same energy transfer ν from the electron side, the struck quark must carry a larger fraction of the proton momentum. As a consequence, in the massive case the hadronic energy flow moves closer to the direction of the incident proton than given by the angle γ_h calculated for the massless case.

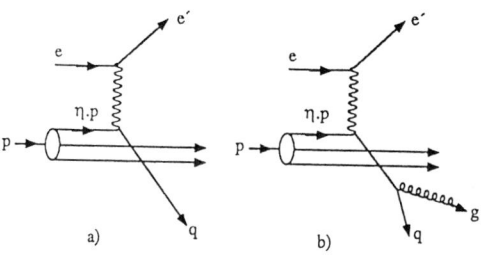

Figure 22. *Diagrams for (a) DIS in lowest order and (b) with gluon bremsstrahlung.*

5.2 Kinematic regime of HERA

Table 2 compares the kinematic ranges accessible at HERA and in previous lepton-nucleon scattering experiments.

	HERA	pre-HERA
s (GeV^2)	10^5	10^3
maximum practical Q^2 (GeV^2)	40,000	400
Δ (cm^{-1})	1×10^{-16}	1×10^{-15}
ν_{max} (GeV)	52,000	500
minimum x at $Q^2 = 10\,\text{GeV}^2$	1×10^{-4}	1×10^{-2}

Table 2. *Kinematic regions accessible at HERA ($E_e = 30$ GeV, $E_p = 820$ GeV) and in previous experiments*

The maximum energy transfer is increased by a factor of approximately 100: HERA is equivalent to a fixed target experiment with an incident electron beam of 52 TeV. The Q^2 domain over which lepton nucleon scattering can be measured is also increased by two orders of magnitude. Since the typical Q^2 values in DIS are much larger than the proton mass the electron interacts with one of the partons (quarks, gluons, ...) rather than with the proton as a whole: HERA is in reality an electron-parton collider.

The correlations between energy and angle of electron and current jet (ignoring gluon emission) are shown in Figure 23 in the x-Q^2 plane. For the purpose of orientation Figure 24 shows in the x-Q^2 plane lines of constant y and W and the kinematic region covered by non-HERA experiments.

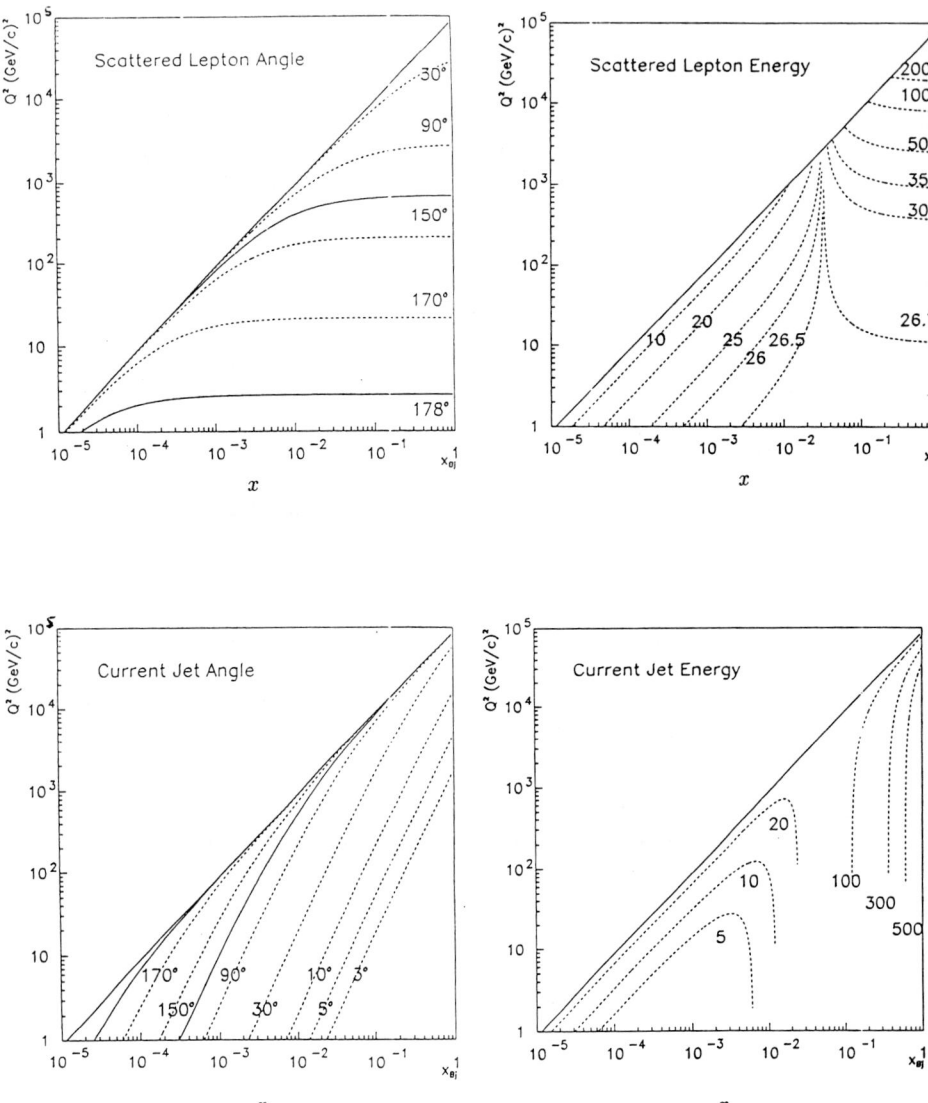

Figure 23. The x-Q^2 dependence of the angle and energy of the scattered electron and the current jet (ignoring gluon radiation) for beam energies of $E_e \times E_p = 26.7\,\text{GeV} \times 820\,\text{GeV}$. The angles are measured w.r.t. the proton direction.

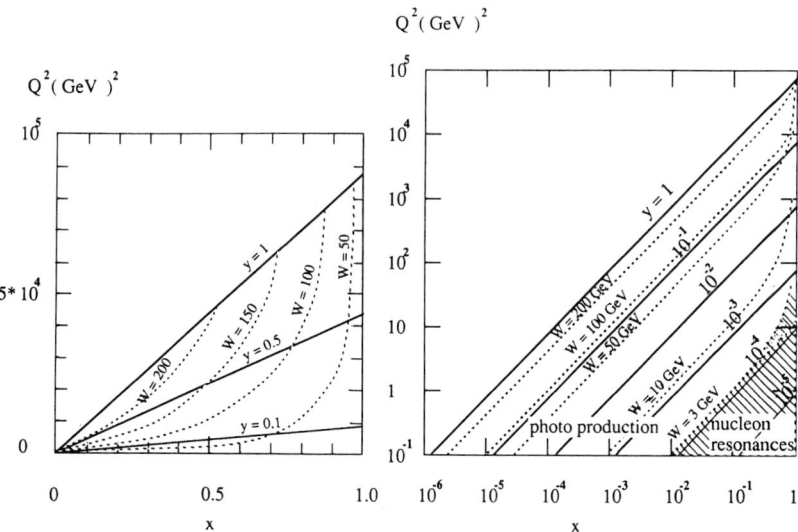

Figure 24. *Lines of constant y in the x-Q^2 plane and the different kinematical regimes as a function of x.*

5.3 Cross sections for DIS

The cross sections for NC and CC scattering are related to the structure functions F_i of the proton (see *e.g.* Ingelman and Rückl 1988, PDG 1992):

Neutral Current, $ep \to eX$:

$$\frac{d^2\sigma(\gamma+Z_0)}{dx\,dy} = \frac{4\pi\alpha^2}{sx^2y^2}\left[\left(1-y+\frac{y^2}{2}\right)F_2(x,Q^2) - \frac{y^2}{2}F_L(x,Q^2) \pm \left(y-\frac{y^2}{2}\right)xF_3(x,Q^2)\right]. \tag{5.10}$$

The upper (lower) sign applies to $e^-(e^+)p$ scattering. The longitudinal structure function is related to F_1, F_2 via

$$F_L(x,Q^2) = F_2(x,Q^2) - 2xF_1(x,Q^2) = 2xF_1(x,Q^2)R(x,Q^2)$$

with

$$R(x,Q^2) = \frac{F_L(x,Q^2)}{2xF_1(x,Q^2)} = \frac{F_2(x,Q^2)}{2xF_1(x,Q^2)} - 1. \tag{5.11}$$

For $Q^2 \gg 1\,\text{GeV}^2$ and x not too small the contribution of the longitudinal structure function F_L is small. The F_3 term measures parity violating contributions which arise from Z^0 exchange. It is significant only when Q is comparable to or larger than the Z^0 mass.

For some applications it is useful to express the photon part of the DIS cross section in terms of the total cross sections σ_T, σ_L for the scattering of transverse and longitudinal photons, respectively, on the proton. Neglecting contributions from Z^0 exchange:

$$2xF_1(x,Q^2) = \frac{Q^2}{(4\pi^2\alpha)}[1-x]\sigma_T(x,Q^2)$$
$$F_2(x,Q^2) = \frac{Q^2}{(4\pi^2\alpha)}[1-x]\sigma_T(x,Q^2) + \sigma_L(x,Q^2) \qquad (5.12)$$

Here, the assumption $4m_p^2 x^2 \ll Q^2$ has been made. A convenient expression for $x \ll 1$ is $\sigma_T(x,Q^2) \approx (112\mu b/Q^2)[2xF_1(x,Q^2)]$, where Q^2 is in GeV^2.

Charged Current, $ep \to nX$:

$$\frac{d^2\sigma(W)}{dx\,dy} = \frac{G_F^2 s}{2\pi}\frac{1}{(1+Q^2/M_W^2)^2}\left[(1-y)F_2(x,Q^2) + y^2 xF_1(x,Q^2) \pm \left(y - \frac{y^2}{2}\right)xF_3(x,Q^2)\right], \qquad (5.13)$$

where G_F is the Fermi coupling constant, $G_F = 1.02\times 10^{-5}/m_p^2$. As before, the upper (lower) sign applies to $e^-(e^+)p$ scattering. For scattering on pointlike spin-1/2 constituents by vector exchange, $F_2(x,Q^2) = 2xF_1(x,Q^2)$ (Callan-Gross 1969).

The expressions for the NC and CC cross sections have very similar structures. This is particularly evident when G_F is expressed in terms of the electroweak coupling,

$$G_F = \pi\alpha/(\sqrt{2}\sin^2\theta_W M_W^2), \qquad (5.14)$$

where θ_W is the electroweak angle, $\sin^2\theta_W = 0.23$. The result is

$$\frac{d^2\sigma(W)}{dx\,dy} = \frac{\pi\alpha^2}{4\sin^4\theta_W sx^2 y^2}\frac{Q^4}{(Q^2+M_W^2)^2}$$
$$\times \left[(1-y)F_2(x,Q^2) + y^2 xF_1(x,Q^2) \pm \left(\frac{y^2}{2} - y\right)xF_3(x,Q^2)\right]. (5.15)$$

The main difference between the NC cross section from photon exchange and the CC cross section from W exchange is due to the different propagator terms,

$$\frac{d^2\sigma(W)}{d^2\sigma(\gamma)} \approx \left[\frac{Q^2}{(Q^2+M_W^2)}\right]^2. \qquad (5.16)$$

The r.h.s. gives 2.4×10^{-8}, 2.4×10^{-6}, 0.02 and 0.4 at $Q^2 = 1, 10, 1000$ and 10^4 GeV^2. It is the heavy W mass which is responsible for the 'weak' CC cross section at small Q^2 values.

5.4 DIS cross sections in the quark-parton model

The structure functions F_i from NC and CC scattering are in principle independent. They can be related via the quark-parton model (QPM) however. The underlying process of the lowest order diagram depicted by Figure 22a is elastic electron-quark scattering, $eq \to eq$. Photon exchange leads to the following cross section for this process:

$$\frac{d\sigma(eq\to eq)}{dQ^2} = \frac{4\pi\alpha^2 e_q^2}{Q^4}\left[1 - \frac{Q^2}{s_{eq}} + \frac{Q^4}{2s_{eq}^2}\right] \qquad (5.17)$$

where s_{eq} is the square of the (eq) c.m. energy, e_q is the electric charge of the quark and the assumption $s \gg m_e^2, m_q^2$ is made. Defining $q(x)\,dx$ as the probability of finding in the proton a quark of type q carrying a momentum fraction between x and $x+dx$, the resulting cross section for electron proton scattering via quark q is

$$\frac{d\sigma^q(ep \to eX)}{dx\,dQ^2} = \frac{4\pi\alpha^2}{Q^4}\left[1 - \frac{Q^2}{s_{eq}} + \frac{Q^4}{2s_{eq}^2}\right]e_q^2 q(x). \tag{5.18}$$

By noting that $s_{eq} = xs$ and $Q^2 = xys$ this can be rewritten as follows:

$$\frac{d\sigma^q(ep \to eX)}{dx\,dy} = \frac{4\pi\alpha^2}{x^2 y^2 s}\left[1 - y + \frac{y^2}{2}\right]e_q^2 x q(x). \tag{5.19}$$

The total ep cross section is obtained by summing over all quark contributions:

$$\frac{d\sigma(ep \to eX)}{dx\,dy} = \frac{4\pi\alpha^2}{x^2 y^2 s}\left[1 - y + \frac{y^2}{2}\right]\sum_q e_q^2 x q(x). \tag{5.20}$$

This is the QPM prediction. The comparison with Equation (5.10) leads to the QPM expression for the structure function F_2 for the case of photon exchange:

$$F_2^{ep \to eX}(x) = \sum_q e_q^2 x q(x) = x[4/9 u(x) + 4/9 \bar{u}(x) + 1/9 d(x) + 1/9 \bar{d}(x) + \cdots]. \tag{5.21}$$

Since only pointlike spin-1/2 constituents contribute, $2xF_1(x) = F_2(x)$ and so $F_L(x, Q^2) = 0$. Inclusion of Z^0 exchange leads to the following expressions:

$$F_2^{ep \to eX}(x) = \sum_q A_q(Q^2) x [q(x) + \bar{q}(x)]$$

$$xF_3^{ep \to eX}(x) = \sum_q B_q(Q^2) x [q(x) - \bar{q}(x)] \tag{5.22}$$

where

$$A_q(Q^2) = e_q^2 - 2e_q v_e v_q \left[\frac{Q^2}{(Q^2 + M_Z^2)}\right] + (v_e^2 + a_e^2)(v_q^2 + a_q^2)\left[\frac{Q^2}{(Q^2 + M_Z^2)}\right]^2$$

$$B_q(Q^2) = -2e_q a_e a_q \left[\frac{Q^2}{(Q^2 + M_Z^2)}\right] + 4 v_e v_q a_e a_q \left[\frac{Q^2}{(Q^2 + M_Z^2)}\right]^2.$$

The corresponding QPM expressions for the CC case are (see e.g. Ingelman and Rückl):

$$F_2^{ep \to \nu X}(x) = x\sum_{i,j}\left\{|V_{u_i d_j}|^2 u_i(x) + |V_{u_j d_i}|^2 \bar{d}_i(x)\right\}$$

$$xF_3^{ep \to \nu X}(x) = x\sum_{i,j}\left\{|V_{u_i d_j}|^2 u_i(x) - |V_{u_j d_i}|^2 \bar{d}_i(x)\right\}, \tag{5.23}$$

where $V_{u_i d_j}$ are the elements of the CKM matrix, u_i and d_j denote up-type and down-type flavours, respectively, and i,j are family indices. At energies well above flavour thresholds Equations (5.23) simplify to

$$F_2^{ep \to \nu X}(x) = x[u(x) + c(x) + \bar{d}(x) + \bar{s}(x) + \cdots]$$

$$xF_3^{ep \to \nu X}(x) = x[u(x) + c(x) - \bar{d}(x) - \bar{s}(x) + \cdots]. \tag{5.24}$$

In QCD, the parton densities become Q^2 dependent, $u(x) \to u(x, Q^2)$, due to gluon radiation or photon-gluon interactions via an intermediate quark. Structure function measurements at a given value of Q^2 can be extrapolated to other Q^2 values by QCD evolution with the help of the GLDAP formalism (Gribov and Lipatov 1972, Lipatov 1974, Dokshitzer 1977, Altarelli and Parisi 1977). In the derivation of the evolution equations, $\ln(1/x)$ terms are neglected over $\ln Q^2$ terms. While this is justified for the range $x > 10^{-2}$ covered by fixed target experiments it may not be so for x values as small as 10^{-4} which are accessible at HERA. This approximation is avoided in the BFKL equations (Kuraev, Lipatov and Fadin 1977, Balitskii and Lipatov 1978).

5.5 Guessing the parton distributions of the proton

It is instructive to estimate the general behaviour of the parton distributions with a toy model. We shall do this for the proton.

Step A: The proton consists of valence quarks only, $p = uud$. Flavour conservation requires:

$$\int_0^1 u(x)\, dx = 2 \qquad \int_0^1 d(x) dx = 1. \tag{5.25}$$

The u and d quarks carry the momentum of the proton which implies

$$x_{u_1} + x_{u_2} + x_d = 1 \tag{5.26}$$

$$\int_0^1 xu(x)\, dx + \int_0^1 xd(x)\, dx = 1 \quad \text{(momentum sum rule)}. \tag{5.27}$$

The assumption of a uniform distribution in the plane (5.26) combined with the normalisation conditions (5.25) leads to

$$u(x) = 4(1-x) \qquad d(x) = 2(1-x) \tag{5.28}$$

with average momentum fractions $\langle x_u \rangle = 2/3$, $\langle x_d \rangle = 1/3$.

Step B: DIS data show that quarks carry only about half of the proton momentum, the other half is carried by gluons:

$$\int_0^1 x[u(x) + d(x)]dx \simeq 0.5 \qquad \int_0^1 xg(x)dx \simeq 0.5, \tag{5.29}$$

where $g(x)$ gives the number of gluons carrying a fraction of the proton momentum between x and $x + dx$. Assuming $\langle x_u \rangle = 1/3$, $\langle x_d \rangle = 1/6$ and making the ansatz $u(x) = a(1-x)^n$ leads to

$$u(x) = 10(1-x)^4 \qquad d(x) = 5(1-x)^4. \tag{5.30}$$

Since gluons carry no electric charge, flavour etc. their number in the proton is not constrained,

$$\int_0^1 g(x)\, dx \geq 0.$$

As a first guess for the gluon x spectrum one may assume one part to be similar to that for u and d quarks, and a second part to arise from gluon bremsstrahlung by quarks, which will depend on Q^2 and on α_s, the strong coupling:

$$g(x) = a(1-x)^4 + bx^{-1}(1-x)^4.$$

The gluon momentum sum rule (5.29) requires $(a/6 + b) = 2.5$. A possible choice is $a = 3$, $b = 2$ (actually, the resulting $g(x)$ does not vary much in $0.1 < x < 0$ for a wide set of a, b values) leading to:

$$g(x) = 3(1-x)^4 + 2x^{-1}(1-x)^4. \tag{5.31}$$

Step C: Gluons can split into a quark-antiquark pair, $g \to q\bar{q}$, called sea quarks (q_s) as distinguished from the valence quarks (q_v) considered previously. The splitting depends on α_s and Q^2. Inclusion of the sea quarks leads to the following constraints:

$$u_s(x) = \bar{u}_s(x) = \bar{u}(x) \qquad d_s(x) = \bar{d}_s(x) = \bar{d}(x)$$
$$u(x) = u_v(x) + u_s(x) \qquad d(x) = d_v(x) + d_s(x)$$
$$\int_0^1 [u(x) - \bar{u}(x)]\, dx = \int_0^1 u_v(x)\, dx = 2$$
$$\int_0^1 [d(x) - \bar{d}(x)]\, dx = \int_0^1 d_v(x)\, dx = 1$$
$$\int_0^1 [s(x) - \bar{s}(x)]\, dx = 0$$
$$\int_0^1 x[u(x) + \bar{u}(x) + d(x) + \bar{d}(x) + s(x) - \bar{s}(x)]\, dx \simeq 0.5$$
$$\int_0^1 xg(x)\, dx \simeq 0.5. \tag{5.32}$$

In view of the splitting process $g \to q_s \bar{q}_s$ it seems natural to assume that a sea quark carries half of the momentum of the parent gluon. Replacing x by $2x$ in the gluon distribution gives for the shape of the sea quark distribution

$$q_s(x) = \text{const.}\,[3(1-2x)^4 + x^{-1}(1-2x)^4] \qquad \text{for } x < 0.5.$$

In order to fix the constant we need to use data. From structure function fits to data, (Martin, Stirling and Roberts 1993) the total momentum fraction carried by sea quarks at $Q^2 = 4\,\text{GeV}^2$ is about 0.18 which determines the constant, provided all sea quark distributions are identical. The valence distributions need now to be corrected for the momentum carried by the sea quarks. With the ansatz $u_v(x) = a(1-x)^n$ the final result is

$$u_v(x) = 16(1-x)^7; \qquad d_v(x) = 8(1-x)^7$$
$$u_s(x) = d_s(x) = s_s(x) = (1-2x)^4 + (3x)^{-1}(1-2x)^4 \qquad \text{for } x < 0.5$$
$$g(x) = 3(1-x)^4 + 2x^{-1}(1-x)^4. \tag{5.33}$$

The resulting parton density for the toy model are shown in Figure 25a. What can one learn from this exercise? The average momentum fraction of the sea quarks is expected to be smaller than that of gluons and the latter to be smaller than that of valence quarks. Furthermore, the proton structure function at $x < 0.01$ will be dominated by the sea quark and gluon contributions.

For comparison, Figure 25b shows the parton distributions obtained from a recent fit to data and evaluated at $Q^2 = 20\,\text{GeV}^2$ (MRS 1993). The qualitative behaviour shown by the toy model is quite similar to that found from the data. A quantitative comparison shows the data to require a wider valence quark distribution, more momentum for the valence u than for the valence d quarks and a steeper gluon distribution.

The parton distributions change with Q^2. Bremsstrahlung of gluons and gluon splitting will become more frequent as Q^2 increases, with the result, that valence quarks will carry less while gluons and sea quarks will carry more of the proton momentum. The predicted evolution of parton distributions with Q^2 is shown in Figure 26.

A recent review of pre-HERA experimental data on the nucleon structure functions has been given by Milsztajn and Virchaux (1993).

5.6 DIS cross sections at HERA energies

The structure functions measured in previous experiments have been extrapolated by QCD evolution with LEPTO (1988) using the parameterisation EHLQ (Eichten et al. 1984, 1986) to the HERA regime. The numbers of events expected from NC and CC scattering are shown in Figure 27 at $Q^2 > 1000\,\text{GeV}^2$ for 500 pb^{-1}. The large NC rates at low Q^2 stem from photon exchange. At $Q^2 > M_Z^2$ contributions from Z-exchange become equally important. The requirement of a minimum of 100 events leads to a maximum Q^2 value of 35,000 GeV^2 up to which NC measurements are feasible.

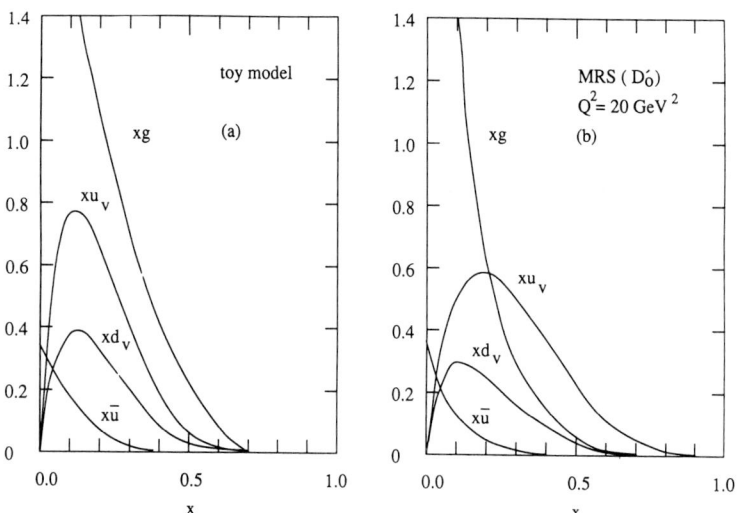

Figure 25. *Parton distributions of the proton (a) from the toy model; (b) calculated for $Q^2 = 20\,\text{GeV}^2$ from a fit to data (MRS 1993).*

The event rate for CC scattering at low Q^2 is much smaller, for the reason discussed above. However, for Q^2 around M_Z^2 the CC and NC cross sections become equal (see Figure 28); weak and electromagnetic interactions are of the same strength. The practical Q^2 limit for CC studies is around 40,000 GeV^2.

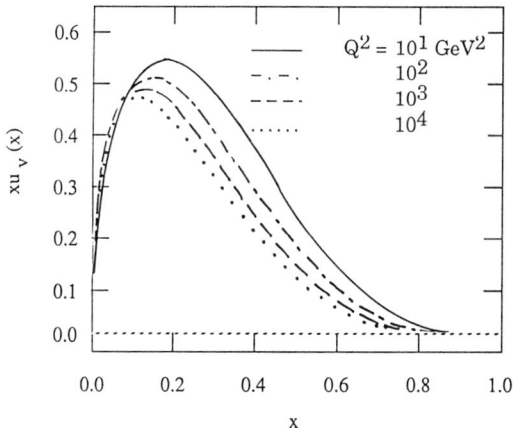

Figure 26. *The evolution of the valence u-quark distribution with Q^2 : $Q^2 = 10$ (full), 10^2 (dash-dotted), 10^3 (dashed) and 10^4 (dotted) GeV^2, from Ingelman and Rückl (1987).*

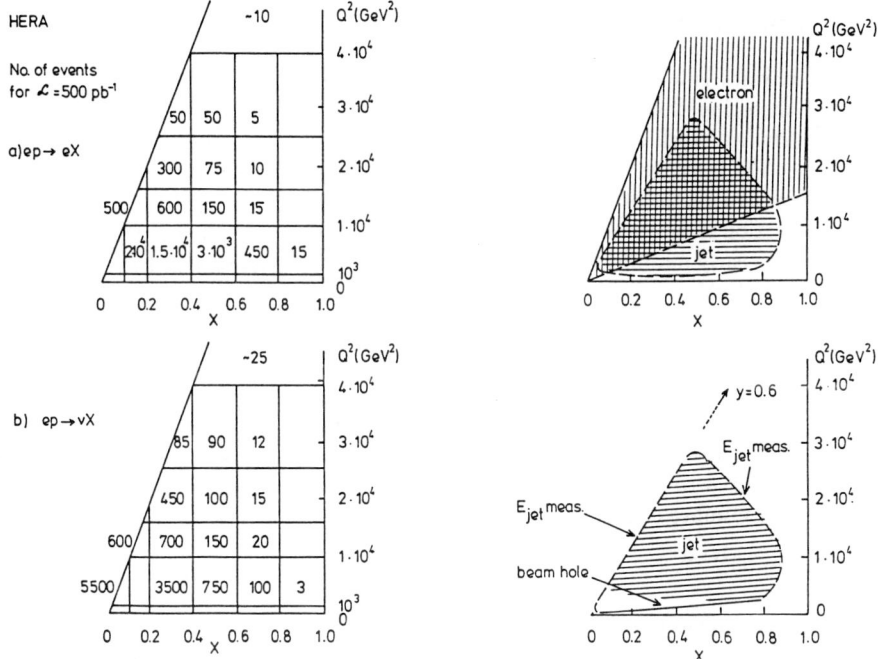

Figure 27. *Event rates for NC and CC scattering at HERA with $L = 500\,\text{pb}^{-1}$ calculated with LEPTO and EHLQ structure functions, and the regions where x and Q^2 can be well measured at HERA for NC scattering, from either the electron or the hadronic system, and for CC scattering from only the hadronic system.*

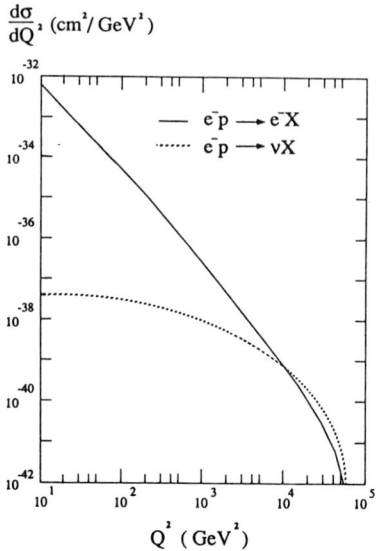

Figure 28. *The cross sections for* NC *and* CC *scattering as a function of* Q^2.

5.7 Small x physics

Small x physics is a new and exciting field of lepton-nucleon scattering. The possibility of accessing this region at HERA has stimulated an intense discussion. Since $x = Q^2/(2m_p\nu)$, small x-values are attained for fixed Q^2 by making the energy transfer ν large. For instance, for $Q^2 = 10\,\text{GeV}^2$, x-values as small as 10^{-4} can be reached at HERA which is a factor of 100 smaller than in previous experiments (see Table 2). The NC cross section is favourably large in this regime, as shown in Figure 29: for instance, the nominal yearly luminosity of 100pb^{-1} should yield 10^6 events with $10^{-4} < x < 10^{-3}$, $10 < Q^2 < 20\,\text{GeV}^2$.

The excitement about the small x region stems from the fact that in perturbative QCD $F_2(x, Q^2)$ grows faster than any power of $\ln(1/x)$ as $x \to 0$ (De Rujula et al. 1974). Intuitively, this can be understood as follows (see Gribov, Levin and Ryskin 1983). Consider scattering (Figure 30a) at small x but not too small Q^2 such that α_s is small, e.g. $Q^2 > Q_0^2 = 10\,\text{GeV}^2$. As $x \to 0$, the numbers of gluons and sea quarks in the proton are expected to grow beyond any limits due to bremsstrahlung and gluon splitting; for instance, the number of gluons with momentum fractions x, $x + dx$ should behave as

$$g(x) \sim x^{-(1+\lambda)}, \qquad \lambda \simeq \alpha_s[12\ln(2)]/\pi \simeq 1/2 \quad \text{(Lipatov 1976 and others)}.$$

Since the transverse size of the partons is fixed ($\sim 1/Q$) and since the partons are confined to the proton and their number grows as $x \to 0$, there must be an $x = x_\text{crit}$ below which partons begin to overlap (Figure 31). This must lead to saturation of the structure functions as $x \to 0$ (Figure 32). A possible mechanism of parton overlap is depicted in Figure 30b : two ladders start from two different partons and begin to interact. A recent review of theoretical developments in small-x physics was given by Kwiecinski (1993).

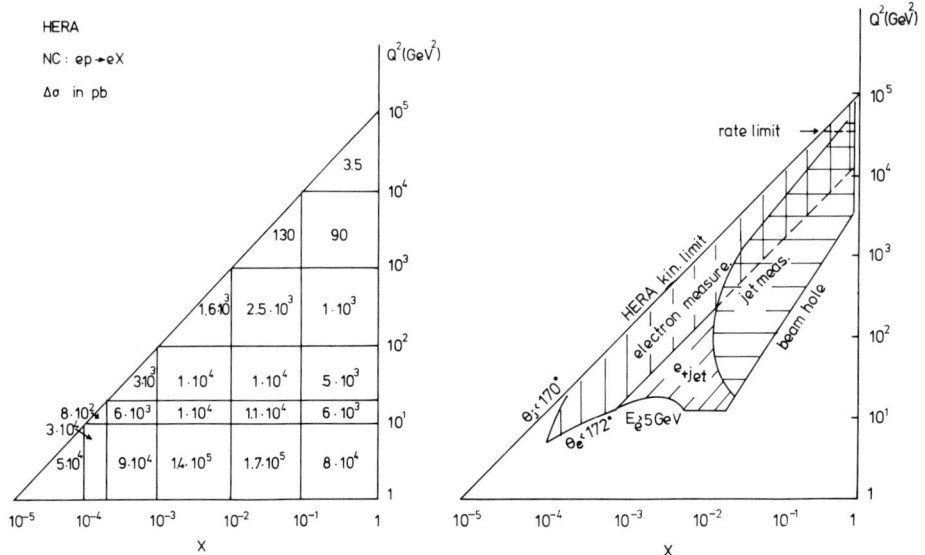

Figure 29. Left: The NC cross section at HERA for $E_e \times E_p = 30\,\text{GeV} \times 820\,\text{GeV}$ calculated with LEPTO and EHLQ structure functions (a). Right: the region where x and Q^2 can be well measured (b).

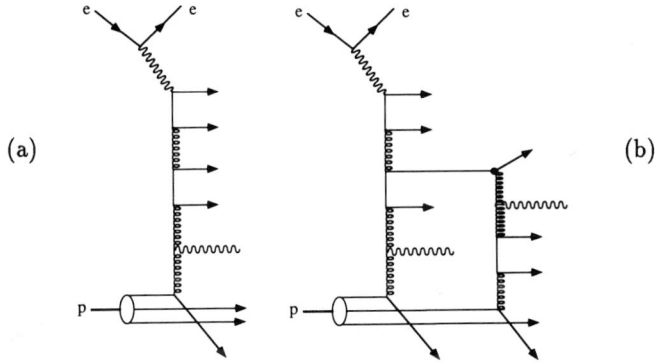

Figure 30. Diagrams for low-x scattering: (a) with a single ladder; (b) with two ladders starting from two different incoming partons and where the two ladders interact.

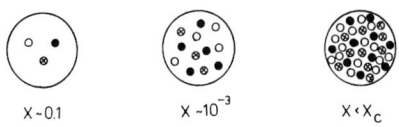

Figure 31. The parton density in the nucleon for different values of x.

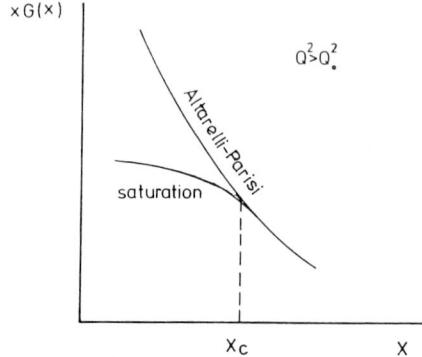

Figure 32. *Expected qualitative behaviour of the gluon structure function at very small x values according to the* **GLDAP** *and* **BFKL** *evolution equations and after inclusion of saturation effects.*

Two estimates for x_{crit} are given in Figure 33 as a function of Q^2 (Kim and Ryskin 1991). The first is characterised by a parameter $R = 5\,\text{GeV}^{-1}$ which may be thought of as the radius of the proton for a uniform parton distribution in the proton. In this case the saturation region is barely within reach at **HERA**. For $Q^2 = 4\,\text{GeV}^2$ the model predicts $x_{\text{crit}} = 10^{-4}$. However, the low-$x$ partons may concentrate, for instance, around the valence quarks and form hot spots (Mueller 1991). Assuming a hot spot radius of $2\,\text{GeV}^{-1}$ observation of saturation effects at **HERA** looks feasible (Figure 33). The amount of saturation one may expect, e.g. for the gluon structure function, is shown in Figure 34 for the two models.

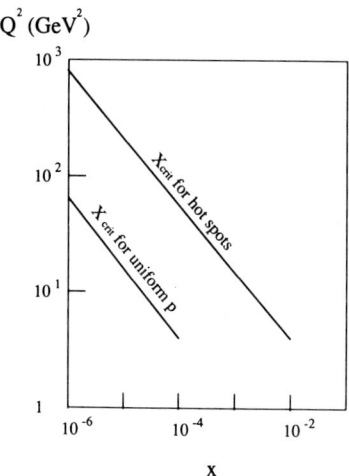

Figure 33. *Model predictions for the x-Q^2 behaviour of x_{crit} for a uniformly populated proton (lower curve) and for the hot spot model (upper curve); taken from Kim and Ryskin (1991).*

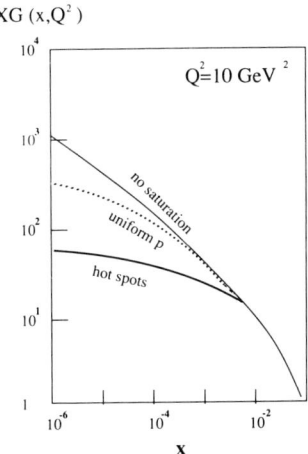

Figure 34. *Model predictions for the x behaviour of the gluon structure function at $Q^2 = 10\,\mathrm{GeV}^2$ assuming no saturation or saturation for a uniformly populated proton, and for the hot spot model (from Kim and Ryskin 1991).*

The smallest-x data for $Q^2 > 5\,\mathrm{GeV}^2$ that were available before HERA were obtained by NMC (1993). Figure 35 shows their recent measurements of F_2 in μp scattering for x-values between 0.008 and 0.5. It is noteworthy that the predictions for F_2 obtained by fitting previous data from higher x-values (Kwiecinski et al. 1990, Morfin and Tung 1992, Glück et al. 1990) fail to fit the NMC data: the NMC data indicate a faster rise of F_2 as x approaches zero. Inclusion of the new NMC data in the structure function fits has resulted in the predictions (Martin et al. 1992) D_0 and D_- shown in Figure 36. The two sets differ in the assumption of whether the gluon structure function is constant or diverges as x goes to zero: $xG(x,Q^2) \sim$ constant ($\sim x^{-0.5}$) for D_0 (D_-). While the two sets give identical results for $x > 0.01$ they make markedly different predictions for F_2 for $x < 10^{-3}$: at $x = 10^{-4}$, F_2 as calculated from D_- is a factor of three larger. Also indicated in Figure 35 is an estimate of the effects of parton saturation on D_-: they are small for a uniform proton but large in the hot spot model for $x = 10^{-4}$.

Deviations from the standard Altarelli-Parisi (GLDAP) evolution at very small values of x are also expected for a 'technical' reason. In the GLDAP evolution for each additional factor of α_s only terms of the order $(\ln Q^2)(\ln(1/x))$ are kept while $(\ln(1/x))$ terms are neglected. This approximation has been avoided in the Lipatov evolution (Lipatov 1976, Kuraev et al. 1977, Balitskii and Lipatov, 1978).

5.8 Experimentally accessible x-Q^2 region

Standard x-Q^2 region: The x-Q^2 region accessible to experiments depends on the structure of the events and on the detector. As discussed above, for NC events, the values of x and Q^2 can be determined from the energy and direction of either the scattered electron or the current jet. For CC events, where the scattered lepton is a

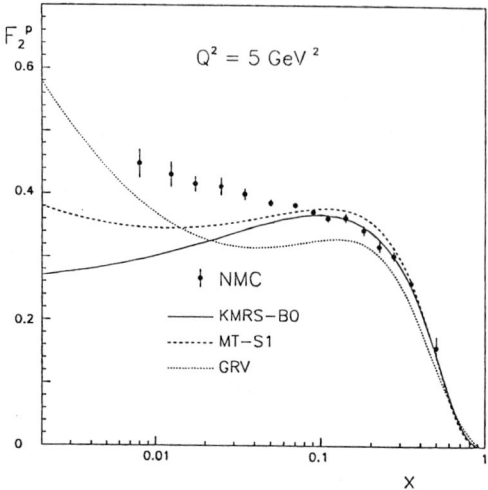

Figure 35. *The structure function F_2 as measured in μp scattering by NMC (1992), with predictions obtained from fits to previous DIS data Kwiecinski et al. 1992, Morfin and Tung 1992, Glück et al. 1990).*

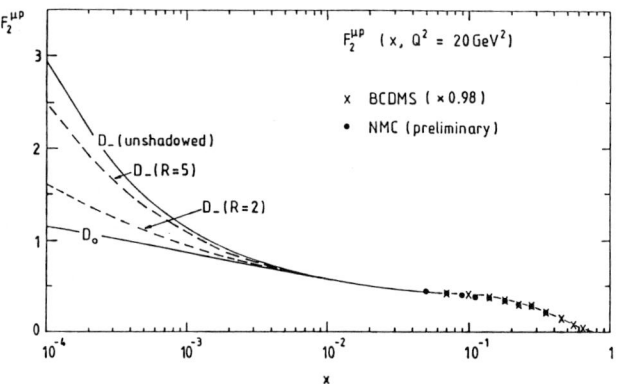

Figure 36. *Prediction for F_2 at very small x-values obtained from a fit (Martin et al. 1992) to new data from NMC (1992) and other experiments assuming as $x \to 0$: $xG(x) \sim x^{-0.5}$ (D_-) and $xG(x) \sim$ constant (D_0).*

neutrino, x and Q^2 can be measured only with the current jet. Figure 27 shows for nominal beam energies and standard x and Q^2 values the regions over which x and Q^2 can be measured well from the electron and the jet parameters, respectively. The main limitations stem from the precision with which the electron and jet energies can be measured, and from the size of the beam holes (see below). For NC scattering, structure function measurements should be feasible for basically the full range of x and Q^2. In the case of CC scattering precise measurements will be difficult for $y > 0.6$ and below $y \sim 0.02$. The well measured region can be extended by operating HERA at lower beam energies.

Small x-region: The major limitation for NC studies at very small x and low Q^2 values comes from the beam holes provided in the forward and rear calorimeters for beam passage. Typical cross sections of these holes are $20 \times 20\,\text{cm}^2$ (ZEUS). The effective hole in the acceptance is somewhat larger since a reliable energy measurement requires the point at which the electron or the jet enters the calorimeter to be some distance away from the cutout. Figure 29b shows an educated guess for the well measured region. It follows from the requirements

$$\theta_{\text{current jet}} < 172°, \quad \theta_{\text{electron}} < 172°, \quad E_{\text{electron}} > 5\,\text{GeV}$$

and from the size of the beam hole. The HERA experiments should be well suited for the region $x > 10^{-4}$, $Q^2 > 10\,\text{GeV}^2$.

6 Photoproduction

The high c.m. energy of HERA allows the study of photoproduction over a wide energy range. An introduction to photoproduction at HERA can be found in the Proceedings of the HERA workshops (*e.g.* Schuler 1991) and in reports by Levy (1992) and Schuler and Sjöstrand (1993).

Besides $p\bar{p}$ interactions, γp scattering is the only other hadronic or hadron-like reaction which can currently be measured at c.m. energies above 100 GeV. In the Vector Dominance Model (VDM) the photon can fluctuate into a low-mass vector meson V (ρ, ω, ϕ ...) which in turn interacts with the proton (Figure 37). As a result, γp scattering can be related to Vp scattering, and via the quark-parton model, to $\pi^\pm p$ and $K^\pm p$ scattering. For instance, the predictions for forward production of vector mesons are:

$$\frac{d\sigma^0}{dt}(\gamma p \to Vp) = \frac{(\alpha/4)}{(\gamma_V^2/4\pi)} \frac{d\sigma^0}{dt}(Vp \to Vp) = \frac{(\alpha/64)}{(\gamma_V^2/4\pi)}(1+\eta_v^2)\sigma_{\text{tot}}^2(Vp)$$

with

$$\begin{aligned}\sigma_{\text{tot}}(\rho^0 p) &= \sigma_{\text{tot}}(\omega p) = 0.5[\sigma_{\text{tot}}(\pi^+ p) + \sigma_{\text{tot}}(\pi^- p)] \\ \sigma_{\text{tot}}(\phi p) &= \sigma_{\text{tot}}(K^+ p) + \sigma_{\text{tot}}(K^- n) - \sigma_{\text{tot}}(\pi^+ p),\end{aligned} \quad (6.1)$$

where γ_V measures the strength of the γ-V coupling, $\gamma_V^2/4\pi = (\alpha^2/12)M_v/\Gamma_{ee}$ = 0.50 ± 0.03, 5.8 ± 0.2, 3.3 ± 0.13 for ρ^0, ω, ϕ respectively (PDG 1992), t is the four momentum transfer squared, and η_v is the ratio of the real-to-imaginary parts of the elastic scattering amplitude; for diffractive scattering $\eta_v = 0$.

As with purely hadronic reactions, photoproduction is expected to have a soft component due to peripheral processes (Figure 37a–c), and a hard component arising from the scattering of a parton from the proton on a parton from a vector meson (Figure 37d). However, in addition to its hadronic features, the photon possesses a property which makes it distinctly different from hadrons: it can couple directly to quarks and the coupling is pointlike. This leads to additional hard scattering processes which become

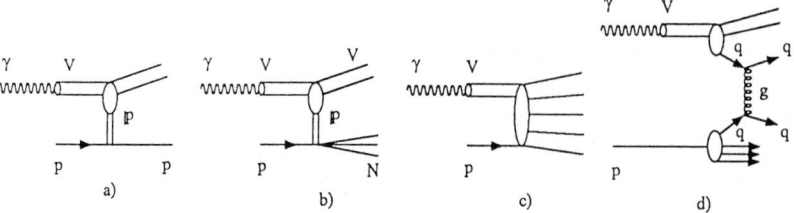

Figure 37. *Diagrams for photoproduction via* VDM *by (a) elastic and (b) inelastic diffractive scattering, (c) by non-diffractive processes and (d) hard scattering.*

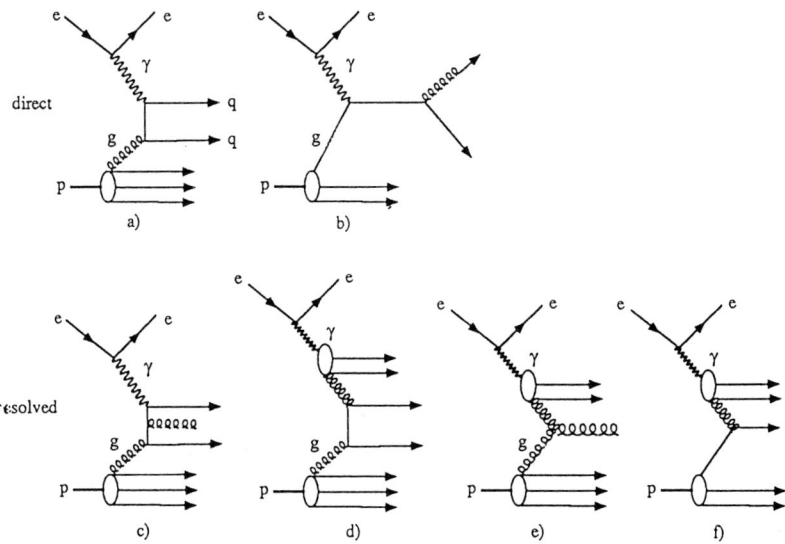

Figure 38. *Diagrams for direct photon processes: (a)* BGF, *(b)* QCDC *and (c–f) for resolved photon processes.*

prominent at high energies and which are not present in hadron-hadron interactions. They are represented by two types of diagrams: the first ('direct photon process') result from photon-gluon fusion into a quark-antiquark pair (Figure 38a) and from photon scattering off a quark in the proton under the emission of a gluon ('QCD Compton process'), see Figure 38b. The quarks coupling to the photon can also emit gluons (*e.g.* Figure 38c), and either a quark or a gluon may participate in the hard scattering (Figure 38d,e). Together with the hard scattering of the hadronic (VDM) photon the hard scattering due to the quark and gluon content of the photon constitute the 'resolved photon processes'.

The need for extra contributions beyond those predicted by the VDM and direct processes was found in the analysis of high p_T hadron production by $\gamma\gamma$-scattering at PETRA (JADE 1981, TASSO 1981,1984, PLUTO 1984 see also Kolanoski 1984) and recently at KEK (AMY 1992, 1993, TOPAZ 1993). AMY and TOPAZ showed that resolved processes can in principle provide for these extra contributions.

The resolved contributions are summed in the photon structure function F^γ which describe the quark and gluon content of the photon. (Note: part if not all the hard scattering contributions from the VDM and from the quark and gluon content of the photon are presumably identical and care must be taken to avoid double counting.) A selection of current predictions for F^γ labelled DG (Drees and Grassie, 1985), LAC1, LAC2, LAC3 (Abramowicz et al. 1991), GRV (Glück et al. 1993) and GS (Gordon and Storrow 1992) is shown in Figure 39.

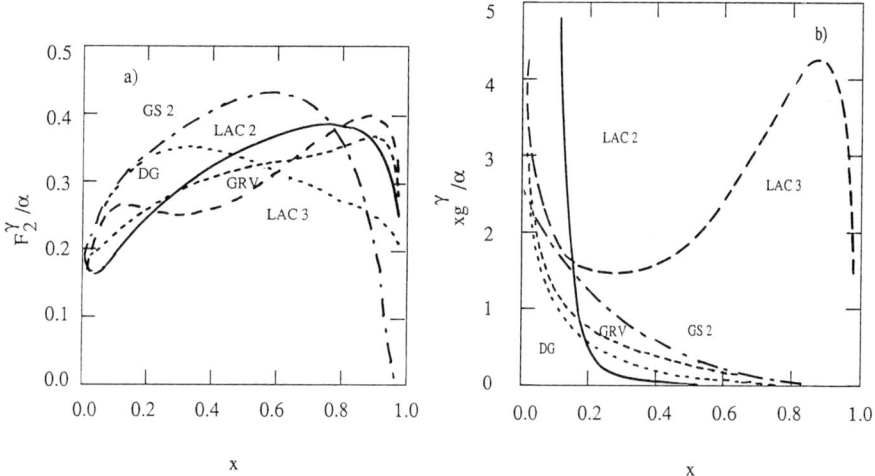

Figure 39. *Parametrisations of the photon structure function (from Drees and Godbole 1992).*

The hard photon processes—direct or resolved—give rise to quark and gluon jets (sometimes called 'mini-jets') with large transverse momenta. Their cross sections have been calculated and found to depend critically on the minimum momentum transfer p_{Tmin} down to which the integration is performed. As will be seen below, at HERA energies hard scattering originating from photon constituents is clearly visible.

6.1 Total photoproduction cross section

In electron-proton collisions quasi-real photons are produced by bremsstrahlung of the electron on the proton. The photon energy E_γ can be determined by measuring the energy of the scattered electron, E'_e, in the electron calorimeter of the luminosity monitor, yielding $E_\gamma = E_e - E'_e$. The photon-proton c.m. energy, W, is given by $W^2 = 4E_\gamma E_p$. For instance, $E_\gamma = 12\,\text{GeV}$ corresponds to $W = 200\,\text{GeV}$.

The photon flux is roughly given by $dN_\gamma \simeq (\alpha/2\pi)\ln(Q^2_{max}/m_e^2)\,dE_\gamma/E_\gamma \simeq 0.015\,dE_\gamma/E_\gamma$, which leads to the following relation between the ep and γp cross sections,

$$\sigma(ep) \simeq 0.015 \int \frac{dE_\gamma}{E_\gamma} \sigma_{tot}(\gamma p).$$

The complete Weizsäcker-Williams approximation gives the following result:

$$\frac{d\sigma(ep)}{dy} = \frac{\alpha}{2\pi}\left\{\frac{[1+(1-y)^2]}{y}\ln\left(\frac{Q^2_{max}}{Q^2_{min}}\right) - 2\left[\frac{1-y}{y}\right]\left(1 - \frac{Q^2_{min}}{Q^2_{max}}\right)\right\}\sigma_{trans}(\gamma p), \quad (6.2)$$

where $y = E_\gamma/E_e$, $Q^2_{min} = m_e^2 y^2/(1-y)$, Q^2_{max} is given by the acceptance for the electron detector and $\sigma_{trans}(\gamma p)$ is the cross section for transverse photons. Note that for typical settings at HERA the second term in (6.2) contributes about 8% (Burow 1993) and should not be neglected.

The event rate for photoproduction is very large at HERA; e.g. selecting photon energies between 10 and 20 GeV and assuming $\sigma_{tot}(\gamma p) = 140\,\mu$b produces 10^5 events for an integrated luminosity of 100 nb^{-1}. The large rate has permitted the two experiments, H1 and ZEUS, to make a measurement of $\sigma_{tot}(\gamma p)$ with a few nb^{-1} of data collected in the first weeks of data taking in 1992. The result were

$$\sigma_{tot}(\gamma p) = 159 \pm 7\,(\text{stat}) \pm 20\,(\text{syst})\,\mu\text{b} \quad \langle W \rangle = 195\,\text{GeV, H1 (1992)}$$
$$154 \pm 16\,(\text{stat}) \pm 32\,(\text{syst})\,\mu\text{b} \quad \phantom{\langle W \rangle = {}} 210\,\text{GeV, ZEUS (1992a)}.$$

The main difficulty in the determination of $\sigma_{tot}(\gamma p)$ lies in the fact that the detector acceptance is strongly process dependent. While for soft and hard processes the acceptance is above 80%, it is only 20–30% for the 'elastic' diffractive reactions, $\gamma p \to Vp$, $V = \rho^0, \omega, \phi$ (Figure 37a) and sensitive to the details of the production mechanism. The truly elastic reaction, $\gamma p \to \gamma p$, is expected to have a cross section which is two orders of magnitude smaller compared to the vector meson processes. The acceptance for inelastic diffraction by dissociation of the photon into higher mass states and/or of the proton into heavier states, $\gamma p \to VX, \to X_1 X_2$, lies in between. Figure 40 shows from H1 for charged particles the distributions of the transverse momentum p_T and the polar angle θ. The data (Figures 40a, c) are well described by the Monte Carlo simulation of the sum of diffractive, soft and hard contributions shown separately in Figures 40b, d.

The H1 and ZEUS experiments repeated the measurement of $\sigma_{tot}(\gamma p)$ with the data from autumn 1992 providing an order of magnitude increase in statistics and allowing for a detailed study of the acceptances.

H1 analysis: From a study of 16,000 photoproduction events obtained with 21.9 nb^{-1}, found (H1 1993d):

$$\sigma_{tot}(\gamma p) = 156 \pm 2\,(\text{stat}) \pm 18\,(\text{syst})\,\mu\text{b}.$$

ZEUS analysis: A sample of about 6000 photoproduction events from 13 nb^{-1} was selected with a tagged electron in the energy range 15.2–18.2 GeV corresponding to W values between 167 and 194 GeV (ZEUS 1993j, l, m). The value of Q^2 was less than 0.02 GeV^2 which implies an expected change of $\sigma_{tot}(\gamma p)$ compared to its value at $Q^2 = 0$ of less than 1%.

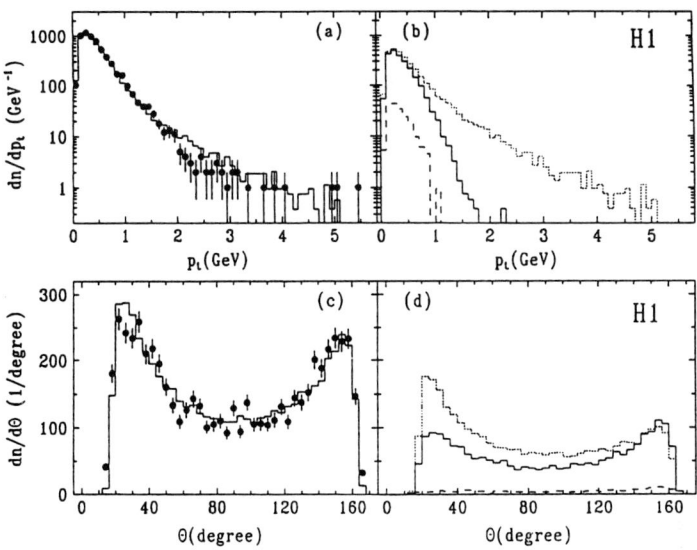

Figure 40. *Inclusive transverse momentum (a) and polar angle (c) distributions of charged particles in photoproduction events (points) compared to the Monte Carlo simulation. In (b) and (d) the contributions of the diffractive (dashed), soft (solid) and hard (dotted) components are shown (from* H1 *1992).*

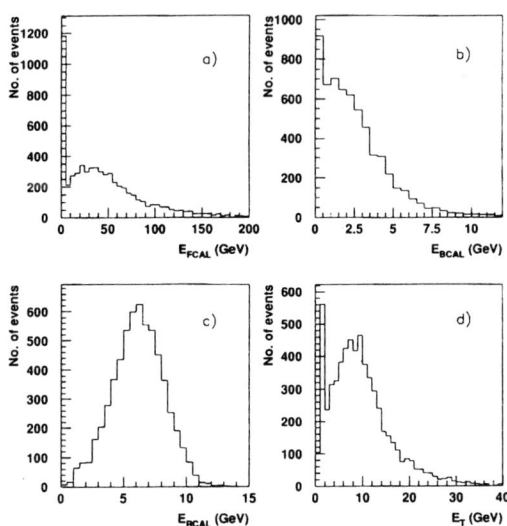

Figure 41. *Distribution in the calorimeter of the energy deposited (a) in the forward, (b) barrel, (c) rear sections and (d) total transverse energy (*ZEUS *1993j, l, m).*

The energy distributions observed for the events in the forward (FCAL), barrel (BCAL) and rear (RCAL) calorimeters are shown in Figure 41. In the E_{FCAL} distribution a distinct class of events is observed with very little energy deposition. These events originate mainly from elastic and inelastic diffractive processes which are characterised by producing very little transverse energy in the forward (i.e. the proton) direction.

From a study of the events with $E_{\text{FCAL}} < 1\,\text{GeV}$ the acceptances and contributions of diffractive processes were determined by comparing the data with various models and Monte Carlo simulations (PYTHIA 1987, 1989, HERWIG 1986, 1992, Nikolaev-Zakharov 1992, 1993). The spatial distribution of the energy deposition in the rear direction permitted the separation of the elastic and inelastic diffractive contributions: photon dissociation into low-mass states deposits energy more closely to the incoming photon direction than high- mass states. The acceptance and contribution from non-diffractive processes were determined using the events with $E_{\text{FCAL}} > 1\,\text{GeV}$. The result for the total cross section is

$$\sigma_{\text{tot}}(\gamma p) = 141 \pm 3\,(\text{stat}) \pm 17\,(\text{syst})\mu b.$$

Figure 42 shows $\sigma_{\text{tot}}(\gamma p)$ as a function of W above the resonance region ($W > 1.75\,\text{GeV}$) as measured in previous experiments up to 18 GeV (S. Alekhin et al. 1987) and by H1 and ZEUS. No dramatic rise is observed between 18 and 200 GeV.

The curve labelled DL in Figure 42 is the prediction of a Regge-type fit to hadron-hadron and pre-HERA photon-proton cross sections of the form

$$\sigma_{\text{tot}} = X s^{\varepsilon} + Y s^{-\eta}, \tag{6.3}$$

where the first term describes pomeron contributions and the second one those from ρ, ω, f and a exchange (Donnachie and Landshoff 1984). The values of ε, η were obtained from fits to pp and $p\bar{p}$ data alone yielding $\varepsilon = 0.0808$ and $\eta = 0.4$ while the coefficients X, Y were determined by fitting the low energy data on $\sigma_{\text{tot}}(\gamma p)$. The curve ALLM (Abramowicz et al. 1991) is also a Regge-type analysis with an ansatz which provides a smooth transition from the photoproduction to the DIS region. The parameters of the model were obtained by a simultaneous fit to the data for $\sigma_{\text{tot}}(\gamma p)$ and DIS at low energy. Both predictions agree rather well with the HERA data.

Observations of 'diffractive' air showers (Chacaltaya and Pamir 1990) and an excess of muons in energetic air showers (Yodh 1990) have suggested $\sigma_{\text{tot}}(\gamma p)$ to rise much faster than $\sigma_{\text{tot}}(p\bar{p})$ at beam energies above of the order 100 TeV. This rise has been attributed to semi-hard scattering involving resolved photon contributions. The other curves shown in Figure 42 are based on the assumption that the total cross section is a sum of a soft part plus the contributions from direct and resolved photons (Drees and Grassie, 1985, Ghandi and Sarcevic, 1991, Forshaw and Storrow, 1991, Schuler and Terron, 1991, Fletcher et al. 1992.). They depend critically on the choice of the photon structure function F^{γ} and on the parameter p_{Tmin}, the lower integration limit for the hard process in QCD, where p_T is the parton transverse momentum. The dashed-dotted lines use the DG parametrisation of F^{γ} with $p_{\text{Tmin}} = 2.0\,\text{GeV/c}$ for the lower (1.4 GeV/c for the upper) line. The dashed (dotted) line uses the LAC1 version with $p_{\text{Tmin}} = 2.0\,(1.4)\,\text{GeV/c}$. Only the dashed-dotted curve—which makes the lowest cross

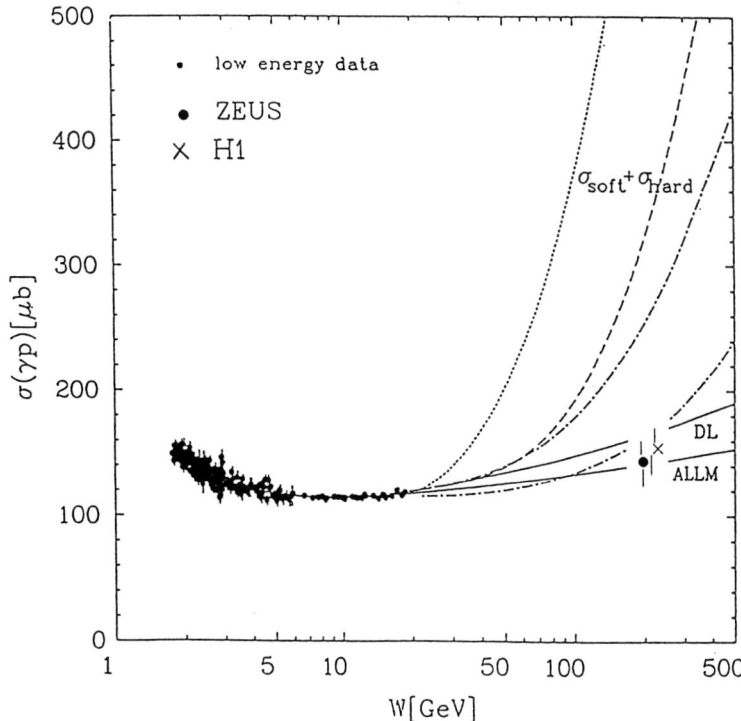

Figure 42. The total γp cross section as a function of the c.m. energy W measured below 18 GeV and by H1 and ZEUS near $W = 200$ GeV. The lower solid curve is the prediction of the ALLM parametrisation and the higher solid curve is that of DL. The dotted (dashed) line uses the LAC1 parametrisation for F^γ with $p_{Tmin} = 1.4$ GeV/c (2 GeV/c). The dashed-dotted lines use the DG parametrisation for F^γ with $p_{Tmin} = 1.4$ GeV/c (upper line) and $p_{Tmin} = 2$ GeV/c (lower).

section prediction—is in agreement with the HERA data. Besides experimental data on the photon structure function higher order calculations are needed in order to assess the behaviour of $\sigma_{tot}(\gamma p)$ above HERA energies. Furthermore, corrections for multiple parton interactions may have to be taken into account.

6.2 Partial photoproduction cross sections

The ZEUS analysis of $\sigma_{tot}(\gamma p)$ also provided measurements of the cross sections for the diffractive and non-diffractive channels:

$$\sigma_{\text{non-diff}} = 91 \pm 11 \mu b$$
$$\sigma_{\text{el diff}} = 17 \pm 8 \mu b$$
$$\sigma_{\text{inel diff}} = 33 \pm 9 \mu b.$$

The cross section $\sigma_{el\ diff}$ for $\gamma p \to V p$, $V = \rho^0 + \omega + \phi$, is shown in Figure 43 together with measurements at low energies. The full curve is a calculation based on VDM, the quark model and the observed energy dependence of $\sigma_{tot}(\gamma p)$. It agrees with the data.

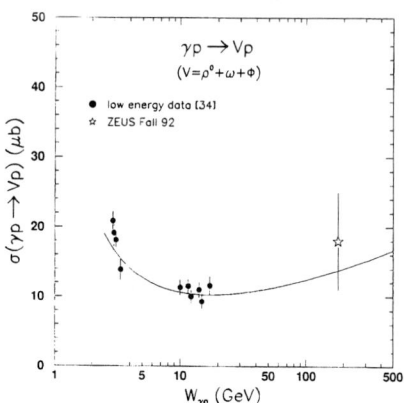

Figure 43. *Cross section for the reaction $\gamma p \to V^0 p$, $V^0 = \rho^0$, ω, ϕ ZEUS (1993j, l).*

Direct observation of $\gamma p \to \rho^0 p$ was reported by both experiments. The $\pi^+\pi^-$ mass distribution as measured with the tracking systems is shown in Figure 44 exhibiting a clear ρ^0 signal.

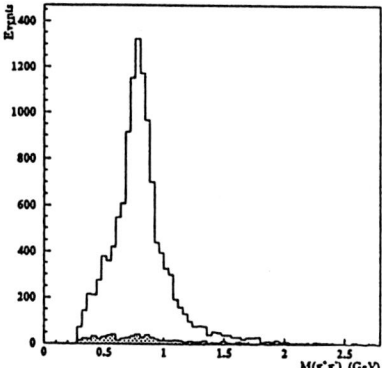

Figure 44. *The $\pi^+\pi^-$ mass distribution from untagged photoproduction (shaded distribution for same sign charged particles); from ZEUS (1993j, l).*

A model of the build-up of the total photon proton cross section by soft processes, diffractive scattering and direct and resolved (called anomalous) photon contributions is shown in Figure 45 (Schuler and Sjöstrand 1993).

6.3 Events with a large rapidity gap

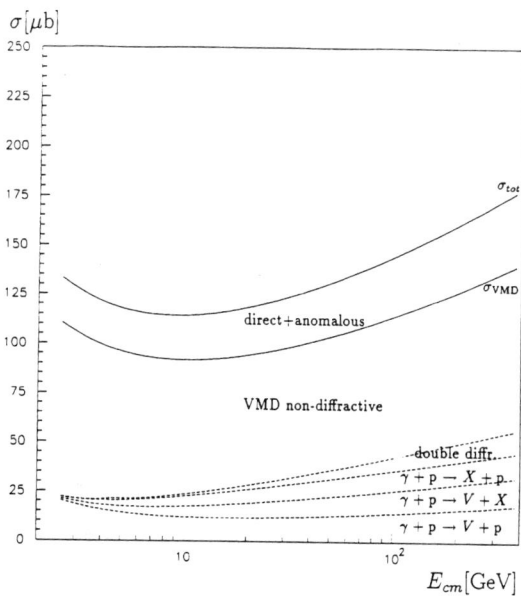

Figure 45. *Build up of the total γp cross section according to Schuler and Sjöstrand (1993).*

In DIS a new class of events was found by ZEUS (1993g) and confirmed by H1(1993j, k) which are characterised by a large rapidity gap between the proton beam direction and the first significant energy deposition away from it. The observation of large rapidity gaps provides a new window for a measurement of the diffractive contribution (elastic and inelastic). It will be discussed in some detail in the section on DIS results.

A search in photoproduction showed the presence of the same type of events (ZEUS 1993j, l). The analysis was performed in terms of the pseudorapidity $\eta = -\ln\tan(\theta/2)$. In Figure 46a the distribution of η_{max} is shown, where η_{max} is the pseudorapidity of the condensate closest to the proton direction—*i.e.* with the maximum η. A condensate is a group of contiguous calorimeter cells with a total energy greater than 0.4 GeV. The data fall into two groups, a large peak at η_{max} values between 3 to 6 consisting mainly of non-diffractive events and described by a Monte Carlo simulation of non-diffractive processes (solid histogram), and a second peak near $\eta_{max} \simeq -2$ arising from diffractive production. The distribution of M_X^2, where M_X is the mass of the diffractive system, shows the bulk of the events to have low masses, $M_X < 10\,\text{GeV}$ (Figure 46b). Note, the data are not corrected for acceptance. The elastic and inelastic diffractive processes discussed before contribute to these large rapidity gap events.

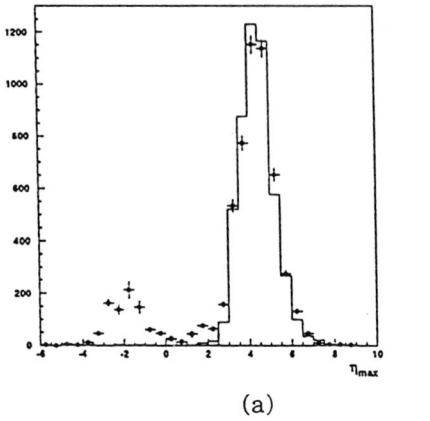 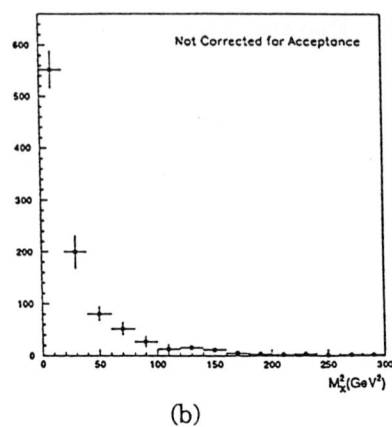

Figure 46. *Tagged photoproduction (ZEUS 1993j, l): (a) Distribution of η_{max} for data (points) and prediction for non-diffractive photoproduction (histogram) (b) Distribution of the mass squared for the hadron system produced in events with a large rapidity gap ($\eta_{max} < 2$).*

6.4 Hard scattering in photoproduction

6.4.1 Considerations of kinematic

While the observed size and energy dependence of the total cross section $\sigma_{tot}(\gamma p)$ do not provide evidence for the direct and resolved photon contributions, the analysis of the final states shows the presence of these terms very clearly in the hard scattering contributions.

The direct and resolved processes (Figure 38) produce quarks and gluons in the final state. At sufficiently high energies the final state partons manifest themselves as jets of hadrons. Let us estimate how high the energy has to be in order to recognise these jets. Consider Compton scattering of a photon on a quark carrying a fractional momentum x as depicted in Figure 47a. The c.m. energy squared of the final state qg system is

$$\bar{s} = 4E_\gamma x E_p = xW^2 \qquad (6.4)$$

or $\bar{s} \sim (0.001\text{–}0.01)W^2$ for typical values of x. In Figure 47b the fragmentation process is sketched in the qg rest system. From e^+e^- data one knows that the average transverse momentum of the outgoing hadrons with respect to the parton direction (Figure 47c) is $\langle p_T \rangle \simeq 0.3\,\text{GeV}$, approximately independent of the parton energy, $E_q^\star = \bar{s}^{1/2}$; the average hadron multiplicity $\langle n \rangle$ grows slowly with $E_q^\star$ ($\langle n \rangle \simeq n_0 + a \exp[b(\ln \bar{s}/\Lambda^2)^{1/2}]$, $n_0 = 3$, $a = 0.4$, $b = 1.93$ for $\Lambda = 0.3\,\text{GeV}$). Therefore the average longitudinal momentum p_L increases rapidly with $E_q^\star$ and the half-opening angle of the jet cone $\langle \delta \rangle$ (Figure 47c,d), defined as $\langle \delta \rangle = \langle p_T \rangle / \langle p_L \rangle$, shrinks with energy: jets become more jetlike as the energy increases. This is illustrated in Table 3 for a few parton energies. This table shows that $(\bar{s})^{1/2} = 2E_q^\star$ needs to be at least 10 GeV and therefore W at least is of order 100 GeV for well collimated jets to emerge.

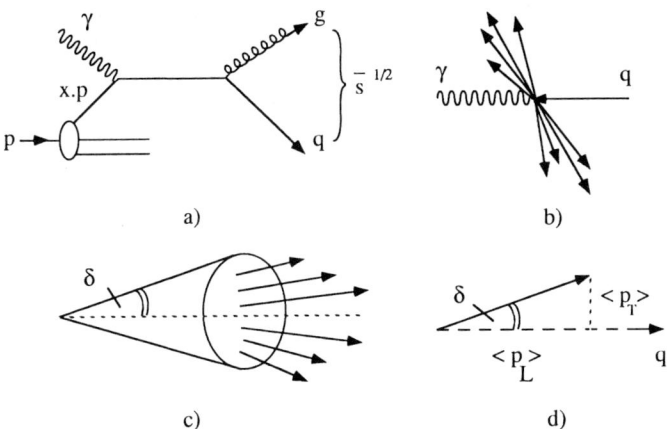

Figure 47. *The* QCD *Compton process (a) producing (b) two back-to-back jets in the photon-quark c.m. system; (c) the jet cone (d) with opening angle δ defined by the average longitudinal and transverse momenta.*

E_q^* (GeV)	$\langle n \rangle$	$\langle \delta \rangle = \langle p_T \rangle / \langle p_L \rangle$
2	3	31°
5	5	19°
20	11	10°

Table 3. *Jet collimation as function of parton energy.*

It is convenient to analyse the jet production at HERA in terms of transverse momentum p_T and pseudorapidity η rather than in terms of p_T and the longitudinal momentum p_z. The pseudorapidity is an approximation to the rapidity Y defined as

$$Y = 0.5 \ln[(E + p_z)/(E - p_z)], \qquad (6.5)$$

where E is the particle energy. A longitudinal transformation into a system which moves with velocity β transforms Y into

$$Y^* = Y + 0.5 \ln[(1 + \beta)/(1 - \beta)]. \qquad (6.6)$$

The transformation causes a shift, $\Delta Y = Y^* - Y$, but does not alter the shape of the rapidity distribution (Figure 48).

The invariant cross section expressed in terms of p_T, Y reads

$$E \frac{d^3\sigma}{d^3p} = \frac{d^3\sigma}{dp_T^2 dY}, \qquad (6.7)$$

which is a particularly useful relation when the cross section factorises into a p_T-dependent and a Y-dependent term. At energies large compared to the particle's mass, $E \gg m$,

$$Y \simeq 0.5 \ln[\{1 + \cos\theta + m^2/(2p^2)\}/\{1 - \cos\theta + m^2/(2p^2)\}]. \qquad (6.8)$$

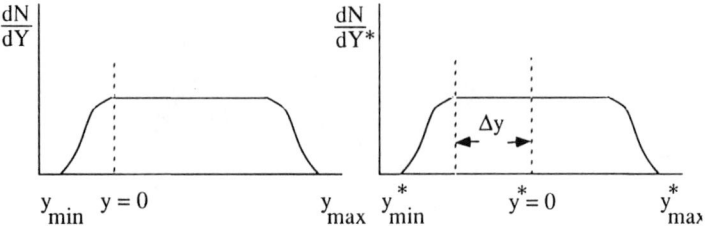

Figure 48. *The rapidity distribution in two Lorentz frames separated by a longitudinal transformation.*

The minimum and maximum rapidities are reached for $\cos\theta = \pm 1$:

$$Y_{\min} \simeq -\ln(2p/m) \qquad Y_{\max} \simeq \ln(2p/m) \qquad \text{for } E \gg m. \tag{6.9}$$

For $1 - |\cos\theta| \gg m^2/(2p^2)$ the rapidity can be approximated by the pseudorapidity,

$$\eta = -\ln\tan(\theta/2). \tag{6.10}$$

In two-to-two parton scattering, such as in direct or resolved photon scattering depicted in Figure 49, the momenta of the incoming partons can be calculated from the two partons observed in the final state. Let x_γ and x_p be the fractions of the photon and proton momenta carried by the interacting partons. The assumption that the photon and proton remnants carry no transverse momenta plus energy-momentum conservation lead to

$$x_p = \sum_{\text{partons}} (E + p_z)_{\text{parton}}/(2E_p); \qquad x_\gamma = \sum_{\text{partons}} (E - p_z)_{\text{parton}}/(2yE_e), \tag{6.11}$$

where E_γ is the initial photon momentum and the sum is over the two final state partons. For the direct process, $x_\gamma = 1$. The expression for x_γ can be rewritten in the form (Barger and Phillips 1987):

$$x_p = \sum_{\text{partons}} E_{\text{Tparton}}[\exp(\eta_{\text{parton}})]/(2E_p); \qquad x_\gamma = \sum_{\text{partons}} E_{\text{Tparton}}[\exp(-\eta_{\text{parton}})]/(2yE_e). \tag{6.12}$$

The final state partons fragment into jets. The jet energies and momenta can be used to estimate the energies and momenta of the final state partons. Using the energy observed by the calorimeter cells assigned to the jet to evaluate the jet energy and longitudinal momentum and since $E_\gamma \simeq yE_e \simeq y_{\text{JB}}E_e$ one can approximate x_γ and x_p as

$$x_p^{\text{meas}} = \sum_{\text{jets}}(E+p_z)_{\text{jet}}/(2E_p); \qquad x_\gamma^{\text{meas}} = \sum_{\text{jets}}(E-p_z)_{\text{jet}}/\sum_h (E-p_z)_h \tag{6.13}$$

or

$$x_p^{\text{meas}} = \sum_{\text{jets}} E_{\text{Tjet}}[\exp(\eta_{\text{jet}})]/(2E_p); \qquad x_\gamma^{\text{meas}} = \sum_{\text{jets}} E_{\text{Tjet}}[\exp(-\eta_{\text{jet}})]/(2E_\gamma), \tag{6.14}$$

where the sum in the denominator runs over all calorimeter cells. Here x_γ is the fraction of $\sum_h (E-p_z)_h$ carried by jets; the energy of the photon remnant is $(1-x_\gamma)E_\gamma$.

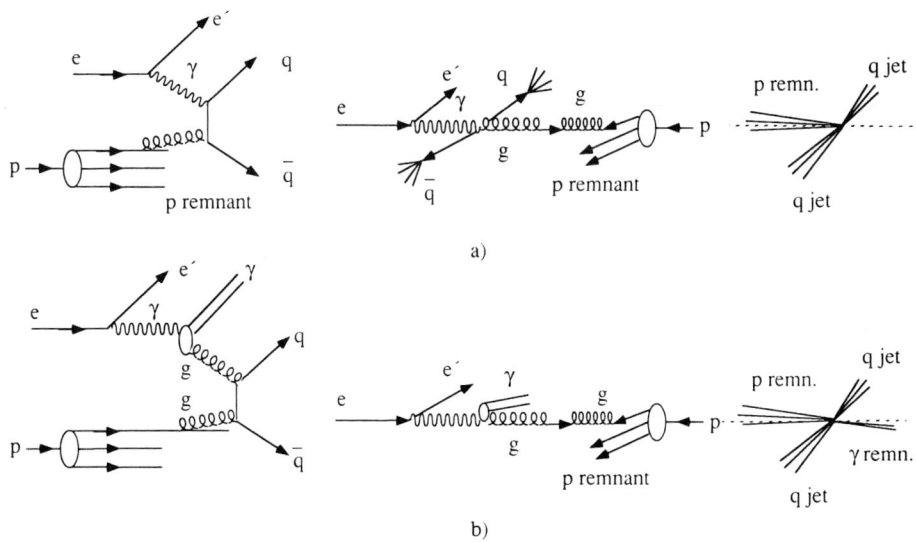

Figure 49. *The topology of hard scattering events from a direct photon process (a) with a proton remnant in the forward direction and two jets, and a resolved photon process (b) with a proton remnant, two jets and a photon remnant in the rear direction.*

6.4.2 Experimental results on hard photoproduction

A first indication for the presence of hard scattering contributions in photoproduction was observed by the Omega-Photon collaboration (1989) when comparing the p_T distribution of charged particles produced by γp and $\pi^\pm p$ or $K^\pm p$ scattering. The data shown in Figure 50 were obtained for beam energies of 110–170 GeV, corresponding to W's of 14–18 GeV. The p_T spectrum produced by photoproduction is somewhat harder than in hadron-hadron scattering for $p_T > 2$ GeV.

H1 analysis (H1 1993e, j, k) The p_T distribution of charged particles measured by H1 for tagged events at an average $W = 197$ GeV is shown in Figure 51a (H1 1993j, k). The photoproduction result from the Omega experiment is included for comparison. The high energy data reach to much larger p_T values (8 GeV/c versus 4 GeV/c) and exhibit a much harder spectrum; e.g. at $p_T = 4$ GeV the cross section for $W = 197$ GeV is a factor of approximately 50 larger compared to 16 GeV. The relative contribution from hard scattering increases with c.m. energy. The p_T spectrum from photoproduction is significantly harder than the one measured by UA1 (1990) for $\bar{p}p$ scattering at the same c.m. energy of 200 GeV (Figure 51b).

Evidence for jet formation (H1 1993j, k) can be seen from Figure 52 where for charged tracks the p_T flow is shown as a function of $\Delta\phi$, the distance in azimuthal angle from the track with the highest p_T. While soft events show no structure in $\Delta\phi$, when the total transverse energy E_T per event detected by the calorimeter is above 10 GeV, particles begin to cluster in opposite cones. A jet search was performed using a cone algorithm (UA1 1983, Huth et al. 1990). Jets were required to have $E_T > 7$ GeV in a cone of $\Delta R = (\Delta\eta^2 + \Delta\phi^2)^{1/2} = 1$. They were selected in the range $-1 < \eta < 1.5$.

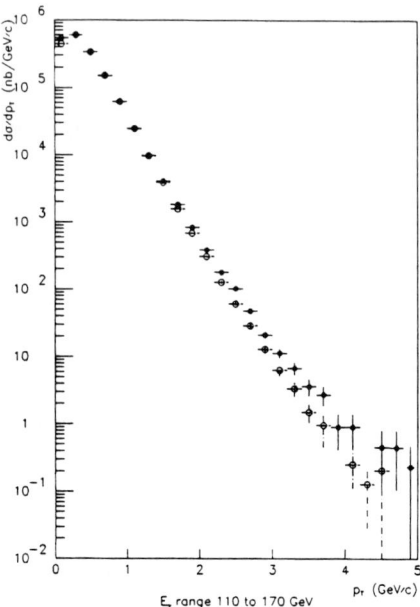

Figure 50. *The transverse momentum spectrum of charged particles produced by photon proton (full points) and hadron proton scattering (open points) for beam energies between 110 and 170 GeV* (OMEGA 1989).

In Figure 53a, the ep cross section, corrected for detector effects to the hadron level, is shown as a function of E_T. A steep drop is observed with E_T. The η dependence of the jet cross section is given in Figure 53b. The curves show leading order (LO) QCD predictions calculated with PYTHIA for different parametrisations of F^γ: LAC2, LAC3, GRV-LO and GRV-LO without gluons. The predictions agree with the measured E_T dependence—with the exception of LAC3—while none of them reproduces well the observed η dependence.

The x_γ distribution was determined from a sample of two-jet events according to Equation 6.14. About 800 two-jet tagged photoproduction events were selected with the requirement that each jet has a transverse energy $E_{Tjet} > 5$ GeV and $|\eta| < 2.5$. The data, not corrected for detector smearing, are shown in Figure 54. The distribution falls rapidly with increasing x_γ. The data were compared with the LO QCD model used already for Figure 53. The curves shown in Figure 54 were calculated by generating events with PYTHIA according to the model and passing them through the H1 detector simulation and reconstruction chain. The choice of a large gluon component in the photon as presented by LAC3 is disfavoured by the data. The dotted curve shows the prediction assuming no gluons from the photon are involved in the interaction. The prediction undershoots the data at low x_γ, where gluons are expected to be important.

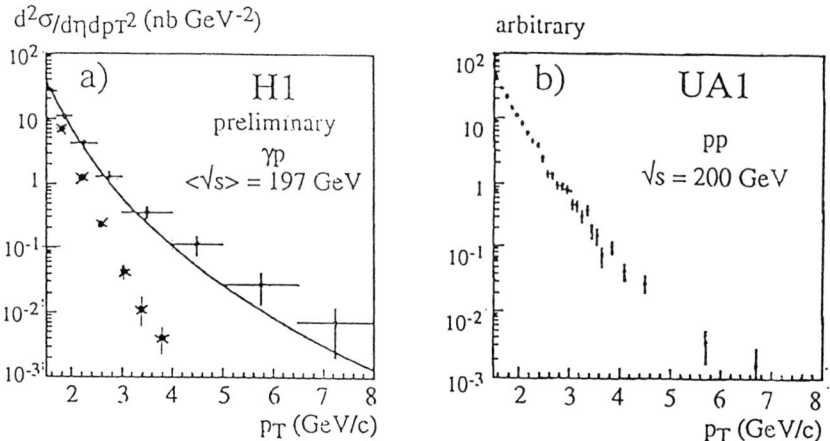

Figure 51. *(a) Inclusive charged track ep cross section for $|\eta| < 1$ at $W = 197$ GeV, $Q^2 \simeq 0$ (H1 1993j, k). Also shown are the photoproduction data from OMEGA (1989) for W around 17 GeV, arbitrarily normalised. The curve shows an arbitrarily normalised spectrum from an NLO QCD calculation of resolved photon interactions (Borzumati et al. 1993). (b) The same spectrum measured at the same c.m. energy in $\bar{p}p$ interactions by UA1 (1990). Figure taken from H1 (1993j, k).*

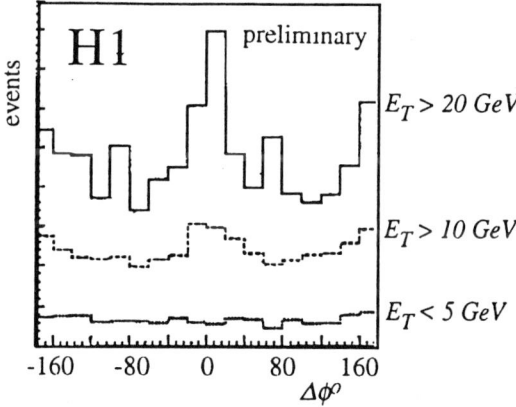

Figure 52. *Distribution of the distance $\Delta\phi$ in azimuth ϕ of charged tracks from the charged track with the largest p_T as a function of the total transverse energy E_T measured in the calorimeter (H1 1993j, k).*

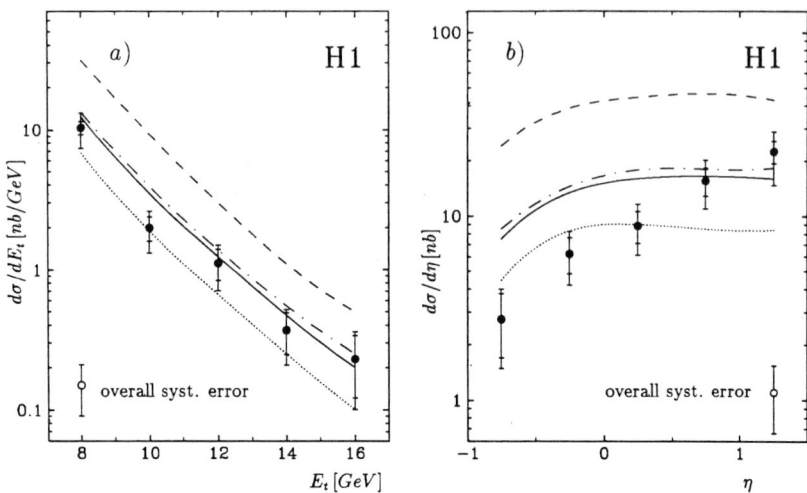

Figure 53. *(a) Inclusive jet E_T spectrum integrated over $-1.0 < \eta < 1.5$. (b) Inclusive jet η spectrum for jets with $E_T > 7\,\text{GeV}$. The curves represent* LO *calculations done with* PYTHIA *using for F^γ* LAC3 *(dashed),* LAC2 *(dashed-dotted),* GRV-LO *(full) and* GRV-LO, *but excluding the gluons originating from the photon (dotted);* (H1 1993e).

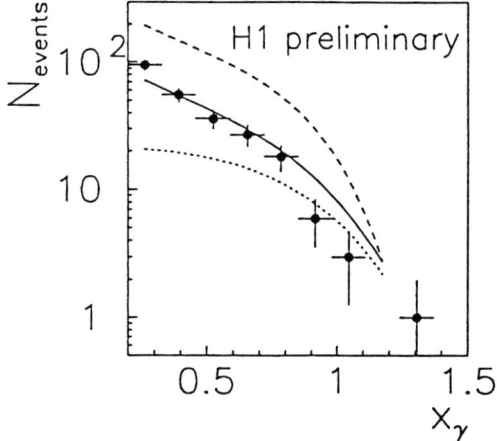

Figure 54. *Momentum fraction of partons from photons. Curves are* LO *calculations as in Figure 53* (H1 1993e).

ZEUS analysis Hard scattering in photoproduction was studied as follows (ZEUS 1992b). About 20% of all photoproduced events have a total transverse energy $E_T > 10\,\text{GeV}$ while soft photoproduction is expected to have $E_T < 10\,\text{GeV}$. A substantial fraction of high E_T events shows two back-to-back jets. A spectacular event is shown in Figure 55, where two jets with about 30 GeV energy each are visible plus a proton remnant. Since no energy is deposited in RCAL, where one would expect the fragments of the photon remnant in a resolved process, this event is a candidate for a direct photon interaction.

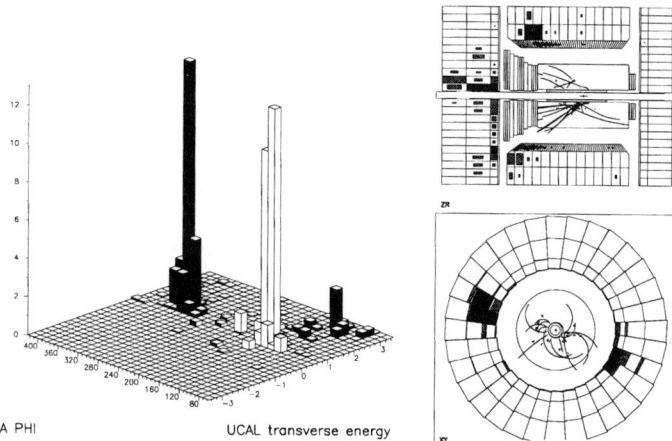

Figure 55. *A tagged photoproduction event with 2 jets. Each jet has an energy of about 30 GeV. This is a candidate for a direct photon interaction since there is no evidence of a photon remnant in the rear calorimeter.*

Clear evidence for resolved photon processes can be seen in the characteristics of the two-jet sample (ZEUS 1992b, 1993i). The kinematics for direct and resolved photon processes is illustrated in Figure 49. In direct processes the final state consists of the two parton jets, the scattered electron close to the electron beam and the proton remnant close to the proton beam. For resolved processes there is in addition the photon remnant which is emitted in the direction of the incident photon (the direction of the incident electron) which may reveal itself as energy deposition in the rear calorimeter. In Figure 56 the energy deposited in RCAL covering η values between -0.75 and -3.8 is plotted against the pseudorapidity, η_{min}, of the most backward jet. If both jets go forward $\eta_{min} > 0$. If direct photon interactions were the sole origin of these events, sizeable energy in RCAL would be expected for events with $\eta_{min} < -0.25$, falling essentially to zero as the jets become more distant from the RCAL region. This trend is indeed observed in Figure 56. However, there is an additional group of events with energy deposition in RCAL as large as 16 GeV even when both jets are far away from RCAL, the nearest jet being as much as three units of pseudorapidity away. This substantial RCAL energy is explained as the remnant of the photon in resolved photon processes.

The distribution of the transverse energy E_{Tjet} of the photoproduced jets is shown in Figure 57a for $\eta_{jet} < 1.6$ (ZEUS 1993i). Jets with E_T up to 19 GeV are seen. The data are well reproduced by a LO QCD calculation performed within the HERWIG Monte Carlo program which considered direct and resolved processes. Initial and final state parton showers were included and the simulated events were passed through the detector and reconstruction chain. According to the model resolved photon contributions are dominant; the direct contributions approach the resolved ones as E_{Tjet} increases beyond 20 GeV. The η_{jet} distribution is shown in Figure 57b. Also shown are the LO QCD predictions. The direct and resolved contributions show quite different distributions: in the resolved case only a fraction of the photon's momentum participates in the hard

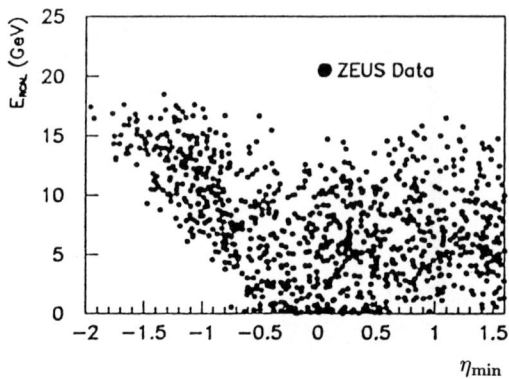

Figure 56. *The energy deposited in RCAL plotted against the pseudorapidity, $\eta_{\min}$, of the most backward jet (ZEUS 1993j, l).*

scatter such that the centre of mass is in general more strongly boosted in the proton direction. The data require a substantial contribution from resolved processes.

Two-jet production was studied requiring both jets to have $\eta_{\text{jet}} < 1.6$. The di-jet mass distribution shows masses up to 40 GeV (Figure 58). The distributions of the proton momentum fraction x_p^{meas} and photon momentum fraction x_γ^{meas} calculated according to (6.13) are displayed in Figure 59 for events with $|\eta_{\text{jet}1} - \eta_{\text{jet}2}| < 1.5$ and $|\phi_{\text{jet}1} - \phi_{\text{jet}2}| > 120°$. The latter cuts were applied in order to improve the resolution for x_γ^{meas}. The bulk of the events have x_p^{meas} values between 3×10^{-3} and 3×10^{-2}. The x_γ^{meas} distribution rises at both low and high values. The Monte Carlo simulations of the direct and resolved processes have very different characteristics. The resolved processes show a rise towards low x_γ^{meas}, as observed in the data but cannot account for the rise at high x_γ^{meas}.

The direct processes predict a sharp rise towards high x_γ^{meas} as observed in the data, and only a small number of events for $x_\gamma^{\text{meas}} < 0.7$. This is an unambiguous signature for the presence of direct processes. The di-jet ep cross sections of quasireal photons in the region $0.2 < E_\gamma/E_e < 0.7$, $E_{\text{Tjet}} > 5$ GeV and $\eta_{\text{jet}} < 1.6$ were found to be

$$21.1 \pm 5.2 \text{ (stat)} \pm 5.7 \text{ (syst)} \text{ nb} \quad \text{for resolved processes,}$$

and

$$9.4 \pm 2.7 \text{ (stat)} \pm 2.7 \text{ (syst)} \text{ nb} \quad \text{for direct proccesses.}$$

6.5 Discussion of the photoproduction results

The first analyses of H1 and ZEUS have shown that photoproduction is accessible at HERA almost up to the kinematic limit. The total photoproduction cross section does not exhibit a dramatic rise in the HERA regime but behaves rather hadron-like. The final states show very clearly the presence of hard scattering manifesting itself in a hardening of the transverse momentum distributions and the production of energetic and large transverse momentum jets. The existence of direct and resolved photon processes have

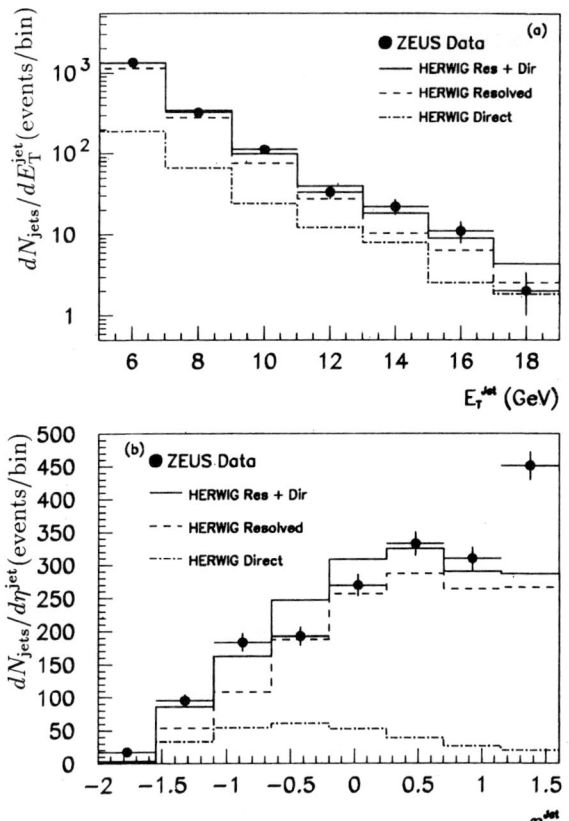

Figure 57. *Inclusive jet distributions for (a) transverse energy of jets, (b) pseudorapidity of jets. The Monte Carlo predictions are normalised to the data in the region $\eta_{jet} < 1.2$. Also shown are the relative contributions of the direct and resolved processes as predicted by Monte Carlo (ZEUS 1993i).*

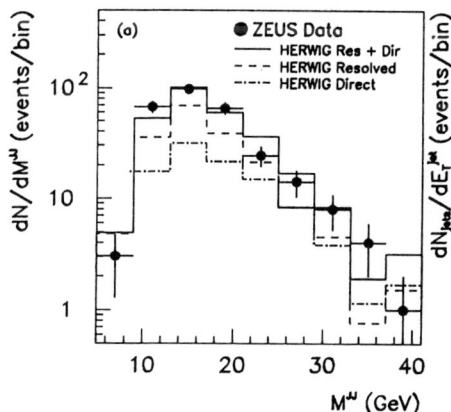

Figure 58. *Two-jet mass for events with two or more jets (ZEUS 1993i).*

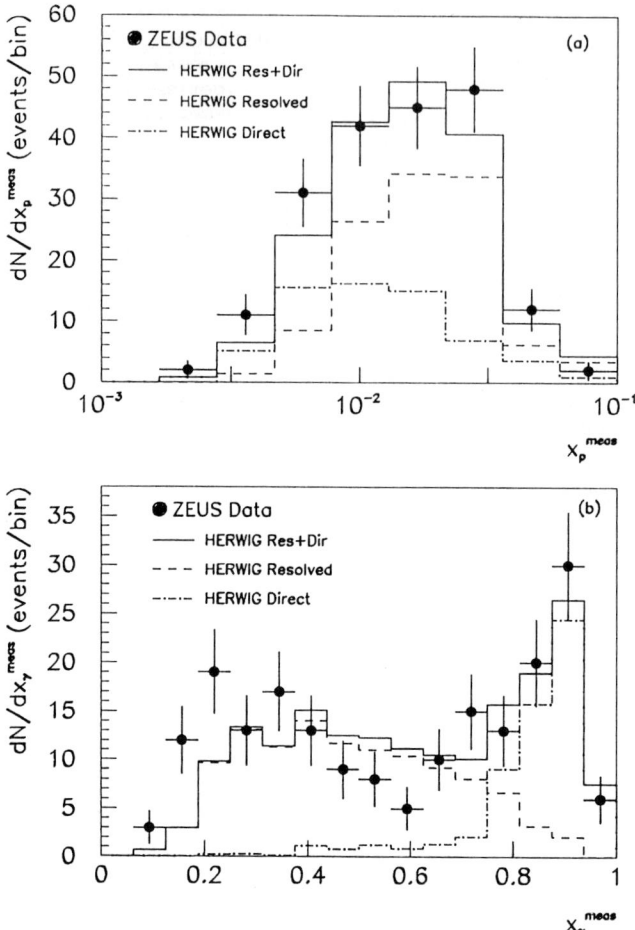

Figure 59. Distributions of (a) x_p^{meas} and (b) x_γ^{meas} for events with two or more jets. The Monte Carlo distributions show the fit of the resolved plus direct contributions to the data in (b) (ZEUS 1993i).

been established unambiguously from the study of two-jet production. At moderate transverse momenta the resolved process is the dominant one. The ensuing picture for the photon in hard scattering (Figure 60) is then either a photon which interacts with a parton (direct process) or a photon which is a bag full of quarks and gluons, of which one interacts with a parton (resolved process).

From the observed features of the data, the determination of the photon structure function via photoproduction of jets looks promising. Important prerequisites for such analyses are next-to-leading order (NLO) QCD calculations. First results of such calculations have become available recently (Gordon and Storrow 1992, Bödeker 1992, Borzumati et al. 1993, Kramer and Salesch 1993, Greco and Vicini 1993). The expected dependence of the jet p_T distribution from resolved processes is shown in Figure 61.

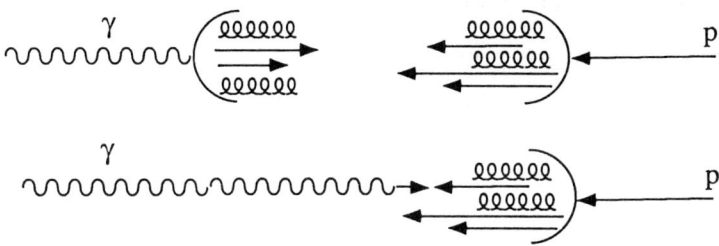

Figure 60. *Hard resolved (top) and direct photon proton scattering (bottom).*

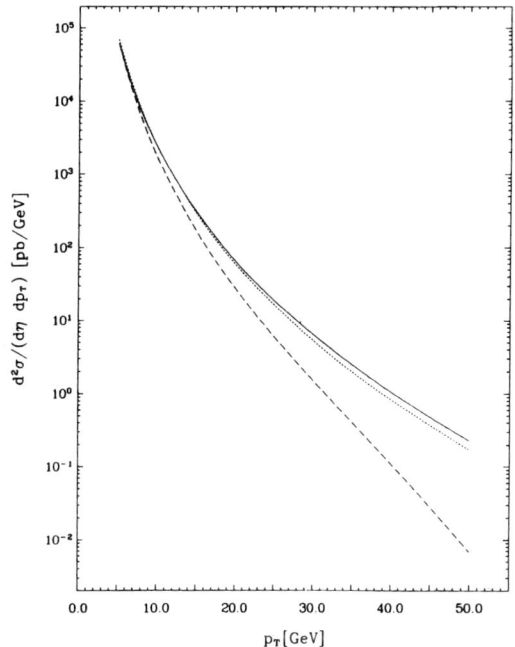

Figure 61. *Dependence on the transverse jet momentum p_T of the full NLO cross section for jet cone radius $R = 1$ and jet rapidities in the laboratory of $\eta = 1$ (full), $\eta = 2$ (dotted) and $\eta = 3$ (dashed); from Kramer and Salesch (1993).*

7 DIS and structure functions of the proton

H1 analysis The inclusive cross section for $ep \to eX$ was determined from autumn 1992 data based on a luminosity of 22.5 nb^{-1} (H1 1993f, i). The scattered electron was detected in the rear (electromagnetic) calorimeter BEMC and by the tracking detectors in front of the BEMC providing the energy E'_e and scattering angle θ'_e. The electron variables were determined from the electron based values of x_e, y_e, Q^2_e according to Equation (5.1). The observed E'_e and θ'_e distributions (Figure 62) are well reproduced by the detector simulation program. The hadron system was measured by a combination

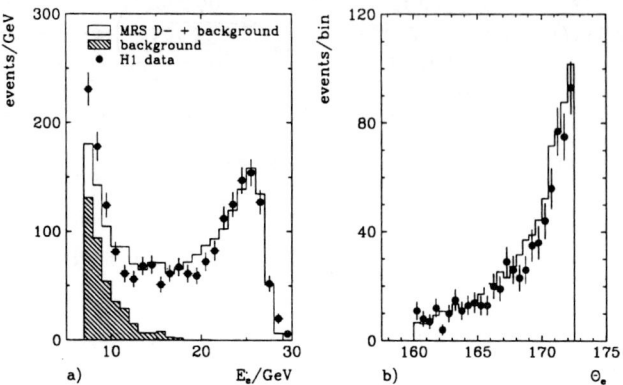

Figure 62. *Distribution of (a) the energy E'_e of the scattered electron before some of the selection cuts; (b) the electron scattering angle θ'_e (H1 1993f, i).*

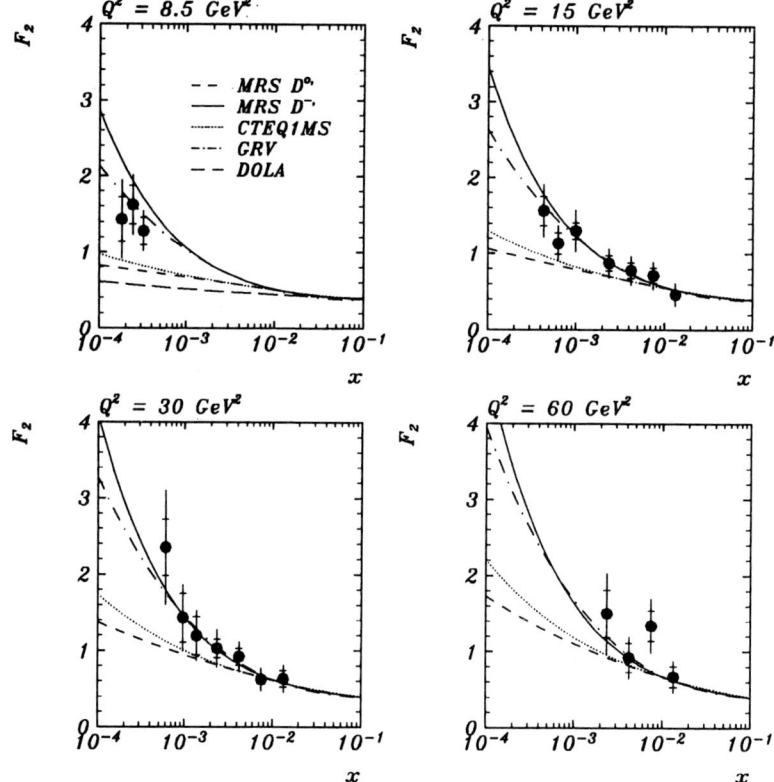

Figure 63. *The structure function $F_2(x, Q^2)$ for different Q^2 intervals. The error bars show statistical and total errors (H1 1993f).*

of calorimetric measurements and reconstructed charged tracks in the central region yielding the hadron based variables x_h, y_h, Q_h^2 according to Equation (5.3). For Q^2, the electron measurement Q_e^2 was used since it has the better resolution. The y value was taken from y_e except for $y_h < 0.3$ where y_h was chosen since it gave a more precise measurement. For a part of the analysis a mixture of electron and hadron information was used to calculate x, $x_m = Q_e^2/sy_h$.

The analysis was restricted to Q^2-values between 5 and 80 GeV^2. For the extraction of the structure function F_2, the contribution of the longitudinal part, F_L, was calculated from QCD. Its contribution to the DIS cross section was found to be small, being at most 8%. Corrections for QED radiative effects were applied using TERAD91 (1991) or HERACLES (1992). The F_2 results obtained from a total of 1026 events are summarised in Figures 63, 64 for fixed bins of Q^2 and x. The systematic errors range from 15–22% to which a global uncertainty from normalisation of 8% has to be added.

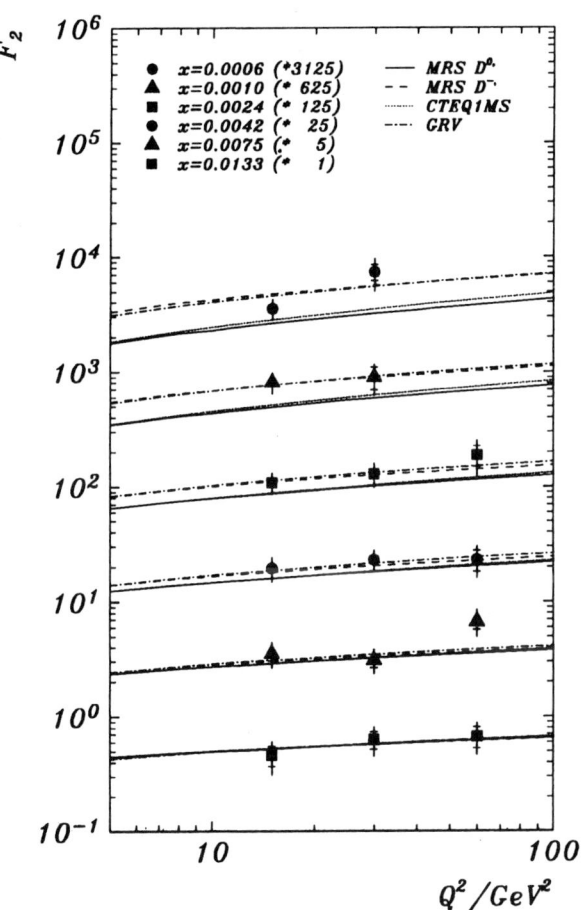

Figure 64. *The structure function $F_2(x, Q^2)$ as in Figure 63 for different x intervals.*

ZEUS analysis The data sample used for the determination of F_2 was obtained in the autumn of 1992 and corresponds to a luminosity of 24.7 nb^{-1} (ZEUS 1993h). The kinematic properties of the scattered electron and produced hadrons were determined with the calorimeter. The values of x and Q^2 were calculated with the double-angle method (Equation 5.6) providing x_{DA}, Q^2_{DA} from the measured angles of the scattered electron and the hadron system. The measured x_{DA} and Q^2_{DA} distributions are in agreement with the expected ones (Figure 65a,b). The distribution of the events in the x_{DA}, Q^2_{DA} plane is displayed in Figure 65c with Q^2_{DA} reaching up to 4,700 GeV^2. Also indicated in Figure 65c is the kinematical area covered by non-HERA experiments.

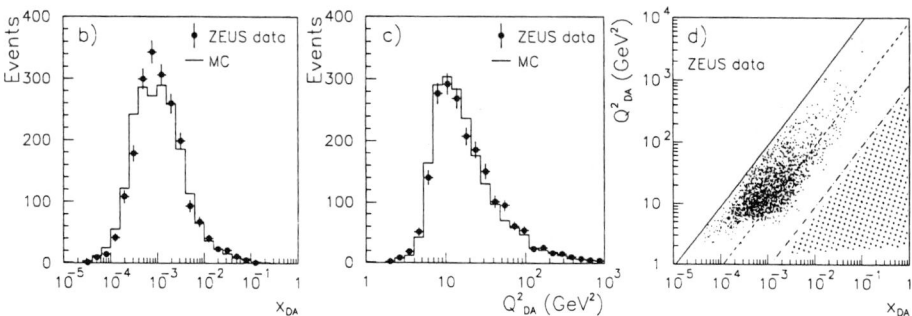

Figure 65. *The measured x and Q^2 distributions and a plot of x v. Q^2 (ZEUS 1993h). The shaded area in (c) shows the x-Q^2 region covered by fixed target experiments.*

The structure function analysis was performed for $Q^2 > 10 \, \text{GeV}^2$ with 1299 events in the chosen x, Q^2 bins. The contribution from F_L was calculated from QCD giving a maximum correction of 12%. Corrections for radiative effects from QED processes were calculated with HERACLES. Figures 66, 67 show F_2 as a function of x and Q^2. The systematic errors range from 7 to 31%; a global normalisation uncertainty of 7% has to be added.

7.1 Discussion of the F_2 results

The results on F_2 from H1 and ZEUS are summarised in Figure 68. Also shown in Figure 68 are measurements from BCDMS (1989) and NMC (1992). The HERA data connect on nicely with the BCDMS and NMC results. The HERA data extend the x-range for F_2 by two orders of magnitude down to $x = 1.6 \times 10^{-4}$. As x decreases from 10^{-2} to 4×10^{-4} the structure function rises by a factor of 2—reported for the first time by H1 (1993i). The Q^2 dependence of F_2 for fixed x is shown to be in good accord with the expected logarithmic violation of scaling (Figure 64, 67).

The data on F_2 are compared in Figure 68 with different extrapolations or predictions. MRS (D'_-), MRS (D'_0) (Martin, Stirling and Roberts 1993) and CTEQ 1MS (CTEQ 1993) are extrapolations of structure function parametrisations obtained from fits to low-energy data at larger x values. For MRS (D'_-) the small x evolution of the gluon density at $Q^2_0 = 4 \, \text{GeV}^2$ is singular [$x \, g(x) \equiv G(x) \sim x^{-0.5}$, Lipatov behaviour], while it

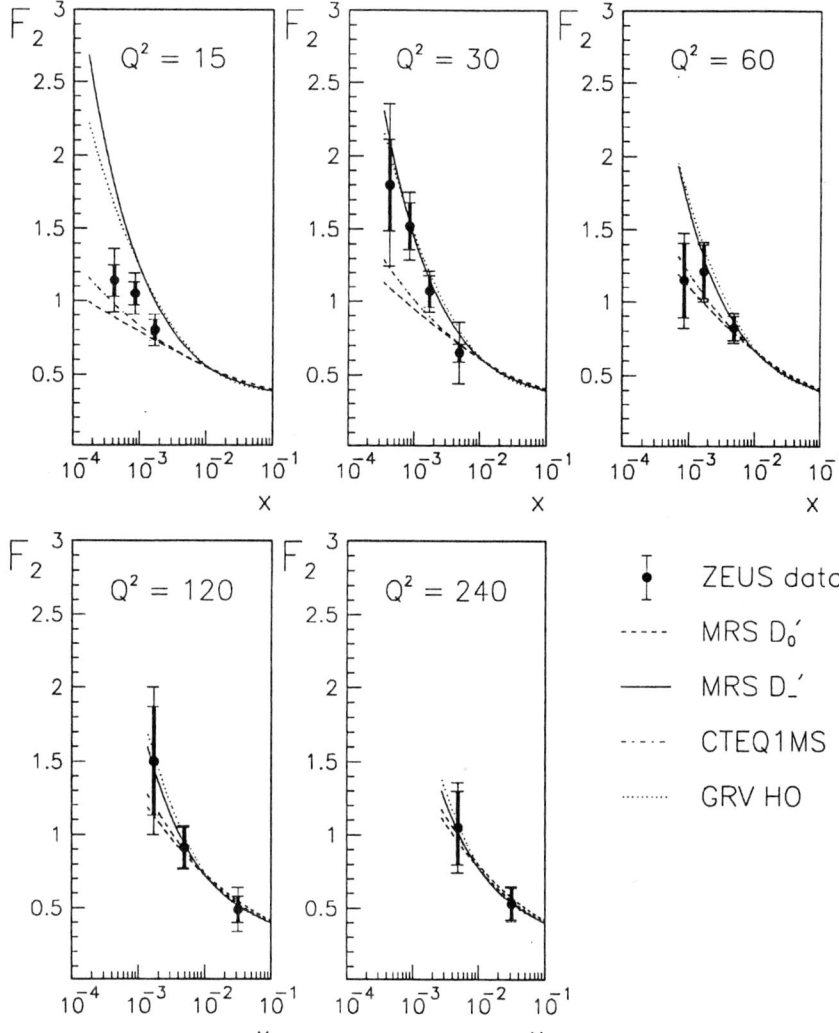

Figure 66. The structure function $F_2(x, Q^2)$ for different Q^2 intervals. The inner error bars are the statistical errors, and the outer error bars show the systematic errors added in quadrature (ZEUS 1993h).

is constant for MRS (D_0') [$G(x)\to$const. for $x\to 0$]. Note that MRS (D_0'), evolved downwards below $Q^2 = 4\,\text{GeV}^2$, would lead to negative parton densities. For CTEQ 1MS the gluon density is singular, but the sea quark distribution is not strongly coupled to the gluon density, leading to a slower rise of F_2 with decreasing x. GRV HO (Glück, Reya and Vogt 1993) is the prediction of a higher-order QCD calculation. Following a proposal by Parisi and Petronzio (1976) the small x partons are radiatively generated according to the GLDAP equations, starting from 'valence like' quark and gluon distributions at $Q_0^2 = 0.3\,\text{GeV}^2$ (see Figure 69). This procedure has been questioned because of the

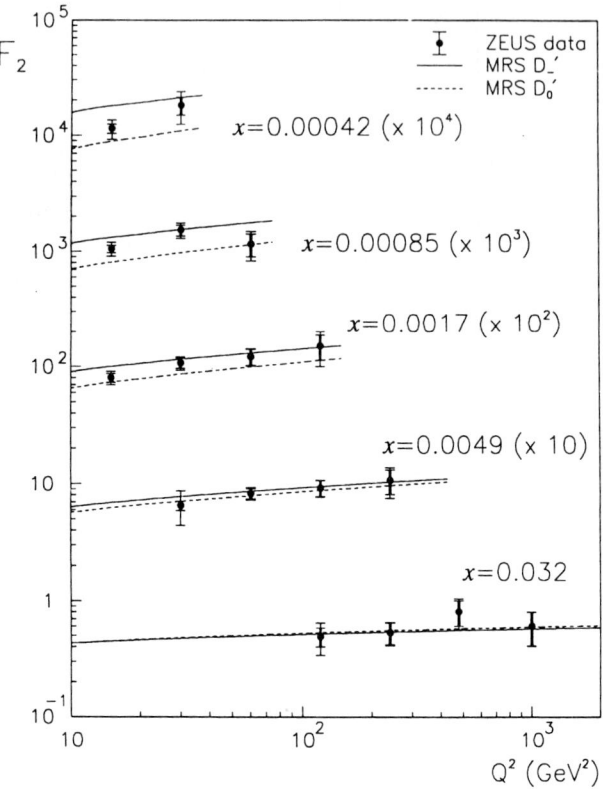

Figure 67. *The structure function $F_2(x, Q^2)$ as in Figure 66 for different x intervals* (ZEUS 1993h).

small value of $Q_0^2 \simeq 5\text{--}10\Lambda^2$ at which the evolution starts (see e.g. Forshaw 1993). The parametrisation DL (Donnachie and Landshoff 1993) is a Regge theory motivated fit, which is applicable for Q^2 values up to about 10 GeV^2. It is worth noting that all five sets of structure functions give a good fit to the non-HERA data, covering x values above approximately 0.04, and that it takes measurements at much smaller x values such as available at HERA to select between the different physics possibilities.

The GRV HO and MRS (D'_-) curves reproduce the F_2 data reasonably well while CTEQ 1MS, MRS (D'_0) and DL fall below the measured points (Figure 68). The conclusion then is that the data prefer a rising gluon density leading also to a rising sea quark density as x decreases. An important message from GRV is that F_2 at low x may change rapidly with Q^2 between approximately 0.5 and 10 GeV^2 and precise data in this region will be of great importance. The Q^2 behaviour expected by GRV (1992) can be seen from Figure 70. The figure shows also that for Q^2 below 5 GeV^2 higher order QCD corrections become large.

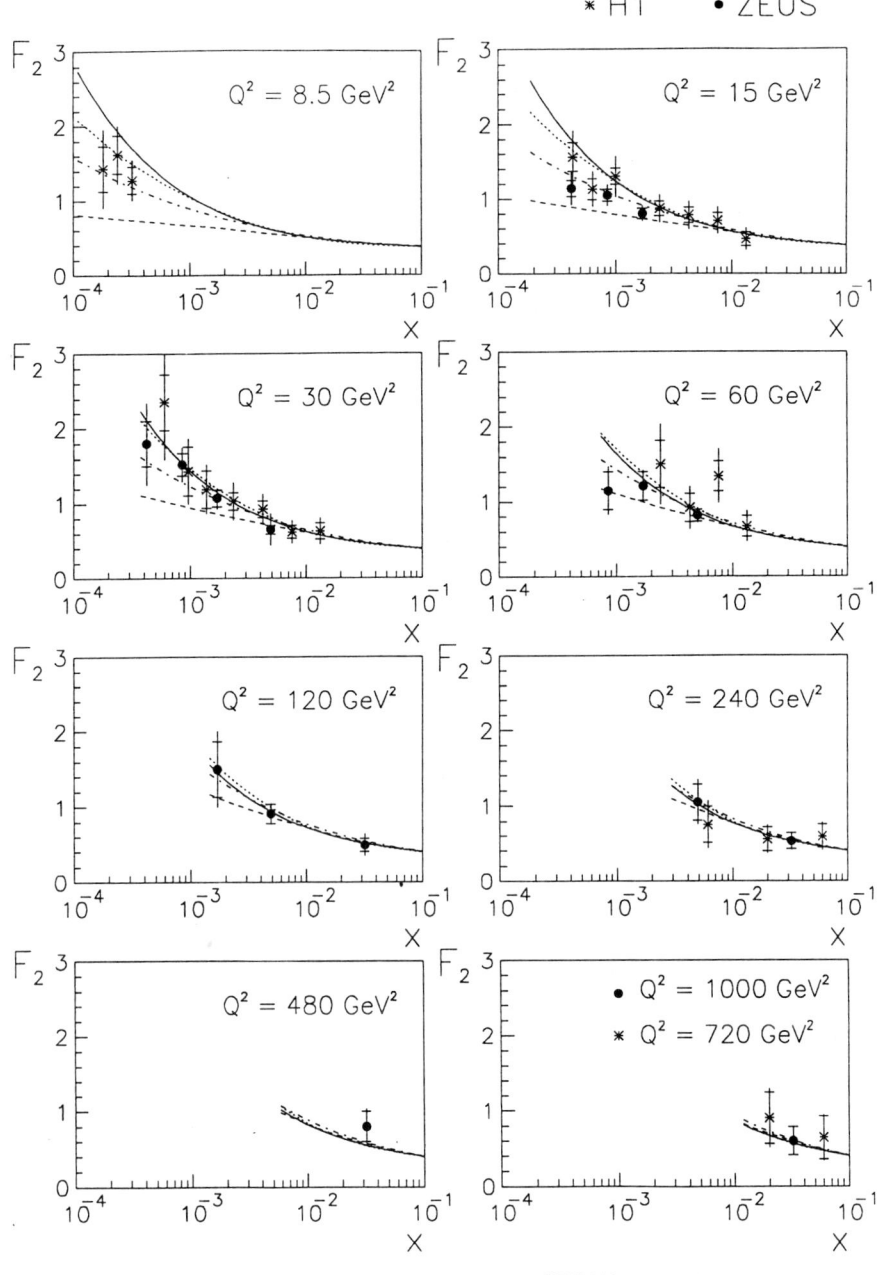

Figure 68. The $F_2(x, Q^2)$ data from H1 and ZEUS for fixed Q^2. The curves show the parametrisations MRS (D'_-) – solid line, MRS (D'_0) – dashed line and CTEQ 1MS – zigzag line and the predictions GRV HO and DL – dotted line.

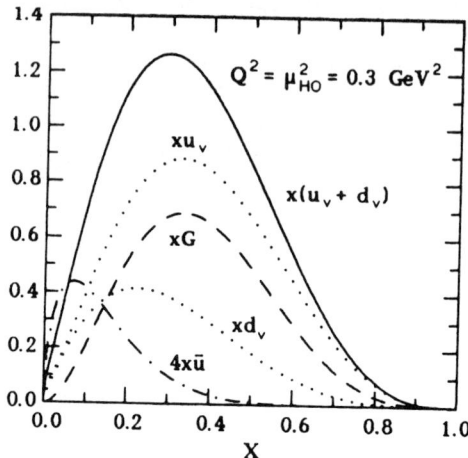

Figure 69. The 'valence' parton distributions of GRV from which the evolution is started at Q_0^2 (GRV 1992).

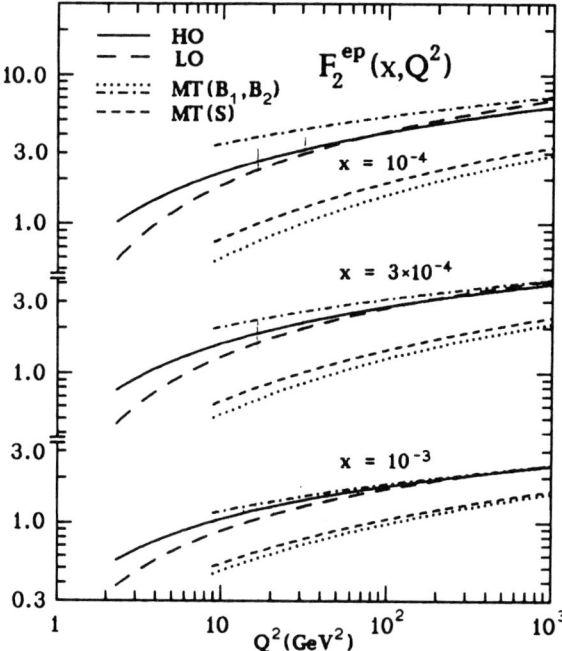

Figure 70. The GRV (1992) prediction for the Q^2 dependence of $F_2(x, Q^2)$ in leading (LO) and higher order (HO). Shown are also parametrisations by Morfin and Tung (1991).

7.2 Scaling violations of F_2 and the gluon structure function

The Q^2 dependence of F_2 at small values of x can be related in a straightforward albeit approximate way with the gluon density $G(x, Q^2)$ of the proton. This was shown by Prytz (1993) following a procedure which had been applied before to derive $G(x, Q^2)$ from the longitudinal structure function $F_L(x, Q^2)$ (Cooper-Sarkar et al. 1988). It is instructive to follow the exercise in some detail.

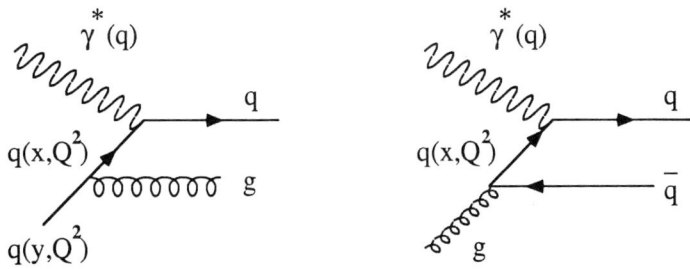

Figure 71. *Lowest order contributions to scaling violations of $F_2(x, Q^2)$.*

Leading-order (LO) QCD diagrams are shown in Figure 71: in the first process a quark carrying a momentum fraction y of the proton emits a gluon, retains a fraction x of the proton momentum and scatters on the virtual photon. In the second process the proton emits a gluon which splits into a $q\bar{q}$ pair one of which, carrying a fraction x of the proton momentum, interacts with the virtual photon. The GLDAP evolution equation for the density of quark i which interacts with the photon reads:

$$\frac{dq_i(x, Q^2)}{d\ln Q^2} = \frac{\alpha_s(Q^2)}{2\pi} \int_0^1 \frac{dy}{y} \left[P_{qq}(x/y) q_i(y, Q^2) + P_{qg}(x/y) g(y, Q^2) \right]. \quad (7.1)$$

$\alpha_s P_{qq}(x/y)$ is proportional to the probability that a quark with momentum fraction x has come from an initial state quark with momentum fraction y which has radiated a gluon.

$\alpha_s P_{qg}(x/y)$ is proportional to the probability that a quark with momentum fraction x has come from a $q\bar{q}$ pair created by a gluon.

In leading order

$$\alpha_s(Q^2) = 12\pi/[(33 - 2N_f)\ln(Q^2/\Lambda^2)], \quad (7.2)$$

where N_f is the number of flavours. Since the typical mass of the $q\bar{q}$ pair is less than 10 GeV, $N_f = 4$ seems a good assumption. From (5.21) and (7.1) follows:

$$F_2(x, Q^2) = x \sum_q e_q^2 q(x, Q^2)$$

$$\frac{dF_2}{d\ln Q^2} = \sum_q e_q^2 \frac{\alpha_s(Q^2)}{2\pi} \int_0^1 \frac{dy}{y} \left[P_{qq}(x/y) q_i(y, Q^2) + P_{qg}(x/y) g(y, Q^2) \right]. \quad (7.3)$$

At very low x the quark contribution can be neglected. Taking the leading order result for P_{qg},

$$P_{qg}(u) = 0.5\{(1-u)^2 + u^2\}, \quad (7.4)$$

making the replacement $y = x/(1-z)$ and defining $G(x, Q^2) = xg(x, Q^2)$ leads to

$$\frac{dF_2}{d\ln Q^2} \simeq \sum_i e_i^2 \frac{\alpha_s(Q^2)}{2\pi} \int_0^{1-x} dz P_{qg}(z) G(x/(1-z), Q^2). \tag{7.5}$$

A Taylor expansion of $G(y, Q^2)$ is made about $z = 1/2$. Noting that P_{qg} is symmetric around $z = 1/2$ and that the second and higher derivatives of $G(y, Q^2)$ are small if $G(y, Q^2)$ is of the form $y^d(1-y)^a$, leads to the final result

$$\frac{dF_2}{d\ln Q^2} \simeq \sum_i e_i^2 \frac{\alpha_s(Q^2)}{2\pi} \frac{1}{3} G(2x, Q^2)$$

$$G(2x, Q^2) \simeq \frac{27\pi}{10\alpha_s(Q^2)} \frac{dF_2(x, Q^2)}{d\ln Q^2}. \tag{7.6}$$

ZEUS (1993 l) and H1 (1993h) analysed their F_2 data according to (7.6) and obtained the results for $G(x, Q^2)$ shown in Figure 72. The data shown from ZEUS were obtained by recalculating them for $Q^2 = 20\,\text{GeV}^2$ and $N_f = 4$, as used by H1. The two experiments agree rather well. Since for both experiments the dominant error is the statistical one, combining the two measurements produces an approximately 30–40% measurement of G at small x. The data indicate that $G(x, Q^2)$ could be very large at small x.

In a recent paper, R.K. Ellis, Kunszt and Levin (1993) presented a procedure for extracting $G(x, Q^2)$ from data on $F_2(x, Q^2)$ without neglecting the quark contribution and including terms up to third order in α_s.

7.3 The energy dependence of $\sigma(\gamma^* p)$

The F_2 data from H1 and ZEUS can be approximated by a function of the form $a(Q^2) + b(Q^2) x^{-1/2}$. Ignoring the logarithmic scaling violations (see above), a reasonable description is obtained with

$$F_2(x, Q^2) \simeq 0.3(1 + 0.1\, x^{-1/2}) \tag{7.7}$$

for the region $x = 4\times 10^{-4}$–1×10^{-2}, $Q^2 = 10$–$40\,\text{GeV}^2$. The $x^{-1/2}$ accounts for the rise of the structure function at small values of x (Lipatov term). Neglecting longitudinal contributions, this leads to the following expression for the total cross section of virtual transverse photon proton scattering (see 5.12):

$$\sigma_{\text{Ttot}}(\gamma^* p) \simeq (4\pi^2\alpha/Q^2) F_2(x, Q^2)/(1-x) \simeq (4\pi^2\alpha/Q^2)[0.3(1 + 0.1 W/Q)] \tag{7.8}$$

For fixed Q^2, the total cross section has a term which is independent of the c.m. energy and a second part, the Lipatov term, which rises with W. Using the optical theorem, $\text{Im} A(\gamma^* p) \sim \sigma_{\text{Ttot}}(\gamma^* p)$, where $A(\gamma^* p)$ is the invariant amplitude for forward elastic $\gamma^* p$ scattering, one sees that at very high values of W and fixed Q^2 the elastic cross section will exceed the total one, $\sigma_{\text{el}}(\gamma^* p) > \sigma_{\text{Ttot}}(\gamma^* p)$. The Lipatov term will violate unitarity unless it is damped at some $x < x_{\text{crit}}$ (see the discussion in Section 5.6).

Comparison with the data on $\sigma_{\text{Ttot}}(\gamma p)$ for *real* photons at $W = 18$ and 200 GeV (see Section 6) shows that in the DIS data at $Q^2 = 10\,\text{GeV}^2$ the contribution from the W-dependent term is much larger than at $Q^2 = 0$, where

$$\sigma_{\text{Ttot}}(\gamma p) \simeq \text{const.}\,[1 + 0.001 W].$$

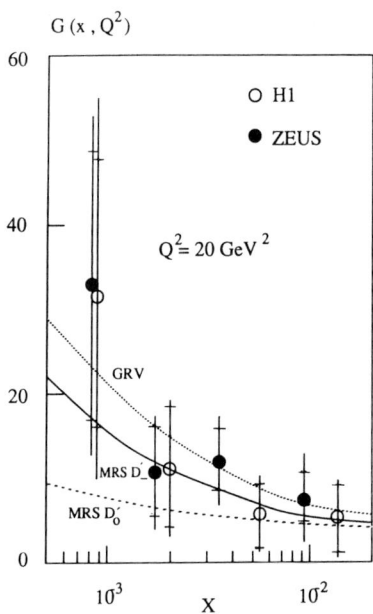

Figure 72. *The gluon distribution function $G(x, Q^2)$ for the proton at $Q^2 = 20\,\text{GeV}^2$ as measured by* H1 *(1993h) and* ZEUS *(1993 l). The inner error bars are the statistical errors, the full have the systematic errors added in quadrature. The curves show different parametrisations or predictions for $G(x, Q^2)$:* CTEQ *(1993) and* GRV *(1993) are calculated in* LO*; the* MRS *curves are given in the* DIS *renormalisation scheme.*

8 Final states in DIS

In lowest order lepton-proton NC scattering, the transverse momentum of the scattered electron is balanced by a single jet associated with the struck quark, the proton remnant carrying relatively little transverse momentum. Higher order QCD processes modify this picture. In particular, a hard gluon can be radiated from the struck quark (QCD Compton scattering, QCDC), or a gluon from the proton can interact with the exchanged boson giving rise to quark-antiquark production (Boson Gluon Fusion, BGF), as illustrated in Figure 73.

One consequence of these higher order processes is the broadening of the transverse momentum distribution of the particles with respect to the exchanged boson direction. This was observed, for example, by EMC (1981, 1987). The production of multi-jet events was reported in a fixed-target muon experiment at c.m. energies up to approximately 30 GeV(E665 1992). By virtue of the large photon-proton c.m. energies, W, available at HERA the contributions from these higher order QCD diagrams become directly visible as multijet events. This was demonstrated by ZEUS and H1.

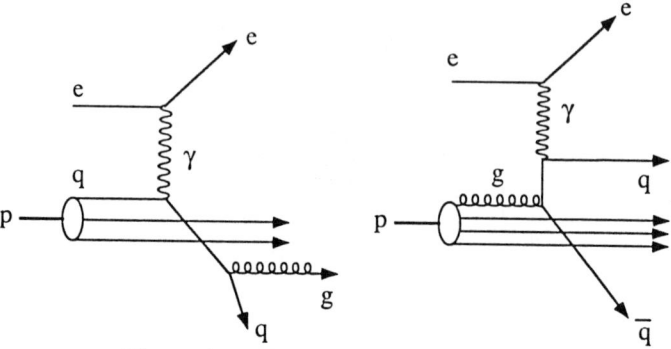

Figure 73. *Diagrams for* QCDC *and* BGF.

8.1 QCD models

For a quantitative comparison of the measurements with QCD predictions, models which describe the QCD processes at the parton level and the fragmentation into hadrons are indispensable. In general, the models are incorporated into Monte Carlo generators which include the simulation of the passage of final state particles through the detector. The features of models which are considered frequently by H1 and ZEUS are briefly summarised.

In LEPTO QCD processes are calculated up to $O(\alpha_s)$ according to exact first order matrix elements (ME). The QPM prediction is obtained by turning $O(\alpha_s)$ matrix elements off. Higher order contributions are simulated in the leading logarithm approximation (parton shower approach, PS). The struck quark can emit partons either before or after the boson vertex (Figure 74). As the quark is radiating the initial state shower—before the boson vertex—it becomes more off-shell or virtual. After the interaction, the quark may again be off-mass-shell and returns to the mass shell by radiating the final state shower. The amount of gluon radiation depends on the scale of virtuality which can be chosen to be either Q^2 or W^2 (PS(Q^2) or PS(W^2)), or a function of both. Since $\langle Q^2 \rangle$ is much smaller than $\langle W^2 \rangle$ in the experiment, PS(W^2) will predict much more gluon radiation than PS(Q^2). The combination of the two approaches (ME+PS) gives the first order parton emission plus the higher-order emissions through parton showers. The probabilities for all partonic subprocesses are matched to avoid double counting. The fragmentation into hadrons is performed with the Lund string model as implemented in JETSET (1986).

In contrast to the bremsstrahlung-like parton shower model PS the colour dipole model CDM does not distinguish between initial and final state radiation and includes interference effects between them. The struck parton and the proton remnant form a colour dipole. When this dipole radiates a gluon, it splits into two radiating dipoles: one between the struck quark and the gluon and the other between the gluon and the remnant. Repeated gluon emission leads to a chain of such dipoles. Since one end of the dipole is considered nonpoint-like—the proton remnant—the maximum p_T^2 in the hadronic c.m. for an emitted gluon varies as $W^{4/3}$. The simulation of QCDC for DIS is only approximate because of the extended proton remnant. CDM is implemented in ARIADNE (1992) which in turn is using LEPTO for generating the hard scattering.

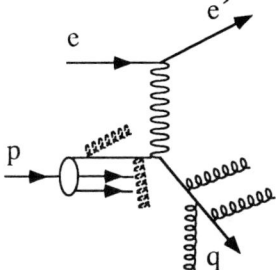

Figure 74. *Diagram with parton showers.*

The HERWIG (1992) generator does not consider explicitly the $O(\alpha_s)$ matrix elements. Rather, leading-log parton showers are considered. The parton shower takes place inside a cone of angular size set by the incoming and outgoing struck quark. In DIS this correlates the initial and final state parton showers. The characteristic scale is given by $2E^2(1 - \cos\Psi)$ where E is the energy of the parton and Ψ is the angle with respect to its colour connected partner. The scale is essentially Q^2. It sets the upper limit for the shower evolution variables. Fragmentation into hadrons is modelled with a cluster fragmentation model.

8.2 Jet production

Jet production in DIS was analysed by ZEUS (1993c) for an integrated luminosity of $27\,\text{nb}^{-1}$. Events were selected with $Q^2 > 4\,\text{GeV}^2$, $x \geq 3\times 10^{-4}$ and $30 < W < 280\,\text{GeV}$, the average being 110 GeV. The transverse momentum distribution was determined in the $\gamma^* p$ rest system with the help of condensates. Condensates are contiguous energy deposits in the calorimeter defined as follows: the calorimeter cell with the highest energy is used as the seed and the cells adjacent to the seed cell are merged. According to simulations, for polar angles above 9°, the condensate multiplicity agrees with that of the final state particles to 90% and about 80% of the generated stable particles are associated with only one condensate. In the analysis condensates were treated as massless particles.

From the condensates the sphericity axis was determined in the $\gamma^* p$ rest system. Figure 75 shows the distribution of the square of the transverse momentum, p_T^2, of the condensates w.r.t. the sphericity axis. The data were restricted to the hemisphere around the γ^* direction, which should contain the struck quark, and where one expects the gluon radiation effects to be most visible. The measured distribution is clearly much broader than the quark-parton model (QPM) expectations. The tail extends up to approximately $20\,\text{GeV}^2$ while the QPM predicts no events above $5\,\text{GeV}^2$.

A search was carried out for two or more jets (in addition to the proton remnant and excluding the scattered electron) in the HERA system. Jets were required to have a transverse energy greater than 4 GeV in a cone of radius $\Delta R = (\Delta\phi^2 + \Delta\eta^2)^{1/2} = 1$ and $\eta_{\text{jet}} < 2$. With this algorithm, 2502 (76%), 662 (20%), 95 (2.9%) and 15 (0.5%) events with zero, one, two or three jets were found. Examples of one-, two-and three-jet events are shown in Figure 76. Note that the jets are isolated and their transverse energies are large, ranging from 5 to 30 GeV for the events shown.

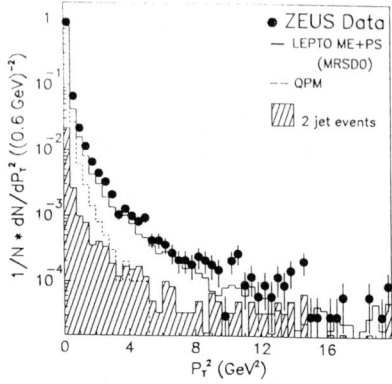

Figure 75. *Distribution of transverse momentum squared of condensates in the $\gamma^* p$ c.m. system, measured w.r.t. the sphericity axis (ZEUS 1993c).*

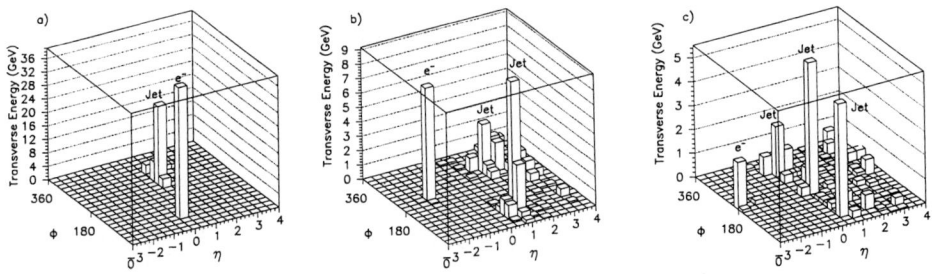

Figure 76. *Transverse energy distributions of events with jets in the (η, ϕ) plane: (a) one-jet event, (b) two-jet event, (c) three-jet event (ZEUS 1993c).*

Two-jet events are candidates for the **QCDC** and **BGF** processes. The shaded area in Figure 75 shows the distribution for two-jet events. The E_{Tjet} distribution extends out to 22 GeV (Figure 77a, b). The pseudorapidity distribution rises towards large η values (Figure 77c). The two jets are preferably back-to-back in the plane transverse to the beams (Figure 77d). The jet energy profile, which is sensitive to fragmentation effects, is shown in Figure 78 where the transverse energy flow into cells within a cone radius of $R = 2$ around the jet axis was computed. The jet width is 0.5–0.6 (**FWHM**) in ϕ and η.

The data were compared with the expectations from **QPM** and **QCD** (Figures 75, 77 and 78). The simple **QPM** fails to describe the data. First order **QCD** matrix elements plus parton showers (**ME+PS**) are in accord with the data: they reproduce the jet rates, describe the p_T^2 distributions as well as the jet E_T distribution and energy profile. It is noteworthy that the calculation based on exact first order matrix elements alone (**ME**), i.e. without parton showers, predicts a narrower jet profile than observed (Figure 78). The agreement with the **QCD** predictions suggests that two-jet production with large transverse jet energies is dominated by the **QCD** Compton and **BGF** processes.

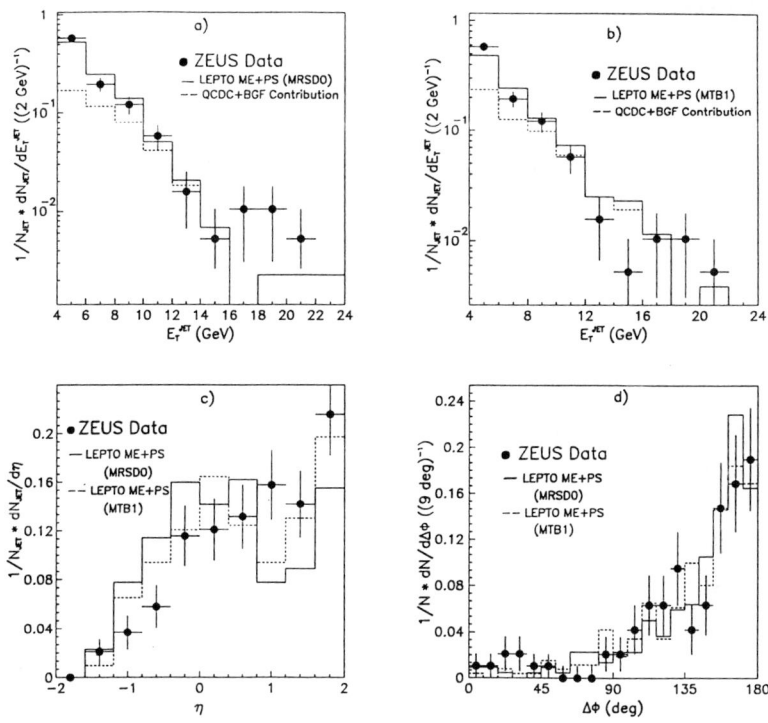

Figure 77. *Distributions of jets for two-jet events measured in the lab. frame w.r.t. the beam direction: (a), (b) transverse jet energy, (c) pseudorapidity and (d) azimuthal angular separation between the two jets* (ZEUS 1993c).

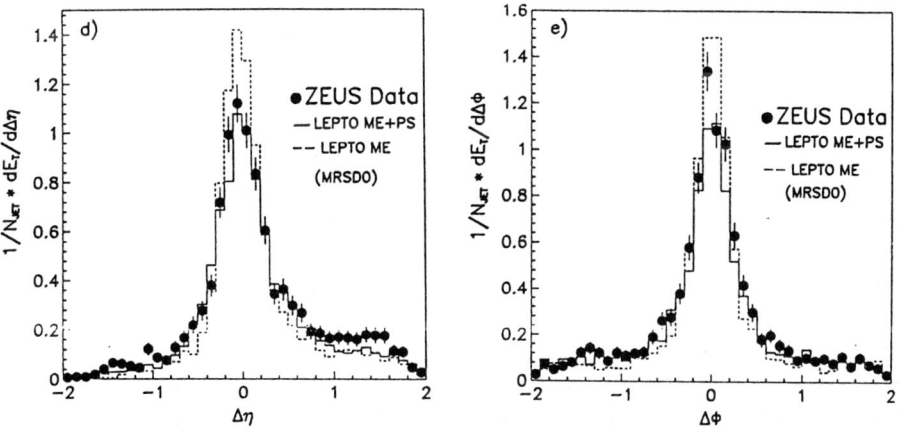

Figure 78. *The profile of jets belonging to the two and three-jet samples* (ZEUS 1993c).

8.3 Jet rates

H1 (1993g) studied the jet multiplicity in NC scattering as a function of the jet resolution with the aim of measuring α_s. The event sample consisted of 769 (47) events with $12 < Q^2 < 80\,\text{GeV}^2$ ($Q^2 > 100\,\text{GeV}^2$) obtained from a luminosity of 22.5 nb^{-1}. Jets were selected with the JADE algorithm (JADE 1986). Two jets were considered to be resolved if their scaled invariant mass satisfied the condition:

$$y_{ij} = m_{ij}^2/W^2 > y_{\text{cut}}. \tag{8.1}$$

Here $m_{ij}^2 = 2E_iE_j(1-\cos\theta_{ij})$, E_i, E_j are the jet energies and θ_{ij} is the angle between them. Events with N jets in addition to a jet from the proton remnant (spectator jet) were classified as $(N+1)$ jet events. The lowest order configuration (QPM) with a current quark and a proton remnant in the final state is expected to produce preferentially $(1+1)$ events; the first order QCD processes should yield predominantly $(2+1)$ events. The fraction of $(2+1)$ to $(1+1)$ events should therefore be sensitive to the value of α_s.

Figure 79 shows the fractional jet rates for the $(1+1)$, $(2+1)$ and $(\geq 3+1)$ configurations as a function of y_{cut} for the low and high Q^2 samples. The data are uncorrected for acceptance and other detector effects. Note, that the data points are highly correlated since the same event sample is used for each value of y_{cut}.

For the test of QCD and the extraction of α_s the optimum value of y_{cut} is one where, firstly, the leading (LO) and all higher order QCD processes and, secondly, the transition from the parton to the hadron level give practically the same (2+1) fraction. Thirdly, the (2+1) fraction should not be too sensitive to detector smearing effects. The curves in Figure 79 show the predictions of various QCD models which were introduced above and which include hadronisation and detector smearing; calculation of the hard scattering part is done in LO only. For the ME+PS model the differences between the predictions at the parton and hadron levels become small near $y_{\text{cut}} \simeq 0.02$ (Figure 79a, c). Detector smearing has a 10–15% effect on the $(2+1)$ fraction near this y_{cut} value. Of various models studied, ME+PS describes the data best (Figures 79b, d).

The Q^2 dependence of the $(2+1)$ fraction measured for $y_{\text{cut}} = 0.02$ is compared in Figure 80 with the predictions (a) of ME+PS for a fixed $\alpha_s = 0.25$, and with a running α_s (with $\Lambda = 200\,\text{MeV}$) (b) for ME alone and (c) for ME + PS. The expected Q^2 behaviour of α_s for (a) and (b) or (c) is different. Because of the Q^2 dependence of α_s, the matrix element for jet production decreases with increasing Q^2. This decrease is not manifest directly in R_{2+1} because of the increase in phase space for jet production at larger Q^2 with the selection cuts applied. With higher statistics and good control of systematic effects from the detector and theory a test for a running α_s and a measurement of Λ may be feasible.

ZEUS (1993k) performed a similar study. The jet rates, corrected for detector effects and initial state QED radiation are shown in Figure 81. The correction factor proceeding from the detector level to the produced hadronic final state varies with y_{cut} between 5 and 20%. The difference between the hadron and the parton levels is below 10% for most of the region. ME+PS provides a good description for the $(0,1+1)$ and $(2+1)$ rates. The conclusion on the prospect for measuring α_s are the same as before.

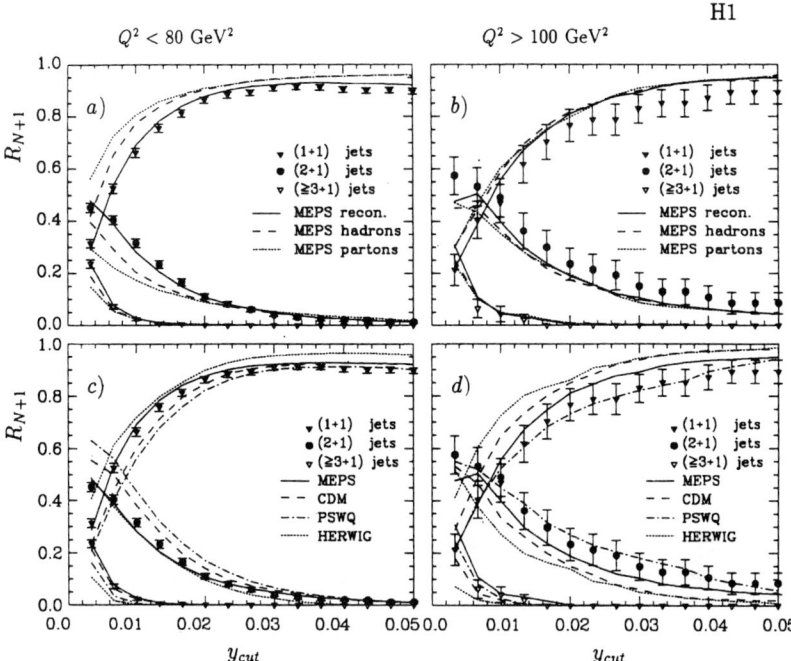

Figure 79. *Fraction of $N+1$ jets (R_{N+1}) versus the cut variable of the jet algorithm for $12 < Q^2 < 80\,\mathrm{GeV}^2$ (a, c) and $Q^2 > 100\,\mathrm{GeV}^2$ (b, d), compared with simulations of ME+PS at the detector level and at the parton level, and in (c, d) to predictions from QCD based models ME+PS, CDM, PS($W \cdot Q$) and HERWIG (H1 1993g).*

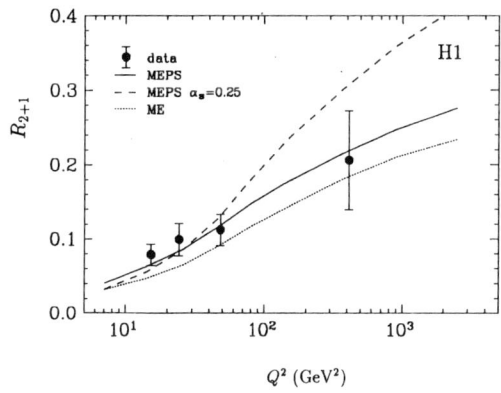

Figure 80. *$(2+1)$ jet fraction R_{2+1} at $y_{\mathrm{cut}} = 0.02$ versus Q^2, corrected for detector effects for $W^2 > 5000\,\mathrm{GeV}^2$ and $y < 0.5$, in comparison with the ME+PS model with a running α_s and a constant $\alpha_s = 0.25$, and also with the matrix elements with no showers (ME) (H1 1993g).*

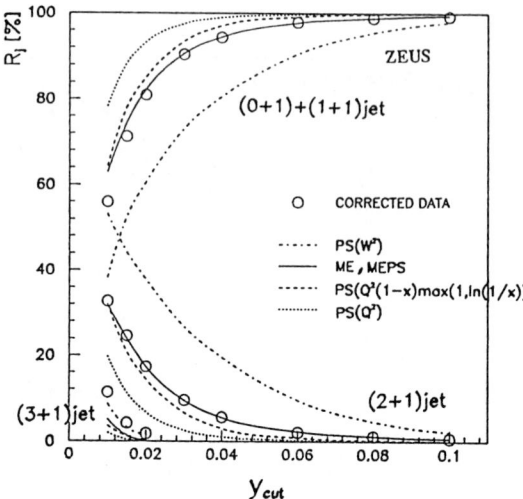

Figure 81. *Fraction of $N+1$ jets (R_{N+1}), corrected for detector effects and initial state radiation, versus the cut variable of the jet algorithm for $x < 10^{-3}$ and $Q^2 > 10\,\text{GeV}^2$. Predictions from* QCD *based models* ME, PS, ME+PS *are also shown (*ZEUS *1993k).*

8.4 Energy flow

In the quark parton model (QPM) the final hadronic state in DIS consists of a jet of particles originating from the struck quark representing the current jet and the particles produced by the proton remnant. Since the proton remnant is assumed to carry very little transverse momentum, the transverse momentum of the scattered electron is balanced by the current jet (Figure 82). The direction γ_h of the current jet can be calculated from the energy and angle of the scattered electron (Equation 5.6). QCD introduces substantial corrections to this simple picture. The phase space between the current jet and the proton remnant is filled with particles materialising from the emission of additional gluons and quarks created by colour transfer between the struck quark and the proton remnant.

The energy flow around the direction of the hypothetical struck quark was found to be very revealing when confronting these ideas with the data from ZEUS 1993e. The analysis was made with DIS events from a luminosity of 30 nb^{-1}. Figure 82 shows the energy flow as a function of the pseudorapidity calculated from the calorimeter cell energies relative to the direction of the struck quark,

$$\Delta\eta = \eta_{\text{cell}} - \eta_h,$$

where $\eta_h = -\ln\tan(\gamma_h/2)$. The events were required to have $Q^2 > 10\,\text{GeV}^2$ and $x < 10^{-3}$. Cells with $\theta < 10°$ ($\eta_{\text{cell}} > 2.4$) were removed to reduce the influence of the proton remnant. Averaged over the event sample, the direction of the struck quark is $\gamma_h \simeq 169°$ or $\eta_h \simeq -2.4$. In the QPM one expects the final state hadrons coming from the struck quark to concentrate around $\Delta\eta = 0$, and those from the proton remnant at large positive $\Delta\eta$ (see the sketch in Figure 82). The energy flow observed for the data

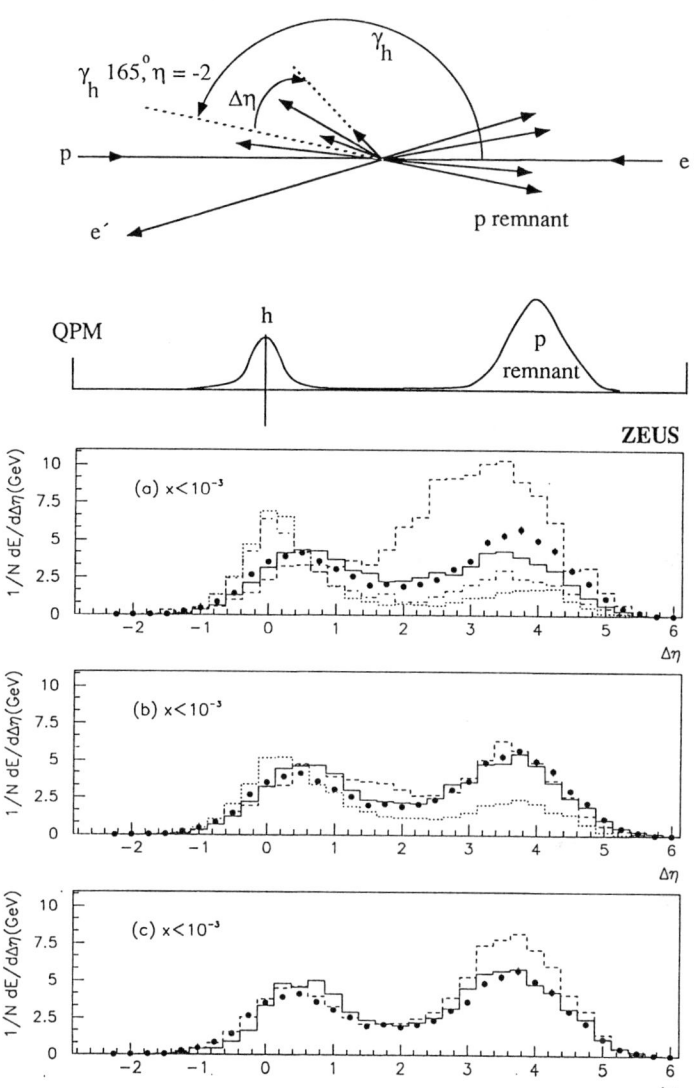

Figure 82. *Deep inelastic neutral current scattering at $x < 10^{-3}$, $Q^2 > 10\,\text{GeV}^2$: the energy weighted pseudorapidity difference, $\Delta\eta$, measured with the calorimeter w.r.t. the struck quark from the quark-parton model (see the sketch at the top). The data (dots) are compared with QCD models: (a) ME+PS (full histogram), ME (dashed-dotted), PS(W^2) (dashed), PS(Q^2) (dotted); (b) CDM+BGF (full), CDM (dashed), PS($Q^2(1-x)$) (dotted); (c) HERWIG (full) without soft underlying events (SUE) and including SUE (dashed); from ZEUS (1993e).*

shows several striking features. There are two peaks, one near $\Delta\eta = 0$, the other one at large $\Delta\eta$ values. The peak at low $\Delta\eta$ values is shifted from the expectation of the QPM by about 0.5 rapidity units in the positive direction to $\Delta\eta \simeq +0.5$. Furthermore, almost all energy appears at positive $\Delta\eta$ values, between the direction of γ_h and the proton remnant.

Qualitatively, a continuous colour flow producing a continuous energy flow between the struck quark direction and the proton remnant at these low-x values (*i.e.* high W values) is expected from QCD radiation when averaging over many events. Also, if the struck quark radiates a gluon in the final state the intermediate quark has a nonzero positive mass. Therefore, the momentum fraction of the proton carried by the struck quark is larger than x, calculated from the electron side (Equation 5.9), causing a shift towards positive $\Delta\eta$ values for the energy flow from the struck quark. Furthermore, initial state radiation will emit particles close to the proton direction and will therefore add to the energy flow in this region. Several QCD models, ME+PS, CDM+BGF (*i.e.* adding the matrix element for BGF in the calculation) and HERWIG, are found to give a quantitative description of the data.

9 Production of events with large rapidity gaps

The standard DIS events (see Figure 83) show energy deposition in the forward region, presumably coming from the fragmentation of the proton remnant, from initial state QCD radiation or from fragmentation of the struck quark. ZEUS (1993g) has observed a class of events which have different characteristics. In Figure 84 one of such events is displayed. It shows a well identified electron from which a Q^2 value 64 GeV^2 is inferred. The special feature of the event is the absence of energy deposition in the forward direction the first significant deposition of energy is found at $\theta > 90°$. The presence of this new type of events in DIS was confirmed by H1 (1993j, k).

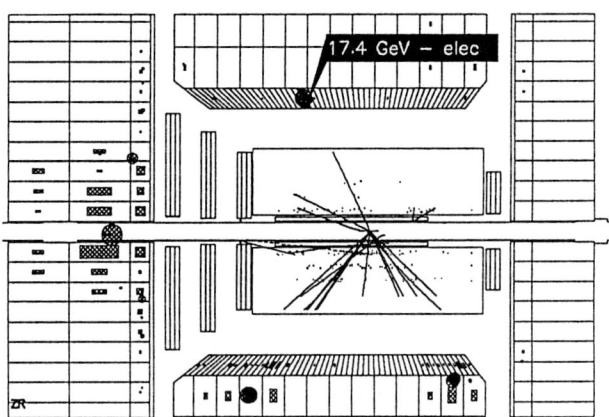

Figure 83. *Typical event of standard deep inelastic neutral scattering.*

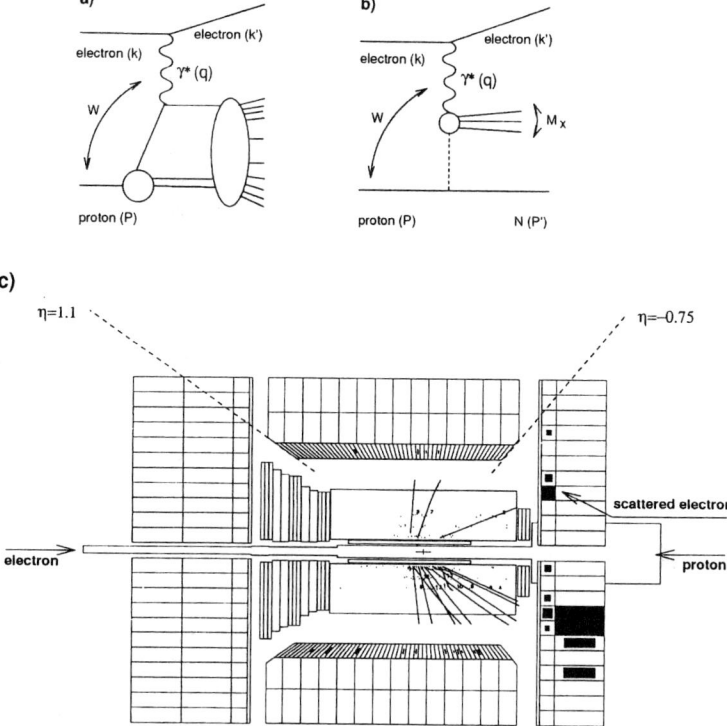

Figure 84. *(a) Schematic diagram describing particle production in* DIS*; (b) the same for diffractive dissociation in* DIS. W *is the c.m. energy of the* $\gamma^* p$ *system and* M_x *the invariant mass of the hadronic system measured in the detector.* N *represents a proton or a low-mass nucleon system. (c) A* DIS *event at* $Q^2 = 64\,\text{GeV}^2$ *with a large rapidity gap in the* ZEUS *detector.*

9.1 Experimental results

In the ZEUS analysis (1993g) the effect was quantified by considering all DIS events (total luminosity 24.7 nb^{-1}) with $Q^2 > 10\,\text{GeV}^2$. A calorimeter cluster was defined as an isolated set of adjacent cells with summed energy above 400 MeV. The pseudorapidity of the cluster closest to the forward direction, *i.e.* with the largest η value, was called η_{max}. The distribution of η_{max} for all DIS events (Figure 85a) shows two groups of events, one concentrated at large η_{max} values and a second one with $\eta_{\text{max}} < 2$. The standard Monte Carlo simulation for DIS scattering (*e.g.* CDM) predicts the shaded distribution which does not account at all for the number of events observed with $\eta_{\text{max}} < 2$. Figure 85b shows that for these events W and η are correlated, which is not observed for the bulk of the events.

For the further study, events with $\eta_{\text{max}} < 1.5$ were denoted as events with a large rapidity gap. An η_{max} of 1.5 corresponds to a rapidity gap of at least 2.8 units. The mass

(a)

(b)

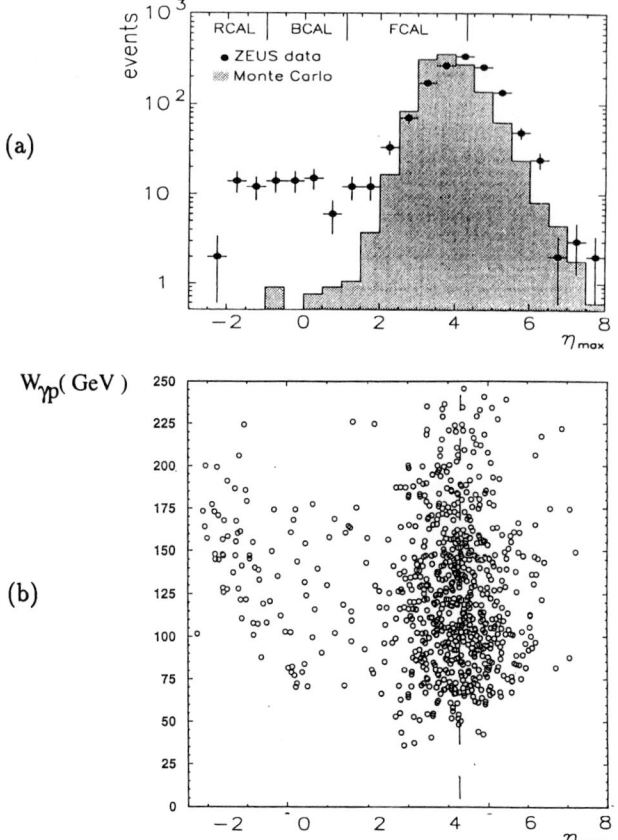

Figure 85. *(a) Distribution for* DIS *events of the maximum rapidity* $\eta_{\max}$ *of a calorimeter cluster in an event, for data (points) and Monte Carlo events (shaded histogram) (b) Correlation between the* γ^*p *c.m. energy* W *and* $\eta_{\max}$ *(from* ZEUS *1993g, j, l).*

M_x of the hadronic system was calculated from the energies detected in the calorimeter cells. Denote by E_H, p_H and q_H the energy, momentum and angle of the hadronic system. By comparing with Section 5.1 one finds:

$$\cos\theta_H = \sum_h p_{zh}/|\sum_h \vec{p}_h|$$
$$E_H - p_H \cos\theta_H = 2E_e y_{DA}$$
$$p_H^2 \sin^2\theta_H = Q_{DA}^2(1 - y_{DA})$$
$$M_x = [E_H^2 - p_H^2]^{1/2} \quad (9.1)$$

Large rapidity gap events have preferentially small M_x values with typical values around 10 GeV in distinction to the events with $\eta_{\max} > 1.5$ (Figure 86a). The M_x^2 distribution falls off rapidly, $dN/dM_x^2 \sim (M_x)^{-n}$, $n \simeq 2-4$, as shown in Figure 87. For $W > 150$ GeV, acceptance corrections have little dependence on W. For $W > 150$ GeV the contribution of large rapidity gap events to the total DIS cross section is, within errors, constant with W (Figure 86b). The contribution of the large rapidity gap events to the DIS cross section as a function of Q^2 is, within errors, also constant with Q^2 (Figure 88). It is worth noting that the data reach Q^2 values as high as $100\,\text{GeV}^2$. The events with $\eta_{\max} < 1.5$ represent 5.4% of the total DIS events.

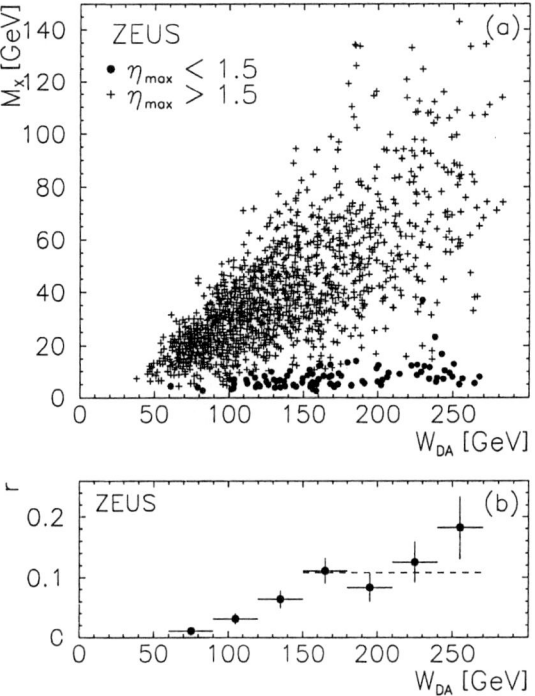

Figure 86. (a) Correlation between M_x and W for events with $\eta_{\max} > 1.5$ (crosses) and $\eta_{\max} < 1.5$ (dots); (b) Fraction of DIS events r with $\eta_{\max} < 1.5$ as a function of W, from ZEUS (1993g).

The H1 findings for large rapidity gap events in DIS are shown in Figure 89. The maximum η value ($\eta_{\max}$) for clusters measured in the calorimeter is 3.8. The distribution of the difference in pseudorapidity between this value and the closest cluster with greater than 400 MeV in an event, $\Delta\eta = 3.8 - \eta_{\max}$, is given in Figure 89a A steep fall-off from zero up to $\Delta\eta \simeq 1.4$ is followed by a long tail extending $\Delta\eta \simeq 6$. Large rapidity gap events are defined to have $\Delta\eta > 2$ ($\eta_{\max} < 1.8$); the DIS events are found to have such a rapidity gap, while a Monte Carlo simulation with LEPTO, which reproduces well the fall-off in the region of small $\Delta\eta$, predicts only 0.1% (see histogram in Figure 89a). The distribution of M_x (Figure 88b) shows that small M_x values are preferred; note, that the distribution is uncorrected and detector acceptance is particularly important for $M_x < 4\,\text{GeV}$. The ratio of events with a large rapidity gap to all events does not depend significantly on Q^2, as shown in Figure 89c.

The results obtained by the two experiments on large rapidity gap events in DIS are in good agreement. What is the origin of these events? The behaviour of the energy flow in the bulk of DIS events (previous section) suggests the absence of energy flow over

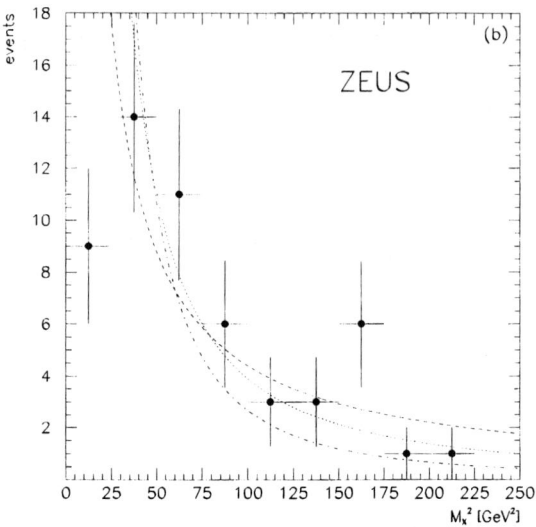

Figure 87. Distribution (uncorrected) of M_x^2 for events with $\eta_{max} < 1.5$ and $W > 150\,\text{GeV}$; the dashed line shows a $1/M_x^2$ dependence, the dotted line $1/M_x^3$, the dashed-dotted $1/M_x^4$; from ZEUS (1993j, l).

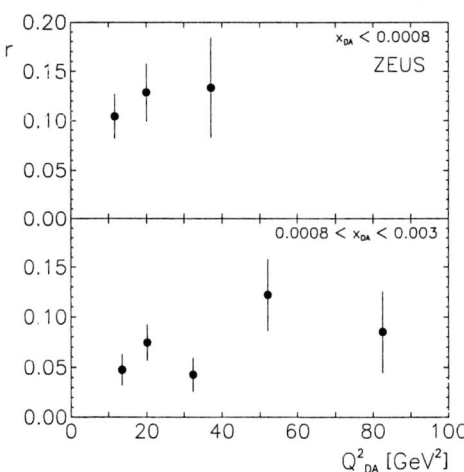

Figure 88. Fraction of DIS events r with $\eta_{max} < 1.5$ as function of Q^2 for two x intervals (ZEUS 1993g).

a large rapidity range be attributed to missing colour flow. The near constancy of the percentage of large rapidity gap events with Q^2 points to a leading twist contribution to the proton structure function or, in other words, since the structure function F_2 shows only logarithmic scaling violations the process leading to large rapidity gap events appears to behave in the same way.

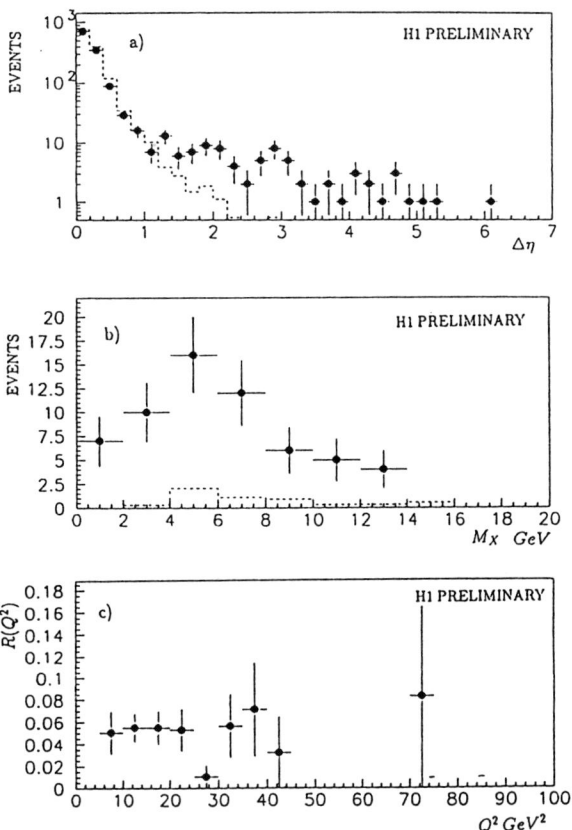

Figure 89. *(a) Distribution of the rapidity gap $\Delta\eta$ for DIS events (b) Effective mass of the visible hadronic system of events with $\Delta\eta > 2$ (c) Ratio $R(Q^2)$ of events with a rapidity gap $\Delta\eta > 2$ over all DIS events as function of Q^2 (H1 1993j, k).*

9.2 Large rapidity gaps by peripheral scattering

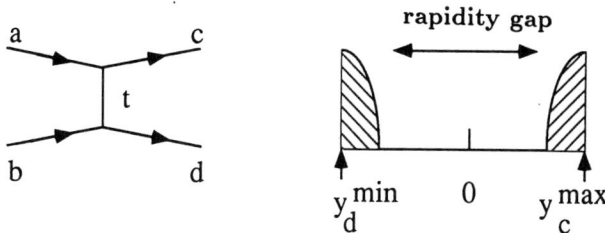

Figure 90. *(a) Diagram for $a+b\to c+d$; (b) rapidity distributions of c, d.*

In soft processes large rapidity gaps can be produced by peripheral scattering of hadrons. Consider two-body scattering $a+b\to c+d$ sketched in Figure 90 in the c.m. system.

The square of the four-momentum transfer at particle momenta $p_a, p_b, \ldots$ large compared to their masses $m_a, m_b, \ldots$ can be approximated by

$$t = (a-c)^2 \simeq -2p_a p_c (1 - \cos\theta) \tag{9.2}$$

where θ is the scattering angle. Peripheral processes are characterised by small $|t|$ values with negligible cross section for, say, $|t|_{max} > 1\,\text{GeV}^2$. For $2p_a p_c \gg |t|_{max}$ this limits the scattering to small angles,

$$\cos\theta > 1 - |t|_{max}/(2p_a p_c) \tag{9.3}$$

and the rapidities Y of c and d to very large (small) values (see Equation 6.8):

$$\begin{aligned}Y_c &\simeq 0.5\ln\{[p_c(1+\cos\theta)]/[p_c(1-\cos\theta)+m_c^2/(2p_c)]\} \\ &> 0.5\ln\{4p_c^2/[m_c^2 + |t|_{max}(p_c/p_a)]\}.\end{aligned} \tag{9.4}$$

The maximum (minimum) rapidities are

$$\begin{aligned}Y_c^{max} &\simeq \ln(2p_c/m_c) \simeq 0.5\ln(s/m_c^2) \\ Y_d^{min} &\simeq -\ln(2p_d/m_d) \simeq -0.5\ln(s/m_d^2),\end{aligned} \tag{9.5}$$

where s is the square of the c.m. energy. The minimum rapidity gap between c and d is (Figure 90b):

$$\Delta Y > 0.5[\ln\{s/(m_c^2 + |t|_{max})\} + \ln\{s/(m_d^2 + |t|_{max})\}]. \tag{9.6}$$

For the example of $\sqrt{s} = 200\,\text{GeV}$, c being a proton, d having a mass of 5 GeV and $|t|_{max} = 1\,\text{GeV}^2$, the rapidity gap between c and d is $\Delta Y \simeq 9$. Suppose d decays into pions. Then for a standard pion multiplicity the pions populate rapidities around the centre value Y_d with an rms spread of about $\pm(1\text{-}2)$ units. The resulting rapidity gap between the proton and the pions from d is 6 to 7 units large. *Peripherality and a c.m. energy which is large compared to the particle masses produce a large rapidity gap.*

For the case that $a + b \to c + d$ proceeds via the exchange of a Regge pole R, the energy dependence of the cross section is given by:

$$\frac{d\sigma}{dt} \simeq f(t) s^{2\alpha_R(t)-2} \tag{9.7}$$

Here $\alpha_R(t)$ is the trajectory for R, $\alpha_R(t) \simeq \alpha_R(0) + a't$, $a' \simeq 1\,\text{GeV}^{-2}$. For the pion trajectory the intercept $a_\pi(0) \simeq 0$; for the rho trajectory $a_\rho(0) \simeq 0.5$. The resulting energy dependence of the forward cross section is

$$\frac{d\sigma}{dt}(t=0) \sim s^{-2} \quad \text{for } \pi \text{ exchange}$$
$$\sim s^{-1} \quad \text{for rho exchange}.$$

The implication is that cross sections for peripheral processes of this type, which are characterised by the exchange of quantum numbers in the t channel (e.g. isospin, charge,...), decrease very fast with s.

There exists a special class of peripheral processes due to diffractive scattering where no quantum numbers are exchanged and which show (almost) energy independent cross sections (for a review of experimental data see e.g. Goulianos 1983, 1990). They are described by the Regge trajectory of a hypothetical particle, the pomeron,

$$\alpha_P(t) \simeq 1 + 0.5t,$$

which leads to $d\sigma/dt(t=0)$ = constant as a function of s. Therefore, of the possible Regge exchanges, only pomeron exchange has a chance of surviving with a significant cross section at high c.m. energies.

9.3 Discussion

The large rapidity gap events observed by ZEUS and H1 in DIS are produced at high γ^*p c.m. energies (W's up to 270 GeV) with large rates (5–20% of the total DIS events). It is therefore suggestive to attribute them to pomeron exchange (Figure 91a). One must keep in mind, however, that in general diffractive production has been studied in soft processes. The large rapidity gap events observed by ZEUS in photoproduction (Section 6.3) are an example of this. In contrast, the events under discussion here from DIS are produced by hard scattering at Q^2 values above 10 GeV^2 and up to 100 GeV^2. The low-x data on the proton structure function F_2 indicate two pieces which build up the total virtual photon proton cross section at large Q^2, one which is independent of the total c.m. energy W, and one whose contribution increases with energy. It is suggestive to associate the constant piece with the soft pomeron and the energy dependent component (Lipatov term) with a hard pomeron.

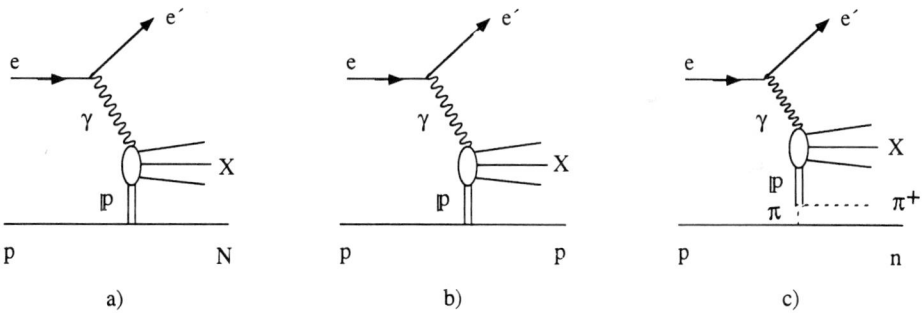

Figure 91. *Diagrams for pomeron exchange in DIS producing a system X plus (a) a nucleon state N, (b) a proton, (c) a neutron plus a mesonic state m via pion exchange, where m is a pion or a higher mass state.*

The concept of the pomeron structure function has been studied in terms of perturbative QCD (Gribov, Levin and Ryskin 1983, Berger et al. 1987, Bartels and Ingelman 1990, Ryskin 1991, Levin and Wüsthoff 1992, Nikolaev and Zakharov 1992, Ingelman and Pryrtz 1993, Collins, Frankfurt and Strikman 1993). It was suggested that the pomeron structure could be probed with a virtual photon at HERA (Ingelman and Schlein 1985, Donnachie and Landshoff 1987, 1992, Streng 1987). Rapidity gaps as a means for detecting new physics was discussed by Bjorken (1992).

On the basis of pp data from R608 (1985) taken at the CERN ISR Ingelman and Schlein (1985) suggested that the pomeron may have a partonic structure. Proton-proton collider data from the UA8 collaboration (1988, 1992) gave strong evidence for high transverse momentum jets in diffractively produced high mass systems suggestive of the hard scattering from partons within the pomeron. Observation of events with large rapidity gaps by $p\bar{p}$ collisions were reported recently also by D0 (Forden 1993). A new study from UA8 (UA8 1993, Schlein 1993) suggests that the pomeron may have a superhard part with one parton carrying almost all of the pomeron momentum in a significant fraction of the events.

Deep inelastic electron proton scattering is well placed for unravelling the properties of the pomeron. By keeping Q^2 fixed and varying x the relative contributions from the conjectured soft and hard pomerons can be changed. The analysis of the (jet) structure of the system X will tell us more about the partonic structure of the pomeron. Important information can also be gained from an analysis of the produced nucleon system N (Figure 91a). For a fraction of the events one expects N to be just a proton (Figure 91b). The momentum of this proton (p_f) will be almost equal to the momentum of the incoming proton (p_{beam}),

$$x_{\text{pom}} = [p_{\text{beam}} - p_f]/p_{\text{beam}} \simeq [M_x^2 + Q^2]/[W^2 + Q^2] \sim O(10^{-2}), \qquad (9.8)$$

where x_{pom} is the momentum fraction of the proton carried by the pomeron. Such protons can be detected in the Leading Proton Spectrometer (LPS) of the ZEUS detector (Section 3.2). Information from the LPS will provide another signature for diffractive events.

Pomeron exchange can also occur as part of a multiperipheral process such as depicted in Figure 91c where the incoming proton emits a virtual pion which scatters on the pomeron emitted at the photon vertex. The contribution of a pomeron-pion ladder to inelastic diffractive ρ^0 production by real photon proton scattering was predicted to increase with photon energy E_γ and to be substantial at high energies (Wolf 1971). The cross section was calculated for $\gamma \to \rho^0 N + i\pi$, $i \geq 1$, taking into account the diagrams of Figure 92. The result is shown in Figure 93 as a function of the photon energy, E_γ, assuming a stationary proton: at $E_\gamma = 300\,\text{GeV}$ ($W = 24\,\text{GeV}$) the cross section amounts to about 5μb which is approximately 50% of the elastic ρ^0 cross section.

A subprocess of inelastic diffractive scattering is $ep \to eXn\pi^+$ where X results from diffraction dissociation of the virtual photon (analogously to Figure 92b). In this case the neutron n carries almost the full momentum of the beam proton and can be tagged with a calorimeter close to the proton beam as pointed out by Levman and Furutani (1992). During the 1993 running ZEUS had a prototype of a Forward Neutron Calorimeter (FNC) installed and found that detection of these neutrons is feasible. A full FNC will complement the LPS information from protons on diffractive production.

10 Concluding remarks

The experimental results presented in this report were extracted from data obtained by H1 and ZEUS during the 1992 running. The 1993 data which are currently under study represent a twenty-fold increase in luminosity and promise a wealth of new information.

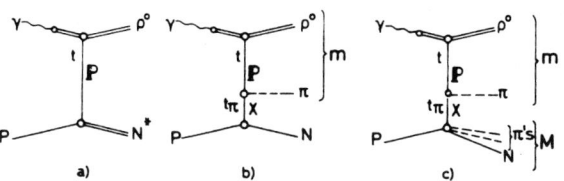

Figure 92. *Diagrams for inelastic diffractive ρ^0 photoproduction: (a) isobar production by pomeron exchange; (b) and (c) double peripheral scattering with pomeron and pion exchange.*

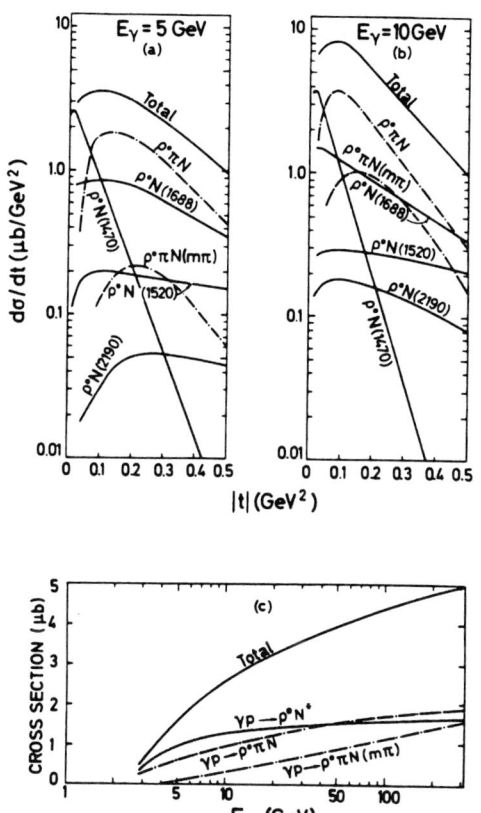

Figure 93. *Cross sections for inelastic diffractive ρ^0 photoproduction via isobar production (diagram (a) of Figure 92) and double peripheral scattering (diagrams (b) and (c) of Figure 92). The curves labelled 'Total' show the incoherent sum of all contributions (from Wolf 1971).*

They will allow, for instance, the Q^2 range in NC scattering to be increased by an order of magnitude and a first study of CC scattering at large Q^2 to be made. The search for exotic particles such as leptoquarks and excited leptons and quarks will also benefit from the increased statistics. The results reported from the 1992 data (H1 1993c, ZEUS 1993d) for some species have already exceeded the sensitivity of previous searches. Running in 1994 will presumably lead to a further tenfold increase in luminosity. The large growth in event statistics will help the studies discussed here to be performed with much improved precision and will bring many more channels within reach of the two experiments.

Acknowledgements

Prof. D. Saxon as director and Drs. A.T. Doyle, P. Negus, K. J. Peach, L. Vick and A. Walker as organisers have provided a most stimulating and enjoyable atmosphere for the participants of the school. I am grateful to Prof. J. Bartel for discussions and to Prof. W. Buchmüller, Dr. R. Klanner, Profs. E. Lohrmann and J. Whitmore for a critical reading of the manuscript and for many comments. S. Gharavi has prepared diagrams and figures. I would like to thank Mrs I. Harm for help with the manuscript.

References

In the following 'HERA Workshop 1991' stands for Physics at HERA, Proc. Workshop 1991, ed. by W. Buchmüller and G. Ingelman, April 1992.

Abramowicz H, Charchula K and Levy A, 1991, *Phys Lett* **B269** 458.
Abramowicz H *et al.* 1991, *Phys Lett* **B269** 465.
Alekhin S I *et al.* 1987, CERN-HERA 87-01 and references given there.
Altarelli G and Parisi G, 1977, *Nucl Phys* **126** 297.
AMY collaboration, Tanaka R *et al.* 1992, *Phys Lett* **B277** 215.
AMY collaboration, Kim B J *et al.* 1993, KEK preprint 93-97.
ARIADNE, Lönnblad L, 1992, *Comp Phys Com* **71** 15, and references given there.
Balitskii Y Y and Lipatov L N, 1978, *Sov J Nucl Phys* **28** 822.
Barber D B *et al.* 1992, DESY Report 92-136; 1993 private communication by Zetsche F.
Barger V D and Phillips R J N, 1987, *Collider Physics, Frontier in Physics Series* (Vol 71), Addison-Wesley.
Bartels J and Ingelman G, 1990, *Phys Lett* **B235** 175.
BCDMS collaboration, Benvenuti A C *et al.* 1989, *Phys Lett* **B223** 485.
Bentvelsen S, Engelen J and Kooijman P, HERA Workshop 1991, Vol 1, p.23.
Berger E *et al.* 1987, *Nucl Phys* **B286** 704.
Bjorken J D, 1969, *Phys Rev* **179** 1547.
Bödeker D, 1992, *Phys Lett* **B292** 164.
Borzumati F M, Kniehl B A and Kramer G, 1993, *Z Phys* **C59** 341.
Buchmüller W and Ingelman G, 1991, Proc. Workshop: Physics at HERA, April 1992, ed. by Buchmüller W and Ingelman G, DESY, Hamburg.
Buon J and Steffen K, 1985, DESY report 85-128.
Burow B 1993, thesis Univ. Toronto.

CTEQ collaboration, Brock R et al. 1993, Handbook of Perturbative QCD.
Chacaltaya and Pamir collaborations, 1990, contrib. VI Int. Symp. on Very High Energy Cosmic Ray Interactions, ICRR-Rpt-216-90-9.
Chang C et al. 1975, *Phys Rev Lett* **35** 901.
Chaves H, Seifert R J and Zech G, 1991, HERA Workshop 1991, Vol 1, p. 57.
Collins J C, Frankfurt L and Strikman M, 1993, *Phys Lett* **B307** 161.
Cooper-Sarkar A M et al. 1988, *Z Phys* **C39** 281.
D0 collaboration, Forden G, 1993, 28th Renc. Moriond, Les Arcs, Savoie, 1993.
Donnachie A and Landshoff P V, 1984, *Nucl Phys* **B244** 322.
Donnachie A and Landshoff P V, 1987, *Phys Lett* **B191** 309; 1992, ibid **B285** 172.
Donnachie A and Landshoff P V, 1994, *Z Phys* **C61** 139.
Dokshitzer Y L, 1977, *Sov Phys JETP* **46** 641.
Drees M and Grassie K, 1985, *Z Phys* **C28** 451.
Drees M and Godbole R M, 1992, Bombay Univ. Report BU-TH 92-5.
E665 collaboration, 1992, Adams M R et al. 1992, *Phys Rev Lett* **69** 1026.
Eichten E et al. 1984, *Rev Mod Phys* **56** 579; 1986, ibid **58** 1047.
Ellis R K, Kunszt Z and Levin E M, 1993, *Fermilab-PUB-93/350-T, ETH-TH-93/41*
EMC collaboration, Aubert J J et al. 1981, *Phys Lett* **B100** 433; Arneodo M et al. 1987, *Z Phys* **C36** 527.
Feynman R P, 1972, *Photon-Hadron Interactions*, Reading, Mass., Benjamin.
Fletcher R S et al. 1992, *Phys Rev* **D45** 377.
Forshaw J R and Storrow J K 1991, *Phys Lett* **B268** 116.
Forshaw J R, 1992, RAL 92-073.
Fox D J et al. 1974, *Phys Rev Lett* **41** 1504.
Fritzsch H and Gell-Mann M, 1972, Proc. 16th Int. Conf. High Energy Physics, Chicago, Vol 2, p. 135, Batavia, Ill., Fermi Nat. Accel. Lab.
Fritzsch H, Gell-Mann M and Leutwyler H, 1973, *Phys Lett* **47** 365.
Ghandi R and Sarcevic I, 1991, *Phys Rev* **D44** R10.
Glück M, Reya E and Vogt A, 1990, *Z Phys* **C48** 471.
Glück M, Reya E and Vogt A, 1992, *Z Phys* **C53** 127.
Glück M, Reya E and Vogt A, 1993, *Phys Lett* **B306** 391.
Gordon L E and Storrow J K, 1992a, *Phys Lett* **B291** 320; 1992b, *Z Phys* **C56** 307.
Greco M and Vicini A, 1993, Frascati LNF-93/017.
Gribov V N and Lipatov L N, 1972, *Sov Journ Nucl Phys* **15** 438, 675.
Gribov L V, Levin E M and Ryskin M G, 1983, *Phys Rep* **100** 1.
Goulianos K, 1983, *Phys Rep* **101** 169; 1990, *Nucl Phys B, Proc Suppl* **12** 110.
Gross D J and Wilczek F, 1973, *Phys Rev Lett* **30** 1343; *Phys Rev* **D8** 3633; 1974, *Phys Rev* **D9** 980.
H1 collaboration, 1986, Technical Proposal for the H1 Detector.
H1 collaboration, Ahmed T et al. 1992, *Phys Lett* **B297** 205.
H1 collaboration, Ahmed T et al. 1993a, *Phys Lett* **B298** 469; 1993b, *Phys Let* **B299** 374.
H1 collaboration, Abt I et al. 1993c, *Nucl Phys* **B396** 3; 1993d, DESY 93-103; 1993e, *Phys Lett* **B314** 436; 1993f, *Nucl Phys* **B407** 515; 1993g, DESY 93-137; 1993h, DESY 93-146.
H1 collaboration, De Roeck A and Klein M, 1993i, DESY 93-014.
H1 collaboration, De Roeck A, 1993j, 1993 EPS Conf. Marseille, DESY 94-005; Dainton J 1993k, 1993 Lepton-Photon Conference at Cornell.
HERA, A Proposal for a Large Electron-Proton Colliding Beam Facility at DESY, DESY HERA 81-10 (1981).
HERA-B collaboration, 1992, Experiment to Study CP Violation in the B System Using an Internal Target at the HERA Proton Ring.

HERACLES, Kwiatkowski A, Spiesberger H and Möhring H-J, 1992, *Comp Phys Com* **69** 155, and references given there.
HERMES collaboration, 1990, Proposal to Measure the Spin-Dependent Structure Functions of the Neutron and the Proton at HERA.
HERWIG, Webber B R, 1986, *Ann Rev Nucl Part Sci* **36** 253; and Marchesini G et al. 1992, *Comp Phys Comm* **67** 465.
Huth J et al. 1990, Proc. 1990 DPF Summers Study on High Energy Physics, Snowmass, Colorado, ed. Berger E L, (World Scientific, Singapore, 1992) p. 134.
Ingelman G and Schlein P, 1985, *Phys Lett* **B152** 256.
Ingelman G and Rückl R, 1987, DESY Report 87-140; 1988, *Phys Lett* **201** 369.
Ingelman G and Prytz K, 1993, *Z Phys* **C58** 285.
Jacquet F and Blondel A, 1979, Proc. Study of an ep Facility in Europe, ed. Amaldi U, 79/48,p. 391.
JADE collaboration, Bartel W et al. 1981, *Phys Lett* **107B** 163.
JADE collaboration, Bartel W et al. 1986, *Z Phys* **C33** 23.
Janssens T, Hofstadter R et al. 1966, *Phys Rev* **142** 922.
JETSET, Sjöstrand T 1986, *Comp Phys Com* **39** 347; Sjöstrand T and Bengtsson M 1987, ibid **43** 367 (1987) and CERN-TH 6488/92 (1992).
Kim V T and Ryskin M G, 1991, DESY 91-064.
Kolanoski H, 1984, *Two-Photon Physics at e^+e^- Storage Rings*, Springer Tracts in Modern Physics **105** .
Kramer G and Salesch S G, 1993, DESY 93-10.
Kuraev E A, Lipatov L N and Fadin V S, 1977, *Sov Phys JETP* **45** 199.
Kwiecinski J et al. 1990, *Phys Rev* **D42** 3645.
Kwiecinski J, 1993, Lecture Notes in Physics, Substructure of Matter as Revealed with Electroweak Probes, Proc. Schladming, 1993, ed. by Mathelitsch L and Plessas W, Springer Verlag 1994, p. 215.
LEPTO, Ingelman G, LEPTO5.2, program manual, unpublished; Bengtsson H, Ingelman G and Sjöstrand T, 1988, *Nucl Phys* **B301** 554.
Levin E M and Wüsthoff M, 1992, DESY 92-166 and FERMILAB-PUB-92/334.
Levman G and Furutani K 1992, ZEUS internal note, unpublished.
Levy A, 1992, Photoproduction at HERA, DESY Acad. Training Progr., 27-29/4/1992.
Lipatov L N, 1974, *Sov J Nucl Phys* **20** 181.
Lipatov L N, 1976, *Sov J Nucl Phys* **23** 338.
Martin A D, Stirling W J and Roberts R G, 1993, *Phys Rev* **D47** 867.
Martin A D, Stirling W J and Roberts R G, 1993, *Phys Lett* **B306** 145.
Martin A D, Stirling W J and Roberts R G, 1993, RAL 93-77/DTP 93/86.
Milsztajn A and Virchaux M, 1993, Lecture Notes in Physics, Substructure of Matter as Revealed with Electroweak Probes, Proc. Schladming, 1993, ed. by Mathelitsch L and Plessas W, Springer Verlag 1994, p. 257.
Morfin J G and Tung W K, 1991, *Z Phys* **C52** 13.
Mueller A H, 1990, Proc. Small x-Workshop at DESY, 1990, ed. by Ali A and Bartels J 1991, *Nucl Phys* **18C** 125.
Nikolaev N N and Zakharov B G, 1992, *Z Phys* **C53** 331, and 1993, LANDAU 16-93.
NMC collaboration, Amaudruz et al. 1992, *Phys Lett* **B295** 159.
OMEGA-Photon collaboration, Apsimon R J et al. 1989, *Z Phys* **C43** 63.
Parisi G and Petronzio R, 1976, *Phys Lett* **B62** 331.
Peccei R D, Proc. HERA Workshop 1987, ed. by Peccei R D, August 1988. DESY, Hamburg.
PDG Particle Data Group, Physical Review **D45** (1992).
PLUTO collaboration, Berger, Ch et al. 1984, *Z Phys* **C26** 191.

Prytz K, 1993, *Phys Lett* **B311** 286.
PYTHIA Bengtsson H-U and Sjöstrand T, 1987, *Comp Phys Com* **46** 43; Sjöstrand T, 1989, *Z Phys* **C42** 301.
R608 collaboration, Smith A M et al. 1985, *Phys Lett* **B163** 267; 1986, ibid **B167** 248; Henkes T et al. 1992, *Phys Lett* **B283** 155.
De Rujula A et al. 1974, *Phys Rev* **D10** 1649.
Ryskin M G, 1991, *Sov J Nucl Phys* **53** 668.
Rutherford E, 1911, *Phil Mag* **21** 669.
Schlein 1993, 1993 EPS Conf. Marseille and UCLA-PPh0061.
Schuler G, 1991, HERA Workshop 1991, Vol 1, p. 131
Schuler G and Terron J, 1991, HERA Workshop 1991, Vol 1, p. 599.
Schuler G and Sjöstrand T, 1993, *Phys Lett* **B300** 169; 1993, *Nucl Phys* **B407** 539.
SLAC-MIT collaboration Riordan E M et al. 1974, *Phys Lett* **B52** 249; see also SLAC-PUB-1634.
SLAC-MIT collaboration Atwood W B et al. 1976, *Phys Lett* **64B** 479.
Sokolov A A and Ternov M, 1964, *Sov Phys Doklady* **8** 1203.
Streng K H, 1987, Proc. HERA Workshop Vol 1, 365, ed. by Peccei R D, and 1988, CERN TH 4949.
Taylor R R, 1969, Proc. 4th Intern. Symp. Electron Photon Interactions at High Energies, p. 251, Daresbury, England, Daresbury Nucl Phys Lab.
TASSO collaboration, Brandelik et al. 1981, *Phys Lett* **107B** 290.
TASSO collaboration, Althoff M et al. 1984, *Phys Lett* **138B** 219.
TERAD91, 1991, Akhundov A et al. 1991, HERA Workshop 1991, 1285.
TOPAZ collaboration, Hayashii H et al. 1993, KEK preprint 93-47.
UA1 collaboration, Arnison G et al. 1983, *Phys Lett* **123B** 115.
UA1 collaboration, Albajar C et al. 1990, *Nucl Phys* **B335** 261.
UA8 collaboration, Bonino R et al. 1988, *Phys Lett* **B211** 239; Brandt A et al 1991, Proc. Joint Intern. Lepton-Photon Symp. and 1991 EPS Conf. High Energy Physics, Geneva, ed. by Hegarty S, Potter K and Quercigh E, p. 741; Brandt R et al. *Phys Lett* **B297** 417.
Voss G-A, 1988, Proc. First Euro. Acc. Conf., Rome, 1988, p. 7.
Watanabe Y et al. 1975, *Phys Rev Lett* **35** 898.
Weinberg S, 1973, *Phys Rev Lett* **31** 494.
Wiik B H, 1982, Electron-Proton Colliding Beams, The Physics Programme and the Machine, Proc. 10th SLAC Summer Institute, ed. Mosher A, 1982, p. 233; Proc. XXVI Int. Conf. High Energy Physics, Dallas, 1992.
Wolf G, 1971, *Nucl Phys* **B26** 317.
Yodh G, 1990, *Nucl Phys* B (**Proc. Suppl.**) **12** 277.
ZEUS collaboration, 1986, The ZEUS Detector, Technical Proposal (1986).
ZEUS collaboration, Derrick M et al. 1992a, *Phys Lett* **B293** 465; 1992b, *Phys Lett* **B297** 404.
ZEUS collaboration, 1993a, The ZEUS Detector, Status Report 1993, ed. by Holm U.
ZEUS collaboration, Derrick M et al. 1993b, *Phys Lett* **B303** 183; 1993c, *Phys Lett* **B306** 158; 1993d, *Phys Lett* **B306** 173, 1993e, *Z Phys* **C59** 231; 1993f, *Phys Lett* **B316** 207; 1993g, *Phys Lett* **B315** 481; 1993h, *Phys Lett* **B316** 412; 1993i, DESY 93-151 and *Phys Lett* B, to be published.
ZEUS collaboration, Klanner R, 1993j, 1993 EPS Conf. Marseille; Park, I, 1993k, ibid and DESY 93-186; Martin J 1993l, 1993 Lepton-Photon Conf. Cornell; 1993m, Abramowicz H, Klanner R and Martin J, 1993, DESY 93-158.

Higgs Phenomenology

W J Stirling

Department of Physics
University of Durham

1 Introduction

In these lectures we discuss the phenomenology of the Higgs boson, both in the Standard Model (SM) and in the minimal supersymmetric extension (MSSM). There will be some overlap with other lecture courses at the School; in particular, the basic theory of the Higgs mechanism has already been covered in detail by Aitchison. Blondel has discussed the rôle of the Higgs boson in electroweak radiative correction physics, Jenni has covered the experimental aspects of Higgs searches at future hadron-hadron colliders, and Peccei has discussed in some detail the shortcomings of the SM and the various ways to avoid these in extended models.

In the next section we will briefly review the fundamental theoretical issues which are relevant for SM Higgs phenomenology, and then proceed to review search strategies at the various types of collider. We then motivate going beyond the SM by introducing supersymmetry. We derive the Higgs spectrum and interactions in the MSSM, and review the associated phenomenology. Finally, we mention some other aspects of MSSM physics.

2 The Standard Model Higgs boson

2.1 The Higgs mechanism

A gauge field theory incorporating a SU(2)×U(1) weak isospin/hypercharge gauge symmetry accounts for all the observed electromagnetic and weak interactions of the quarks and leptons:

$$\mathcal{L}_{EW} = \sum_f \left[\bar{f}_L \gamma^\mu (i\partial_\mu - \tfrac{1}{2} g \tau_i W^i_\mu - \tfrac{1}{2} g' Y B_\mu) f_L + \bar{f}_R \gamma^\mu (i\partial_\mu - \tfrac{1}{2} g' Y B_\mu) f_R \right] \\ - \tfrac{1}{4} W^i_{\mu\nu} W^{\mu\nu}_i - \tfrac{1}{4} B_{\mu\nu} B^{\mu\nu}, \quad (1)$$

with representations

$$\begin{pmatrix} \nu_e \\ e^- \end{pmatrix}_L, \quad \begin{pmatrix} u \\ d \end{pmatrix}_L, \quad e_R^-, \quad u_R, \quad d_R, \quad \cdots \quad \text{etc.} \quad (2)$$

However, as formulated in this way, the theory contains *massless* gauge bosons and fermions, in contradiction with what is known about the masses of the $W^\pm$, Z^0, τ, b, ... particles. Any attempt to incorporate masses for these particles by introducing explicit mass terms of the form

$$\tfrac{1}{2} M^2 V_\mu V^\mu, \quad -m \bar{f} f \quad (3)$$

in $\mathcal{L}_{EW}$ breaks the SU(2)×U(1) gauge symmetry and destroys the renormalizability of the theory. The Higgs mechanism (Higgs 1964, Brout and Englert 1964) of spontaneous symmetry breaking appears to be the simplest way of incorporating masses into the theory while preserving renormalizability. A complex SU(2) doublet of scalar fields is introduced, with a potential

$$V(\phi) = \lambda (\phi^\dagger \phi)^2 - \mu^2 \phi^\dagger \phi. \quad (4)$$

With the parameters $\lambda, \mu^2 > 0$, this potential has a minimum which is not at $\phi = 0$. In fact there is a circle of degenerate minima, Figure 1, corresponding to

$$|\phi| = \sqrt{\frac{\mu^2}{2\lambda}} \equiv \frac{v}{\sqrt{2}}. \quad (5)$$

The field acquires a non-zero vacuum expectation value at a particular point on the circle of minima away from $\phi = 0$ and the symmetry is broken (*spontaneous symmetry breaking*); three of the four scalar field components get 'eaten' by the SU(2) gauge particles to form massive vector bosons ($W^\pm, Z^0$), and there remains a physical neutral scalar boson—the Higgs boson (H). This is the 'minimal model'; many extensions of the Standard Model (supersymmetry, technicolor, grand unification, ...) retain the Higgs mechanism as the primary method for mass generation for gauge bosons, albeit with more complicated Higgs sectors and more physical scalars. In this review we will concentrate on the minimal model with a single Higgs boson, although many of the remarks also apply to more complicated models. An excellent review of all aspects of Higgs physics in the Standard Model and in more complicated extensions can be found in the *Higgs Hunter's Guide* (Dawson et al. 1990).

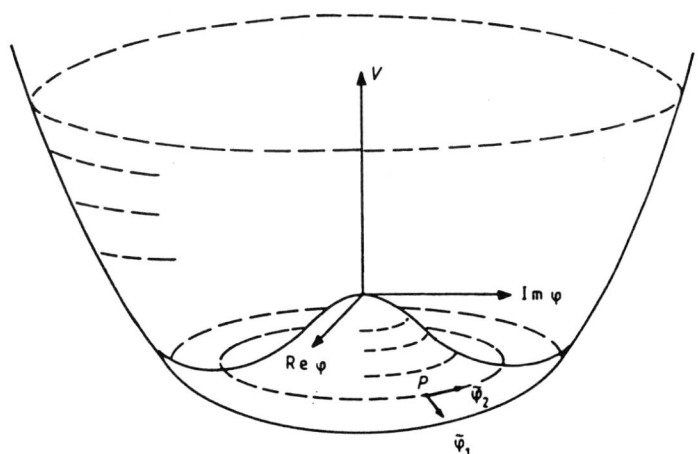

Figure 1. *The scalar potential which gives rise to spontaneous symmetry breaking.*

The precise way in which the Higgs mechanism is realized in the full electroweak Weinberg-Salam theory is complicated by the proliferation of gauge and representation indices. The details are contained in the lectures of Aitchison (these Proceedings). However, the basic principles can be understood by studying a simpler model. Let us consider, therefore, a theory with a single massless 'photon' field, invariant under local U(1) gauge transformations. To this theory we add a complex scalar field ϕ:

$$\mathcal{L} = -\frac{1}{4} F_{\mu\nu} F^{\mu\nu} + (D_\mu \phi)^* (D^\mu \phi) + \mu^2 \phi^* \phi - \lambda \left(\phi^* \phi \right)^2 . \tag{6}$$

The second term on the RHS is the gauge-invariant kinetic term for the ϕ field, and the third term is just the potential introduced above in Equation (4). This Lagrangian defines a theory which (i) has local U(1) invariance and is renormalizable, and (ii) has a potential energy which is bounded below and has a minimum at $|\phi| = \sqrt{\mu^2/2\lambda}$. To examine the physical content of this theory we expand the scalar field about a particular minimum of the potential, chosen to be $\phi = v/\sqrt{2} = \sqrt{\mu^2/2\lambda}$:

$$\phi(x) = \frac{1}{\sqrt{2}} [v + H(x)]. \tag{7}$$

Substituting into the Lagrangian (6) gives

$$\mathcal{L} = -\frac{1}{4} F_{\mu\nu} F^{\mu\nu} + \frac{1}{2} g^2 v^2 A_\mu A^\mu - \lambda v^2 H^2 + \left\{ H^3, H^4, HAA, H^2 AA \right\}. \tag{8}$$

Now we have a theory with (i) a *massive* vector field A^μ, with $M_A = gv$, (ii) a real physical scalar particle H, with mass $M_H = \sqrt{2\lambda} v$, and (iii) a set of interactions involving the H and A particles, shown schematically in Equation (8). Note that the number of degrees of freedom has been preserved. The original two spin states of the massless gauge boson and the two components of the scalar field have reappeared as the three spin states of the massive vector field and the single scalar field H.

In the SM, a doublet of complex scalar fields is introduced

$$\phi = \begin{pmatrix} \phi^+ \\ \phi^0 \end{pmatrix} = \frac{1}{\sqrt{2}} \begin{pmatrix} \phi_1 + i\phi_2 \\ \phi_3 + i\phi_4 \end{pmatrix}. \tag{9}$$

Then, with

$$\langle 0|\phi_3|0\rangle = v, \quad \langle 0|\phi_i|0\rangle = 0, \ (i = 1, 2, 4), \tag{10}$$

we find gauge and scalar boson mass terms of the form

$$\begin{aligned}
M_W &= \tfrac{1}{2}gv & \Rightarrow v = 246 \text{ GeV}, \\
M_Z &= \tfrac{1}{2}\sqrt{g^2 + g'^2}\, v \equiv M_W/\cos\theta_W, \\
M_\gamma &= 0, \\
M_H &= 2\mu^2 = 2\lambda v^2.
\end{aligned} \tag{11}$$

Evidently, one of the parameters in the Higgs potential is fixed by the measured parameters of the electroweak theory. The other, which is related to the physical Higgs mass, is essentially arbitrary.

By introducing Yukawa interactions between the scalar field ϕ and the fermion fields, the Higgs mechanism can also be used to give masses to the fermions. A Yukawa interaction of the form $g_f \bar{f} f \phi$ leads to a fermion mass $m_f = g_f v/\sqrt{2}$, and residual $\bar{f}fH$ interactions of strength $g_f = \sqrt{2} m_f/v$.

The fundamental vertices by which the Higgs couples to the fermions and weak gauge bosons of the SM are shown in Figure 2. Note that the Higgs couples to the W and Z with a standard weak coupling strength, whereas the coupling to the fermions is proportional to the fermion mass. In other words, the Higgs prefers to couple to the heaviest available fermions. This is manifest in the partial decay widths (see below), which are readily obtained from the diagrams of Figure 2.

2.2 Limits on the Higgs mass

Higgs boson phenomenology is the study of the various ways in which the particle can be identified experimentally. It is a subject which has received an enormous amount of interest, principally in connection with proposals to build new machines. One of the main difficulties is that the arbitrariness of the mass leads *a priori* to a proliferation of different production and decay scenarios—from low-energy nuclear physics to the very highest energy hadron colliders. However, the situation has been significantly clarified in the last few years by experimental lower bounds on the Higgs mass from LEP. From the absence of any signal from the process $e^+e^- \to Z^*H \to f\bar{f}H$, the limit

$$M_H > 62.5 \text{ GeV} \quad (95\% \text{ cl}) \tag{12}$$

has been derived (Carter 1993). There are, in addition, the following theoretical arguments which attempt to put limits on the Higgs mass.

As the Higgs mass increases, the amplitude for high-energy $W_L W_L$ scattering (*i.e.* the scattering of longitudinal W bosons), which includes s- and t-channel Higgs exchange diagrams, Figure 3, becomes large. Schematically, the leading contributions at

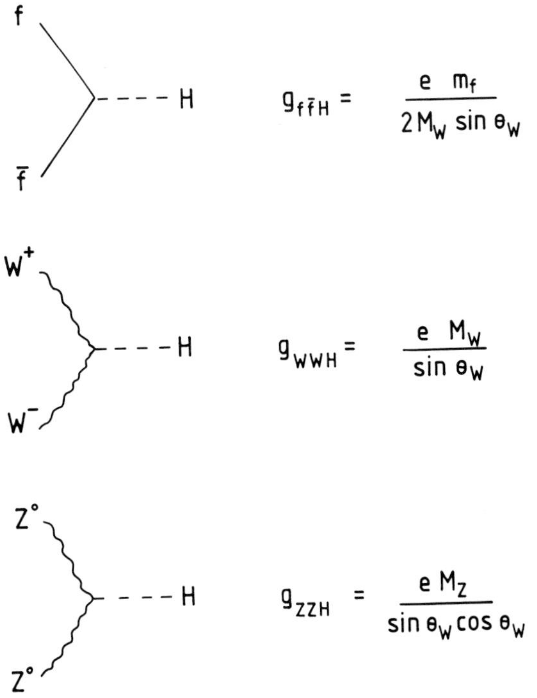

$$g_{f\bar{f}H} = \frac{e\, m_f}{2 M_W \sin\theta_W}$$

$$g_{WWH} = \frac{e\, M_W}{\sin\theta_W}$$

$$g_{ZZH} = \frac{e\, M_Z}{\sin\theta_W \cos\theta_W}$$

Figure 2. *Standard Model couplings of the Higgs boson to fermions and W and Z bosons.*

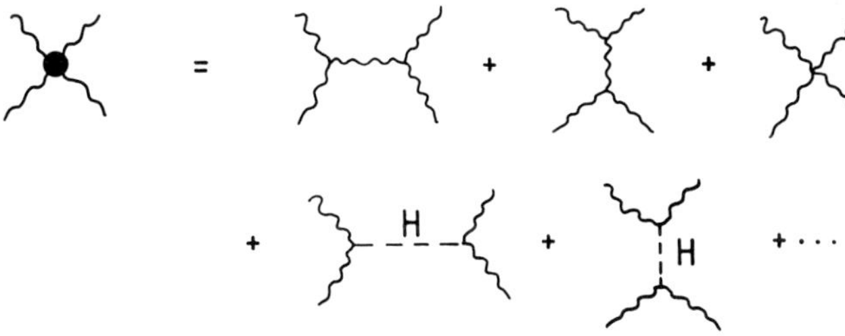

Figure 3. *Feynman diagrams for WW scattering in the Standard Model.*

high energy are

$$\lim_{s\to\infty} \mathcal{M}_{W_L W_L \to W_L W_L} = g^2 s - g^2 \frac{s^2}{s - M_H^2} + \cdots, \qquad (13)$$

where the first term on the RHS comes from the three non-Higgs diagrams in Figure 3. Note that without the Higgs exchange contributions this amplitude would grow linearly with s and eventually violate the unitarity bound. This is another manifestation of the non-renormalizability of the electroweak theory if the weak boson masses are simply introduced by hand. The effect of the Higgs is to cancel the bad high energy behaviour, at an energy scale $\sqrt{s} \sim M_H$. Now one may argue that this cancellation should take place *before* the unitarity bound is reached, in which case an upper limit on the Higgs mass

$$M_H < \left(\frac{4\pi\sqrt{2}}{3G_F}\right)^{\frac{1}{2}} \sim 700\,\text{GeV} \tag{14}$$

can be derived. We should emphasize however that it is *perturbative* unitarity rather than unitarity that is being violated here; non-perturbative effects could restore unitarity for Higgs masses which exceed the above bound. It is true in any case that $VV \to VV$ ($V = W^\pm, Z$) scattering becomes strong in the heavy Higgs limit (*i.e.* $M_H \gtrsim 1\,\text{TeV}$), and new non-perturbative effects presumably become important. A recent review, which describes some possibilities and their experimental signatures, can be found in Bagger et al. (1993).

Another tentative upper limit comes from consideration of the renormalized λ parameter which appears in the potential (4). Like all couplings in the theory, this parameter 'runs' with the renormalization scale μ. Explicit calculation gives

$$\frac{\partial \lambda(\mu)}{\partial \ln \mu} = \frac{3}{2\pi^2}\left[\lambda^2(\mu) - g_t^2 + \cdots\right], \tag{15}$$

where g_t is the Higgs-top quark Yukawa coupling

$$g_t = 2^{1/4}\sqrt{G_F}\,m_t. \tag{16}$$

On the RHS of Equation (15) we have displayed two potentially important contributions. It the top quark is not too heavy, the first term dominates the evolution of λ when the parameter is large:

$$\lambda(\mu) = \frac{\lambda(v)}{1 - \frac{3\lambda(v)}{2\pi}\ln\left(\frac{\mu}{v}\right)}. \tag{17}$$

Note that the coupling increases with the scale and eventually a singularity (the Landau pole) is reached. If this Landau pole is to be avoided before some new physics scale Λ (for example the Planck scale or the scale of some Grand Unification) is reached, then the coupling at low energy $\lambda(v)$ cannot be too large. This in turn translates, via Equation (11), into an upper limit on the Higgs mass. In particular, for $\Lambda = \Lambda_{\text{planck}}$ we find $M_H < 175\,\text{GeV}$, but the limit increases as Λ decreases. For example, if new physics sets in at the $O(1\,\text{TeV})$ scale then the upper limit on the Higgs mass is of the same order.

Another problem arises if the top mass is large, in which case the Yukawa coupling g_t in Equation (15) becomes important. This contribution has the opposite sign from the first term, and can drive λ to negative values. If this happens, the potential of Equation (4) is no longer bounded from below, and the Higgs vacuum becomes unstable. The avoidance of this can be translated into a *lower* limit on the Higgs mass which depends on m_t:

$$M_H > 1.7\,(m_t - 85\,\text{GeV}), \tag{18}$$

where we have approximated the results of Lindner et al. (1989). Both the above limits have been studied quantitatively in Lindner (1986). The results are summarized in

Figure 4, which shows allowed regions in the $M_H - m_t$ plane for different new-physics scales Λ (Lindner 1986).

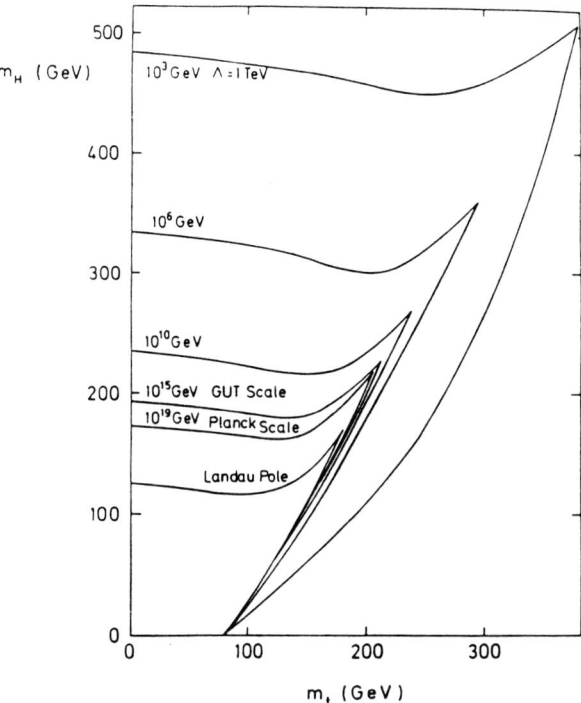

Figure 4. *Allowed regions in the $M_H - m_t$ plane, from Lindner (1986).*

Information on M_H can also be obtained from electroweak radiative corrections. Full details can be found in the lectures of Blondel (these Proceedings), and so only the general ideas are presented here. In the SM, there are several important tree-level relations between the couplings and masses. An example is

$$1 - \frac{M_W^2}{M_Z^2} = \sin^2 \theta_W = \frac{1}{2}\left[1 - \left(1 - \frac{4\pi\alpha}{\sqrt{2}G_F M_Z^2}\right)^{\frac{1}{2}}\right], \quad (19)$$

which allows the W mass to be predicted once the basis set of parameters (α, G_F, M_Z) is known precisely. Such a relation is, however, only valid at lowest order in electroweak perturbation theory. Nowadays experimental measurements are so accurate that higher-order corrections must be taken into account. For example, Equation (19) becomes

$$1 - \frac{M_W^2}{M_Z^2} \equiv \sin^2 \theta_W = \frac{1}{2}\left[1 - \left(1 - \frac{4\pi\alpha}{\sqrt{2}G_F M_Z^2(1 - \Delta r)}\right)^{\frac{1}{2}}\right], \quad (20)$$

where the radiative corrections have been incorporated into a correction term Δr. Now there are one-loop contributions to Δr which involve virtual top and Higgs particles.

Examples, which arise in the calculation of the renormalized W propagator, are shown in Figure 5.

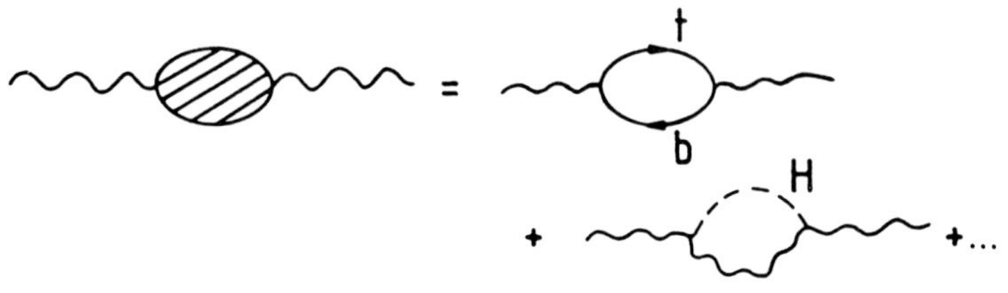

Figure 5. *One-loop contributions to the W propagator involving the top quark and Higgs boson.*

An explicit calculation gives (Veltman 1977)

$$\Delta r \simeq \Delta\alpha + \frac{\sqrt{2}G_F}{16\pi^2}\left[-3\cot^2\theta_W\, m_t^2 + \tfrac{11}{3}M_W^2 \ln\frac{M_H^2}{M_W^2}\right] + \cdots \qquad (21)$$

where ... denotes less important contributions. Thus we see that, given input information on α, G_F and M_Z, a precise measurement of M_W can yield information on m_t (quadratic dependence) and to a lesser extent on M_H (logarithmic dependence). There are many other quantities (partial Z widths, Z decay asymmetries, ν scattering cross sections, *etc.*) which can be analysed in a similar way. As discussed by Blondel, the most precise information comes from global fits to all such quantities, with m_t and M_H unknown parameters to be fitted. A recent example of the result of such a fit is shown in Figure 6, which shows χ^2 as a function of m_t for $M_H = 60\,\text{GeV}$, $300\,\text{GeV}$, $1000\,\text{GeV}$. In these fits the strong coupling is held fixed at the value measured in LEP event shape analyses: $\alpha_s(M_Z) = 0.123 \pm 0.006$ (LEP 1993). Evidently, there is as yet no statistically significant information on M_H from such fits.

If one takes the same three values of M_H but allows α_s to vary, the m_t values in Table 1 are obtained. The table also gives the corresponding χ^2 and $\alpha_s(M_Z)$ values. Note that smaller M_H corresponds to smaller m_t and α_s. If we recall (from the lectures of Bethke—these Proceedings) that the world average for α_s, including the very precise deep inelastic scattering measurements and event shapes from LEP, is

$$\alpha_s(M_Z) = 0.118 \pm 0.007, \qquad (22)$$

we might be tempted to conclude from Table 1 that lighter Higgs masses are slightly preferred. Figure 6 also shows the impact of a direct top mass measurement from the Fermilab $p\bar{p}$ collider. If the top quark is found at the lower end of the indicated mass range ($\sim 140\,\text{GeV}$) then we would conclude that M_H is at most of order a few hundred GeV. Obviously this would have far-reaching consequences for future search strategies.

We can therefore conclude that future Higgs boson phenomenology will focus on the mass range $O(60\,\text{GeV}) < M_H < O(1\,\text{TeV})$. For reasons which will become

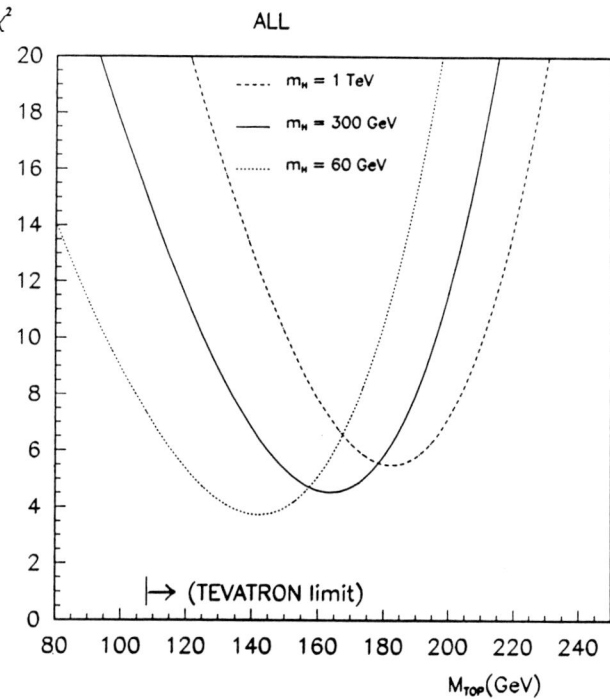

Figure 6. χ^2 curves for a Standard Model global fit to electroweak data, from LEP 1993, for three fixed Higgs masses.

M_H (GeV)	m_t (GeV)	χ^2	$\alpha_s(M_Z)$
60	143^{+17}_{-18}	3.7	0.118 ± 0.006
300	166^{+16}_{-17}	4.4	0.120 ± 0.006
1000	182^{+15}_{-16}	5.4	0.122 ± 0.006

Table 1. Fitted m_t and α_s values for various fixed M_H, from Blondel.

clear below, it is convenient to identify three distinct Higgs mass regions: a 'heavy' Higgs is defined as one with $M_H > O(200\,\text{GeV})$, an 'intermediate-mass' Higgs has $O(90\,\text{GeV}) < M_H < O(200\,\text{GeV})$, and a 'light' Higgs has $M_H < O(90\,\text{GeV})$.

2.3 Decay channels

The 'leading-order' Higgs partial decay widths are readily obtained from the diagrams of Figure 2:

$$\Gamma(H \to f\bar{f}) = \frac{CG_F m_f^2 M_H}{4\pi\sqrt{2}} \left(1 - \frac{4m_f^2}{M_H^2}\right)^{3/2}$$

$$\Gamma(H \to W^+W^-) = \frac{G_F M_H^3}{8\pi\sqrt{2}} \left(1 - \frac{4M_W^2}{M_H^2}\right)^{1/2} \left(1 - \frac{4M_W^2}{M_H^2} + \frac{12M_W^4}{M_H^4}\right)$$

$$\Gamma(H \to ZZ) = \frac{1}{2} \Gamma(H \to W^+W^-)|_{M_W \leftrightarrow M_Z} \quad (23)$$

where C is a colour multiplicity factor: $C = 3$ for quarks and $C = 1$ for leptons. Note that these results are only valid *above* the two-particle threshold in each case.

A complicating feature of present-day Higgs phenomenology is the lack of precise knowledge of the top quark mass. This not only effects the dominant decay modes, but also influences the production cross sections (see below). For example, given the preference of the Higgs to couple to the heaviest available fermions, the decay modes of, say, a 140 GeV Higgs are completely different depending on whether m_t is greater or less than half this value, *viz.* 70 GeV. However, as the lower limits on the top mass, directly from searches and indirectly from electroweak radiative corrections, increase above 100 GeV, the picture begins to simplify. The reason is that for $m_t > M_W, M_Z$, as suggested nowadays by searches at the Fermilab $p\bar{p}$ collider, the decay width for $H \to t\bar{t}$ is always smaller than those for $H \to W^+W^-$ and $H \to ZZ$, and so the $t\bar{t}$ decay mode is never dominant. In this scenario, below the WW threshold (and assuming of course that $M_H > 2m_b$) the Higgs will decay predominantly into $b\bar{b}$, as in Figure 7.

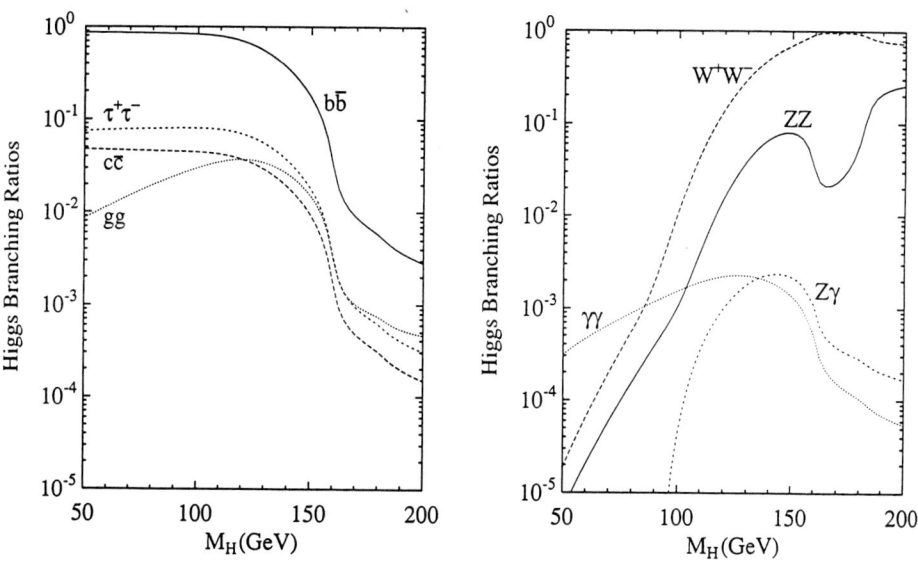

Figure 7. *Branching ratios of the intermediate-mass Higgs boson.*

The next most important decay mode is $\tau^+\tau^-$, with a branching ratio suppressed by a factor of $3(m_b^2/m_\tau^2)$. Above the WW threshold the Higgs decays predominantly into WW and ZZ pairs (apart from a small mass window where only the former is kinematically allowed), Figure 8.

Away from the thresholds, the branching ratios are approximately in the ratio 2:1, Equation (23). There is a slight dip in the WW, ZZ branching ratios around the $t\bar{t}$

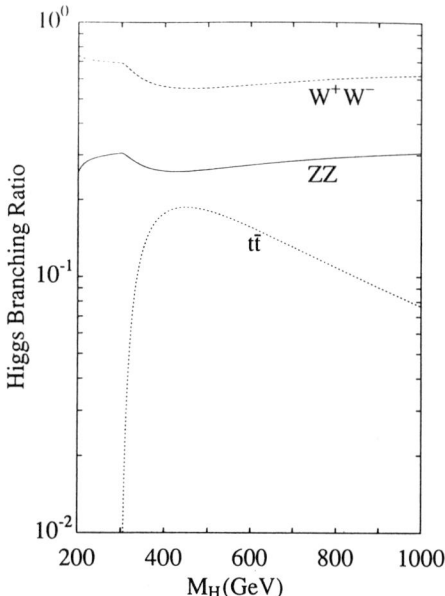

Figure 8. *Branching ratios of the heavy Higgs boson, for $m_t = 150\,\text{GeV}$.*

threshold. Of course the W and Z bosons into which the Higgs decays are themselves unstable, and so the actual final states consist of leptons and quark jets. Table 2 gives the branching ratios for the various channels, ignoring small perturbations from $H \to t\bar{t}$. As we shall see, final states containing jets have large backgrounds at hadron colliders. A large fraction of the decays are therefore not useful, and it is the much rarer purely leptonic final states that yield the main hope for discovery.

channel	$H \to W^+W^-$	$H \to Z^0 Z^0$	total
4 jets	0.56	0.52	0.55
2 jets + $l\nu$	0.37		0.25
2 jets + l^+l^-		0.14	0.05
2 jets + $\nu\bar{\nu}$		0.26	0.09
$l^+\nu l^-\bar{\nu}$	0.06		0.04
$l^+l^-l^+l^-$		0.010	0.003
$l^+l^-\nu\bar{\nu}$		0.036	0.012
$\nu\bar{\nu}\nu\bar{\nu}$		0.032	0.011

Table 2. *Branching ratios of a heavy SM Higgs boson decaying into quark and lepton ($l = e,\,\mu,\,\tau$) final states.*

In addition to the 'leading-order' decay channels discussed above, certain rare decay channels which appear beyond leading order can be important. Paramount among these is the $H \to \gamma\gamma$ decay, which is mediated by intermediate W and quark triangle loops:

$$\Gamma(H \to \gamma\gamma) = \frac{\alpha^2 G_F M_H^3}{128\sqrt{2}\pi^3} \left| \sum_q 3e_q^2 I_q \left(\frac{m_q^2}{M_H^2}\right) + I_W \left(\frac{M_W^2}{M_H^2}\right) \right|^2 \quad (24)$$

where I_q and I_W are dimensionless functions given by:

$$\begin{aligned}
I_q(x) &= 4x[2 + (4x - 1)F(x)] \\
I_W(x) &= -2[6x + 1 + 6x(2x - 1)F(x)] \\
F(x) &= \frac{1}{2}\theta(1 - 4x)\left[\ln\left(\frac{1 + \sqrt{1 - 4x}}{1 - \sqrt{1 - 4x}}\right) - i\pi\right]^2 \\
&\quad - 2\theta(4x - 1)\left[\sin^{-1}\left(1/2\sqrt{x}\right)\right]^2 .
\end{aligned} \quad (25)$$

The corresponding branching ratio is shown in Figure 7.

A further point is worth mentioning in connection with the Higgs branching ratios. Higher-order QCD corrections to the $b\bar{b}$ width are very important. The main effect is to 'run' the b-quark mass down to a lower value, and since $\Gamma \propto m_b^2$ the partial width is correspondingly decreased. This in turn boosts the branching ratios of certain important rare decay modes (e.g. $H \to \gamma\gamma$ and $H \to ZZ^*$) above what a lowest order calculation would naïvely give. A complete discussion of the all various Higgs partial decay widths including higher-order corrections—together with the relevant formulae—can be found in Dawson et al. (1990).

A final property of the Higgs boson which is important for hadron-collider phenomenology is the fact, already evident in Equation (23), that the total Higgs decay width grows rapidly with M_H to become comparable with the mass at around 1 TeV. This is illustrated in Figure 9, which shows Γ_H as a function of M_H for $m_t = 150$ GeV.

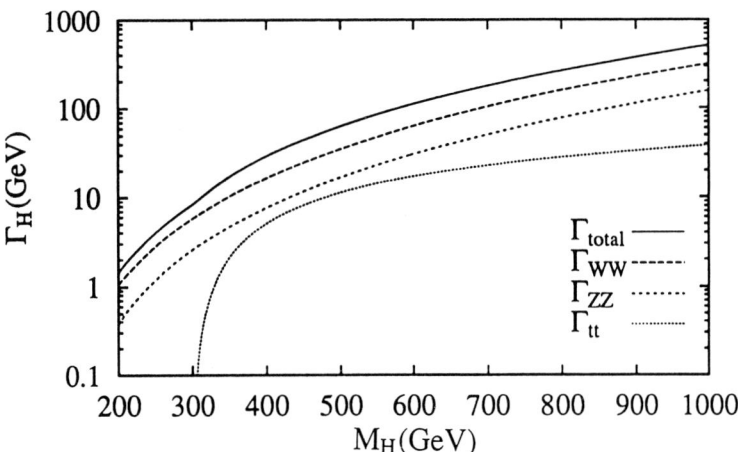

Figure 9. *Total decay width of the Standard Model Higgs boson as a function of M_H, for $m_t = 150$ GeV.*

3 Higgs at LEP

LEP-I and SLC are primarily 'Z factories'. Higgs phenomenology at these machines is therefore the study of rare Z decays involving a Higgs in the final state. There is, in fact, only one relevant decay mechanism:

$$Z \to HZ^* \to Hf\bar{f}, \qquad (26)$$

with $f = e, \mu, \nu$. At leading order, the differential partial decay width $\Gamma(Z \to Hf\bar{f})$, normalized to $\Gamma(Z \to f\bar{f})$, is

$$\frac{1}{\Gamma(Z \to f\bar{f})} \frac{d\Gamma(Z \to Hf\bar{f})}{dx} = \frac{\alpha(x^2 - 4r^2)^{\frac{1}{2}}}{48\pi \sin^2\theta_W \cos^2\theta_W} \frac{12 - 12x + x^2 + 8r^2}{(x - r^2)^2 + (\Gamma_Z/M_Z)^2}, \qquad (27)$$

where $x = 2E_H/M_Z$ is the scaled Higgs energy, with kinematical limits $2r \leq x \leq 1+r^2$, and $r = M_H/M_Z$. The total rate can be obtained by numerical integration. Figure 10 shows the partial decay width for $Z \to H\mu^+\mu^-$, normalized to that for $Z \to \mu^+\mu^-$, as a function of M_H.

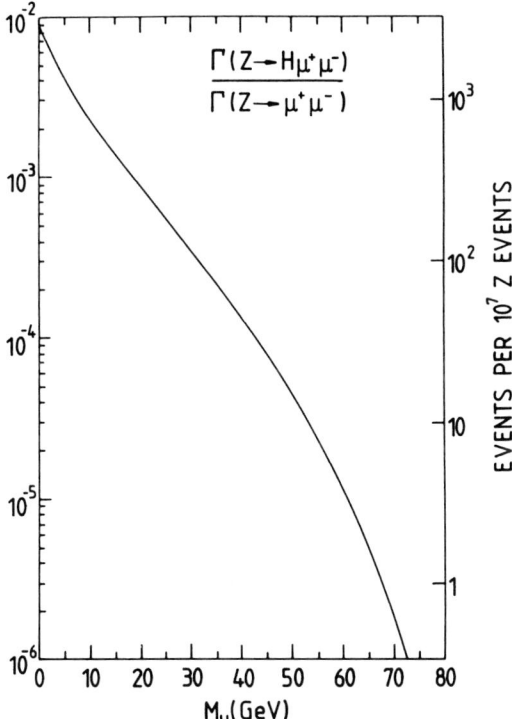

Figure 10. *Partial decay width for the $Z \to H\mu^+\mu^-$ process, normalized to the $\mu^+\mu^-$ decay width.*

At LEP, various search strategies have been devised to rule out a very light SM Higgs, so that the absence of signal events can be interpreted as a *lower* bound on M_H. For example, if M_H is so small that the only (leading order) kinematically allowed decay channel is $H \to e^+e^-$, then the lifetime is so long that the Higgs will usually decay *outside* the detector. In this case one looks for events of type $e^+e^- \to e^+e^-+$ missing momentum, with the final state e^+e^- almost back-to-back and directed away from the beam direction. As M_H increases, the lifetime becomes shorter, the Higgs decays inside the detector, and final states such as $H\nu\bar\nu$ with $H \to e^+e^-, \mu^+\mu^-, \pi^+\pi^-, K^+K^-, D\overline{D}$, ... according to the available energy, become observable. Finally, for $M_H > 2m_b$, the Higgs decays into two jets (which may appear as a merged monojet if the mass is not too high) and final states of the form $(e^+e^-, \mu^+\mu^-, \nu\bar\nu)+$ jet(s) provide a signature. An example of a search covering the low-mass region in the ALEPH detector at LEP is shown in Figure 11. The number of events expected in various channels is shown as a function of M_H, and the absence of observed events rules out the whole range down to $M_H = 0$. Similar results have been obtained by the other LEP experiments.

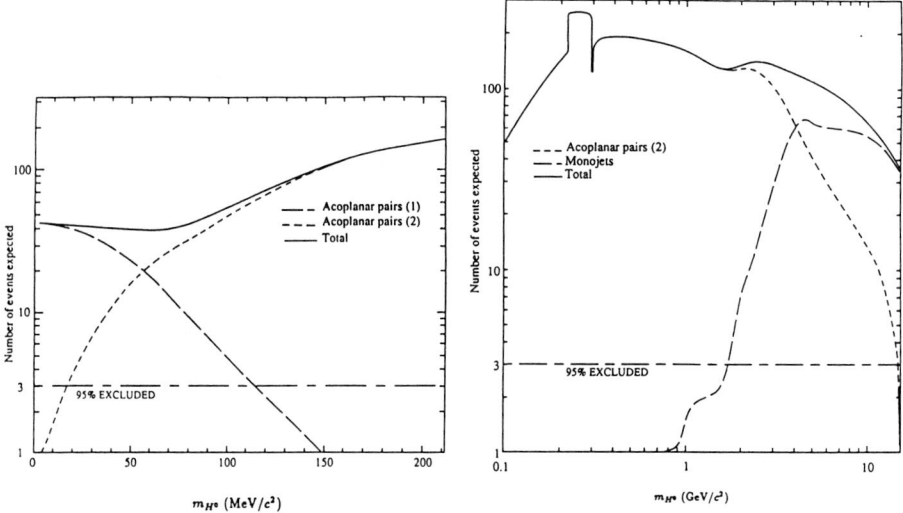

Figure 11. *Search for a light SM Higgs in the ALEPH detector at LEP (ALEPH 1990).*

A recent comprehensive review of SM Higgs searches at LEP can be found in Gross and Yepes (1993). No events have been seen so far, and the resulting lower limit on the Higgs mass, combining the limits from the four experiments ALEPH, DELPHI, L3 and OPAL, is currently (Carter 1993)

$$M_H > 62.5\,\text{GeV} \qquad (95\%\ \text{cl}) \qquad (28)$$

from a total sample of approximately 3.4 million hadronically decaying Z's. It is clear from the steepness of the curve in Figure 10 for large Higgs masses that it becomes progressively harder to improve the above lower limit at LEP-I: $O(70\,\text{GeV})$ is a realistic ultimate limit.

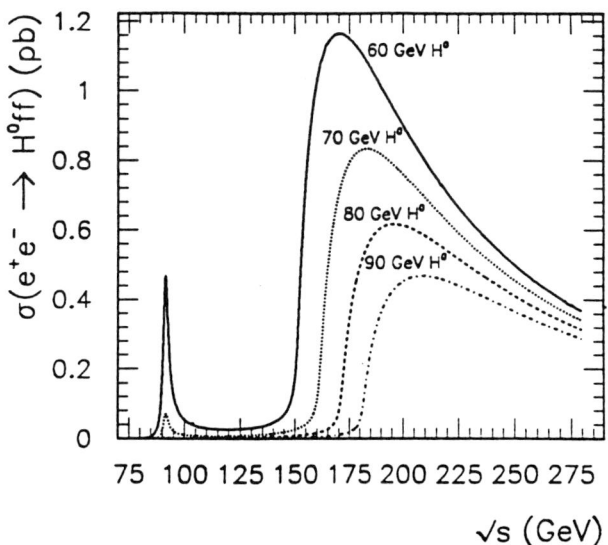

Figure 12. *Total cross section for $e^+e^- \to f\bar{f}H$ as a function of the centre-of-mass collision energy, for various values of M_H, from Gross and Yepes (1993).*

With LEP-II, the e^+e^- collision energy will approach 200 GeV. The dominant Higgs production process is then $e^+e^- \to ZH$, from the same Feynman diagram as for the LEP-I process (26) but now with a *real* rather than a *virtual* Z in the final state (*i.e.* the opposite situation to the real Z decay discussed above). The cross section is, at leading order,

$$\sigma(e^+e^- \to ZH) = \frac{\pi\alpha^2[1+(1-4\sin^2\theta_W)^2]P(P+3M_Z^2)}{24\sin^4\theta_W \cos^4\theta_W \sqrt{s}(s-M_Z^2)^2}, \tag{29}$$

where P is the three-momentum of the Z:

$$P = \frac{1}{2s}\left[s^2 - 2s(M_Z^2 + M_H^2) + (M_Z^2 - M_H^2)^2\right]^{\frac{1}{2}}. \tag{30}$$

These cross sections are quite sizeable—Figure 12 shows the total cross section as a function of the collision energy for various Higgs masses. If we assume that the Higgs in this mass range decays predominantly into $b\bar{b}$ pairs (*i.e.* two jets), then the largest background is from ZZ pair production: $e^+e^- \to Z(\to q\bar{q}) + Z(\to f\bar{f})$. Note that the Z which decays into two jets and simulates $H \to b\bar{b}$ is off-mass-shell and therefore gives a continuous distribution of masses falling steeply away from the Z pole. The background is therefore only important for Higgs masses near to M_Z. However, the ability to efficiently tag b quarks using vertex detectors allows, with sufficient statistics, a separation of real Z's decaying only 15% of the time to $b\bar{b}$, from Higgs bosons with $M_H \sim M_Z$ decaying 90% of the time to $b\bar{b}$. Enhancement of Higgs signals at LEP-II using b-tagging has been studied by all four LEP experiments.

Figure 13 shows a typical result from the L3 experiment (Alcaraz et al. 1993), using the 2 jet + $\nu\bar{\nu}$ channel. For $M_H = 80$ GeV there is little difficulty in separating the Higgs peak from the Z background. For $M_H = 90$ GeV the peaks coincide, but the requirement that the jets originate from $b\bar{b}$ makes the ZH signal stand out clearly above the ZZ background.

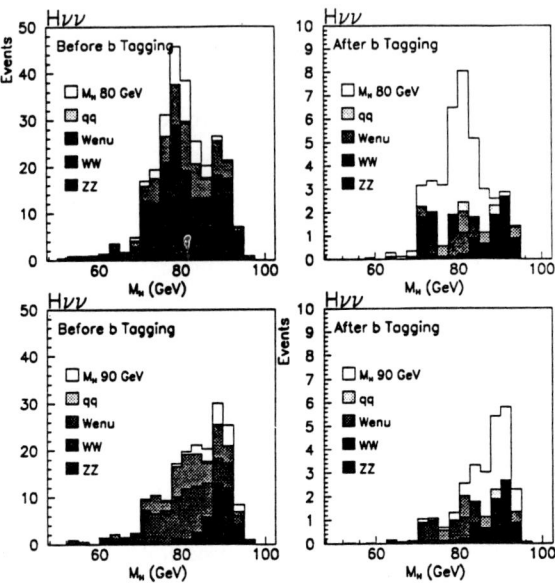

Figure 13. *A simulation of Higgs production at LEP-II with and without b-tagging, from the L3 collaboration (Alcaraz et al. 1993).*

Ultimately, the discovery limit at LEP-II depends on the maximum e^+e^- centre-of-mass energy attainable. As an approximate guide, with 500 pb^{-1} of integrated luminosity, the sensitivity of LEP-II extends to $M_H \simeq \sqrt{s} - 100$ GeV.

4 Higgs at hadron colliders

In this section we consider the most important search strategies at high energy hadron colliders. We consider in particular the SSC (pp collisions at $\sqrt{s} = 40$ TeV) and the LHC (pp collisions at 16 TeV). When these machines come into operation, either the Higgs boson will have been discovered, or a lower mass limit of $O(90\,\text{GeV})$ will have been set. Higgs boson phenomenology at hadron colliders focusses therefore on intermediate-mass and heavy Higgs bosons. There has been a vast amount of work on this topic in recent years, and we are only able to give a brief summary here. More detailed discussions can be found, for example, in Dawson et al. (1990) and Froidevaux et al. (1990).

There are only a limited number of production mechanisms which give cross sections large enough to be relevant at these machines. Each makes use of the Higgs' preference to couple to heavy particles:

(i) gluon-gluon fusion: $gg \to H$,

(ii) vector boson fusion: $qq \to Hqq$ via $W^+W^-, ZZ \to H$,

(iii) associated production with vector bosons: $q\bar{q} \to WH, ZH$,

(iv) associated production with top quarks: $gg, q\bar{q} \to t\bar{t}H$.

The Feynman diagrams for these are shown in Figure 14.

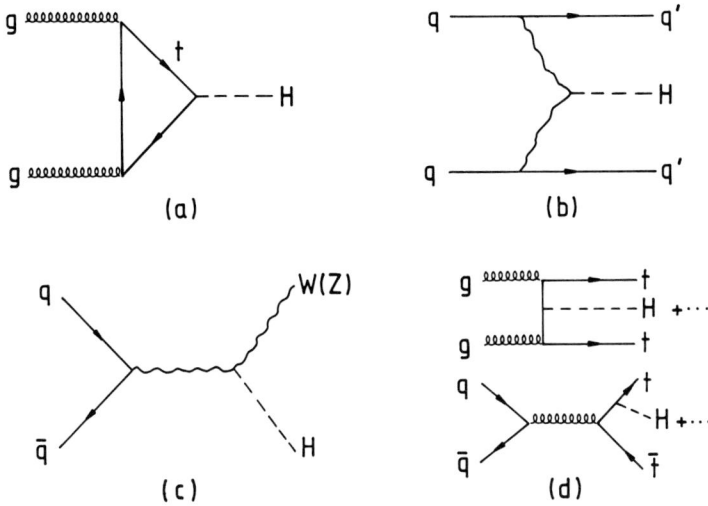

Figure 14. *The dominant Higgs production mechanisms at hadron colliders: (a) gluon-gluon fusion, (b) vector-boson fusion, and associated production with (c) W bosons and (d) $t\bar{t}$.*

Except possibly for very heavy Higgs, the dominant production mechanism is gluon fusion, $gg \to H$, via an intermediate top quark loop, Figure 14a. The matrix element summed and averaged over spins and colours is (Georgi et al. 1978)

$$|M|^2(gg \to H) = \frac{\alpha_s^2(M_H^2) G_F M_H^4}{288\sqrt{2}\pi^2} \left| I\left(\frac{m_t^2}{M_H^2}\right) \right|^2 \qquad (31)$$

where $I(x)$ is a dimensionless function given by:

$$I(x) = 3x[2 + (4x - 1)F(x)]. \qquad (32)$$

The function $F(x)$ is defined in Equation (25). The factor $|I(x)|^2$ vanishes as $x \to 0$, attains a maximum value of about 3.3 at $x \sim 0.2$ and then decreases to the constant value of 1 as $x \to \infty$.

The cross sections for Higgs production via the gluon fusion mechanism in 16 TeV and 40 TeV pp collisions are shown in Figures 15 and 16 respectively, as a function of M_H for the three values $m_t = 100, 150, 200$ GeV. The cross section increases with increasing

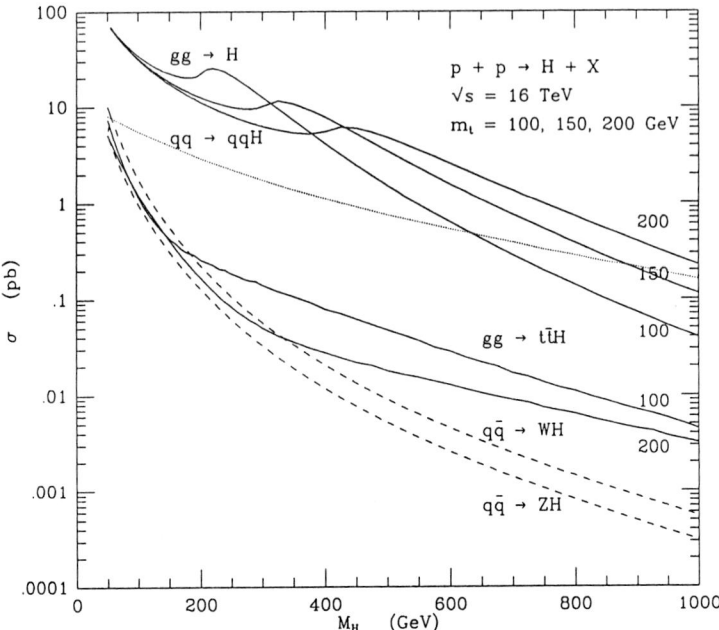

Figure 15. *Higgs production cross sections in pp collisions at* LHC, $\sqrt{s} = 16\,\text{TeV}$.

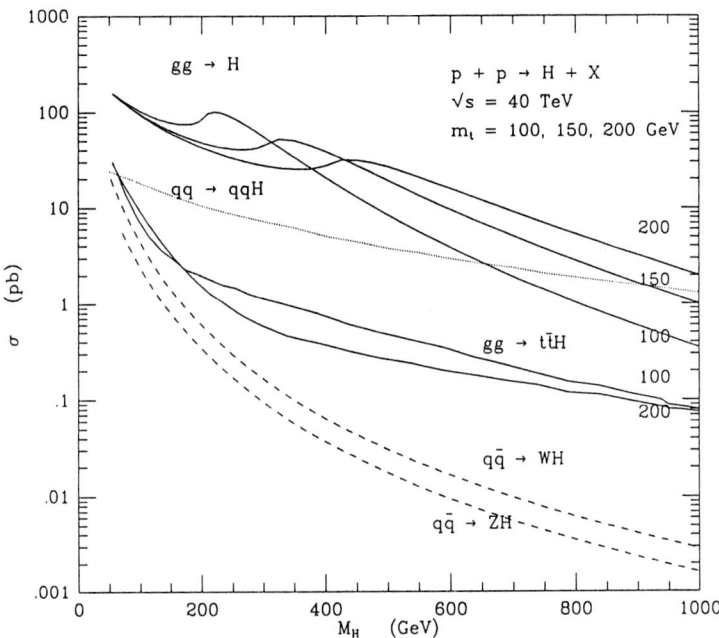

Figure 16. *Higgs production cross sections in pp collisions at* SSC, $\sqrt{s} = 40\,\text{TeV}$.

m_t for fixed large M_H, is approximately independent of m_t for $M_H \sim 400\,\text{GeV}$, and is larger for smaller top masses in the intermediate-mass Higgs region.

When the top quark was believed to be relatively light, the mechanism that received the most attention was Higgs production by WW or ZZ fusion, Figure 14b. If the Higgs width is not too large compared to its mass, the cross section is calculated from the $2 \to 3$ process: $qq \to Hqq$. The matrix element squared for WW fusion is (Cahn and Dawson 1984)

$$|M|^2(q(p_1)q(p_2) \to q(p_3)q(p_4)H) = \frac{512 M_W^8 G_F^3}{2\sqrt{2}}$$
$$\times \frac{(p_1 \cdot p_2)(p_3 \cdot p_4)}{((p_3 - p_1)^2 - M_W^2)^2((p_4 - p_2)^2 - M_W^2)^2}$$

$$|M|^2(q(p_1)\bar{q}(p_2) \to q(p_3)\bar{q}(p_4)H) = \frac{512 M_W^8 G_F^3}{2\sqrt{2}}$$
$$\times \frac{(p_1 \cdot p_4)(p_2 \cdot p_3)}{((p_3 - p_1)^2 - M_W^2)^2((p_4 - p_2)^2 - M_W^2)^2}, \quad (33)$$

and for ZZ fusion (Cahn and Dawson 1984):

$$|M|^2(q(p_1)q'(p_2) \to q(p_3)q'(p_4)H) = (v_q + a_q)^2(v_{q'} + a_{q'})^2 \frac{M_W^8 G_F^3}{8\cos^8\theta_W \sqrt{2}}$$
$$\times \frac{(p_1 \cdot p_2)(p_3 \cdot p_4)}{((p_3 - p_1)^2 - M_Z^2)^2((p_4 - p_2)^2 - M_Z^2)^2}$$

$$|M|^2(q(p_1)\bar{q}'(p_2) \to q(p_3)\bar{q}'(p_4)H) = (v_q + a_q)^2(v_{q'} - a_{q'})^2 \frac{512 M_W^8 G_F^3}{8\cos^8\theta_W \sqrt{2}}$$
$$\times \frac{(p_1 \cdot p_4)(p_2 \cdot p_3)}{((p_3 - p_1)^2 - M_Z^2)^2((p_4 - p_2)^2 - M_Z^2)^2}, \quad (34)$$

with the quark vector and axial couplings given by

$$v_u = \tfrac{1}{2} - \tfrac{4}{3}\sin^2\theta_W, \qquad a_u = \tfrac{1}{2}$$
$$v_d = -\tfrac{1}{2} + \tfrac{2}{3}\sin^2\theta_W, \qquad a_d = -\tfrac{1}{2}. \quad (35)$$

Convolution with the appropriate quark distribution functions gives the cross sections shown in Figures 15 and 16. As all the electroweak parameters are well-determined, and the quark distributions are sampled in a x range where they are constrained by deep inelastic scattering data, there is very little uncertainty in these predictions.

The cross sections for Higgs production in association with W and Z bosons are not large, but may be useful for detecting an intermediate-mass Higgs especially if high luminosity is available, since the Higgs can be 'tagged' by triggering on the weak boson. The production mechanisms are $q\bar{q} \to WH$, $q\bar{q} \to ZH$, Figure 14c, with matrix elements (Glashow et al. 1978):

$$|M|^2(q\bar{q} \to WH) = \frac{\pi^2\alpha^2}{9\sin^4\theta_W \hat{s}} \frac{\hat{s}^2 + M_W^4 + M_H^4 - 2\hat{s}(M_H^2 - 5M_W^2) + M_W^2 M_H^2}{(\hat{s} - M_W^2)^2}$$

$$|M|^2(q\bar{q} \to ZH) = (v_q^2 + a_q^2) \frac{\pi^2 \alpha^2}{9 \sin^4 \theta_W \cos^4 \theta_W \hat{s}}$$
$$\times \frac{\hat{s}^2 + M_Z^4 + M_H^4 - 2\hat{s}(M_H^2 - 5M_Z^2) + M_Z^2 M_H^2}{(\hat{s} - M_Z^2)^2}. \qquad (36)$$

The cross sections at LHC and SSC are shown in Figures 15 and 16. The WH cross section is uniformly a factor of about 2 higher than the ZH cross section.

In the same spirit as the previous mechanism, a Higgs boson can be radiated off a heavy top quark produced in gluon-gluon or quark-antiquark fusion: $gg \to t\bar{t}H$, $q\bar{q} \to t\bar{t}H$ (Kunszt 1984, Gunion et al. 1987, Dicus and Willenbrock 1989), Figure 14d. Not surprisingly, the cross section depends quite sensitively on the top quark mass. Although the top quark-Higgs coupling is large for large m_t, this is counterbalanced by the parton distribution suppression of the correspondingly heavier $t\bar{t}H$ final state. The cross sections for $m_t = 100$ GeV and 200 GeV are shown in Figures 15 and 16. The lighter top quark cross section is larger over most of the M_H range, by a factor between one and three.

Signals and backgrounds for a wide variety of production and decay channels have been studied (Dawson et al. 1990, Froidevaux et al. 1990). Three channels, however, stand out from the others in that they appear to provide the 'best' chance for discovery of the Higgs at LHC and SSC. These are

(i) $gg \to H \to \gamma\gamma$,

(ii) $gg \to H \to ZZ^{(*)} \to l^+l^-l^+l^-$, the so-called 'gold-plated' channel,

(iii) $q\bar{q} \to WH \to l\bar{\nu}_l\gamma\gamma$, $gg\,(q\bar{q}) \to t\bar{t}H \to WH + X \to l\bar{\nu}_l\gamma\gamma + X$.

Note that in each case the branching ratios involved are extremely small. For example the $H \to 4l^\pm$ ($l = e, \mu$) decay mode of a heavy Higgs has a branching ratio of only 0.15%, Table 2 (see page 235).

The total production rates are readily obtained by combining the branching ratios in Figures 7 and 8 and Table 2 of Section 2.3 with the cross sections of Figures 15 and 16. Thus Figure 17 shows the cross section times branching ratio for 'direct' $pp \to H(\to \gamma\gamma) + X$ production, at $\sqrt{s} = 16$ TeV and 40 TeV, and Figure 18 shows the corresponding rate for the four-charged-lepton channel ($l = e, \mu$). Note that for the former, the effect of a rising branching ratio combined with a falling cross section yields a signal which is approximately constant as a function of M_H in the intermediate-mass region. Only for Higgs masses of order 140–150 GeV and above does the signal start to fall away rapidly. The $H \to 4l^\pm$ rate as a function of M_H has an interesting structure. At the low mass end, the rate falls very sharply to zero below about $M_H = 150$ GeV—a reflection of the behaviour of the branching ratio shown in Figure 7. At the high mass end, the dominant feature is the strong dependence on the top quark mass.

Of course, none of the above channels is free of background. The main concern is with SM background processes which are *irreducible*, i.e. contain exactly the same set of particles in the final state as the signal. For the three classes of signal process defined above, the irreducible backgrounds come from (i) $q\bar{q}, gg \to \gamma\gamma$, (ii) $q\bar{q} \to ZZ^{(*)} \to l^+l^-l^+l^-$, and (iii) $q\bar{q} \to W\gamma\gamma \to l\bar{\nu}_l\gamma\gamma$, $gg\,(q\bar{q}) \to t\bar{t}\gamma\gamma \to l\bar{\nu}_l\gamma\gamma + X$.

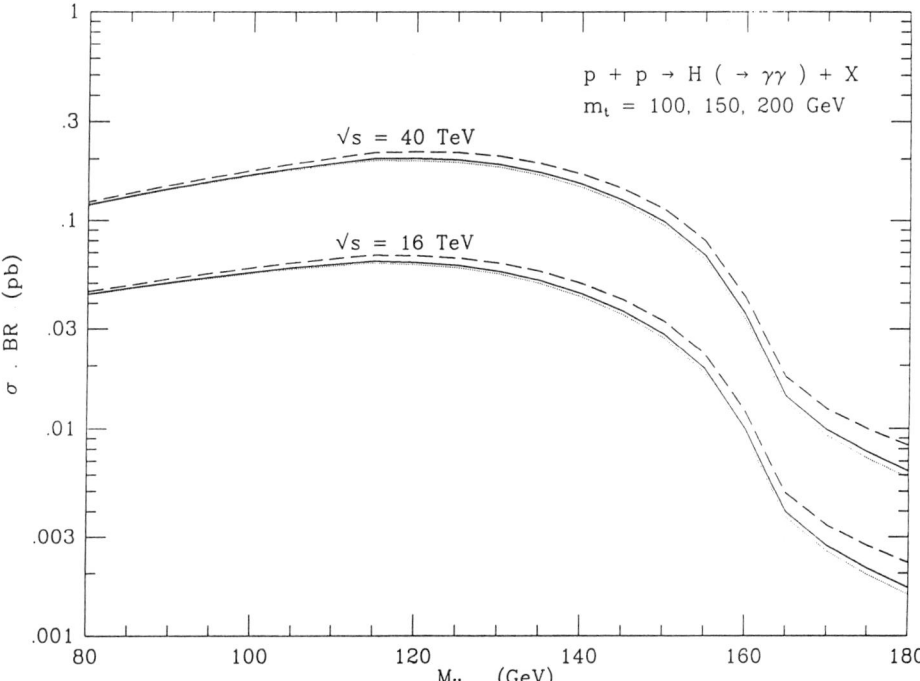

Figure 17. *Cross section times branching ratio for* $pp \to H(\to \gamma\gamma) + X$ *as a function of* M_H, *at* LHC *and* SSC.

Reducible backgrounds, for example where the charged lepton comes from the decay of a b quark and not a W, can usually be suppressed by additional (*e.g.* isolation) cuts. Particularly troublesome in this respect is hadronic $t\bar{t} \to bW^+\bar{b}W^-$ production, which is not only an irreducible background for $H \to W^+W^-$, but which also can mimic $H \to 4l^\pm$ when both b quarks decay semileptonically. It has been shown, however, that in most cases all backgrounds can be suppressed to an acceptable level by suitable cuts on the final state particles, while still leaving the bulk of the signal intact. In particular, the four-lepton channel offers the cleanest possibility of discovery, in a Higgs mass region bounded from above *and* below, the exact window depending on the available luminosity, the top mass, and the experimental acceptance. As a rough guide, assuming integrated luminosities $\int \mathcal{L} = 10^5 \text{pb}^{-1}$ at $\sqrt{s} = 16\,\text{TeV}$ (LHC) and $\int \mathcal{L} = 10^4 \text{pb}^{-1}$ at $\sqrt{s} = 40\,\text{TeV}$ (SSC) gives a Higgs discovery window in this channel of approximately

$$130\,\text{GeV} < M_H < 800\,\text{GeV}. \tag{37}$$

If only $\int \mathcal{L} = 10^4 \text{pb}^{-1}$ is available at LHC, the upper limit drops to around 600 GeV. An example of the detection of a heavy Higgs in this channel in the ATLAS detector at LHC is shown in Figure 19.

The $pp \to H(\to \gamma\gamma) + X$ channel allows the mass region between the LEP-II limit and 130 GeV to be searched. The problem here is the large irreducible continuum

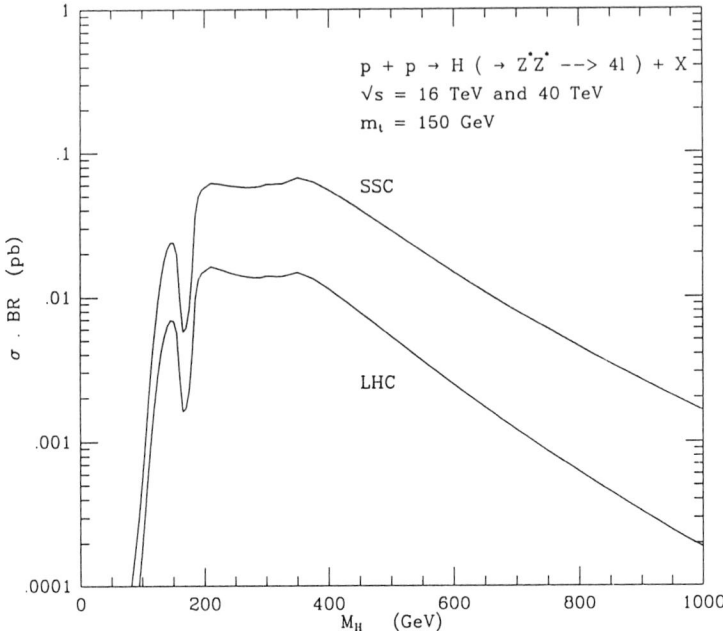

Figure 18. *Cross section times branching ratio for* $pp \to H(\to l^+l^-l^+l^-) + X$ $(l = e, \mu)$ *as a function of* M_H, *at* LHC *and* SSC.

background from gg, $q\bar{q} \to \gamma\gamma$ production, which rises steeply as a function of $M_{\gamma\gamma}$. The signal can only be observed above the continuum background if the experimental $M_{\gamma\gamma}$ resolution is very high. To give an example, Table 3 (from Jenni) shows the result of a simulation of this process in the ATLAS detector at the LHC.

M_H (GeV)	80	90	110	130	150
$\sigma B(H \to \gamma\gamma)$ (fb)	51	57	68	70	35
acceptance (%)	23	30	41	46	51
mass resolution (%)	1.45	1.40	1.22	1.19	1.12
signal in mass bin	600	876	1430	1650	915
bkgd. in mass bin	36k	34k	25k	20k	13.5k
stat. significance	3.2	4.8	9.0	11.7	7.9

Table 3. *Observability of* $H \to \gamma\gamma$ *in* ATLAS *at* LHC *as a function of* M_H, *for* $\int \mathcal{L} = 10^5 \text{pb}^{-1}$.

Evidently, there appears to be no problem for $M_H > O(100 \text{ GeV})$: Figure 20 shows the expected $M_{\gamma\gamma}$ spectrum for the signal and irreducible background for $M_H = 110 \text{ GeV}$ and $\int \mathcal{L} = 10^5 \text{pb}^{-1}$. The main concern is with a possible window of unobservability in

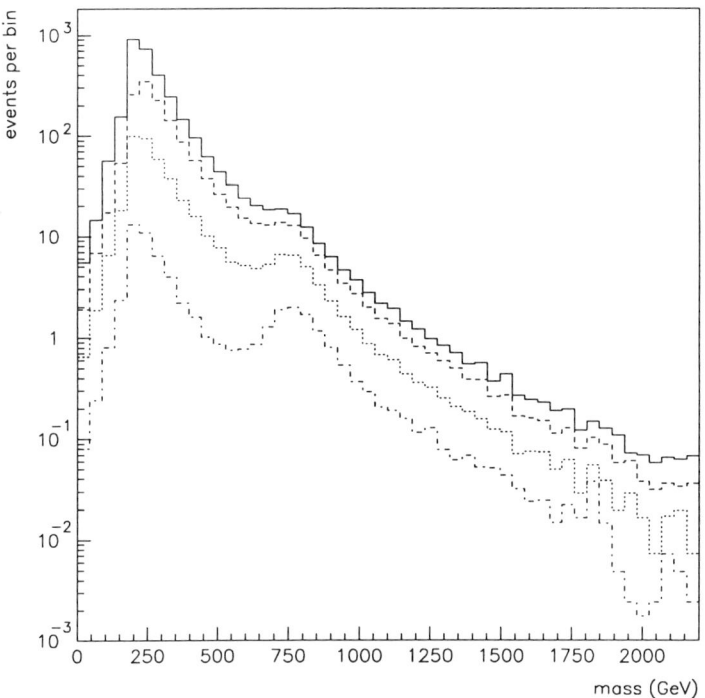

Figure 19. *Expected M_{ZZ} signal and background spectrum as simulated in the* ATLAS *detector at* LHC, *for* $M_H = 800 \text{ GeV}$ *and* $\int \mathcal{L} = 10^5 \text{pb}^{-1}$, *from Jenni. Successive cuts applied: all events in acceptance (solid), events with $P_T(Z) > \frac{1}{4} M_{ZZ}$ (dashed), events with one (dotted) and two (dot-dashed) reconstructed tag jets (see text).*

this channel at the low mass end.

The WH, $t\bar{t}H$ production processes also allow the light intermediate-mass Higgs region to be covered. The idea here is to use the decay channels $WH \to l\nu\gamma\gamma$, which gives a clean final state with manageable backgrounds. High $M_{\gamma\gamma}$ resolution is not required—the main problem is that with the anticipated machine luminosity, the number of signal events is rather small. Figure 21 shows a simulation of the direct WH channel at LHC (Summers 1992), with the signal for three different Higgs masses and the most important backgrounds. The $t\bar{t}H$ channel provides an additional source of events, particularly at the SSC where the event rate approximately doubles. Further experimental details can be found in Seez et al. (1990).

The above channels, therefore, provide coverage of the Higgs mass range up to $O(800 \text{ GeV})$ at both the LHC and SSC. A considerable amount of effort has gone into investigating whether the search can be extended to $M_H \sim 1 \text{ TeV}$. Attention has focussed on the heavy Higgs decay channels which have larger branching ratios (see Table 2), in particular $H \to WW \to l\nu jj$ and $H \to ZZ \to l^+l^-\nu\bar{\nu}$.

The $H \to ZZ \to l^+l^-\nu\bar{\nu}$ signal can, in principle, be identified as a Jacobian peak in the missing transverse energy distribution, at $E_T^{\text{miss}} \sim M_H/2$. Aside from the irre-

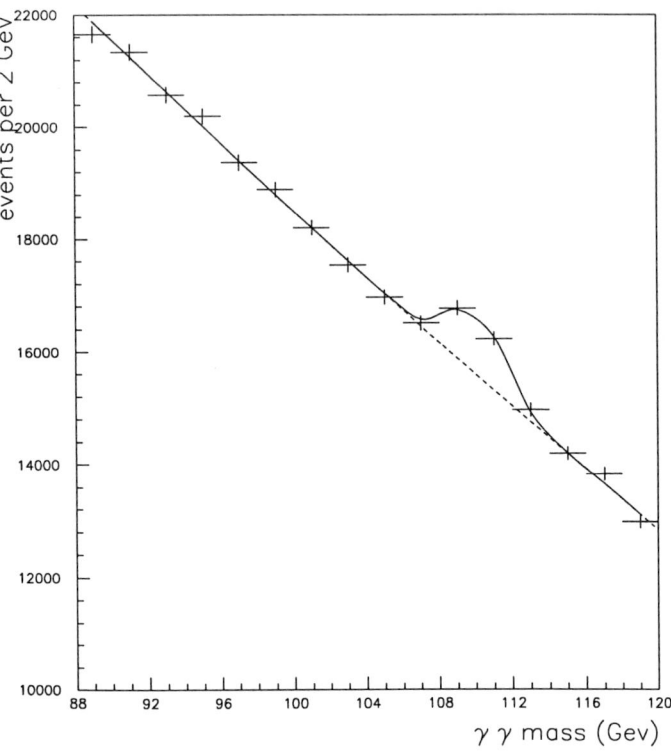

Figure 20. *Expected $M_{\gamma\gamma}$ spectrum as simulated in the* ATLAS *detector at* LHC, *for $M_H = 110$ GeV and $\int \mathcal{L} = 10^5 \text{pb}^{-1}$, from Jenni (these Proceedings).*

	signal	$t\bar{t}$ background
central cuts and reconstruction	640	240000
single jet tag $E^j > 600$ GeV $p_T^j > 25$ GeV	390	4500
double jet tag $E^j > 600$ GeV $p_T^j > 25$ GeV	90	200

Table 4. *Expected rates for the $H \to WW \to l\nu jj$ signal and $t\bar{t}$ background for a 1 TeV Higgs, in* ATLAS *at* LHC *with $\int \mathcal{L} = 10^5 \text{pb}^{-1}$, from Jenni (these Proceedings).*

ducible ZZ continuum background, the main problem is with the much larger $Z+$ jets background. This can only be suppressed if the detector has a large acceptance in rapidity for hadronic jets, and has a high hermiticity.

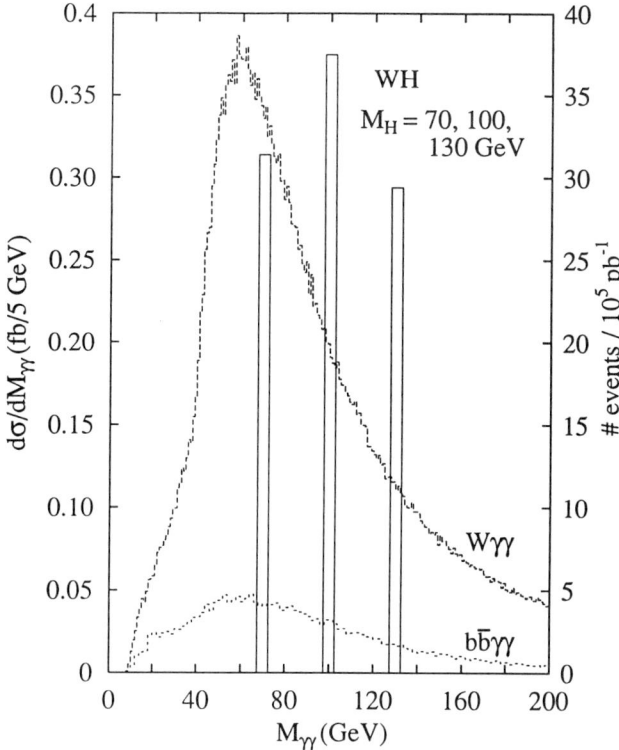

Figure 21. *The signal $pp \to WH + X \to l\nu\gamma\gamma + X$ with the main backgrounds at LHC, from Summers (1992).*

The key to utilizing the $H \to WW \to l\nu jj$ channel is 'jet tagging' (Cahn et al. 1987). The idea is to use the WW and ZZ fusion production processes, which are comparable with the gg fusion process at large M_H (see Figures 15 and 16) and to tag the quark jets which are produced in association with the Higgs. These jets are energetic and are produced at large rapidity, typically $E^j \sim 1\,\text{TeV}$ and $p_T^j \sim M_W/2$. The probability of finding such jets in background $W+$ jets and $t\bar{t}$ production is small, and so requiring one or two tag jets in an event provides a powerful background rejection. Of course this procedure requires high forward calorimeter performance, in both coverage and granularity. Table 4, which shows signal and background rates for a $M_H = 1\,\text{TeV}$ Higgs in the ATLAS detector at LHC, indicates that jet tagging does indeed yield a statistically significant signal in this channel. Note that jet tagging can also enhance the significance of the signal in the 'gold-plated' $H \to ZZ \to l^+l^-l^+l^-$ channel, see Figure 19.

5 Higgs and Supersymmetry

The Standard Model with a single, fundamental Higgs scalar is a well-defined, mathematically consistent theory. However, as discussed in detail by Aitchison, it suffers from a severe *naturalness* or *fine-tuning* problem, which we will summarize below. The most elegant solution to this problem is to make the theory supersymmetric. In fact, as we shall see, there is a well-defined minimal supersymmetric version of the Standard Model (MSSM) which has many theoretically attractive features apart from solving the naturalness problem: it is perturbatively calculable, readily embeddable in Grand Unified Theories, supergravity *etc.* , while retaining all the successes of the SM at low energy. From an experimental point of view, the attraction of the MSSM is that there is a reasonably well-defined spectrum of new supersymmetric particles to look for.

There are, of course, other ways of invoking new physics to solve the SM naturalness problem. Some of the alternatives (technicolor, $t\bar{t}$ condensates, ...) are discussed in detail by Peccei (these Proceedings). In general, the Higgs boson is not a fundamental scalar in these models. What distinguishes the MSSM is that not only are fundamental scalar fields retained, but also that there are *two* complex doublets, compared with one in the SM. This leads to five Higgs bosons (two charged and three neutral), one of which is expected to be light, *i.e.* with mass comparable to M_Z or less. It follows that the discovery of a light Higgs boson, or more than one Higgs boson, would lend very strong support to the minimal supersymmetric model. In other words, the Higgs sector could provide the key to supersymmetry!

5.1 The rationale for SUSY

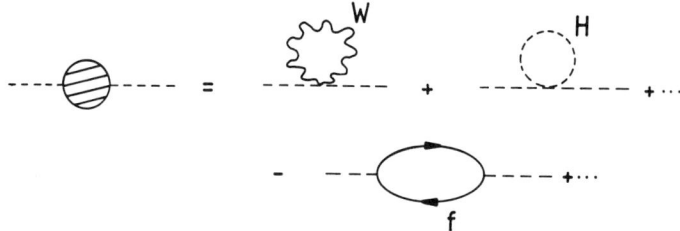

Figure 22. *One-loop contributions to the renormalized Higgs propagator in the* SM.

To understand how the naturalness problem arises in the SM, consider the one-loop corrections to the Higgs propagator which arise in the calculation of the renormalized Higgs mass, Figure 22. These diagrams are quadratically divergent,

$$\delta M_H^2 \sim g^2 \int^\Lambda \frac{d^4k}{(2\pi)^4 k^2} \sim \frac{\alpha}{\pi} \Lambda^2 , \tag{38}$$

where an ultra-violet cut-off has been introduced. This is not in itself a problem—the theory is renormalizable and such divergences can be cancelled in a consistent way. However, suppose that the theory is embedded in a Grand Unified Theory (GUT) characterized by a large mass scale M_X. For example,

$$\text{SU}(5) \xrightarrow{M_X} \text{SU}(3) \times \text{SU}(2) \times \text{U}(1) \xrightarrow{M_W} \text{SU}(3)_c \times \text{U}(1)_{em}. \tag{39}$$

Now there are contributions to δM_H^2 from interactions between heavy Higgs bosons ($M_\Phi \sim M_X$), responsible for the GUT symmetry breaking, and the SM light Higgs, mediated via light gauge-boson exchange. These effectively replace the cut-off Λ in (38) by $g^2 M_X$. If M_X is just above the limit allowed by proton decay, then

$$\delta M_H^2 \sim g^4 M_X^2 \sim 10^{24}. \tag{40}$$

This violates the 'naturalness' principle, which states that quantum corrections to a parameter in the theory should be no larger than its physical value ('t Hooft 1980). In this case, in order to retain a light Higgs boson which breaks the electroweak symmetry at the W mass scale, the bare mass parameter of the theory must be tuned to one part in 10^{12}!

vector supermultiplet		chiral supermultiplet	
$J=1$	$J=\frac{1}{2}$	$J=\frac{1}{2}$	$J=0$
gluons g	gluinos $\tilde{g}$	quarks q_L, q_R	squarks $\tilde{q}_L, \tilde{q}_R$
$W^\pm, Z^0$	winos, zino $\tilde{W}^\pm, \tilde{Z}^0$	leptons l_L, l_R	sleptons $\tilde{l}_L, \tilde{l}_R$
photon γ	photino $\tilde{\gamma}$	Higgses H_1, H_2	Higgsinos $\tilde{H}_1, \tilde{H}_2$

Table 5. *The MSSM particle spectrum.*

In the supersymmetric Standard Model, the fine-tuning problem is solved by introducing a new super-particle (sparticle) for each particle which has the same gauge quantum numbers, but a spin which differs by $\frac{1}{2}$; see Table 5. In this case there is a cancellation between the loop contributions in Figure 22 involving the fermion and boson particles and sparticles,

$$\delta M_H^2 \sim \frac{\alpha}{\pi} \left[(\Lambda^2 + O(m_B^2)) - (\Lambda^2 + O(m_F^2)) \right] \sim \frac{\alpha}{\pi} \left| m_B^2 - m_F^2 \right|. \tag{41}$$

The residual correction will then be naturally small, $\delta M_H^2 \lesssim M_H^2$, providing

$$\left| m_B^2 - m_F^2 \right| < O(1\,\text{TeV})^2, \tag{42}$$

i.e. the supersymmetric particles should be no heavier than $O(1\,\text{TeV})$.

There is some supporting evidence for $O(1\,\text{TeV})$ sparticles from coupling constant unification. The three gauge couplings are now known so accurately that simple 'grand unification' at some large energy scale can be ruled out, see Figure 23. Invoking supersymmetry restoration at an intermediate scale M_{SUSY} allows the three couplings to evolve to a single value at a scale M_{GUT} consistent with proton lifetime limits, see for example Ellis et al. (1991a), Amaldi et al. (1991), Langacker and Luo (1991), Anselmo et al. (1992). Numerical studies show that $M_{\rm SUSY}$ must be of order 1 TeV, see Figure 23. Note, however, that in realistic models the sparticles have a range of masses and the β-functions change in small increments as the various thresholds are crossed, see for example Roberts and Ross (1992), making the analysis more complicated. Nevertheless, it does appear to be true that $O(1\,{\rm TeV})$ or less *is* a typical scale for the sparticle masses in this unification scenario.

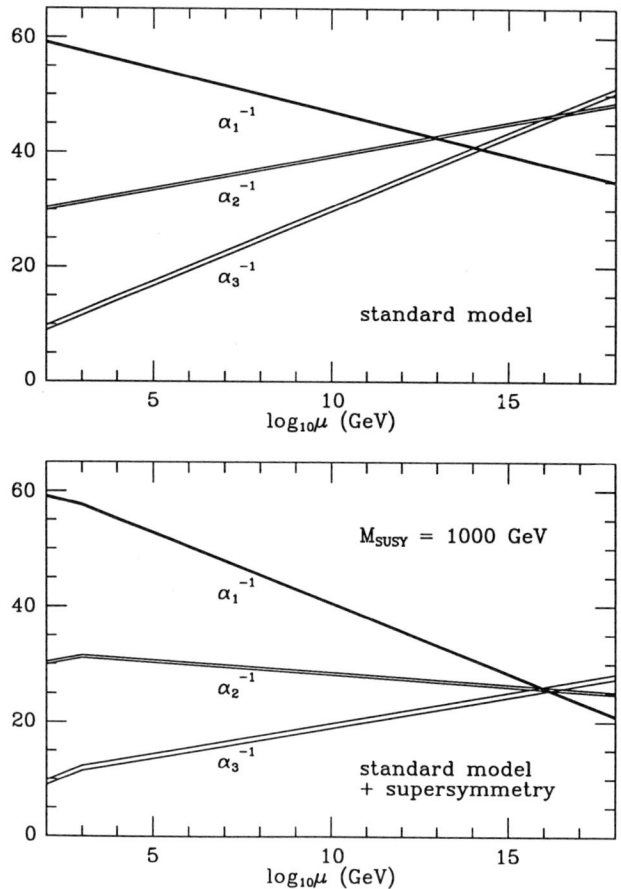

Figure 23. *Unification of the three couplings of the Standard Model. In the upper figure the couplings are evolved using the standard β-functions of the theory, while in the lower figure the complete set of supersymmetric partners are introduced at a single scale of 1 TeV. The changes in the β-functions thus induced allow the couplings to meet at a single point.*

5.2 The MSSM Higgs sector

We turn now to the Higgs sector in the MSSM, leaving a discussion of the sparticle spectrum and phenomenology to a later section. The precise details of how the minimal supersymmetric version of the Standard Model is constructed are beyond the scope of these lectures. A summary is given in the lectures of Peccei, and there are excellent reviews in the literature, for example Nilles (1984), Haber and Kane (1985), Barbieri (1988).

Our starting point will be the form of the MSSM Higgs potential, the analogue of Equation (4):

$$V(H_1, H_2) = m_1^2|H_1|^2 + m_2^2|H_2|^2 + m_3^2(H_1 H_2 + \text{h.c.})$$
$$+ \frac{g^2}{8}\left(H_2^\dagger \vec{\sigma} H_2 + H_1^\dagger \vec{\sigma} H_1\right)^2 + \frac{g'^2}{8}\left(|H_2|^2 - |H_1|^2\right). \quad (43)$$

We make the following observations.

(i) H_1 and H_2 are complex SU(2) doublets of scalar fields. Note that there are *two* such doublets in the MSSM. The reason is that separate scalar fields are needed to give masses to the up-like and down-like fermions, in contrast to the SM where a single doublet (and its complex conjugate) could do both. Even in the absence of such Yukawa interactions, two doublets are needed to avoid anomalies appearing in loop diagrams: the potential problem here is caused by the spin-$\frac{1}{2}$ Higgsino partners. In the SM, the electric charge assignment in the SU(2) doublet is $(+1, 0)$, and a second doublet with opposite charge assignment $(0, -1)$ is needed in the MSSM to cancel the anomaly arising from the charged fermion field.

(ii) The mass parameters m_1^2, m_2^2 and m_3^2 are essentially arbitrary, and have their origins in the explicit SUSY-breaking terms added to the supersymmetric potential. They are the analogue of the parameter $-\mu^2$ in the SM potential (4).

(iii) A key feature of the MSSM potential is that the quartic terms are *not* arbitrary—cf. the $\lambda(\phi^\dagger \phi)^2$ term in (4)—but are fixed by the SU(2)×U(1) gauge couplings g and g'. These terms originate in the supersymmetrized gauge interactions of the scalar multiplets.

(iv) Some restrictions on the m_i^2 parameters are necessary to ensure that the potential V is bounded below and has a minimum with $\langle H_1 \rangle$, $\langle H_2 \rangle \neq 0$ and real. Without loss of generality one can choose $m_3^2 < 0$, whence a sufficient condition is $m_1^2 > 0$, $m_2^2 < 0$. It is interesting that these latter inequalities arise naturally when the corresponding equal and positive quantities are evolved downwards from the scale of grand unification (see Section 5.5 and the lectures of Peccei).

The physical scalar spectrum and interactions are made explicit by expanding the H_i fields about their vacuum expectation values:

$$H_1 = \begin{pmatrix} v_1 + \frac{1}{\sqrt{2}}(S_1 + iP_1) \\ H_1^- \end{pmatrix}, \quad H_2 = \begin{pmatrix} H_2^+ \\ v_2 + \frac{1}{\sqrt{2}}(S_2 + iP_2) \end{pmatrix}. \quad (44)$$

where $\langle H_1 \rangle = v_1$ and $\langle H_2 \rangle = v_2$. In the SM, the two original parameters μ and λ in the potential were replaced by v and M_H, with the former fixed by the W mass. Similarly, in the MSSM, the three parameters can be chosen as v_1, v_2 and M_A, where M_A is the mass of one of the physical scalars (see below). The combination $v_1^2 + v_2^2$ is fixed by the weak boson mass,

$$\begin{aligned} M_W^2 &= \tfrac{1}{2} g^2 \left(v_1^2 + v_2^2 \right) \\ M_Z^2 &= \tfrac{1}{2} (g^2 + g'^2) \left(v_1^2 + v_2^2 \right), \end{aligned} \quad (45)$$

while the ratio

$$\frac{v_2}{v_1} \equiv \tan \beta \quad (46)$$

is arbitrary. The gauge coupling constants g and g' also appear in the (tree-level) expressions for the MSSM Higgs masses and interactions, but these can be replaced by the weak boson masses M_W and M_Z.

The Goldstone bosons which get 'eaten' by the $W^\pm$ and Z^0 are

$$\begin{aligned} G^+ &= -\cos\beta \, (H_1^-)^* + \sin\beta \, H_2^+ \\ G^- &= (G^+)^* \\ G^0 &= -\cos\beta \, P_1 + \sin\beta \, P_2. \end{aligned} \quad (47)$$

There are five physical scalars, two charged ($H^\pm$) and three neutral (the CP-even h, H and the CP-odd A):

$$\begin{aligned} H^+ &= \sin\beta \, (H_1^-)^* + \cos\beta \, H_2^+ \\ H^- &= (H^+)^* \\ A &= \sin\beta \, P_1 + \cos\beta \, P_2 \\ h &= -\sin\alpha \, S_1 + \cos\alpha \, S_2 \\ H &= \cos\alpha \, S_1 + \sin\alpha \, S_2. \end{aligned} \quad (48)$$

The angular ranges are, by convention,

$$0 \leq \beta \leq \frac{\pi}{2}, \quad -\frac{\pi}{2} \leq \alpha \leq 0. \quad (49)$$

The charged and neutral physical scalar masses are obtained by substituting the fields defined in (44) and (48) into the potential (43). They can be expressed, for example, in terms of the parameters β, M_A, M_W and M_Z. For the charged Higgs bosons we obtain

$$M_{H^+}^2 = M_{H^-}^2 = M_W^2 + M_A^2. \quad (50)$$

The mass-squared matrix of the CP-even neutral Higgs bosons is

$$\mathcal{M}_0 = \sin 2\beta \left[\tfrac{1}{2} M_Z^2 \begin{pmatrix} \cot\beta & -1 \\ -1 & \tan\beta \end{pmatrix} + \tfrac{1}{2} M_A^2 \begin{pmatrix} \tan\beta & -1 \\ -1 & \cot\beta \end{pmatrix} \right], \quad (51)$$

with eigenvalues

$$\begin{aligned} M_h^2 &= \tfrac{1}{2} \left[M_A^2 + M_Z^2 - \sqrt{(M_A^2 + M_Z^2)^2 - 4 M_A^2 M_Z^2 \cos^2 2\beta} \right], \quad (52) \\ M_H^2 &= \tfrac{1}{2} \left[M_A^2 + M_Z^2 + \sqrt{(M_A^2 + M_Z^2)^2 - 4 M_A^2 M_Z^2 \cos^2 2\beta} \right]. \quad (53) \end{aligned}$$

From these relations a set of important inequalities can be derived:

$$M_W, M_A < M_{H^\pm}, \tag{54}$$
$$M_h < |\cos 2\beta| M_Z < M_Z < M_H, \tag{55}$$
$$M_h < M_A < M_H. \tag{56}$$

We must stress, however, that these results are valid only at tree-level. Radiative corrections can lead to significant violations of the above bounds, which have important implications for phenomenology. This will be discussed in more detail below.

Note that the mixing angle α is not an independent parameter—it can be expressed in terms of the physical scalar masses, for example by

$$\cos 2\alpha = -\cos 2\beta \left(\frac{M_A^2 - M_Z^2}{M_H^2 - M_h^2} \right). \tag{57}$$

As was the case for the SM, the MSSM Higgs sector can also be used to generate masses for the fermions, through Yukawa interactions of the form $g_f \bar{f} H_i f$. It is straightforward to show that these lead, via spontaneous symmetry breaking, to

$$m_e = g_e v_1, \quad m_d = g_d v_1, \quad m_u = g_u v_2, \quad \text{etc.} \tag{58}$$

The same Yukawa couplings are also responsible for interactions between the fermions and the three neutral Higgs bosons. These are similar to those for the SM Higgs, except that the mixing angles α and β appear at the vertices when the H_i fields are re-expressed in terms of h, H and A. Table 6 lists the Higgs-fermion couplings relative to the SM couplings of Figure 2. Also shown are the couplings to W and Z bosons. Note again the dependence on the mixing angles, and also that the CP-odd A Higgs does not have tree-level couplings to W^+W^- or ZZ. Thus, the dominant decay of a heavy A is to fermion pairs, in contrast to the SM heavy Higgs. Another phenomenologically important trilinear vertex, not shown in Table 6 is ZAh, which has the same strength as ZZh.

	$d\bar{d}, s\bar{s}, b\bar{b}$; $e^+e^-, \mu^+\mu^-, \tau^+\tau^-$	$u\bar{u}, c\bar{c}, t\bar{t}$	W^+W^-, ZZ
h	$-\sin\alpha/\cos\beta$	$\cos\alpha/\sin\beta$	$\sin(\beta-\alpha)$
H	$\cos\alpha/\cos\beta$	$\sin\alpha/\sin\beta$	$\cos(\beta-\alpha)$
A	$-i\gamma_5 \tan\beta$	$-i\gamma_5 \cot\beta$	0

Table 6. *Couplings of the* MSSM *neutral Higgs bosons to fermion and gauge boson pairs, relative to the* SM *couplings of Figure 2.*

Finally, notice that in the limit where $M_A \to \infty$ with β fixed, four of the five MSSM Higgs bosons become heavy. Only the h boson remains light, with $M_h < M_Z$. Furthermore, it follows from Equation (57) that in this limit

$$\alpha \to \frac{\pi}{2} - \beta, \tag{59}$$

and the couplings of the h to fermions and weak gauge bosons become *identical* to those of the SM Higgs.

5.3 The importance of radiative corrections

Perhaps the most striking feature of the MSSM Higgs mass spectrum is the prediction that there is a least one scalar lighter than the Z^0 boson. However this is a *tree-level* result only—recent calculations of the radiative corrections to the mass spectrum show that in fact the bound $M_h < M_Z$ can be violated by a large amount (Okada *et al.* 1991, Ellis *et al.* 1991b, Brignole *et al.* 1991, Haber and Hempfling 1991). This has, not surprisingly, profound implications for phenomenology, since the radiative corrections may push the lightest Higgs mass beyond the reach of LEP-II.

The dominant one-loop radiative corrections to the neutral Higgs mass matrix come from interactions with heavy quarks, in particular the bottom and top quarks. Thus

$$\mathcal{M} = \mathcal{M}_0 + \begin{pmatrix} \Delta_1^2 & 0 \\ 0 & \Delta_2^2 \end{pmatrix}, \qquad (60)$$

where $\mathcal{M}_0$ is the tree-level mass matrix defined in Equation (51), and

$$\Delta_1^2 \simeq \frac{3G_F}{\pi\sqrt{2}} \frac{m_b^4}{\cos^2\beta} \log\frac{m_{\tilde{b}}^2}{m_b^2}, \quad \Delta_2^2 \simeq \frac{3G_F}{\pi\sqrt{2}} \frac{m_t^4}{\sin^2\beta} \log\frac{m_{\tilde{t}}^2}{m_t^2}. \qquad (61)$$

From these one can readily derive the corrections to the masses M_h and M_H, and to the mixing angle α. The dominant effect clearly comes from Δ_2^2, which has a quartic dependence on m_t. The residue of the quark-squark cancellation in the renormalization of the Higgs mass shown in Equation (41) is seen here as the suppression of the correction when the quarks and squarks are degenerate in mass.

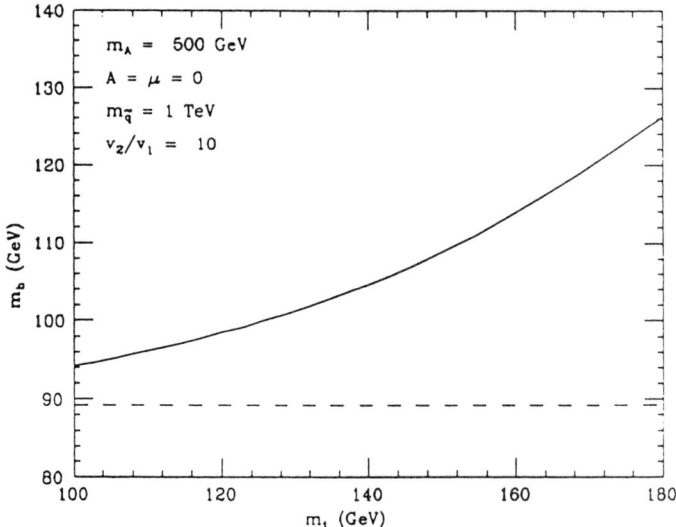

Figure 24. *The mass of the lightest CP-even* SUSY *Higgs boson, as a function of m_t, before (dashed line) and after (solid line) radiative corrections, from Ellis et al. (1991b).*

The radiatively-corrected masses of the neutral Higgs bosons now depend on the quark and squark mass spectrum, as well as on M_A and β. There is no longer a formal upper limit on M_h, but it is reasonable to assume 'natural' values of the other parameters when calculating the allowed ranges of the Higgs masses. For example, the assumption $m_{\tilde{q}} = O(1\,\text{TeV})$ is usually made. An example of the impact of the radiative corrections on M_h is shown in Figure 24, from Ellis et al. (1991b). The choice of parameters $M_A = 500\,\text{GeV}$, $\tan\beta = 10$ would correspond to a tree-level M_h slightly less than M_Z (dashed line). The corrected mass is shown as a function of m_t, with $m_{\tilde{q}} = 1\,\text{TeV}$ assumed for the quark masses. The maximum value of M_h is attained in the limit $M_A \to \infty$ (as at tree level). Figure 25 from Kunszt and Zwirner (1992) shows contours of $M_h^{\max}$ as functions of (a) $(m_{\tilde{q}}, m_t)$ for fixed $\tan\beta = m_t/m_b$ (a typical value in SUSY-GUTS), and (b) $(\tan\beta, m_t)$ for fixed $m_{\tilde{q}} = 1\,\text{TeV}$. The conclusion is that a reasonable upper limit is $M_h < O(130\,\text{GeV})$.

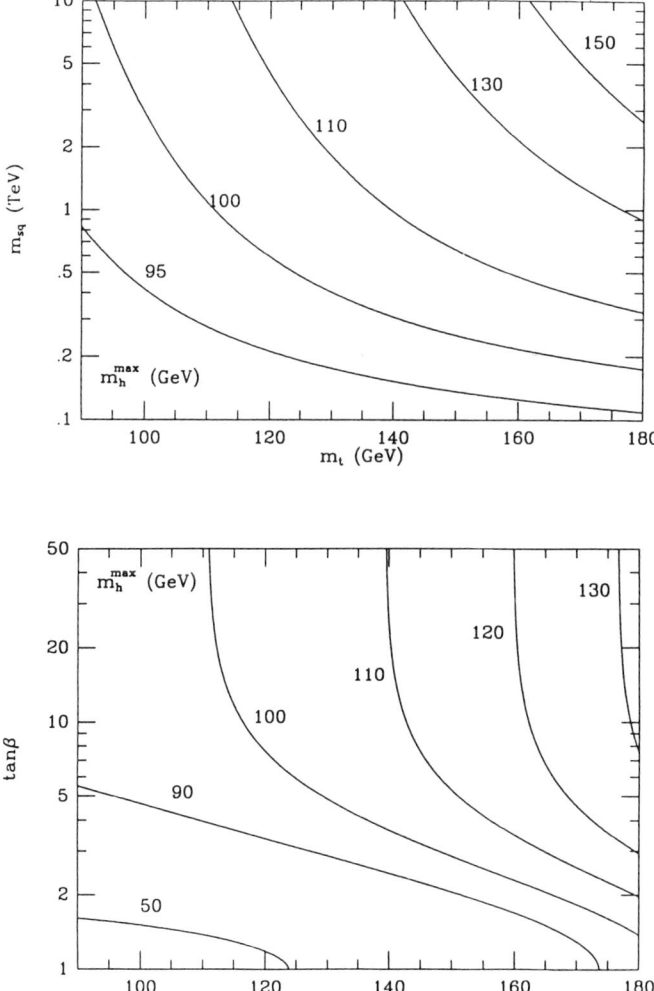

Figure 25. *Contours of $M_h^{\max}$ in the $(m_{\tilde{q}}, m_t)$ and $(\tan\beta, m_t)$ planes, from Kunszt and Zwirner (1992).*

Radiative corrections to the other Higgs masses and to the Higgs–fermion, Higgs–gauge-boson and scalar-interaction vertices have also been computed. The effects are in general less dramatic than on M_h.

5.4 MSSM Higgs phenomenology

SUSY-Higgs phenomenology is the search for the five scalar bosons h, H, A and $H^\pm$. The properties of these can be parametrized in terms of the unknowns $(\tan\beta, M_A, m_t)$. Through radiative corrections, there is also some dependence on the squark masses (see Figure 25), but it seems appropriate to assume $m_{\tilde{q}} = O(1\,\text{TeV})$. In practice, the three parameters are varied over a 'reasonable' range of values. The same qualitative arguments used to derive an upper limit on the SM Higgs mass also apply in the MSSM case, and so only Higgs masses (in particular M_A) less than $O(1\,\text{TeV})$ need be considered. The top mass is usually varied over the range 100–200 GeV. Finally, although $\tan\beta$ is unknown, *a priori*, the embedding of the supersymmetric SM in a GUT gives $1 \lesssim \tan\beta \lesssim m_t/m_b$, see for example Haber and Kane (1985).

The goal of most phenomenological studies has been to investigate whether at least one of the MSSM Higgs bosons can be discovered at present and future colliders, in particular at LEP-I, LEP-II, LHC or SSC, whatever the values of $(\tan\beta, M_A, m_t)$. This is the so-called 'no-lose theorem'. The procedure in practice is to (i) fix a set of parameter values, (ii) compute the mass spectrum and coupling strengths, (iii) compute the branching ratios and production cross sections, (iv) compute event rates for signals and backgrounds at the various colliders, and (v) establish that at least one of the Higgs gives a statistically significant signal. As we shall see below, the standard approach is to explore which regions of the $(\tan\beta, M_A)$ plane are accessible for a given m_t. Several such phenomenological analyses have recently been performed (Kunszt and Zwirner 1992, Barger *et al.* 1992a, Baer *et al.* 1993, Gunion and Orr 1992), with broadly similar conclusions. We shall only be able to show some representative figures here—all the details can of course be found in the literature.

We begin by considering the Higgs mass spectrum. Figure 26 shows the masses of the h, H and $H^\pm$ bosons as a function of M_A and $\tan\beta$, for $m_t = 150\,\text{GeV}$, from Gunion and Orr (1992). Note that, as discussed above, the H and $H^\pm$ become heavier with increasing M_A, while the h remains light. The next step is to compute the branching ratios of the dominant decay channels. A large part of the calculation can be taken over directly from the SM Higgs analysis of Section 2.3. The main differences are: (i) the mixing angle factors of Table 6 must be included in the appropriate channels, (ii) additional decay channels of heavier Higgs into lighter Higgs bosons, for example $H \to hh$, $A \to Zh$, ..., can be important, and (iii) if the sparticles are sufficiently light, then sparticle–anti-sparticle final states might also be accessible. Most phenomenological studies have assumed that this is not the case.

A detailed analysis of all the Higgs branching ratios as a function of the various parameters is beyond the scope of the present review. A very clear discussion is given, for example, in Kunszt and Zwirner (1992). We give one phenomenologically-important example by way of illustration. As we have seen above, it is possible that the h boson lies just above the reach of LEP-II. It will then have to be looked for at the SSC or LHC, and as for the SM Higgs, the important decay channel will be $h \to \gamma\gamma$. The partial

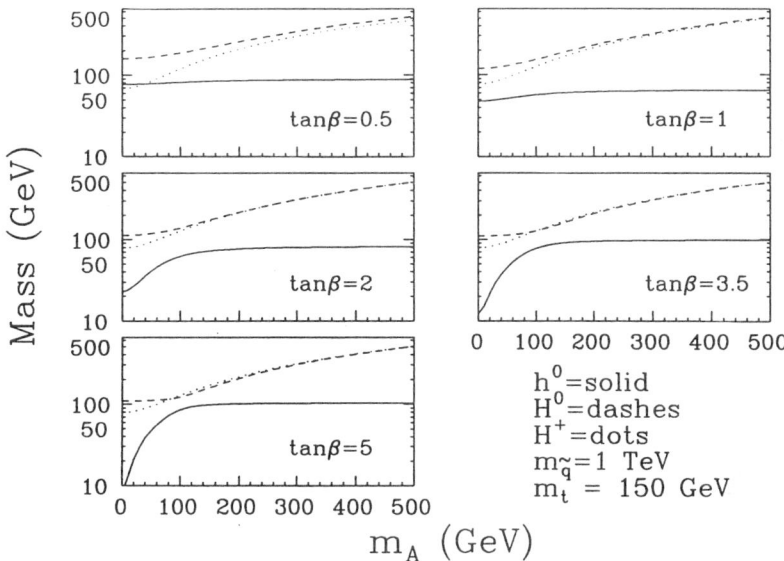

Figure 26. *An example of a MSSM Higgs mass spectrum, from Gunion and Orr (1992).*

width for this is obtained by adding the appropriate factors from Table 6 to the SM Higgs result (24).

$$\Gamma(h\to\gamma\gamma) = \frac{\alpha^2 G_F M_h^3}{128\sqrt{2}\pi^3}$$
$$\times \left| -\frac{1}{3}\frac{\sin\alpha}{\cos\beta} I_q\left(\frac{m_b^2}{M_h^2}\right) + \frac{4}{3}\frac{\cos\alpha}{\sin\beta} I_q\left(\frac{m_t^2}{M_h^2}\right) + \sin(\beta-\alpha) I_W\left(\frac{M_W^2}{M_h^2}\right) \right|^2. \quad (62)$$

The total h decay width is again dominated by the $h \to b\bar{b}$ channel (above threshold), which is proportional to the SM partial width:

$$\Gamma(h\to b\bar{b}) = \frac{3 G_F m_b^2 M_h}{4\pi\sqrt{2}} \frac{\sin^2\alpha}{\cos^2\beta} \left(1 - \frac{4m_b^2}{M_h^2}\right)^{3/2}. \quad (63)$$

Figure 27 (Gunion and Orr 1992) shows the $\gamma\gamma$ branching ratios for the h, H and A bosons, as a function of their mass and for various $\tan\beta$. The $h \to \gamma\gamma$ branching ratio shows a significant variation with $\tan\beta$, originating in the explicit α and β dependence in Equations (62) and (63).

The relevant processes for MSSM searches at LEP-I are

$$\left. \begin{array}{c} Z \to Z^* h \\ Z \to hA \end{array} \right\} \text{ with } Z^* \to l\bar{l} \text{ and } h, A \to b\bar{b}, \tau^+\tau^-. \quad (64)$$

These channels are complementary, in the sense that the rates are proportional to $\sin^2(\beta-\alpha)$ and $\cos^2(\beta-\alpha)$ respectively. The SM Higgs search also applies to the Z^*h

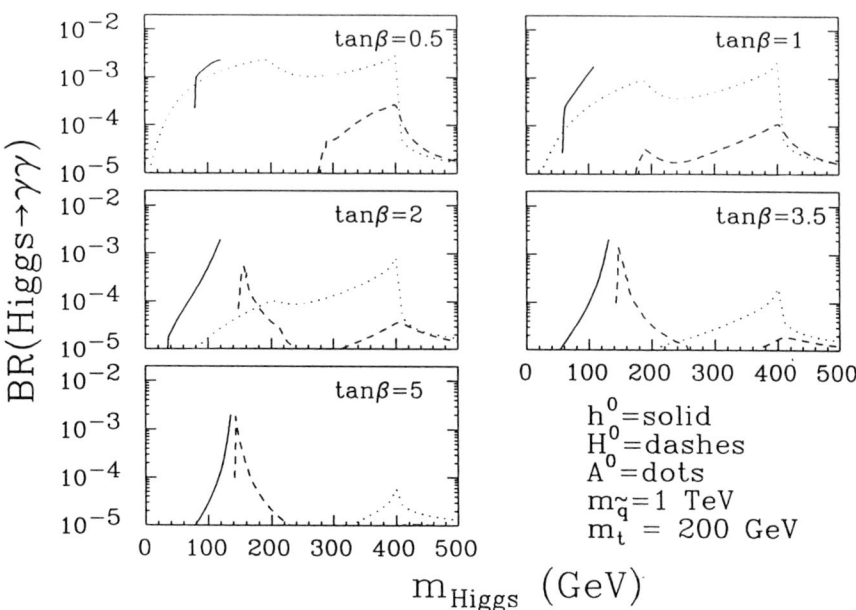

Figure 27. *The $\gamma\gamma$ branching ratios for the h, H and A MSSM Higgs bosons, as a function of their mass and for various $\tan\beta$, from Gunion and Orr (1992).*

channel, with no observed events corresponding to either a lower limit on m_h or an upper limit on the factor $\sin^2(\beta - \alpha)$. In the latter case, the $Z^* \to hA$ channel is more dominant, providing of course that the A boson is sufficiently light. Note that hA events are tagged by $b\bar{b}b\bar{b}$, $b\bar{b}\tau^+\tau^-$, and $\tau^+\tau^-\tau^+\tau^-$ final states.

With no Z^*h or hA events seen so far, LEP-I is able to exclude h and A masses less than $O(40\text{--}50)$ GeV. The exact values depend on the assumed values of $\tan\beta$ and m_t, and it is more appropriate to translate the experimental limits into an excluded region in the $(\tan\beta, M_A)$ plane, for a fixed m_t. Note that if M_A is large, a lower limit on M_h from LEP-I is equivalent to a lower limit on $\tan\beta$, the exact value depending on m_t. However, if m_t is too large, then radiative corrections push the h mass above the LEP-I kinematic limit, and there is no limit on $\tan\beta$. This is illustrated in Figure 28, from Kunszt and Zwirner (1992), which summarizes the current LEP-I limits. Note that parameter values below and to the left of the solid lines are excluded.

At LEP-II, the relevant processes are

$$e^+e^- \to Z^* \to Zh, \; ZH, \; hA, \; HA, \tag{65}$$

corresponding to l^+l^-jj, $\nu\bar{\nu}jj$ and $jjjj$ final states. Once again, many features of the SM Higgs search discussed in Section 3 are also relevant here. As at LEP-I, the Zh and hA channels are complementary, and the LEP-I contour shown in Figure 28 can be pushed further out into the $(\tan\beta, M_A)$ plane, the exact region excluded depending on the available collider energy and luminosity. Two scenarios are illustrated in Figure 28: the dashed line corresponds to a total event rate (*i.e.* adding the cross sections

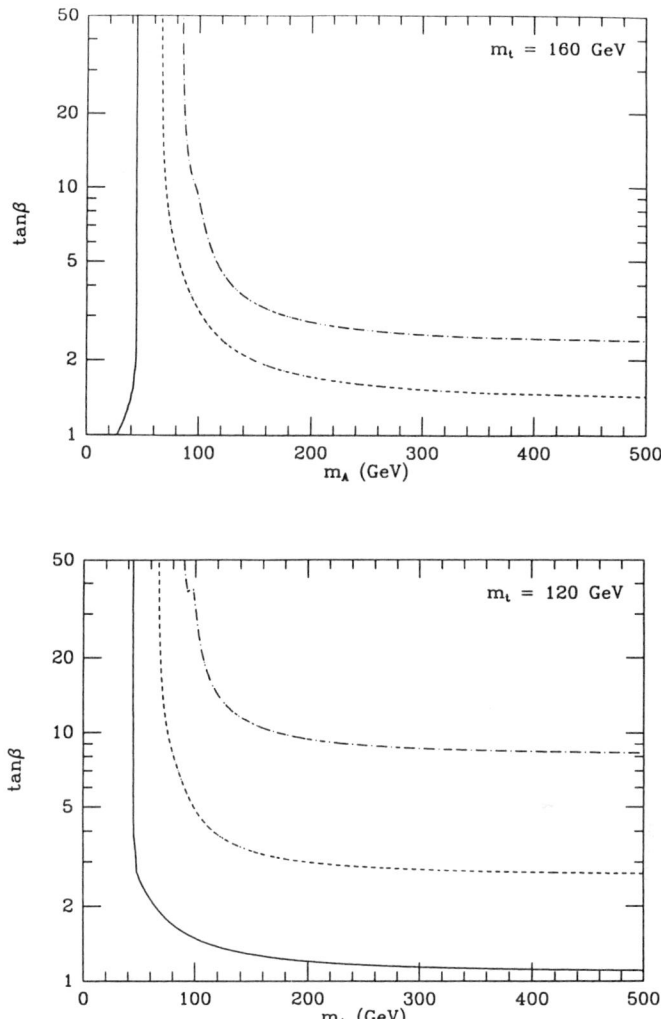

Figure 28. Regions in the $(\tan\beta, M_A)$ plane excluded by LEP-I (solid lines) and by LEP-II (dashed and dash-dotted lines—see the text) for two m_t values, from Kunszt and Zwirner (1992).

for the processes in (65)) of 100 per 500 pb^{-1} of luminosity at $\sqrt{s} = 175$ GeV—a conservative scenario, while the dash-dotted curve corresponds to 25 events per 500 pb^{-1} at $\sqrt{s} = 190$ GeV—an optimistic scenario. As expected, higher m_t values reduce the region which can be excluded, as radiative corrections push M_h beyond the kinematic limit.

At the high-energy hadron colliders, MSSM Higgs searches are broadly similar to those discussed already for the SM Higgs (Kunszt and Zwirner 1992, Barger et al. 1992a, Baer et al. 1993, Gunion and Orr 1992). The key decay channels are again $(h, H, A) \to \gamma\gamma$, with the Higgs bosons either produced directly or in association with $t\bar{t}$ or $W^{\pm}$, and $(h, H) \to 4l^{\pm}$ with $l = (e, \mu)$. The rates are generally smaller than those for the corresponding SM channels, due to the (α, β) suppression factors at the vertices (Table 6). In terms of coverage of the $(\tan\beta, M_A)$ plane, a key rôle is played by $h \to \gamma\gamma$. We have already seen how large values of these parameters correspond to h masses at the top end of the allowed range (i.e. $M_h > M_Z$), with properties similar to the intermediate-mass SM Higgs. This then leads to 'straightforward' detection at LHC or SSC. As an example, Figure 29 shows the accessible region for this channel at SSC, assuming $20\,\text{pb}^{-1}$ of integrated luminosity. The question is then whether the region in $(\tan\beta, M_A)$ accessible in this way extends down to overlap with the region accessible at LEP-II, Figure 28. The answer appears to be 'almost, but not quite'. Before looking at this in detail, we note that the decays $(H, A) \to \gamma\gamma$ and $(H, h) \to 4l^{\pm}$ give additional coverage in relatively small regions of the $(\tan\beta, M_A)$ plane; see, for example, Figure 29 for the region covered by $(H, A) \to \gamma\gamma$ (Barger et al. 1992a).

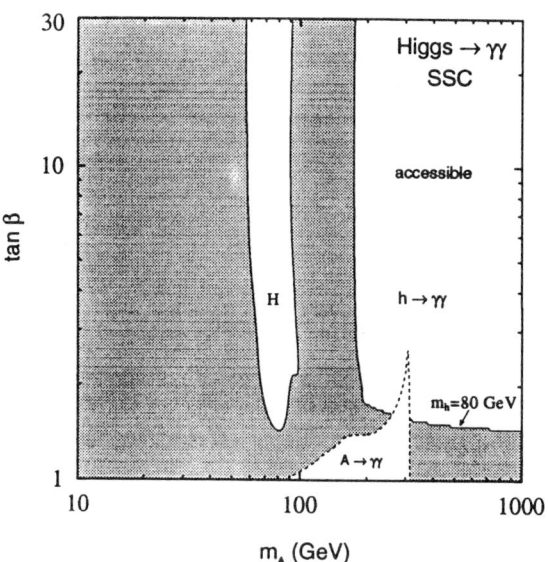

Figure 29. *Regions in the $(\tan\beta, M_A)$ plane accessible at the SSC (assuming $\int \mathcal{L} = 20\,\text{pb}^{-1}$ and $m_t = 150\,\text{GeV}$) via the direct channels $(h, H, A) \to \gamma\gamma$, from Barger et al. (1992a).*

Although neutral scalar searches at LEP and LHC/SSC cover most of the MSSM parameter space, *charged Higgs* production can also yield important signals, providing $m_t > M_{H^{\pm}}$ so that $t \to bH^+$ is an allowed decay channel. Since $M_{H^{\pm}}$ increases with M_A (Equation (50)), for fixed m_t the relevant region of parameter space is M_A not too large. This is illustrated in Figure 30 (Gunion and Orr 1992), which shows $t \to bH^+$ branching ratio contours in the $(\tan\beta, M_A)$ plane for $m_t = 150\,\text{GeV}$. As m_t increases, the curve move proportionally to the right. For $\tan\beta \gtrsim 1$ and $M_{H^\pm} < m_t$, the dominant charged Higgs decay mode is $H^+ \to \tau^+\nu_\tau$, and so a clean signature is provided by $pp \to t\bar{t} + X$ followed by $t \to bW^+ \to bl^+\nu_l$ and $\bar{t} \to \bar{b}H^- \to \bar{b}\tau\bar{\nu}_\tau$. The net result is a region of accessibility given approximately by $M_A < m_t - O(40\,\text{GeV})$, independent of $\tan\beta$.

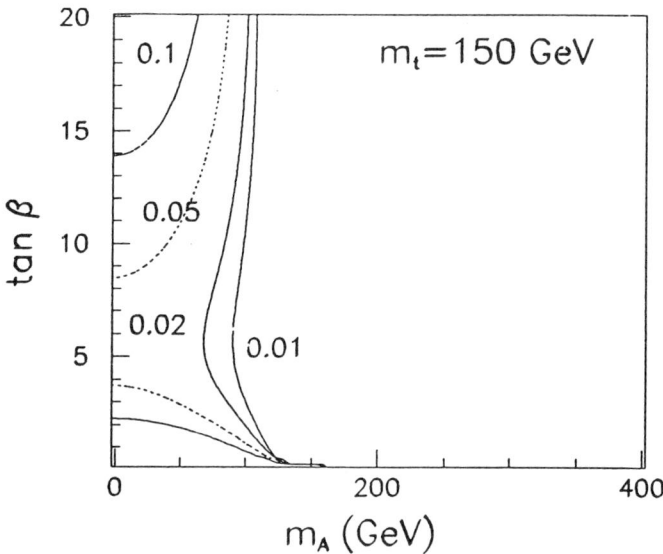

Figure 30. $t \to bH^+$ *branching ratio contours in the* $(\tan\beta, M_A)$ *plane for* $m_t = 150\,\text{GeV}$, *from Gunion and Orr (1992).*

Figure 31 (Barger et al. 1992a) shows the result of putting together *all* the possibilities for detecting at least one MSSM Higgs at either LEP or SSC. For the latter, the parameters $\sqrt{s} = 40\,\text{TeV}$ and $\int \mathcal{L} = 20\,\text{pb}^{-1}$ have been used. The lower energy of the LHC is largely compensated by higher luminosity, so that assuming $\sqrt{s} = 16\,\text{TeV}$ and $\int \mathcal{L} = 100\,\text{pb}^{-1}$ gives comparable results. Broadly similar conclusions have been reached by other groups (Kunszt and Zwirner 1992, Baer et al. 1993, Gunion and Orr 1992). Note the region of inaccessibility in the middle of the plot, *i.e.* for large $\tan\beta$ and M_A between about 100 and 180 GeV. This corresponds to $m_t = 150\,\text{GeV}$. If m_t is larger, the increased coverage provided by $t \to bH^+$ decreases the problematical region, so that by $m_t = 200\,\text{GeV}$ it has largely disappeared. The window of inaccessibility is also closed by increasing the energy of LEP-II, as was already apparent from Figure 28. In connection with this, a e^+e^- linear collider (NLC) with parameters

$$\sqrt{s} > O(300\,\text{GeV}); \qquad \int \mathcal{L} > O(10\,\text{fb}^{-1}) \qquad (66)$$

would close this window by allowing the detection of h bosons with mass up to the theoretical MSSM limit, see for example Barger et al. (1992b). Another way to close the window of inaccessibility partially is to try to use the $\tau^+\tau^-$ decay of the H and A bosons at hadron colliders, which would give a non-negligible signal for very large $\tan\beta$

and moderately large M_A. In Kunszt and Zwirner (1992), for example, it was estimated that the part of the window above $\tan\beta \sim 15\text{--}20$ could be closed using this channel. However more work needs to be done on the feasibility of the $\tau^+\tau^-$ mode in a realistic detector simulation before any firm conclusion can be drawn (Pauss 1991).

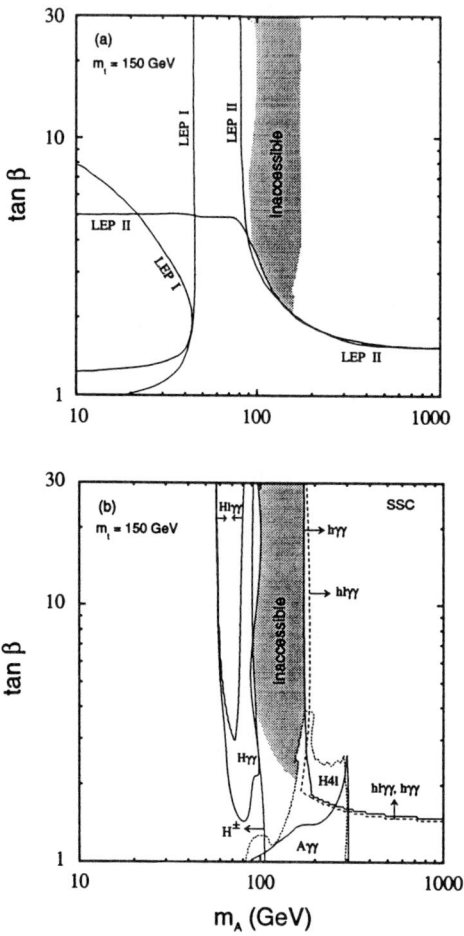

Figure 31. *Combined* LEP *and* SSC *discovery regions for* $m_t = 150\,\text{GeV}$, *from Barger et al. (1992a)*.

It is interesting to ask how many of the MSSM Higgs bosons can actually be discovered in the various regions of the $(\tan\beta, M_A)$ plane. A detailed answer is given, for example, in Barger *et al.* (1992a). Roughly speaking, to the right of the window of inaccessibility only the h boson can be detected (at LHC/SSC), while to the left of and below the window, the bosons are light enough that at least two can generally be detected. There does not, however, appear to be any significant region where all five can be detected (Barger *et al.* 1992a).

In summary, the MSSM has a rich Higgs structure (five scalars), whose properties are parametrized by $\tan\beta$, M_A and m_t. At least one light neutral scalar with mass less than about 130 GeV is expected. A combination of LEP-I, LEP-II, LHC and SSC will give complementary coverage of *almost all the parameter space*. The small window of inaccessibility would close if (a) the top is heavy, (b) a high-energy e^+e^- linear collider was available, or, possibly, (c) if a way could be found of utilizing the $\tau^+\tau^-$ decay channels of the heavier neutral Higgs bosons. It is important, however, to remember that all statements about coverage of the parameter space depend quite sensitively on the exact values assumed for the luminosity, energy and detection efficiency at the various machines.

Finally, we should also mention the possibility of constraining the MSSM Higgs parameters by using the experimental bound on $b \to s\gamma$ decays (Hewett 1993, Barger et al. 1993). A light H^- would, at the one-loop level, give an additional contribution to the SM W^-–mediated decay rate in excess of the experimental upper limit. This may already have excluded part of the otherwise inaccessible region in the $(\tan\beta, M_A)$ plane.

5.5 Other SUSY signatures

Aside from the discovery of one or more of the SUSY Higgs bosons, the most compelling and direct evidence for supersymmetry would come from the discovery of the *sparticles* themselves. In this section we give a very brief overview of sparticle phenomenology. More details can be found in the many excellent reviews in the literature, for example Ellis (1992), Baer and Tata (1992), Barger and Phillips (1993).

Supersymmetric particle spectra are usually discussed in the context of a SUSY-GUT; typically the particle and sparticle masses are well-constrained once the common scalar and gaugino masses at the $M_{\rm GUT}$ scale, m_0 and $m_{1/2}$, are specified. These are chosen such that the physical sparticle masses, obtained by evolving downwards from $M_{\rm GUT}$, are of order 1 TeV or less. The imposition of coupling constant unification $(g_1, g_2, g_3) \to g_{\rm GUT}$ leads to a further constraint on m_0 and $m_{1/2}$.

In a particular class of models, including the MSSM, the discrete quantum number

$$R = (-1)^{2S+L+3B} \quad \begin{cases} S = \text{spin} \\ L = \text{lepton number} \\ B = \text{baryon number} \end{cases}, \qquad (67)$$

known as R-parity, is conserved. Thus, particles have $R = 1$ while sparticles have $R = -1$ and must be produced in pairs in ordinary particle collisions, e.g. $qq \to \tilde{q}\tilde{q}$. Heavy sparticles will be unstable, decaying into lighter particles and sparticles. However by conservation of R-parity the lightest sparticle (LSP) must be stable. In many models, the LSP is the photino so that, for example, $\tilde{g} \to q\bar{q}\tilde{\gamma}$ is a possible gluino decay mode. Since the LSP escapes undetected, sparticle decays are typically characterized by missing energy in the final state.

Figure 32, from Roberts and Ross (1992), shows an example of a realistic MSSM sparticle spectrum. The lines correspond to running masses $m_i(\mu)$, with the physical masses $\overline{m}_i$ defined by $m_i(\overline{m}_i) = \overline{m}_i$. The masses m_1 and m_2 in the MSSM Higgs potential (43) satisfy $m_1 = m_2 = \sqrt{(m_0^2 + \mu_0^2)}$ at $\mu = M_{\rm GUT}$. Note that spontaneous symmetry

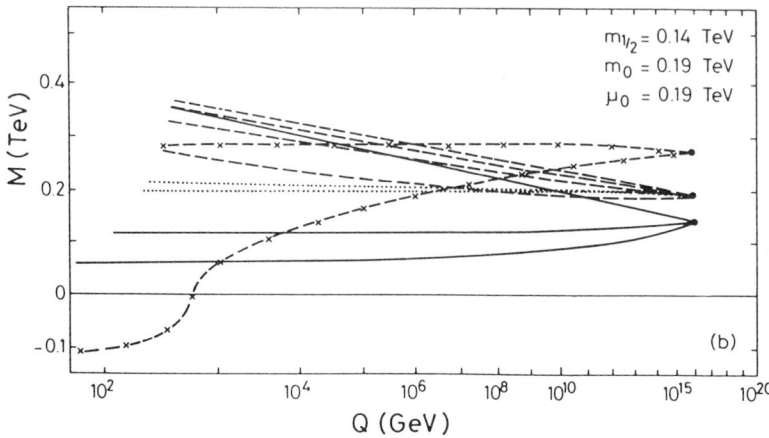

Figure 32. *A realistic* MSSM *particle spectrum consistent with* SM *coupling unification and electroweak symmetry breaking (Roberts and Ross 1992). The lines correspond to gauginos (solid), squarks (dashed), sleptons (dotted) and Higgs (dash-crossed). Other parameter values are* $m_t = 160$ GeV, $\alpha_s = 0.118$ *and* $\tan\beta = 21$.

breaking happens automatically in this model! The large Yukawa coupling of the H_2 Higgs field to the top quark (Equation (58)) drives the mass parameter m_2^2 negative at small scales. The top quark mass is chosen to be 160 GeV, the strong coupling is $\alpha_s(M_Z) = 0.118$, and the SUSY mass parameters at the GUT scale are chosen so that the correct electroweak symmetry-breaking scale is reproduced. In this scenario $\tan\beta = 21$, and the (tree-level) Higgs masses are $(91, 264, 264, 276)$ GeV for $(h, H, A, H^\pm)$ bosons respectively. Note that the physical 'neutralinos' $\tilde{\chi}_i^0$ $(i = 1, 4)$ are, in general, mixtures of the four neutral $\tilde{Z}^0$, $\tilde{\gamma}$, $\tilde{H}_1^0$, $\tilde{H}_2^0$ fields, with $\tilde{\chi}_1^0$ the LSP by convention. Likewise, the two physical 'charginos' $\tilde{\chi}_i^\pm$ $(i = 1, 2)$ are mixtures of the $\tilde{W}^\pm$, $\tilde{H}_1^-$, $\tilde{H}_2^+$ fields.

LEP provides a clean environment in which to search for light sparticles:

(i) Light sneutrinos are excluded by Z^0 line-shape measurements, which rule out contributions to the total decay width from

$$Z^0 \to \tilde{\nu}_i + \tilde{\bar{\nu}}_i\,. \tag{68}$$

(ii) Null searches for acoplanar lepton pairs exclude light sleptons and charginos:

$$Z^0 \to \tilde{l}^+ + \tilde{l}^- \to l^+ \tilde{\chi}_1^0 + l^- \tilde{\chi}_1^0 \to l^+ l^- + \text{missing energy}, \tag{69}$$
$$Z^0 \to \tilde{\chi}_i^+ + \tilde{\chi}_i^- \to l^+ \nu \tilde{\chi}_1^0 + l^- \bar{\nu} \tilde{\chi}_1^0 \to l^+ l^- + \text{missing energy}\,. \tag{70}$$

For these pair production processes, the lower mass limits are approximately

$$m_{\tilde{\nu},\tilde{l},\tilde{\chi}_i^\pm} \gtrsim \tfrac{1}{2} M_Z\,. \tag{71}$$

(iii) Light neutralinos are excluded by null searches for processes

$$Z^0 \to \tilde{\chi}_i^0 + \tilde{\chi}_i^0 \to f\bar{f}\tilde{\chi}_1^0 + f\bar{f}\tilde{\chi}_1^0 \to f\bar{f}f\bar{f} + \text{missing energy}, \qquad (72)$$

which would correspond to final states with leptons and/or jets and missing energy. Here the lower limits on the $m_{\tilde{\chi}_i^0}$ depend on the specific mix of gauginos and Higgsinos assumed.

Significantly higher mass limits for squarks and gluinos can be achieved at hadron colliders. Being strongly interacting, these particles can be copiously pair produced in collisions of ordinary quarks and gluons, *e.g.*

$$\begin{aligned} gg &\to \tilde{g}\tilde{g}, \tilde{q}\tilde{q} \\ qg &\to \tilde{q}\tilde{g} \\ qq &\to \tilde{q}\tilde{q}, \text{ etc.} \end{aligned} \qquad (73)$$

If the squarks and gluinos are not too heavy, then the dominant decay modes are either

$$\begin{aligned} \tilde{q} &\to q\tilde{\chi}_1^0 \\ \tilde{g} &\to q\bar{\tilde{q}} \to q\bar{q}\tilde{\chi}_1^0, \end{aligned} \qquad (74)$$

if $m_{\tilde{q}} < m_{\tilde{g}}$, or

$$\begin{aligned} \tilde{q} &\to q\tilde{g} \to qq\bar{q}\tilde{\chi}_1^0 \\ \tilde{g} &\to q\bar{q}\tilde{\chi}_1^0, \end{aligned} \qquad (75)$$

if $m_{\tilde{q}} > m_{\tilde{g}}$. In addition to these 'direct' decays, indirect 'cascade' decays (Baer *et al.* 1991) can also occur if the squarks and gluinos are sufficiently heavier than the charginos and neutralinos, for example

$$\begin{aligned} \tilde{g} &\to q\bar{q}\tilde{\chi}_i^0 \to q\bar{q}q\bar{q}\tilde{\chi}_1^0 \\ \tilde{q} &\to q\tilde{\chi}_i^\pm \to ql\nu\tilde{\chi}_1^0. \end{aligned} \qquad (76)$$

The characteristic signature is therefore 'jets + missing transverse energy' for direct decays, with the additional possibility of 'jets + leptons + missing transverse energy' for cascade decays. On average the missing energy will be smaller in the latter, because there are more intermediate stages in the decay. Of course such events can also arise in the Standard Model, and in practice care must be taken to account for all possible background processes. Figure 33 shows the region in the $(m_{\tilde{q}}, m_{\tilde{g}})$ plane currently excluded by the CDF collaboration (CDF 1992) at the Fermilab $p\bar{p}$ collider, assuming direct decays only. Including also the possibility of cascade decays lowers these bounds by some 20 GeV (Baer and Tata 1992). Note the light squark/gluino window in this figure—very light squarks and gluinos produce so little missing transverse energy that they are impossible to distinguish from the normal QCD jet background.

The fact that no sparticles have been discovered so far can be interpreted as a lower limit on the fundamental MSSM mass parameters m_0 and $m_{1/2}$. In the scenario illustrated in Figure 32, the sparticle masses are not too far above their experimental lower bounds, and detection of at least some of them would be guaranteed at LHC and SSC.

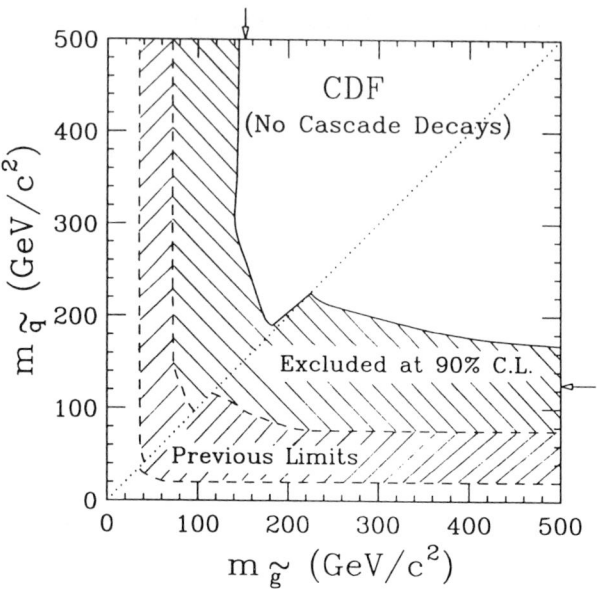

Figure 33. *Limits on squark and gluino masses (assuming direct decays only) from the CDF collaboration (CDF 1992) at $\sqrt{s} = 1.8\,\text{TeV}$.*

6 Summary

The Higgs boson holds the key to one of the most pressing problems in particle physics today: understanding the origin and values of the particle masses. In these lectures we have discussed the theory and phenomenology of the Higgs sector, in both the Standard Model and in its minimal supersymmetric extension. We have seen that present and future colliders—in particular, LEP, LHC and SSC—are well placed to search for the SM Higgs boson if, as expected, its mass is less than $O(1\,\text{TeV})$. Already, LEP-I has ruled out a Higgs boson lighter than about 60 GeV. LEP-II will extend this search towards 100 GeV, at which point the pp colliders will take over. A high-energy linear e^+e^- collider would in many ways be complementary to a pp collider, offering similar coverage in mass but utilising a different set of decay modes (Zerwas et al. 1992). Such a machine would also be ideal for covering the difficult intermediate-mass range. Unfortunately, the technical problems in designing this type of machine with the required luminosity appear overwhelming at present.

The Higgs mechanism, while it has many attractive features, is not without difficulties. There is, for example, a severe fine-tuning ('naturalness') problem associated with preventing the Higgs mass being renormalized up to some very high new-physics scale. Supersymmetry is one way of circumventing this difficulty while retaining the essential features of the SM Higgs sector. In the minimal supersymmetric model, the masses and couplings of the five Higgs bosons are expressible in terms of just two parameters, $\tan\beta$ and M_A. The key feature is at least one Higgs with a mass less than $O(130\,\text{GeV})$, the exact upper bound depending on m_t. Present and future experiments will provide sensitivity to essentially all of the MSSM parameter space.

Theorists have come up with various ways of going beyond the Standard Model, but it seems clear that the only way to really understand the mechanism of electroweak symmetry breaking and the origin of particle masses is to *search* for scalar bosons, and to measure their properties and interactions with the other particles. To quote the *Higgs Hunter's Guide*:

> "As has happened so often in the past, once the data informs us as to the path Nature has chosen, understanding should soon follow."

References

Alcaraz J, Felcini M, Pieri M and Zhou B, 1993, preprint CERN-PPE/93-28.
ALEPH collaboration: Decamp D et al., 1990, *Phys Lett* **B236** 233; 1990, **B246** 306.
Amaldi U, de Boer W and Furstenau H, 1991, *Phys Lett* **B260** 447.
Anselmo F et al., 1992, preprints CERN-PPE/92-103, CERN-TH.6429,6543, *Nuovo Cimento* **105A** 1335, 1357.
Baer H, Bisset M, Dicus D, Kao C and Tata X, 1993, *Phys Rev* **D47** 1062.
Baer H and Tata X, 1992, Experimental Consequences of Supersymmetry, *Workshop on Decay Properties of SUSY particles, Erice, Italy, 1992.*
Baer H, Tata X and Woodside J, 1991, *Phys Rev* **D44** 207.
Bagger J et al., 1993, preprint FERMILAB-Pub-93/040-T.
Barbieri R, 1988, *Riv Nuovo Cimento* **11** 1.
Barger V, Berger M S and Phillips R J N, 1993, *Phys Rev Lett* **70** 1368.
Barger V, Cheung K, Phillips R J N and Stange A L, 1992a, *Phys Rev* **D46** 4914.
Barger V, Cheung K, Phillips R J N and Stange A L, 1992b, *Phys Rev* **D47** 3041.
Barger V and Phillips R J N, 1993, Search for Physics Beyond the Standard Model, *International Symposium on 30 Years of Neutral Currents - 1993.*
Brignole A, Ellis J, Ridolfi G and Zwirner F, 1991, *Phys Lett* **B271** 123.
Brout R and Englert F, 1964, *Phys Rev Lett* **13** 321.
Cahn R N and Dawson S, 1984, *Phys Lett* **B136** 196.
Cahn R N, Ellis S D, Kleiss R and Stirling W J, 1987, *Phys Rev* **D35** 1626.
Carter A A, 1993, in *Proc IoP Nuclear and Particle Physics Conference, Glasgow.*
CDF collaboration: Abe F et al., 1992, *Phys Rev Lett* **69** 3439.
Dawson S, Gunion J F, Haber H E and Kane G L, 1990, *The Higgs Hunter's Guide* (Addison-Wesley, Reading).
Dicus D A and Willenbrock S S D, 1989, *Phys Rev* **D39** 751.
Ellis J, 1992, Ten Years of SUSY Confronting Experiment, preprint CERN-TH.6707.
Ellis J, Kelley S and Nanopoulos D V, 1991a, *Phys Lett* **B260** 131.
Ellis J, Ridolfi G and Zwirner F, 1991b, *Phys Lett* **B257** 83; 1991, *Phys Lett* **B262** 477.
Froidevaux et al., 1990, Report of the Higgs working group, in *Proc Large Hadron Collider Workshop, Aachen*, eds Jarlskog G and Rein D, ECFA 90-133 II 427.
Georgi H, Glashow S L, Macahek M E and Nanopolous D V, 1978, *Phys Rev Lett* **40** 692.
Glashow S L, Nanopolous D V and Yildiz A, 1978, *Phys Rev* **D18** 1724.
Gross E and Yepes P, 1993, *Int J Mod Phys* **A8** 407.
Gunion J F, Haber H E, Paige F E, Tung W-K and Willenbrock S S D, 1987, *Nucl Phys* **B294** 621.
Gunion J F and Orr L H, 1992, *Phys Rev* **D46** 2052.

Haber H E and Hempfling R, 1991, *Phys Rev Lett* **66** 1815.
Haber H E and Kane G L, 1985, *Phys Rep* **117** 75.
Hewett J L, 1993, *Phys Rev Lett* **70** 1045.
Higgs P W, 1964, *Phys Lett* **12** 132; 1964, *Phys Rev Lett* **13** 508; 1966, *Phys Rev* **145** 1156.
't Hooft G, 1980, Recent Developments in Field Theories, eds 't Hooft G *et al.* (Plenum Press, New York).
Kunszt Z, 1984, *Nucl Phys* **B247** 339.
Kunszt Z and Zwirner F, 1992, *Nucl Phys* **B385** 3.
Langacker P and Luo M, 1991, *Phys Rev* **D44** 817.
LEP, 1993, *The LEP Electroweak Working Group*, informal note.
Lindner M, Sher M and Zaglaeur H, 1989, *Phys Lett* **B228** 139.
Lindner M, 1986, *Z Phys* **C31** 295.
Nilles H-P, 1984, *Phys Rep* **110** 1.
Okada Y, Yamaguchi M and Yanagida T, 1991, *Prog Theor Phys* **85** 1; *Phys Lett* **B262** 54.
Pauss F, 1991, CERN Academic Training Programme lectures, and references therein.
Roberts R G and Ross G G, 1992, *Nucl Phys* **B377** 571.
Seez C *et al.*, 1990, Report of the Higgs working group, in *Proc Large Hadron Collider Workshop, Aachen* eds Jarlskog G and Rein D, ECFA 90-133 **II** 474.
Summers D J, 1992, *Phys Lett* **B277** 366.
Veltman M, 1977, *Acta Phys Pol* **B8** 475.
Zerwas P M *et al.*, 1992, *Proc Workshop on e^+e^- Collisions at* 500 GeV: *the Physics Potential*, eds Zerwas P M *et al.*, DESY 92-123A/B 1.

Physics and Experimental Challenge of Future Hadron Colliders

Peter Jenni

CERN, Geneva, Switzerland

1 Introduction

The Standard Model (SM) of particle physics has been tested to unprecedented precision by the experimental results obtained from the four experiments at LEP. The results are impressive. The SM proves to be a mathematically consistent theory capable of precise predictions at the microscopic level. However it is universally accepted that the SM cannot be the last word, although no universally accepted extension of the SM has emerged.

There remain parts of the SM still to be discovered, most notably the top quark and the Higgs boson, the scalar which remains after the generation of mass. Some of the proposed extensions of the SM (e.g. SUSY) bring a rich spectrum of particles which await discovery. The experimenter is confronted with a promise of new discoveries, but no precise predictions of mass to guide the experimentation. In this situation high energy hadron colliders, which offer a broad spectrum of mass, represent the best means for future discovery.

There are two projects, the the Large Hadron Collider (LHC) in CERN and Superconducting Supercollider (SSC) in the USA. (These lectures were given and prepared before the cancellation of the SSC project; we have not altered them to take this into account.) The technology for realising the machines appears to exist, but they present extreme experimental challenges, which will be addressed in these lectures.

The outline of the lectures will be as follows. First we will discuss the LHC and the SSC, the physics goals and the experiments; then we will discuss calorimetry, electron identification and the inner detector; finally we will discuss muon identification, the trigger and data acquisition (DAQ), and the simulated physics. Most of the detailed experimental examples will be taken from the LHC *Letter of Intent* of the ATLAS Collaboration.

2 LHC and SSC

The LHC (Large Hadron Collider) is the project planned for CERN. This collider is to be located in the existing LEP tunnel. The colliding particles are protons with a centre-of-mass energy $\approx 16\,\text{TeV}$. A second mode of operation for the LHC as a heavy ion collider will be possible. The location of this collider within existing facilities allows the optimal use of the already operating accelerators in the injector chain. With the LHC in the same tunnel as LEP there exists also the future possibility of collisions between LHC and LEP as an *e-p* collider. The constraint of the diameter of the LEP ring places a high demand on the guide field of the bending magnets required for LHC operation at 16 TeV. This magnetic field must be in the region of 10 T.

	p-p	*e-p*	Pb-ions
Max cm. energy (TeV for $B = 9.5\,\text{T}$)	15.4	1.36	1262
Luminosity (cm^{-2} s^{-1})	1.6×10^{34}	2.8×10^{32}	1.8×10^{27}
Number of bunches	4725	508	800
Bunch spacing (m, ns)	4.5, 15	49.4, 164.7	31.5, 105
Particles/bunch	10^{11}	$9.2 \times 10^{10} - 3.0 \times 10^{11}$	6.2×10^{7}
Particles/beam	4.7×10^{14}	$9.2 \times 10^{13} - 1.5 \times 10^{14}$	5.0×10^{10}
Number of experiments	3	1	2
β at interaction point (m ($\beta_x \times \beta_y$))	0.5	(0.85, 0.26)–(32.7, 3.05)	0.5
r.m.s. radius at interaction point (μm (x, y))	15	(120, 37)	12.4
r.m.s. collision length (cm)	5.3	3.8	5.3
crossing angle (μrad)	200	0	200

Table 1. LHC *main parameters.*

	LHC	SSC
Beam energy (TeV)	7.7	20
Maximum Luminosity cm^{-2}s^{-1}	1.7×10^{34} (*)	10^{33}
bunch spacing (ns)	15	16
stored energy per beam (MJoule)	583	485

* - at each of 3 interaction points

Table 2. LHC *and* SSC *design parameters.*

The SSC project is also a proton-proton collider which is planned to operate at a centre-of-mass energy of 40 TeV. This project is located in Texas and is being built as a new facility. The tunnel for SSC is considerably larger in diameter than LHC and this allows the field in the bending magnets to be smaller; about 6.6 T. This scale of field, using superconducting magnets is readily achievable with technology in use in current machines e.g. HERA, Tevatron.

Each of these machines is designed to run at high luminosity, because the cross sections for the most sought after physics signals are very small.

3 Physics goals

3.1 The Higgs particle

The benchmark process for these hadron colliders is the detection of the Standard Model Higgs particle, the scalar remnant after the symmetry breaking mechanism of the electro-weak sector. The total cross section for various masses of Higgs at LHC and SSC energies is given in Figure 1. The two main production processes for the Higgs

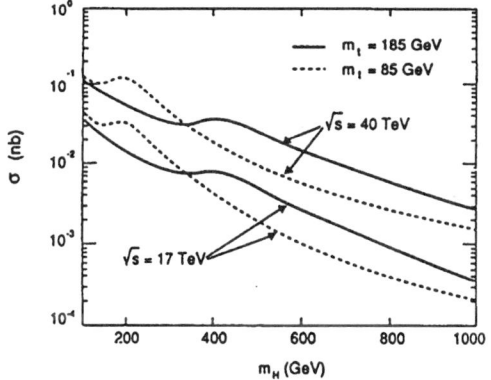

Figure 1. *Total Higgs cross section*

particle are predicted to be those of gluon fusion, through top quark loops (gg→H^0), and intermediate vector boson fusion (WW→H^0 q$\bar{q}$). The production cross sections for these two processes at LHC energy are given in Figure 2.

The mass m_H of the anticipated Higgs particle plays a major role in determining the decay modes which may be used for its detection. At low to intermediate m_H of 50 to 200 GeV, one has the H^0→$\gamma\gamma$ as the practical decay mode, up to about 130 GeV, when the decay modes of H^0 →W*W or Z*Z become important. This behaviour is roughly independent of the top mass variation from 90 to 200 GeV. A heavy Higgs particle of 200 to 1000 GeV has H^0 →W$^+$W$^-$ and H^0 →ZZ as the dominant decay modes. A top mass in the range from 90 to 200 GeV has an effect for m_H less than 400 GeV, in that

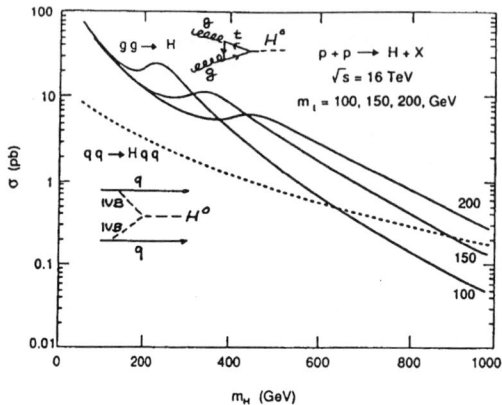

Figure 2. *Higgs production cross section from IVB fusion and gluon fusion.*

the decay mode $H^0 \to t\bar{t}$ has significance at the higher top mass. The H^0 width in the low and intermediate mass range is narrow, but it becomes broad at high masses. This is important for considerations on instrumental resolutions.

The detection of a heavy H^0 depends on the ability to detect e, μ and ν (via missing P_T) and jets under high luminosity conditions, as outlined below in Table 3.

Process and comments	Indicative branching ratio
• $H^0 \to WW$ or $ZZ \to 4$ jets – not accessible due to huge QCD background	56%

• $H^0 \to WW \to \ell\nu + 2$ jets
  – very difficult because of $t\bar{t} \to WWb\bar{b}$
  – forward jet tagging for $qq \to qqH^0$
	16%

• $H^0 \to ZZ \to \ell^+\ell^- + 2$ jets(M_Z) – background : Z + jets continuum	3%
• $H^0 \to ZZ \to \ell^+\ell^- \nu\bar{\nu}$	0.7%
• $H^0 \to ZZ \to \ell^+\ell^-\ell^+\ell^-$ – lowest rates but cleanest signatures	0.1%

Table 3. *Heavy H^0 detection.*

The possibility of H^0 above a mass of 1 TeV can also be considered. If m_H is above 1 TeV then the Higgs sector can be explored via a measurement of $W_L W_L$ scattering. This will allow investigation of the symmetry breaking sector, $\mathcal{L}_{SB}$. If the effect of $\mathcal{L}_{SB}$ becomes strong then the event rate in Table 4 is expected. For these events it is not easy to establish a clear signal and further work is needed on theoretical calculations for this regime.

	16 TeV	40 TeV
W^+W^+ or W^-W^- into $\ell^\pm \nu \ell^\pm \nu$	50	53
background (high-order EW processes)		
$P_T^\ell > 50\,\text{GeV}$, $|\eta| < 2$, $M_{WW} > 800\,\text{GeV}$	70	34
$\int \mathcal{L}\, dt$ (pb^{-1})	10^5	10^4

Table 4. *Number of events attributable to strong $\mathcal{L}_{SB}$ above 1 TeV.*

3.2 Top quark physics

The other missing particle of the SM, namely the top quark, can be studied at LHC and SSC. Analyses within the SM, together with LEP precision data, and M_W/M_Z from $p\bar{p}$ colliders, predict that $m_t \approx 100\text{--}200\,\text{GeV}$. At the start up of LHC and SSC the top quark may remain undiscovered, or even if discovered, its mass may not be measured to a precision of better than 10 GeV. The electroweak measurements at LEP, combined with an accurate measurement of m_t can be used to set limits on m_H. The uncertainty on the prediction of m_H receives roughly equal contributions from LEP errors and a 5 GeV uncertainty on m_t. The first observation and mass measurement may occur at LHC.

At the LHC energy the cross section for top production, $\sigma_{t\bar{t}} = 3\,\text{nb}$ for m_t of 140 GeV, thus large numbers of $t\bar{t}$ pairs will be produced even at luminosity $10^{32}\,\text{cm}^{-2}\text{s}^{-1}$. At LHC and SSC large values of m_t will cause high rates of WW final states ($t \to Wb$) giving backgrounds to other physics searches ($H^0 \to WW$, $Z' \to WW$).

The possible signatures of the top quark include $t\bar{t} \to WW + b$-jets with one W decaying to $\ell\nu$ and the other to $q\bar{q}'$. This decay type has a background from the production of W + jets. Another possible signal is $t\bar{t} \to WW + b$-jets, with one W decaying to $\mu\nu$ and the other to $e\nu$. Here the backgrounds are expected to be small. In both these signals the ability to tag b-jets is essential as is lepton tracking and momentum resolution.

A direct method for the determination of the top mass is possible for the above decays through the reconstruction of 2 and 3-jet masses in association with a lepton tag. The determination of the top mass is also possible through the indirect method of a fit of the dilepton mass spectra from decays $t\bar{t} \to WW b\bar{b}$ where the W's decay to $\ell\nu$. The isolated $e\mu$ channel can be used to extract a very clean signal since the dominant background is $b\bar{b}$ production followed by b and $\bar{b}$ decays to e and μ. This method will provide the most accurate measurement of m_t, but requires integrated luminosities greater than $10^4\,\text{pb}^{-1}$. It leads to an expected statistical error on m_t of 1.2 (1.9) GeV with a systematic error of 4.0 (5.0) GeV for $m_t = 140$ (200) GeV.

If a charged Higgs exists it may be produced through the decay mode $t \to bH^+$ where the H^+ decays to either $c\bar{s}$ or $\tau \nu_\tau$.

The $\tau \nu_\tau$ channel can be studied by searching for a clean sample of $t\bar{t}$ events with a high P_T lepton (e or μ). In this sample the decay $t \to bH^+$ will yield an excess of events with one isolated τ, compared to events with an additional lepton ($e\mu$). For the search $t \to bH^+$, $H^+ \to c\bar{s}$ the top events are selected, and then a H^+ mass peak is sought in the 2-jet mass distribution.

3.3 Supersymmetry

Strongly interacting supersymmetric (SUSY) particles have large production cross sections at hadron colliders of the energy of LHC and SSC. Two types of signatures have to be considered, depending on the decay chain.

- Direct decays into the $\tilde{\gamma}$, the photino, assumed to be the stable lightest SUSY particle escaping detection, of the form

$$\tilde{g} \to q\bar{q}\tilde{\gamma}$$

This decay exhibits a signature of jets and missing transverse momentum P_T^{miss}.

- Cascade decays involving charginos and neutralinos of the form

$$\tilde{g} \to q\bar{q}'\,\tilde{x}^{\pm}, \text{ where } \tilde{x}_i^{\pm} \to \tilde{x}_j^0 W^{\pm} \quad \text{or} \quad \tilde{g} \to q\bar{q}\tilde{x}^0, \text{ where } \tilde{x}_i^0 \to \tilde{x}_j^0 Z^0.$$

The cascade type decays also exhibit characteristics of jets and missing P_T in addition to the appearance of intermediate vector bosons in the final state.

The common characteristics of the SUSY decays is the combination of jets and missing P_T. This signature remains the most promising for the detection of gluinos and squarks ($\tilde{g}$ and $\tilde{q}$). The irreducible background for such a signature consists mainly of $t\bar{t}$ and W+jet production followed by leptonic W decay, and Z+jet production followed by $Z \to \nu\bar{\nu}$.

3.4 Heavy vector bosons

Searches for heavy Z' and W' can be made at the new hadron colliders where a mass range of several TeV can be explored. These particles appear in a number of models, e.g. minimal extensions of the Standard Model and models for electroweak symmetry breaking through compositeness. The Z' can decay through $Z' \to e^+e^-$ or $\mu^+\mu^-$ and these modes should be easy to detect, depending on the electromagnetic and muon energy resolution. Another probable decay channel, using the Extended Gauge Model, is $Z' \to q\bar{q}$. This channel is more difficult to observe due to the large QCD continuum, but the rates are much larger than those of the previous two-lepton channel. The possible observation of this channel depends on the constant term present in the resolution of the hadronic calorimetry.

3.5 Parton substructure

The existence of parton substructure can be tested by measuring the inclusive jet cross section. A composite nature of quarks would show up as a deviation from the standard QCD expectation at high transverse momenta, where valence quark scattering dominates, as shown in Figure 3. The jet cross section and angular distribution can be studied to test the strength of the compositeness scale (Λ_c). The result will depend not only on the quality of the measurement, but also the theoretical knowledge of the expected cross section.

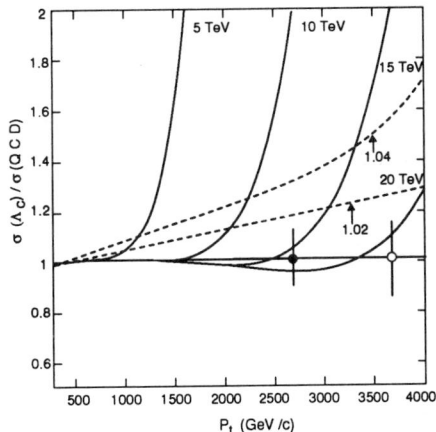

Figure 3. *Deviation from* QCD *for different* Λ_c *values. The dashed lines show the effect on the QCD expectation of possible uncorrected non-linearity of 2% and 4% in the calorimeter response. The statistical sensitivities for* 10^4pb^{-1} *(full circle) and* 10^5pb^{-1} *(open circle) are also shown.*

4 Experiments at LHC and SSC

At these colliders lepton identification is crucial, in that the final state lepton rate is often expected to be small for many processes, but will give the cleanest signals. With luminosities of around $10^{34} \text{cm}^{-2}\text{s}^{-1}$ there will be very large background rates which have the ability to mask these lepton signatures. Muons are easy to identify due to their non-interactive nature, and can be separated from jets, but are hard to measure well. Electrons are relatively easy to measure but hard to identify at a luminosity of $2 \times 10^{34} \text{cm}^{-2}\text{s}^{-1}$ due to the ability of jets (pions) to fake this signal. The requirement of lepton isolation is necessary to solve this problem, as shown in Figure 4.

Characteristically the experiments at LHC and SSC are large detectors with precise inner tracking, followed by electromagnetic then hadronic calorimetry, and finally muon detectors. The calorimetry must be as hermetic as possible in order to make the detection of missing transverse energy (E_T^{miss}) efficient. These detectors must be physically large to contain the high energy interaction products. They must have a fast response time to cope with the high luminosity and low bunch crossing times of the colliders.

4.1 SSC

4.1.1 SDC detector

The Solenoidal Detector Collaborations (SDC) detector typically uses proven technology to achieve the physics detection goals. The main features are, working outward from the interaction point, a silicon and straw tracker for vertex reconstruction and charged

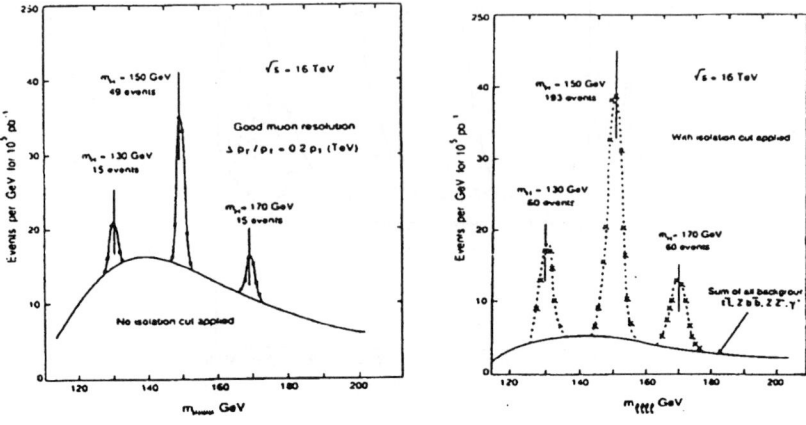

Figure 4. 4μ mass without isolation; e and μ with isolation.

particle tracking, a scintillator based calorimeter for both electromagnetic and hadronic energy measurements and then an iron toroid system for muon detection. The inner detector is contained inside a solenoidal magnetic field of 2 T to provide charged track momentum resolution.

4.1.2 GEM detector

The GEM detector employs somewhat more novel and newer technology. The calorimetry in the GEM detector is split up into separate electromagnetic and hadronic calorimeters. The electromagnetic calorimeter uses a sampling accordion geometry design with liquid krypton as the active material in the barrel region. The accordion geometry provides a coverage for the calorimeter which is hermetic and the use of liquid krypton increases the resolution. The hadronic calorimetry choice is liquid krypton in the barrel region and liquid argon in the end-caps. A scintillator tail catcher completes the barrel calorimeter. The magnetic field is provided by a large solenoid that envelopes the tracking, calorimetry and muon chambers. Since the shape of the magnetic field inside this volume over all η is important, forward field shapers are included at high η to help in momentum resolution.

4.2 LHC

The main detector proposals for LHC are (as of October 1992) ATLAS (a general purpose, balanced approach to e, γ, μ, jets and E_T^{miss}), CMS (a Compact Muon Solenoid, general purpose, optimised for μ and precise e, γ) and L3P (dedicated to μ, e or e, γ, μ options).[1]

---

[1] *Editors' Note:* ATLAS and CMS were requested to prepare a full Technical Proposal during 1993.

4.2.1 CMS

The CMS detector is optimised for muon detection which should remain relatively easy even at a luminosity of 2×10^{34}cm^{-2}s^{-1}. Muons can be identified inside jets, unlike electrons, and efficient muon detection will also allow for more control on isolation cuts. It is possible to trigger on and identify muons down to very low transverse momenta (≈ 5 GeV). Since muon detection and measurement is of primary importance for this detector the magnet design influences the rest of the detector. The magnet selected is a long (14 m) solenoid, of large inner radius (2.9 m), with a 4 T field. Three independent measurements of muon momentum will be made; in air before the absorber, after the coil, and after the return yoke. These last two measurements should be guaranteed at any luminosity.

The central tracker of CMS consists of an inner silicon tracker with a resolution of 15 μm surrounded by micro-strip gas counters with a resolution of 50 μm.

The electromagnetic calorimeter is designed for good electron identification when used in conjunction with the inner tracker. The general design of the calorimeter is a series of crystals, two samplings in depth, with fine lateral granularity achieved by placing the crystals at large radius. With this arrangement the energy and position of charged tracks can be determined well for two shower separation ($\pi^0 \to \gamma\gamma$ rejection). Also the shower direction can be determined when the barycentre of the second sampling is used. In order to get good $\gamma\gamma$ mass resolution, good energy and angular resolution is required. The design aim for the electro-magnetic calorimeter is

$$\frac{\sigma}{E} \approx \frac{2\%}{\sqrt{E}} \oplus \frac{0.15\%}{E} \oplus 0.5\% \quad ; \quad \sigma_\theta \approx \frac{40}{\sqrt{E}} \text{ mrad.}$$

The hadronic calorimetry is realised by a technique where plastic scintillator plates are read out by a WLS (Wave Length Shifting) fibre. The calorimeter consists of 7 interaction lengths in 3 or 4 longitudinal samplings, lateral granularity of ($\Delta\eta\times\Delta\phi$) of 0.1×0.1, copper absorber thickness of 3.75 cm, and scintillator plate thickness of 4 mm. Copper was selected as the absorber for this calorimeter, since high Z materials will degrade muon momentum resolution in the multiple scattering dominated regime. Of the suitable low Z materials, copper has the shortest interaction length.

4.2.2 L3P

The L3P detector proposed for LHC is based on the design principle of placing the detector components far away from the interaction point. This strategy is used to reduce the rate of neutral and charged particles on the detector elements. The nearest charged particle detector is at a radius of 1.6 m and the nearest crystals of the calorimeters are at a radius of 2.9 m from the interaction point.

Muon detection is made with several independent measurements in conjunction with the super conducting solenoidal magnet.

The closest calorimeter elements of the CeF$_3$ calorimeter are placed 2.9 m from the interaction point which ensures a low rate of neutrons and γ's incident on the crystals. This distant placement means that photon 'pointing' is not necessary. There is good

π^0 rejection, and the calorimeter is insensitive to pile-up background. To reduce noise in the calorimeter a low capacitance photodiode readout is required and this implies a low magnetic field in the immediate vicinity.

4.2.3 ATLAS

The ATLAS detector is designed as a general purpose detector with a balanced approach to detection of electrons, muons, jets and missing transverse energy. The detector is optimised to provide as many signatures as possible of new physics at the highest LHC luminosity while at the same time retaining good performance at the lower initial luminosities. The magnet choice consists of an outer muon toroid, complemented by an inner solenoid. This choice of magnets places almost no constraints on the calorimetry and inner detector allowing flexibility for technological choices.

5 Calorimetry

At LHC one of the major considerations for calorimetry and overall detector performance is the effect of pile-up due to the high luminosity and low bunch crossing time. For a luminosity of 2×10^{34}cm^{-2}s^{-1} and bunch crossing time of 15 ns, 18 events on average are expected per bunch crossing. Slow detectors may integrate over several bunch crossings. Shower development time sets a lower limit on the integration time of ~40 ns. It is necessary to have a baseline subtraction because of the effective pile-up.

The general calorimeter design needs to be optimised according to the luminosity of the collider. Since the speed of the calorimeter has a lower limit of 2 bunch crossings, the effect of ~40 events in the calorimeter has to be studied. The size of a calorimeter cell determines the probability that the cell contains less than a certain pile-up energy which will act as noise. The granularity of the EM calorimeter is also set by consideration of position and isolation measurements and not only by pile-up considerations. The effect of 40 minimum bias events in a cluster of $\Delta\eta\times\Delta\phi = 0.08\times0.08$ is shown in Figure 5.

For hadron calorimetry which is largely concerned with detecting jets, the effect of pile-up is Gaussian and can be subtracted by using, for example, bi-polar shaping. The size of the jet cone $\Delta R = (\Delta\eta^2+\Delta\phi^2)^{1/2}$ and the sensitive time have a direct effect on the pile-up noise present in the hadronic calorimeter. The effects of the opening half-angle of the jet cone, and number of minimum bias events, on the mass resolution and reconstructed mass, can be seen in Figure 6.

At the above luminosity, a hadronic calorimeter with a 45 ns sensitive time is affected by pile-up in that a 100 GeV jet has a resolution of $60\%/\sqrt{E_T}$ whereas for a 1000 GeV jet the effect is negligible. At high luminosity operation, the useful energy resolution of the hadronic calorimeter is affected greatly by this fact.

The environment of the collider demands high performance from the calorimeter. This is especially true for intermediate mass Higgs searches which entail the decay modes H$^0 \to \gamma\gamma$ and H$^0 \to$Z*Z, where each Z decays to e^+e^- and good resolution helps the observation of M_Z and Γ_Z. High performance does not only imply high resolution, also it is important to have a very large rejection factor ($\sim 10^4$), in conjunction with inner

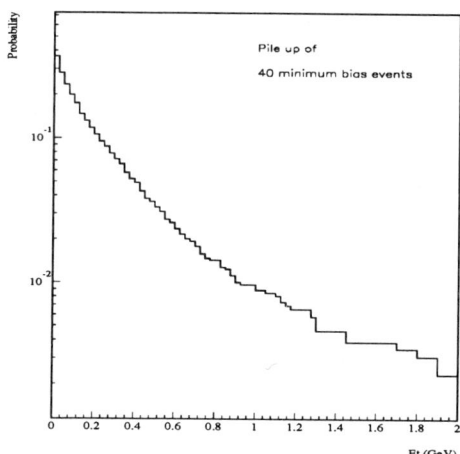

Figure 5. *Probability that transverse energy piled up in EM cluster is larger than E_T.*

tracking) against jet background which can fake a γ as shown in Figure 7. A calorimeter capable of making isolation cuts can identify γ's as opposed to jet fragments, and a high position resolution will help to distinguish single γ's from $\pi^0 \to \gamma\gamma$.

The directional resolution is also an important consideration in the performance of the calorimeter. In the cases of $H^0 \to \gamma\gamma$, where the the event vertex is not known, the H^0 mass resolution depends strongly on the directional resolution of the γ's. In order to achieve an adequate mass resolution, a directional resolution of at least $\sigma_\theta \sim 100\,\mathrm{mrad}/\sqrt{E}$ is required.

The hadronic calorimetry performance requirements are driven mainly by the needs of jet spectroscopy and jet physics. Some examples of processes and calorimeter characteristics are:–

- top quark physics, where the calorimeter resolution has direct impact on multi-jet mass resolution;
- heavy Z' decays into two jets, where the sampling and mainly the constant term have an effect;
- jet cross sections, demand an understanding of possible calorimeter non-linearities;
- high P_T W→jet+jet decays from heavy Higgs, need high granularity to resolve nearby jets.

Bearing these characteristics in mind the ATLAS collaboration has adopted design goals for the hadronic calorimetry of

$$\frac{\sigma_E}{E} = \frac{50\%}{\sqrt{E}} \oplus 3\% \quad ; \quad \text{granularity } \Delta\phi \times \Delta\eta = 0.1 \times 0.1 \text{ (or better)}.$$

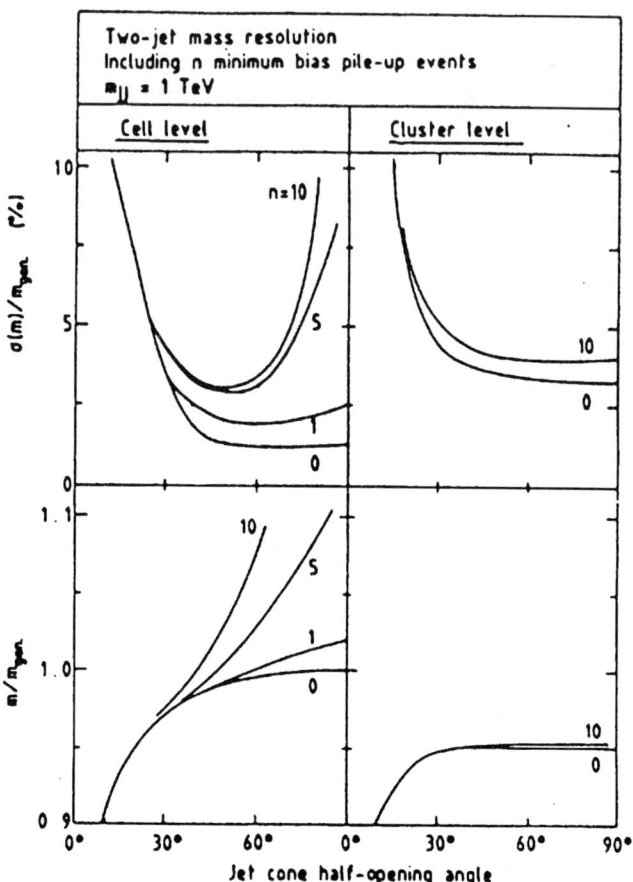

Figure 6. *Effects of the opening half angle of the jet cone (horizontal axis) and the number of piled up minimum bias events (labeled on curves) on the reconstructed two-jet mass (lower plots) and its resolution (top plots). Cell level assumes $\Delta\phi \times \Delta\eta = 10° \times 0.1$. Cluster level assumes adjacent cells with $E > 400$ MeV, $E_T^{cluster} > 10$ GeV.*

Forward calorimetry ($|\eta| > 3$) is required to extend the coverage as close as possible to the complete solid angle. In the high η regions the main challenge is to find a technology which can withstand high radiation doses (1 MGy for 10^5 pb^{-1} integrated luminosity). One attractive option is to profit from a replaceable read out media. Good examples are high pressure gas or liquid scintillator which can be recirculated. The performance requirements for this calorimeter are based on complementing missing E_T measurements of the barrel calorimeters and the need for jet tagging in a heavy Higgs search. Since the forward calorimeter covers a significant rapidity range a vast number of particle products from minimum bias events will hit the device. The requirements

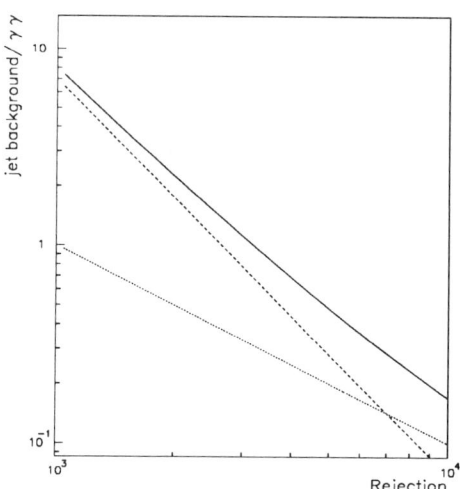

Figure 7. *Expected ratio of jet-jet to $\gamma\gamma$ (dashed) and γ-jet to $\gamma\gamma$ (dotted) rates versus rejection against jets. The solid line is the sum of both ratios.*

for particle identification and directional capability must be reduced, due to this large flux of particles. The design parameters of the forward calorimetry, encompassing a fiducial coverage of $3.0 < |\eta| < 4.5$ are

$$\frac{\sigma_E}{E} = \frac{100\%}{\sqrt{E}} \oplus 7\% \quad ; \quad \text{granularity } \Delta\phi \times \Delta\eta \sim 0.15 \times 0.15.$$

The range of the rapidity coverage of the entire calorimetry has direct implications on the measurement of E_T of an event. Figure 8 shows the effect of η coverage in the $H^0 \to \ell^+\ell^-\bar{\nu}\nu$ channel. The effect of extending the η coverage past $|\eta| = 4$ is minimal and hence the adopted η coverage, is restricted to $|\eta| = 4.5$.

For the case of an intermediate mass Higgs particle decaying to four leptons, electrons have to be measured down to low P_T and over a large rapidity range in order to collect a significant number of events as shown in Figure 9.

5.1 Accordion geometry liquid argon calorimeter

The baseline design for the ATLAS barrel electromagnetic calorimeter is the 'accordion' geometry liquid argon sampling calorimeter. This calorimeter utilises a novel arrangement of the absorber plates and read out media to eliminate cable connections to form towers in depth, increase read out speed, improve hermeticity, and allow flexible granularity. The difference between a 'classical' and 'accordion' calorimeter can be seen in Figure 10.

The location of read out boards on the front and rear of the calorimeter removes the need for cabling running along the side of tower elements. The read out boards are

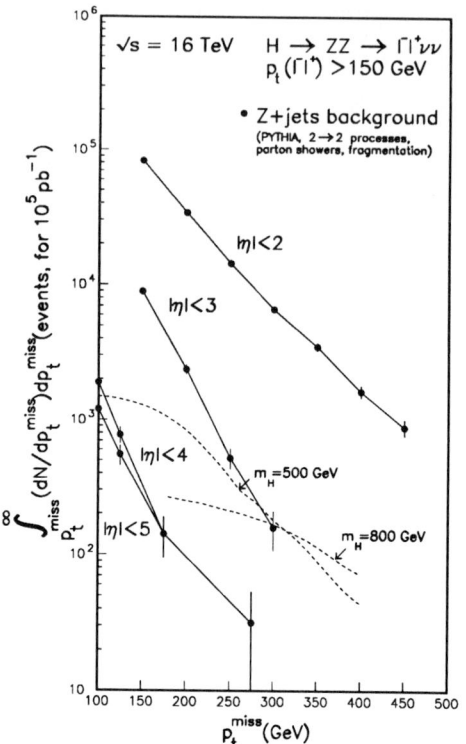

Figure 8. Required η-coverage for missing E_T measurements in $H^0 \to ZZ \to \ell^-\ell^+\nu\bar{\nu}$. For $|\eta| > 4$ the remaining Z+jet background is dominated by $Zb\bar{b}$ with $b \to \nu + X$.

segmented in depth to create the towers, and also segmented in η to create the lateral cells. The elements of accordion geometry can be seen in Figure 11.

Liquid argon, with its low drift velocity, represents a readout problem at speeds required to reduce pile-up at the hadron collider. The solution to this problem can be found in the use of fast bipolar shaping of the signal. The current signal received by a cell rises instantaneously, and is proportional to the amount of charge deposited. A fast bipolar shaping is applied to this signal and a useful signal is produced as shown in Figure 12. The speed of the bipolar shaping has an effect on the calorimeter performance as shown in Figure 13. Increased speed reduces the effect of pile-up, but leads to an increase in electronic noise.

The radiation resistance and reliability of the electronics for this detector are under investigation. Both Si and GaAs preamplifiers have survived thermal cycles, with problems only at the few per mil level. A dedicated programme for the definition of long term behaviour of elements in cold will be set up and will form part of the choice of technology for preamplifiers. Programmes for radiation tests of preamplifiers in cold and warm conditions have been set up with partial results shown in Figure 14.

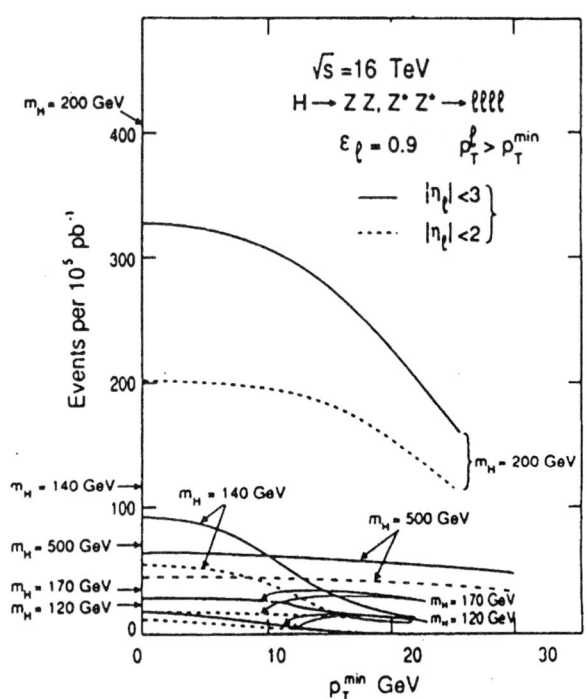

Figure 9. *Effect of η coverage on various mass Higgs signals.*

Uninstrumented material in front of the calorimeter is a potential source of unknown energy loss, so an integrated pre-shower detector is envisaged in front of the ATLAS calorimeter. This will effectively instrument the dead material in front of the calorimeter (solenoid coil and cryostat) enabling energy corrections, providing high granularity for $\gamma - \pi^0$ separation up to $P_T \sim 80$ GeV, providing γ direction measurements for $m_{\gamma\gamma}$ and enhancing electron identification. The position resolution of the pre-shower is critical for the identification of photons from a $H^0 \to \gamma\gamma$ decay. High energy muons have been used to determine position resolution of the layers of the presamples as shown in Figure 15. The energy and angular resolutions of the prototype EM calorimeter and pre-shower are shown in Figure 16 and Figure 17.

The energy resolution from Figure 16 can be parameterised as (with E in GeV),

$$\frac{\sigma_E}{E} = \frac{(10.0 \pm 0.5)\%}{\sqrt{E}} \oplus (0.5 \pm 0.1)\% \oplus \frac{(0.343 \pm 0.028\%)}{E}$$

and the angular resolution parameterised as

$$\sigma_\gamma = \frac{(25 \pm 4)}{\sqrt{E}} \oplus \frac{(76 \pm 28)}{E} \text{ mrad.}$$

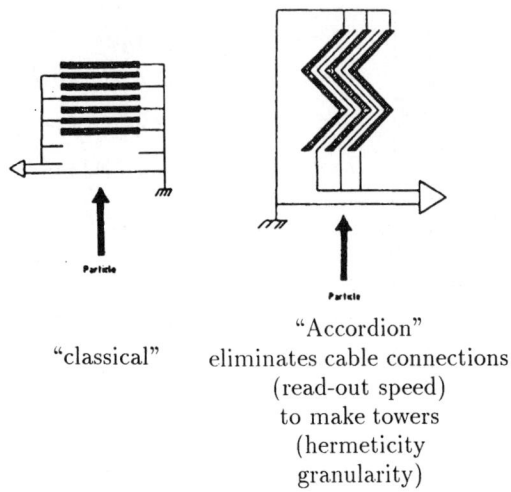

Figure 10. *The classical vs. accordion geometry calorimeter.*

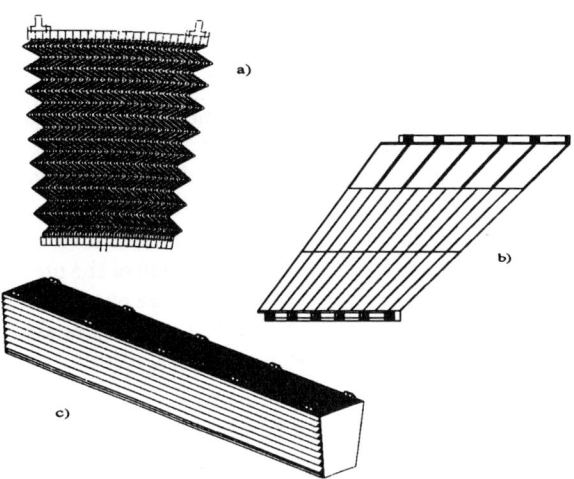

Figure 11. *(a) Transverse view of the* EM *prototype module, with radial dimensions of 1.43–1.95 m. (b) A readout kapton electrode with segmentation in depth and η visible. (c) Perspective view of 2 m long module.*

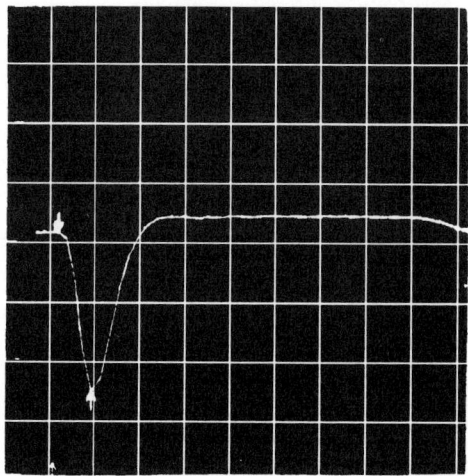

Figure 12. *Fast shaped signal from* 60 GeV *electron hitting one calorimeter cell. The* 400 ns *drift time is clearly visible. (*50 ns×50 mV *per square)*

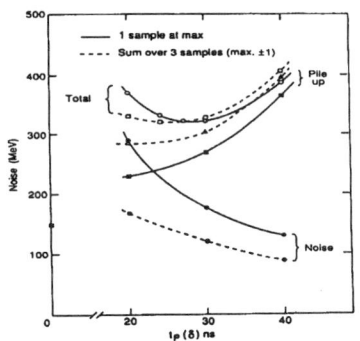

Figure 13. *Calorimeter noise level vs. bipolar shaping pulse speed.*

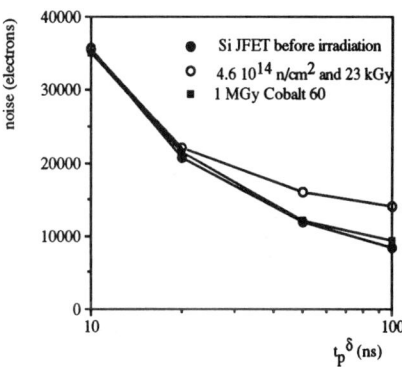

Figure 14. *Electronics noise increase under severe irradiation as a function of shaping time. 'Warm' irradiation results.*

The performance of the combined pre-shower and calorimeter system when used to search for the decay $H^0 \to \gamma\gamma$ has been studied both with Monte Carlo and test beam data. This channel is important for the design of the calorimeter system (energy resolution, γ direction measurement, π^0 rejection). The expected photon energy resolution

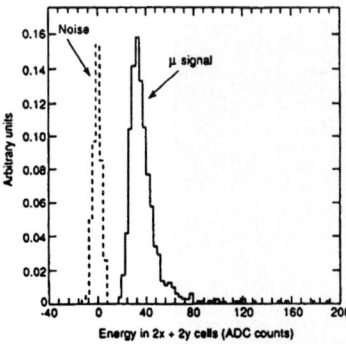 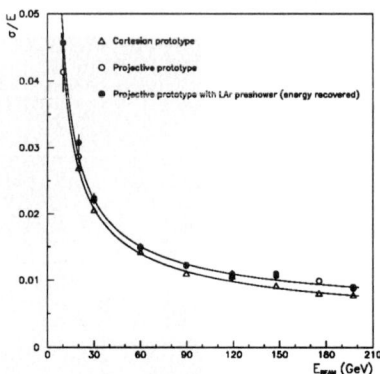

Figure 15. 180 GeV *muon signal in both pre-shower layers. Position resolution of* $340 \pm 40\,\mu m$ *in both coordinates.*

Figure 16. *Energy resolution of Accordion prototypes versus electron energy.*

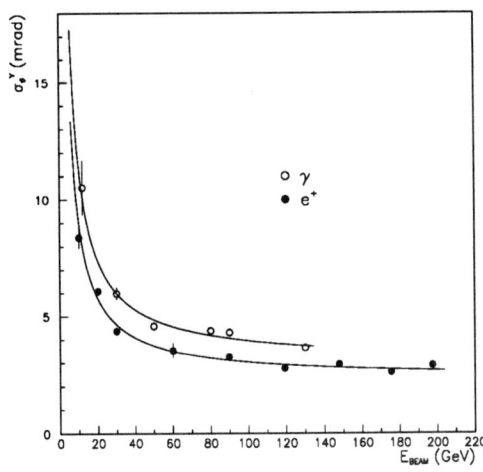

Figure 17. *Angular resolution of the combined pre-shower-calorimeter system versus electron energy (in the η direction).*

for the barrel calorimeter as a function of η is shown in Figure 18. The rejection factor against pions, using the shower shape distributions in the pre-shower, and matching between pre-shower and calorimeter, is shown in Figure 19.

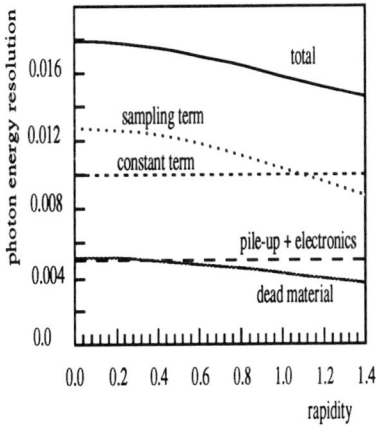
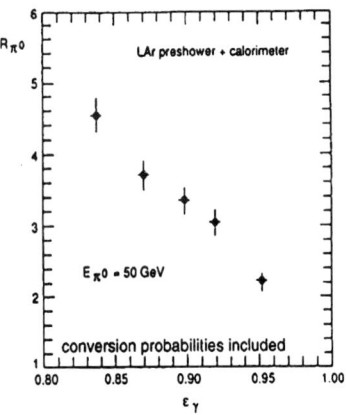

Figure 18. *Energy resolution for a 60 GeV E_T photon as a function of rapidity.*

Figure 19. *Rejection factor against 50 GeV π^0 of the combined pre-shower-calorimeter system. Simulated by overlapping real data from tagged photon beam.*

6 Electron identification

The identification of electrons at LHC and SSC will be crucial for recognising signals of new physics. The main task and difficulty is to identify electrons against the background from fake electrons due to the huge jet rates. Particles that have to be considered in the rejection scheme are:–

- π^0, γ;
- overlaps of π^0 and $\pi^\pm$;
- e^+ and e^- from converted γ's.

The separation of isolated high P_T electrons from $\pi^\pm$ is less of a problem. However the typical calorimeter rejection of 10^3 against jet background is not sufficient to identify single electrons as can be seen in Figure 20. Large rejections against jet and π^0, γ backgrounds are needed to reach the level of real electrons from $W \to e\nu$ and b leptonic decays.

A strategy has been devised for electron identification at LHC using the calorimeter and pre-shower system. The spatial energy deposition in the calorimeter and isolation around the electromagnetic shower are two criteria which can be used to identify electrons. Detailed simulations, including pile-up effects, at a luminosity of $2 \times 10^{34} \text{cm}^{-2}\text{s}^{-1}$ have shown that a rejection factor against jets (R^{jet}) of $\sim 10^3$ with $\varepsilon_e > 90\%$ is possible, provided the electromagnetic calorimeter has fine granularity, see Figure 21. Larger rejections against single $\pi^\pm$ can be achieved, but the calorimeter alone is not sufficient to distinguish signal over the background. To achieve the rejection factors needed use must be made of an inner tracker for matching charged tracks to calorimeter hits.

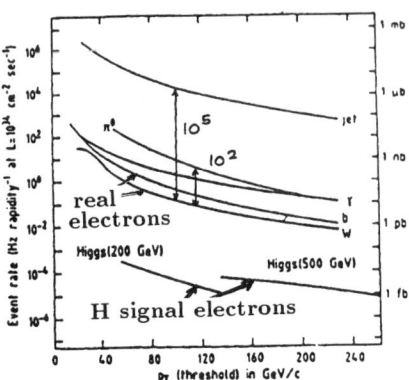

Figure 20. *Event rate vs. P_T threshold. A total rejection factor of 10^5 is needed to observe real electrons from W decays.*

7 Inner detector

The inner detector in ATLAS is placed inside the solenoid magnet which provides a constant 2 T magnetic field over a range of $|\eta| = 1.9$ (the momentum resolution is degraded by ~20% for $1.9 < |\eta| < 2.5$). The inner detector has a tracking volume of length 6.8 m and radius 1.06 m. The desired performance can be split up into high and low luminosity goals, since different types of physics will be studied in each region. At high luminosity the inner tracker should:–

- provide highly efficient reconstruction of isolated high P_T charged tracks, in particular electrons and μ's (above 7 GeV, for $|\eta| < 2.5$);

- enhance electron identification against jet backgrounds, (mainly in the case of π^0) which have passed the initial calorimeter selection, where a rejection of 10^2 to 10^3 is required, and a part of this rejection power should be available at level-2 trigger (of order 50 for $P_T > 40$ GeV);

- provide lepton momentum measurements, with charge sign determination up to ~500 GeV, needed for example to reject opposite sign dileptons from $t\bar{t}$ background in the search for high mass same sign WW boson pairs, for which accurate e and μ momentum measurements at low P_T are required;

- yield reconstruction of low P_T tracks near a high P_T lepton candidate for background rejection; isolation criteria can be used to reduce $Zb\bar{b}$ and $t\bar{t}$ 4-lepton backgrounds in Higgs searches, and the recognition of partner electrons serves to reject Dalitz decays or γ conversions.

The low luminosity goals arise from different physics needs, and include:–

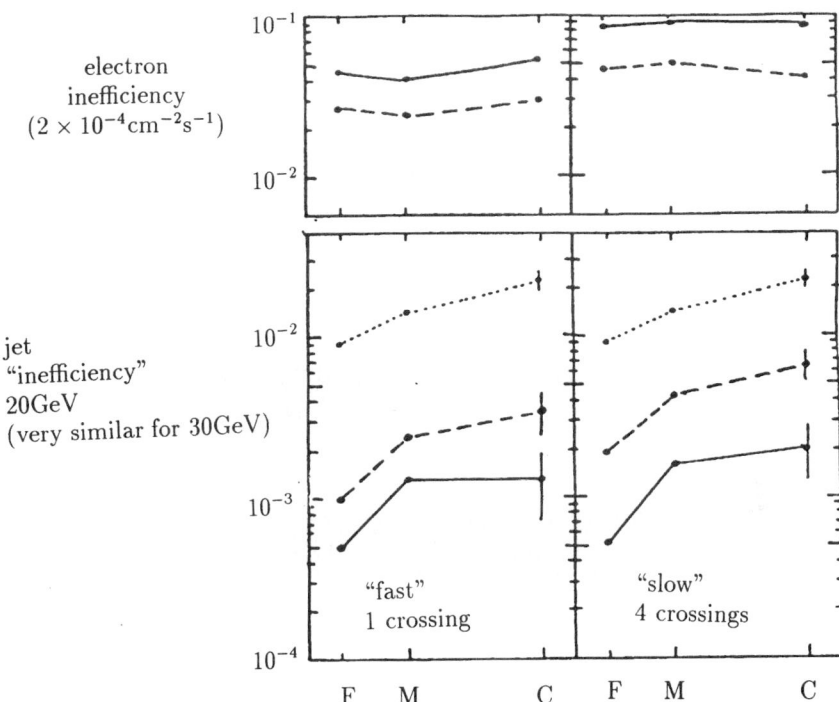

Figure 21. *Electron inefficiency and jet efficiency vs. EM granularity. ($\Delta\phi \times \Delta\eta$ F:0.02^2 M:0.04^2 C:0.06^2). Dotted line, trigger level 2×2 or 3×3 cells. Dashed line, hadron leakage cut. Solid line, isolation cut 4×4 or 9×9 cells.*

- the identification of jets originating from b-quark decays by impact parameter measurements, which reduces combinatorial background especially in t-quark physics;
- the identification of hadronic τ decays, which will be useful in SUSY charged Higgs searches;
- the reconstruction of CP-violating B-decays (for example $B_d^0 \to J/\Psi + K_s^0$).

The required momentum resolution of 20–30% for $P_T = 500\,\text{GeV}$ implies $r\phi$ measurement precisions of less than $20\,\mu\text{m}$ at small radii ($r < 30\,\text{cm}$) and less than $60\,\mu\text{m}$ at outer radii ($r \sim 100\,\text{cm}$). No single tracking detector satisfies all the performance goals so the best features of several detector technologies are used in an integrated design.

The innermost tracker consists of a Silicon Tracker/Vertex detector (SITV) made up from double sided silicon microstrip detectors placed at radii 20 and 30 cm and $|\eta| < 1.5$. An additional layer of pixel detectors is placed at radius 10 cm to ensure good pattern recognition for primary and secondary vertex finding. The innermost forward region for $|\eta| > 1.8$ is covered by GaAs detector rings. The end-caps ($0.9 < |\eta| < 2.5$) are covered by a combination of TRD/tracker arranged as a number of $r\phi$ layers of 4 mm straw drift tubes with radiator foils and MSGC (Micro-Strip Gas Counter) rings. The MSGC rings are interleaved with the TRD/tracker 'wheels'. For the outer barrel region $|\eta| < 0.9$ two concepts are considered. Concept A consists of a silicon tracker (SIT) with axial strip orientation for ϕ measurements and pad layers to provide space point and pattern recognition. Concept B utilises a barrel TRD/tracker with axially oriented straw drift tubes, with six layers of the 'concept A' silicon tracker outside. A general view of the inner tracker with both options can be seen in Figure 22. The characteristics

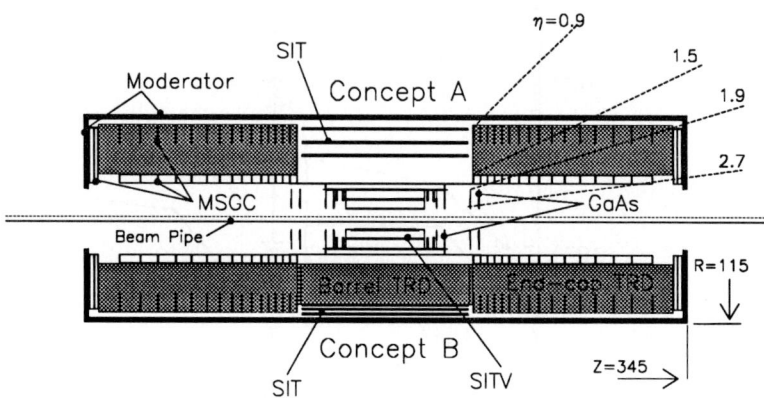

Figure 22. Layout of the inner detector with two design concepts; concept A above and B below the beam line (dimensions in cm).

of the inner detector as it exists in the above two options are given in Table 5.

	SITV	GaAs forward	SIT (A)	TRD barrel	MSGC	TRD
					end-caps	
elements (10^6)	4.2 strips 100 pixels	0.8	2.7	0.11	5	0.26
hits/track	6	4	6	40	4–6	40
% radiation length	6	5	8	12	7.5	12
% normal inc. resolution in $r\phi$ (μm/measurement)	15	20	60	150	45	150
resolution in η (mm in z for barrel in r for forward)	1 strips 0.06 pixels	2	0.3	-	3	30

Table 5. Characteristics of the inner detector.

The components of the inner tracker represent material which sits in front of the calorimeter. This is an important consideration, since for example high P_T electrons can emit bremsstrahlung leading to degraded energy measurement, or γ's can convert. The distribution of material in the inner detector has been studied and is shown in Figure 23.

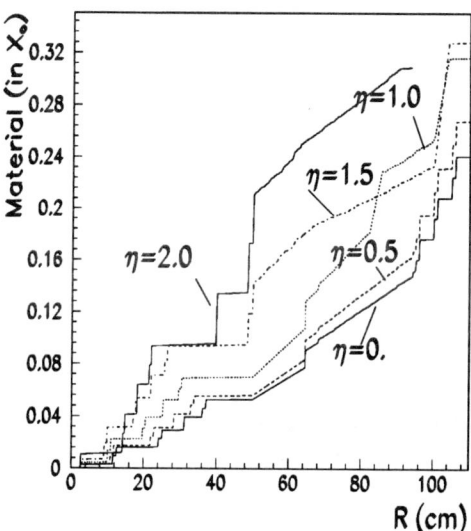

Figure 23. *Radial distribution of the material density for the inner tracker (concept B).*

The potential occupancy of the inner detector has to be considered very seriously because of the projected high luminosity at LHC. The estimation has been made with a realistic amount of material representing the beam pipe in order to take account of conversions, the effects of 'loopers' (low P_T particles that travel on helical trajectories through the inner detector), δ-rays and secondary interactions. In the estimation of the occupancies of the inner detector components, for the SITV and SIT integration was made over 1 bunch crossing, and for the TRD and MSGC integration was made over two bunch crossings due to their distance from the interaction point. The occupancies found from this calculation are shown in Figure 24. Results for the forward regions of the inner detector were found using the same calculation. The occupancy rates are show in Table 6.

inner MSGC	$\sim 1.0\%$
outer MSGC	$\sim 0.4\%$
TRD (all hits)	$\sim 20\%$
TRD (TR hits)	$\sim 2\%$

Table 6. *Occupancy rates for the forward regions of the inner detector.*

The momentum resolution is determined by the high resolution track vector detectors measuring the sagitta and end-point. The intrinsic resolutions of each superlayer

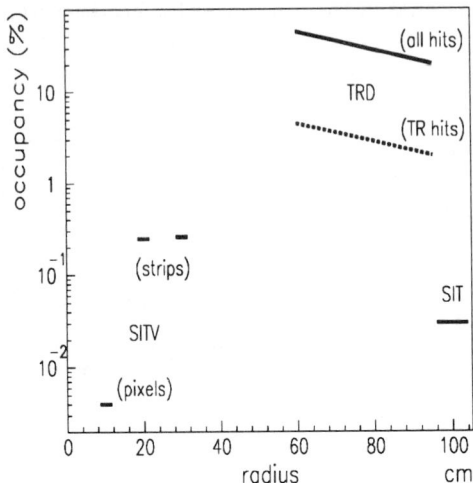

Figure 24. *Occupancy for the barrel detectors at $\eta = 0$ as a function of radius with 20 minimum bias events per bunch crossing.*

are slightly degraded to account for alignment uncertainties. The SITV and GaAs detectors are assigned an $r\phi$ resolution of 20 μm per superlayer, the inner MSGC 45 μm per superlayer, the SIT and outer MSGC 60 μm per superlayer and the transverse beam position is assumed to have an uncertainty of 20 μm. The scaled transverse momentum resolution, $\Delta P_T/P_T^2$ is found to be less than 5×10^{-4} GeV^{-1} for $|\eta|$ less than 2 and degrades to $\sim 10^{-3}$ at $|\eta| = 2.5$, see Figure 25. This degradation at high η is obtained using a realistic magnetic field map of the solenoidal field. At low momenta multiple scattering dominates, contributing to $\Delta P_T/P_T$ a factor of 1–1.5%.

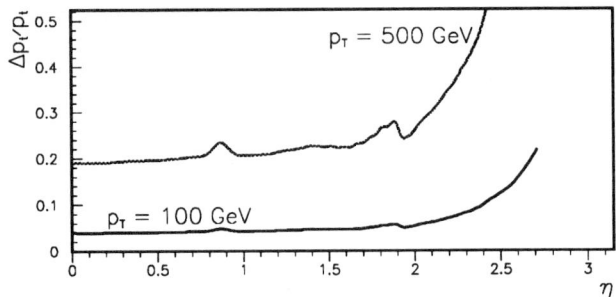

Figure 25. $\Delta P_T/P_T$ *for inner detector as a function of η at high P_T.*

An important function of the inner tracker is its capability to find tracks in conjunction with the electromagnetic calorimeter or muon detector. The inner detector searches for tracks on 'roads' defined by either or both of the other two devices. The typical size of a road is $\Delta\eta\times\Delta\phi = 0.03\times 0.04$ and 0.20×0.06 for a 10 GeV muon and electron respectively. In the simulation 60 minimum bias events are superposed on the high P_T events in an attempt to account for pile-up effects. The performance of the barrel detector with SITV and SIT is shown in Table 7 and with the TRD tracker in Table 8.

$|\eta|$	0		1.2	
Momentum (GeV)	10	100	10	100
Efficiency e (%)	96	97	94	96
Efficiency μ (%)	98.5	98.5	98.5	98.5
Fake tracks* (%)	< 0.2	< 0.2	< 0.2	< 0.2

* within road explained in text

Table 7. *Performance of barrel detector with SITV and SIT. (Lum. 1.7×10^{34}cm^{-2}s^{-1}).*

	barrel		end-cap	
Momentum (GeV)	20	100	20	100
Efficiency e (%)	96	98	96	99
Efficiency μ (%)	> 99	> 99	> 99	> 99
Fake tracks* (%)	< 0.3		< 0.05	

* within road explained in text

Table 8. *Performance of TRD/T pattern recognition. (Lum. 1.7×10^{34}cm^{-2}s^{-1}.)*

Particles with measurable lifetime are of interest, and the precise vertex measurements of the SITV make possible the tagging of such particles. A study of impact parameter resolution has been made assuming a SITV detector which yields three pairs of points between the radii 10 and 30 cm with 20 μm point resolution in $r\phi$. The impact parameter resolution for $\eta = 0$ is given in Figure 26 and as a function of η in Figure 27.

An important component of either concept (A or B) for the inner detector is the TRD/Tracker device. The main functions of the TRD are:–

- efficient pattern recognition over $|\eta| < 2.5$ for $P_T > 0.5$ GeV with low 'ghost' rate even at highest luminosity;
- enhancement of electron identification;
- second level triggering for electron candidates;
- reconstruction of complicated event topologies at low luminosity eg. $B_d^0 \to J/\Psi\, K_s^0$.

The TRD/T is composed of 4 mm diameter kapton straw drift tubes spaced 8 mm apart with the intervening spaces filled with polypropylene foils or foam which produce transition radiation photons. Two different geometries are being studied; the

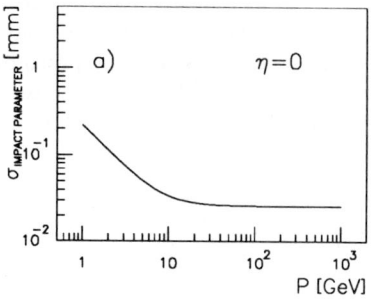
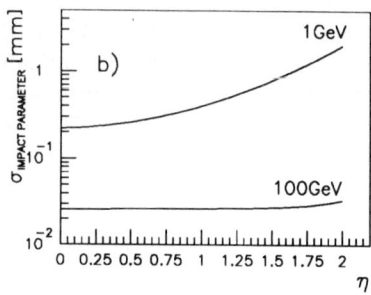

Figure 26. *Impact parameter resolution as a function of particle momentum for $\eta = 0$.*

Figure 27. *Impact parameter resolution as a function of η for 1 GeV and 100 GeV tracks.*

barrel configuration which consists of 100 to 200 cm long axial straws, and the end-cap design which uses 50 cm long radial straws. The straws are read out using a two-threshold scheme; a 0.2 keV threshold read out for tracking drift time measurements and a 5 keV threshold measurement for transition radiation detection. Using drift time measurements a track position can be measured with a resolution of 140 μm. The 5 keV transition radiation hits are used to identify electrons ($E_e > 0.5$ GeV) and high energy pions and muons ($E_{\pi/\mu} > 100$ GeV).

The proximity of the TRD to the interaction point has prompted studies of the radiation hardness of its components. These studies have shown that the straw and radiator components will function for more than 20 LHC years. A full scale prototype end-cap wheel is under construction by the RD6 collaboration.

Hadron rejection can be obtained from the TRD. A simulation of the ability of the TRD to reject hadrons at LHC luminosities was made by using test beam results, see Figure 28. A global study of QCD jet rejection has been performed using a full GEANT simulation of the inner detector (concept B), the pre-shower, and the calorimeter system. The starting point was 500,000 QCD jets with $P_T > 15$ GeV and $|\eta| < 0.5$. Events were analysed using pattern recognition from the combined tracking system and the electron efficiency was checked with 20 GeV electrons including a pile-up of 40 minimum bias events. The results, expressed as cross sections, are shown in Table 9.

8 Muon detection

Muons are considered to be the most 'robust' signature of future hadron collider physics. A thick absorber will 'filter out' other particles even at the highest luminosity, and leave muons as a clean signal. Nevertheless the muon systems were the most controversial topic in current LHC/SSC detector discussions. The questions to understand and resolve include: backgrounds, physics requirements, magnet concepts and design of chambers. For example, the design criteria selected for the ATLAS muon spectrometer are:

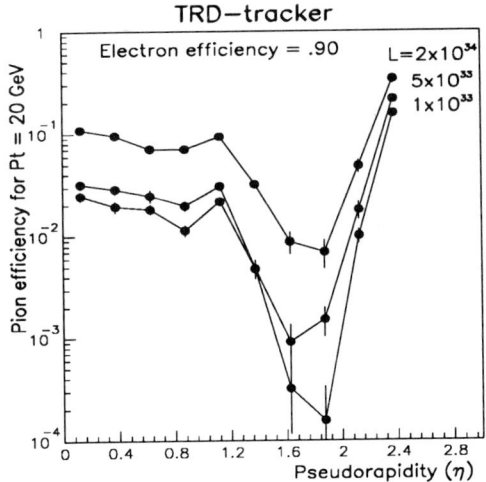

Figure 28. *Hadron rejection of* TRDT *detector versus* η.

real electron sources	isolated e from $W \to e\nu$	3 nb
	e from c, b decays ($\varepsilon = 15\%$)	~ 10 nb
fake electron sources	conversions, Dalitz	~ 0.1 nb
after calo+E/p+pre-shower	hadrons	~ 0.4 nb
$\to$rejection$\sim 10^5$, signal (isolated e)/background ~ 0.6		
fake e-sources	conversions, Dalitz	< 0.1 nb
including TRD	hadrons	< 0.4 nb
(and partner search)		
$\to$rejection$\sim 10^6$, signal(isolated e)/background ~ 6		
Large uncertainties on background estimates (detector modelling, fragmentation) $\to$ TRD brings useful additional rejection		
$\to$ ATLAS detector reaches electron identification goals (a rejection of $\sim 10^6$ against QCD jets with $\varepsilon_e^{tot} \sim 75\%$)		

Table 9. TRD *performance.*

- very good momentum resolution in the range 10–3000 GeV;
- hermetic coverage and momentum measurement to $|\eta| = 3$;
- safe stand-alone operation at luminosities of 10^{34}cm^{-2}s^{-1} and above;
- robust pattern recognition;
- capability of providing an efficient level 1 trigger.

The muon backgrounds that occur after an absorber (the calorimeters) come from two main sources: primary decays from the inner detector volume, whose rates are easy to calculate, and punch-through events, which arise from secondary decays in the hadronic shower, and hadron leakage. More experimental data is required to understand in detail the hadronic and muonic components of the punch through events; their momentum spectrum, and their dependence on absorber material and thickness, and on the magnetic field.

The physics requirements of the muon system have been discussed using the test case of $H^0 \to 4\mu$. These decays are compared to the backgrounds from heavy flavour decays such as $t\bar{t} \to 4\mu+X$ and $Zb\bar{b} \to 4\mu+X$ and continuum decays $ZZ \to 4\mu$.

The detection of muons and the measurement of momentum is simple in principle. After sufficient material to filter out non-muons, several layers of detectors are placed both within and outside a magnetic field, and each detector layer records a track with a certain resolution. The presence of the magnetic field causes charged tracks to bend due to the Lorentz force and from this information the muon momentum can be deduced. The requirements on muon resolution will influence the detector lay-out in a large way. Conventional magnet choices yield the following general resolutions:-

- conventional iron toroids, $\Delta P/P \sim$ 10–20%;
- very high field solenoid or toroids with magnetised iron at 2–6 T, $\Delta P/P \sim$ 5–8%;
- very large open geometry solenoid or air toroids, $\Delta P/P \sim$ 2–4%.

This last choice is somewhat controversial, since only the intermediate mass H^0 ($\to Z^*Z \to 4\mu$) may profit from such a good muon resolution. Each magnet choice has its pros and cons due to their basic design and engineering constraints, summarised in Table 10.

Solenoids	Toroids
Good features	
transverse bending of particles	large η acceptance
stable magnet construction	constant P_T resolution
two momentum measurements (& inner $\vec{B}$ field)	closed field
Bad features	
difficult to cover large range of η	large (and heavy if Fe)
(high) inner field compromises calorimeter operation	limited resolution for warm Fe
very large if in air	difficult access for chambers.

Table 10. *Magnet design considerations.*

The muon chambers detect particles at the outer radii of the detector. Due to their location they will have to cover a very large area (several thousands of m²) so mass production at low cost will be an important issue. Good timing is needed to relate a muon track to a particular beam crossing and provide an efficient trigger. The precision of the chamber alignment is most important. The precision requirements depend on the accuracy aimed at with the muon magnet system, and translate to less than 200 μm for an iron toroid and less than ≤ 50 μm for a high precision muon system. Two approaches to the chamber question are proposed. One approach uses many stacks of 'cheap' drift chambers to provide many track measurements, and the other method uses a few high precision chambers that provide fewer, but higher quality measurements.

As an example, one type of chamber being considered for the ATLAS detector muon system is the honeycomb strip chamber (HSC). An HSC consists of a stack of honeycomb drift tubes as shown in Figure 29 made of folded polyester foil with a thin copper layer on the inside. Two folded foils are combined in a template and form a layer of hexagonal cells that can be glued together to form a light rigid self supporting block of honeycomb. The measured spatial resolution of the chambers using cosmic rays is shown in Figure 30. If small mechanical tolerances are achieved then a resolution of 100 μm per superlayer could be obtained. The resolution near the wire is worse, so the chamber resolution improves with chamber size, but the maximum drift time is proportional to the square of the cell radius. This limits the practical radius due to the counting rate and track multiplicity. Thus the optimal cell radius decreases with $|\eta|$. The muon

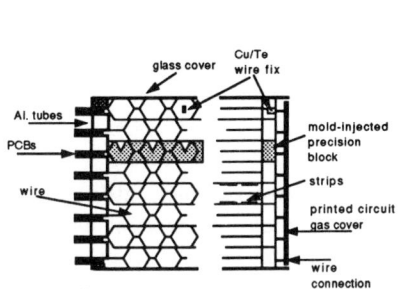
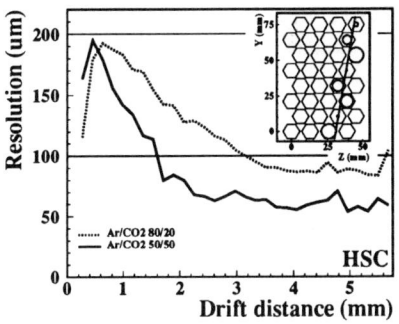

Figure 29. *Cross sections of the Honeycomb Strip Chamber. The left and right hand part of the figure are views perpendicular to the wires and the strips, respectively.*

Figure 30. *Measured spatial resolution of the drift coordinate as a function of the distance between the track and the wire, for two different gas mixtures.*

momentum resolution is determined by successive muon stations, that consist of two superlayers, providing two measurements each of resolution 100 μm per track. For the air-core barrel of ATLAS as shown in Figure 31 there are three such stations providing a sagitta measurement except in the region of the coils (~7% of azimuth) where a point-angle measurement will be made with ~2.5 times worse resolution. In the air-core end cap region three stations are arranged to make point-angle measurements only. The performance of the air-core toroid magnet muon system has been simulated to obtain the muon P_T momentum capability, shown in Figure 32.

The particle rate on the muon system has to be studied. The total absorption length in front of the toroid muon system allows the first muon station, which occurs

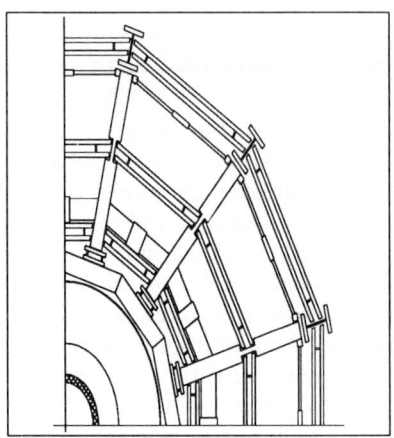 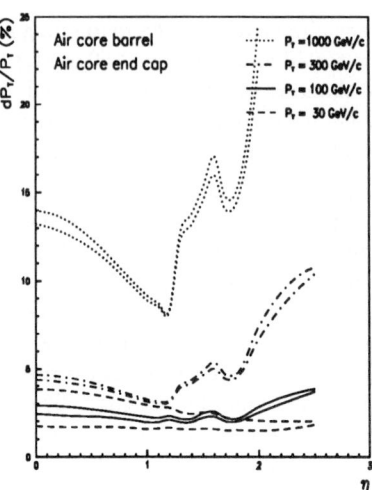

Figure 31. *Transverse view of possible chamber layout in the air-core barrel toroid.*

Figure 32. *Transverse momentum resolution in the air toroid barrel and warm iron end cap systems, as a function of η. All of the curves in this figure are for no inner tracking.*

after sufficient thickness of absorber material, to reduce fake muon rates appreciably. The simulations that were used to study these rates took into account decays of π's and K's in the inner detector, decay muons from hadronic showers (μ-punchthrough), heavy quark decays, IVB decays and Drell-Yan pairs. The results of this simulation are shown in Figure 33.

9 Trigger and data acquisition

The triggering of the ATLAS detector or any detector at a future hadron collider must face several major requirements:-

- selection of rare events in the presence of huge hadronic cross sections;
- overlapping and pile-up of minimum bias events with the 'interesting' event ($\sim$40 inelastic events/bunch crossing at $4\times10^{34}\mathrm{cm}^{-2}\mathrm{s}^{-1}$);
- multi-level trigger and pipeline electronics to handle rates and 15 ns bunch spacing.

The main problem may not be the 'soft' interactions but the large QCD multi-jet background which may fake the interesting physics signatures. The existence of reliable Monte Carlo tools to generate QCD processes is a question that has to be examined. As an example, at LHC energy, the total cross section for the production of a 500 GeV Higgs

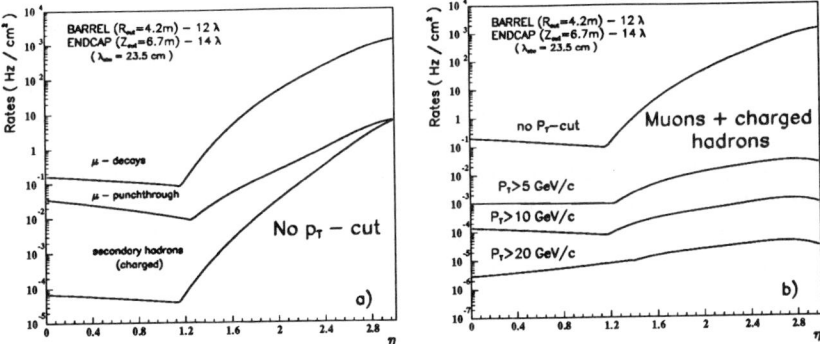

Figure 33. *Charged particle rates on the outer calorimeter surface: (a) all momenta, (b) for different cuts in P_T.*

is ~1pb whereas the total cross section for 250 GeV jet production is ~1μb. Large jet rejection is needed for a single electron trigger at low P_T which might come from an IVB.

General considerations for the trigger/data acquisition system (DAQ) at this time lead to the following conclusions:-

- It is too early for a top-down design due to:-
  - inadequacy of present technology;
  - poor definition of selection procedure and system requirements;
- At present the concentration on system elements appears more profitable:-
  - R & D projects for front-end electronics, trigger processors, readout and data acquisition;
  - more physics simulation of physics for trigger rates;
  - system modelling and behavioural simulation studies;

The event selection chain for the ATLAS detector is based on the following steps:-
- a level-1 trigger with negligible dead time and the shortest possible latency (less than 2μs), making an unambiguous identification of the bunch crossing containing the event of interest;
- a level-2 trigger with programmable algorithms based on local data;
- a level-3 trigger for which the full detector information will be used.

The trigger system is required to reduce a 67 MHz event rate, read from 10^7 detector channels (based on luminosity 1.7×10^{34} cm^{-2}s^{-1} and 20 events per bunch crossing) to 10–100 Hz at the final stage. The data rates are expected to decrease from 1 MB per event (at 67 MHz) to 10–100 MB/s at the final stage.

9.1 Level 1 trigger

The electromagnetic and hadronic calorimeters are included in the level-1 trigger. This trigger also uses information from forward regions. The trigger utilises reduced granularity ($\Delta\eta \times \Delta\phi \sim 0.1 \times 0.1$) and will sum over all samplings in depth for the calorimeters. Events with high P_T electrons, photons and jets, and large E_T^{miss} will be retained. The algorithm under study looks at a 2×1 or 1×2 window in η–ϕ that gives a sharp threshold even if the shower falls on the boundary between two cells, see Figure 34(a). An isolation option of 12 EM calorimeter cells plus 16 hadronic calorimeter cells will also exist, with cells of $E_T < 1$ GeV being ignored to reduce the effect of pile-up and noise. A jet trigger is constructed by using a reduced granularity in the hadronic calorimeter

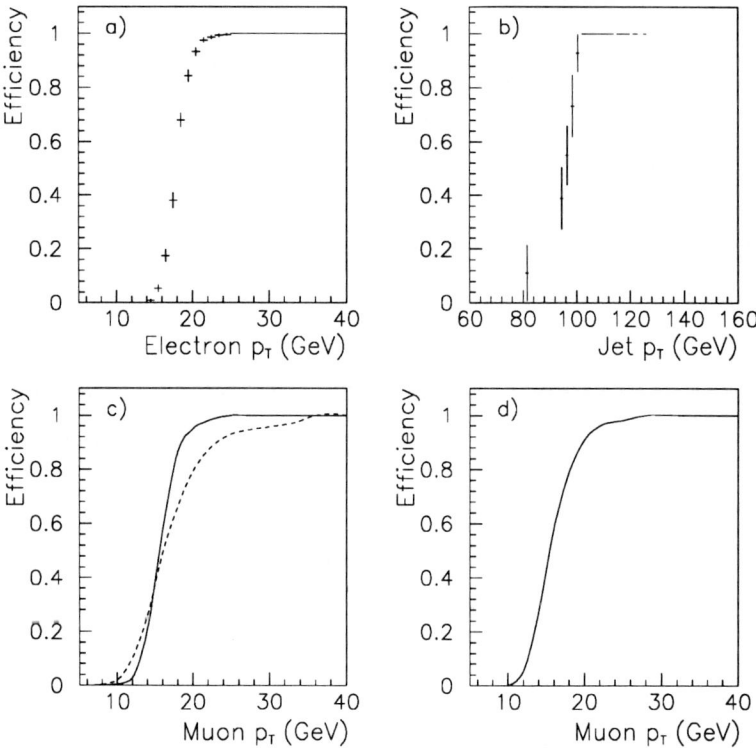

Figure 34. *Trigger efficiency versus P_T for nominal thresholds of: (a) $P_T = 20$ GeV electrons; (b) $P_T = 100$ GeV jets; (c) $P_T = 20$ GeV muons (iron toroid) – $|\eta| < 2.0$ and $2.0 < |\eta| < 2.5$ (dashed); (d) $P_T = 20$ GeV muons (barrel air-core toroid).*

of $\Delta\eta \times \Delta\phi \sim 0.3 \times 0.3$ and scanning with two overlapping windows of 2×2 cells to search for jet signals above a certain threshold. The results for the jet trigger efficiency for a 100 GeV threshold are shown in Figure 34(b).

A muon trigger is also included in the level-1 system. This trigger will provide the unambiguous identification of the bunch crossing with an intrinsic time resolution better than 5 ns. The muon trigger searches for tracks that point to the interaction region. The second and third measurements of the sagitta when compared to the first will reveal the particle bending angle. Using these measurements the particle track can be compared to the interaction region and a bunch crossing identified, see Figure 35. The muon trigger efficiency is shown in Figure 34(c) and (d).

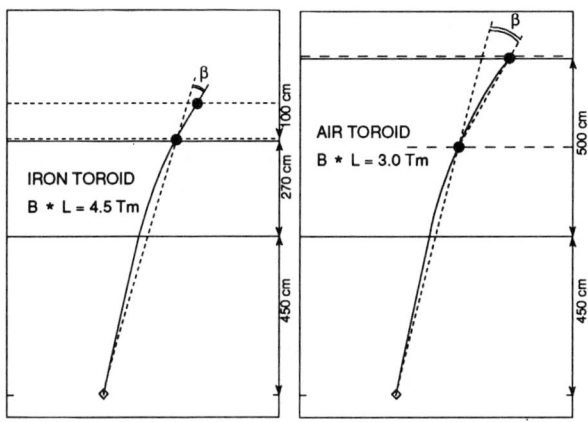

Figure 35. *Concept of triggering with toroids: β is the muon exit angle used in the trigger.*

The level-1 trigger rates, computed for luminosity $1.7 \times 10^{34} \text{cm}^{-2}\text{s}^{-1}$ including pile-up effects are shown in Figure 36. Examples of level-1 trigger rates for certain thresholds and signatures are given in Table 11.

Trigger	Rate
≥ 1 isolated EM cluster with $P_T > 40$ GeV (isolation not required for clusters with $P_T > 65$ GeV)	31 kHz
≥ 2 isolated EM clusters, each with $P_T > 20$ GeV (loose isolation cut)	16 kHz
$\geq 1\mu$ with $P_T > 20$ GeV	8 kHz
$\geq 2\mu$'s, each with $P_T > 20$ GeV	67 Hz
≥ 2 jets, each with $P_T > 200$ GeV	5 kHz

Table 11. *Example of level-1 trigger thresholds and associated rates.*

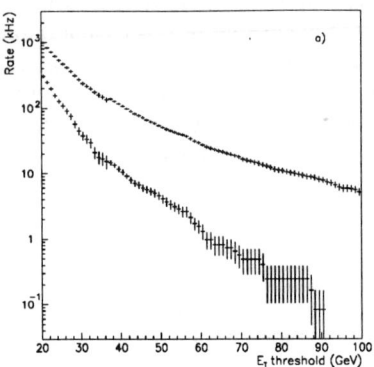 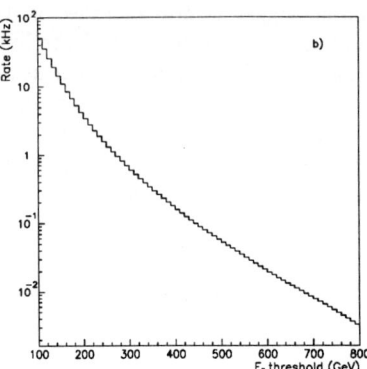

Figure 36. *Inclusive trigger rates as a function of P_T for (a) the e/γ trigger (with and without the isolation requirement), (b) the di-jet trigger.*

9.2 Level 2 trigger

The regions of interest of the detector identified by level-1 will be passed from the first to second level trigger, for example the calorimeter clusters and muon 'roads' to the inner tracking. The level-2 trigger is envisaged to consist of two kinds of processors, 'global' processors that act on the interesting 'features' passed by the 'local' processors. Cost effectiveness and ease of processing allow the use of buffers, that in turn allow the local level-2 processors to operate at about 1% of the rate of the level-1 trigger. The data throughput is greatly reduced by feeding local data to local processors only. These then extract the 'feature' of the local event and pass it to the global level-2 processor.

The level-2 track trigger will act on calorimeter and inner detector and improve the rejection of faked electrons (mainly π^0's) which pass the first level trigger. The rates of real electrons from W and Z particles and both c and b quark decays are of the order 200 Hz. This exceeds the capabilities of the final DAQ, so physics decisions have to be made at level-2. The strategy in this task is to search for a high P_T track in a narrow 'road' defined by the calorimeter cluster and match the track in the calorimeter space with E/p from the central tracking. The results of a simulation of the level-2 track trigger are shown in Figure 37.

The results of the simulations for the level-2 rejection for jets that have passed the level-1 electron cuts ($\varepsilon_e = 90\%$) are are shown in Table 12.

9.3 Level 3 trigger

The level-3 trigger is the final stage in the event selection chain. It reads out data over optical links from all detectors in local memories for events accepted by the level-2 trigger. The data from level-2 is fed into a system of switching networks for event building.

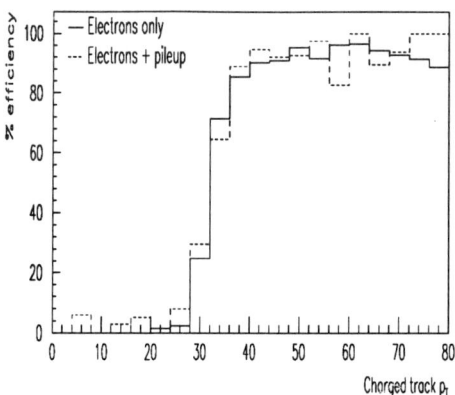

Figure 37. *Trigger efficiency versus electron P_T, with (dashed line) and without (solid line) pile-up background.*

	Rejection factor	
SIT track:	$R = 100$	(events with no high P_T charged hadron)
	$R = 35$	(all events)
TRD track:	$R = 40$	('tracking' requirement)
	$R = 150$	('electron' requirement)
Lvl-2 calo+Pre-shower +SIT track:	$R = 150\text{--}250$	

Table 12. *Expected level 2 rejection rate for jets passing the level 1 electron cuts.*

The level of event building required for ATLAS rules out existing and near-future bus-based event builders, and products currently being developed in telecommunications will provide an adequate solution.

After the events have been 'built' they are sent through a 'farm' ($\sim 10^6$ Mips) of RISC based processors that can run very sophisticated algorithms for further event selection. The final events are then sent for mass storage at an anticipated rate of 10–100 MB/s. Some of the features of the processing system under study are: the operating system (real time features and standardisation e.g. Unix), software development (software engineering, program language), the DAQ system (parallelism and scalability).

10 Simulated physics

Simulations of the response of the ATLAS detector to various physics signals have been made in order to understand the performance and sensitivity to detector parameters. The event generator generally used is PYTHIA except for the case of supersymmetry which uses ISASUSY. The decays gg→$Zb\bar{b}$ and W or Z+4 jets are simulated through matrix calculations, partly to check on PYTHIA. The detector is simulated through the use of SLUG, interfaced to GEANT giving a good description of the geometry and materials. The results of the simulations are quoted using the following assumptions:-

- $\int \mathcal{L} \mathrm{d}t = 10^5 \mathrm{pb}^{-1}$ for nominal high luminosity operation;
- lepton (e or μ) reconstruction efficiency ~90% (including trigger);
- signal is observable for a significance $> 5\sigma$;
- pile-up effects are included separately.

10.1 Top quark simulations

Top quark physics can be studied at LHC with low-luminosity from the beginning. Simulations include the top quark mass measurement and non-standard top decays. The rate of top quark production at LHC energy is expected to be very large even at low luminosity. For $\mathcal{L} = 10^{32} \mathrm{cm}^{-2}\mathrm{s}^{-1}$ and $m_t = 140 \,\mathrm{GeV}$ it is expected that there will be about 20 $t\bar{t}$ pairs produced per minute, 300 reconstructed $t\bar{t} \to \ell\nu$ and jjb events per day, and 30 clean isolated $e-\mu$ pairs produced per day.

A typical set of cuts for the reconstruction of hadronic top decays comprise:-

- A lepton, with $P_T > 40 \,\mathrm{GeV}$;
- At least three reconstructed jets within $|\eta| < 2$ with $P_T > 50 \,\mathrm{GeV}$ and $P_T^{2,3} > 40 \,\mathrm{GeV}$ in the hemisphere opposite to the lepton;
- Two of these jets are required to have an invariant mass within $\pm 20 \,\mathrm{GeV}$ of M_W.

The reconstructed mass is shown in Figure 38. The top signal can be extracted in this channel for an integrated luminosity of 30 pb^{-1}. For 10^3 pb^{-1} the statistical error on m_t is approximately 1 GeV and the expected systematic error is ±6 GeV.

The search for a charged Higgs particle using the decay $H^{\pm} \to c\bar{s}$ has been simulated and the result is shown in Figure 39. This channel is sensitive to the hadron calorimeter resolution, especially if $m_{H^{\pm}} \sim M_W$. Tagging of the b quark via impact parameter requirements is always useful in these searches (top, and H^+) to reject the non-top backgrounds. Non-standard top decay simulations include a search for $t \to bH^+$ at $\mathcal{L} = 10^{33} \mathrm{cm}^{-2}\mathrm{s}^{-1}$. This decay is observed mainly through a search for an excess of τ-leptons in top quark decays ($H^{\pm} \to \tau\nu$ in addition to $W^{\pm} \to \tau\nu$). The selection of $\tau \to \pi\nu$ entails decays with narrow hadronic jets and only one high P_T track. The signal expected, for $m_t = 140 \,\mathrm{GeV}$ and $m_{H^{\pm}} = 100 \,\mathrm{GeV}$, would be an excess of ~400 τ events above a background of ~400 W→τ and ~900 fake τ's. The statistical significance of τ excess is shown in Figure 40.

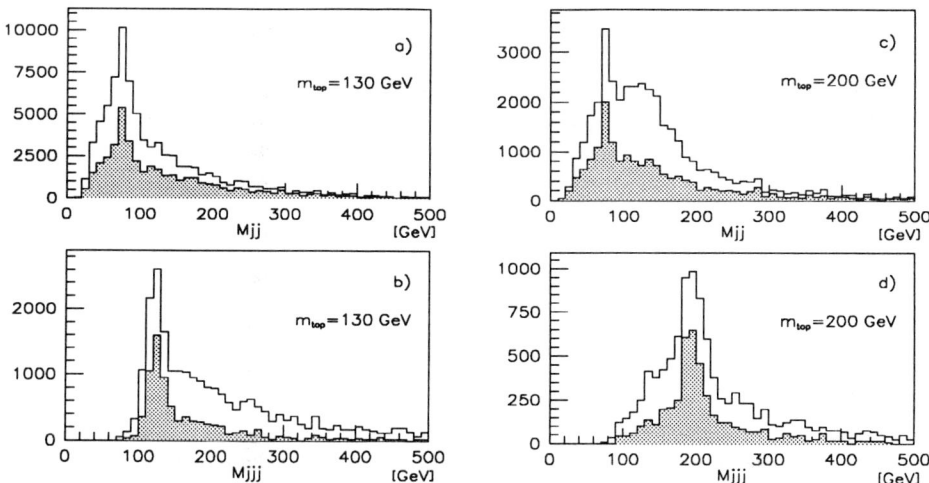

Figure 38. *Reconstructed 2 and 3-jet masses above the combinatorial background. Shaded (white) area shows $t\bar{t}$ events with (without) b-tagging. For $m_t = 130$ GeV (a,b) and $m_t = 200$ GeV (c,d).*

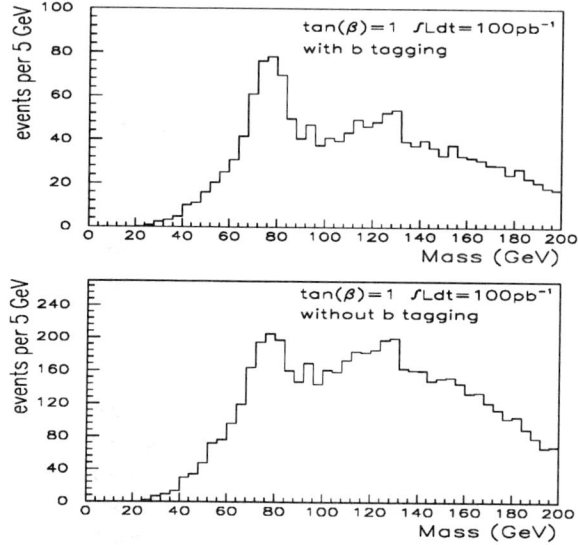

Figure 39. *Distribution of the reconstructed 2-jet mass for charged $H^+ \to c\bar{s}$ with $m_{H^+} = 130$ GeV, above the combinatorial background from $t\bar{t}$ events.*

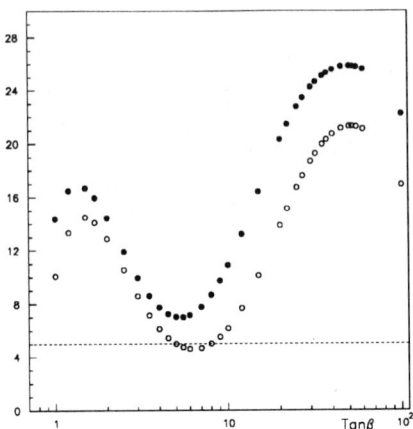

Figure 40. *Statistical significance of expected excess of τ-leptons from charged Higgs decay in $t\bar{t}$ events versus $\tan\beta$, for $m_{H^\pm} = 100\,\text{GeV}$, $m_t = 140\,\text{GeV}$ (full circles) and $200\,\text{GeV}$ (open circles) and an integrated luminosity of $10^4\,\text{pb}^{-1}$.*

10.2 Higgs simulations

A GEANT simulation of $H^0 \to \gamma\gamma$ has been made using 150,000 jets with $P_T > 40\,\text{GeV}$, with the following results. The background from jets is mainly of the form 'jet$\to\pi^0\to$fake γ-candidate'. The rejection, using calorimeter cuts alone, (~ 3000 for $P_T > 40\,\text{GeV}$) results in a fake $\gamma\gamma$ background which is about equal to the continuum $\gamma\gamma$ background. There is a clear need for additional separation of single γ from π^0 using a pre-shower detector (factor ~ 3).

For the Higgs mass range $80 < m_H < 100\,\text{GeV}$, and if $m_H = M_Z$ a significant background arises from the process $Z \to e^+e^-(\gamma)$. Since $\sigma_Z \cdot B(Z \to e^+e^-)/\sigma_H \cdot B(H^0 \to \gamma\gamma)$ is $\sim 25,000$, the resonant $Z \to e^+e^-$ background must be vetoed by a factor $\sim 250,000$. A particularly dangerous topology arises from internal or external bremsstrahlung with hard γ's. The required veto efficiency (per Z-leg) is $\sim 99.8\%$. Low energy electrons down to 1-2 GeV have to be reconstructed in order to approach the desired veto efficiencies. The observability of $H^0 \to \gamma\gamma$ in the ATLAS detector is summarised in Table 13 and the predicted observed signal shown in Figure 41.

The channel $H^0 \to ZZ^* \to 4$ charged leptons for the intermediate Higgs mass range has been studied. ATLAS has good sensitivity for $m_H \geq 130\,\text{GeV}$ with 60 to 180 reconstructed events for $10^5\,\text{pb}^{-1}$, after selection cuts, with an overall efficiency of $\sim 60\%$. The Higgs in this mass range should be narrow ($\Gamma_H < 300\,\text{MeV}$) so the signal significance will depend strongly on the detector resolution. The expected mass resolution σ_m in GeV, for $m_H = 150\,\text{GeV}$, and various detection methods in the ATLAS detector is shown in Table 14.

The backgrounds are large and uncertain, but the main one is expected to be $t\bar{t} \to 4$ charged leptons. This background was examined by generating without bias

	Higgs mass (GeV)				
	80	90	110	130	150
$\sigma_H B(H^0 \to \gamma\gamma)$ (fb)	51	57	68	70	35
acceptance (%)	23	30	41	46	51
mass resolution (%)	1.45	1.40	1.22	1.19	1.12
signal in mass bin	600	876	1430	1650	915
background in mass bin	36k	34k	25k	20k	13.5k
statistical significance	3.2	4.8	9.0	11.7	7.9

Table 13. *Observability of $H^0 \to \gamma\gamma$ in ATLAS (10^5pb^{-1}), direct production.*

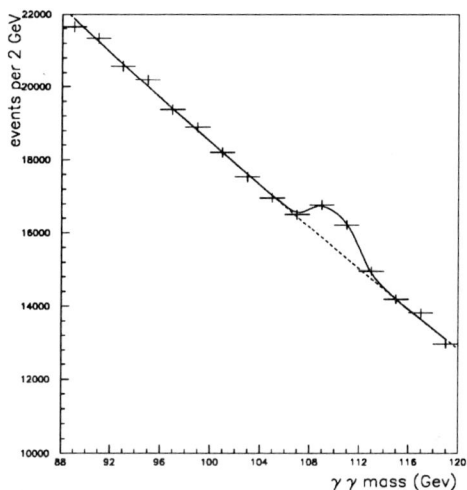

Figure 41. *Expected $m_{\gamma\gamma}$ spectrum for $H^0 \to \gamma\gamma$ signal above irreducible $\gamma\gamma$ background for $m_H = 110 \, \text{GeV}$ and 10^5pb^{-1}.*

5×10^6 $t\bar{t}$ events and studying the systematics due to top fragmentation in PYTHIA. This method extracted the best possible estimate of the $t\bar{t}$ background. This $t\bar{t}$ background was reduced by demanding one dilepton mass combination to be compatible

	GeV
electrons	1.8
muons stand-alone full air core	1.9
inner detector combined with muon system	1.9

Table 14. *The expected mass resolution σ_m in GeV, for $m_H = 150 \, \text{GeV}$, and various detection methods in the ATLAS detector.*

with the Z boson mass. The $Zb\bar{b}$ background was found to be underestimated by the parton shower model and so was generated using a $gg \to Zb\bar{b}$ matrix element calculation.

In order to establish a signal over the full intermediate mass range (120–180 GeV) lepton isolation is vital. The lepton isolation performance was estimated using a full detector simulation of the signal and background processes. From this simulation, isolation (using calorimeter and inner detector) achieves rejections of 30 (15) against $t\bar{t}$ ($Zb\bar{b}$) for $\varepsilon_H \sim 50\%$ including pile-up from 40 minimum bias events. The resulting simulated signal is shown in Figure 42.

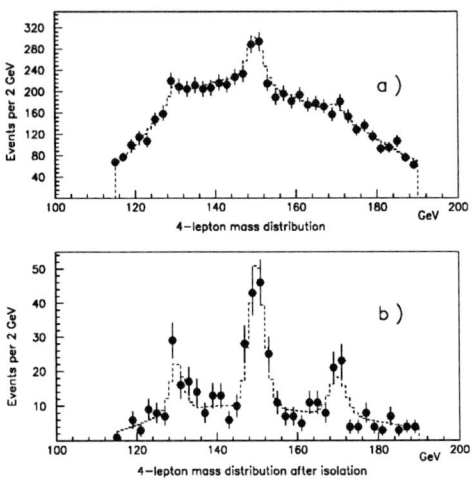

Figure 42. *Reconstructed 4-lepton invariant mass above background without (a) and with (b) isolation cuts, for m_H = 130, 150, 170 GeV. The dashed histogram represents the expected summed signal plus background and the dots show the result of a single experiment with an integrated luminosity of $10^5 pb^{-1}$.*

From this simulation we conclude that for $10^5 pb^{-1}$ the signal is observable over the full intermediate mass range ($m_H \geq 130$ GeV); with isolation, a signal is observable independently in the $H^0 \to 4\mu$ and $H^0 \to e^+e^-e^+e^-$ channels over part of the mass range; and for $10^4 pb^{-1}$ the signal is already observable over part of the mass range.

The response of the ATLAS detector to the heavy Higgs sector ($m_H > 2M_Z$) has been studied. The dominant decays in this mass range are $H^0 \to WW$ and ZZ with $ZZ \to 4\ell$, $2\ell\nu\bar{\nu}$ or $2\ell jj$ and $WW \to \ell\nu jj$. The sensitivity to these channels for large m_H is crucially dependent on the forward calorimetry and luminosity. Forward jet tagging to measure outgoing quark jets in $qq \to qqH^0$ production is considered in the simulation, and the coverage for E_T^{miss} is essential for $H^0 \to ZZ \to \ell\ell\nu\bar{\nu}$.

The potential of the ATLAS detector for this decay has been studied in full, especially the forward calorimeter performance (granularity and coverage), pile-up effects and realistic estimates of background rejection for the above channels. The results show that:–

- jet tagging will be a powerful tool to isolate WW, ZZ fusion processes and reduce the background;
- calorimeter crack effects increase the E_T^{miss} tail by ~10%;
- resolution of the calorimeters has small effect on the E_T^{miss} tail;
- η-coverage is crucial to reduce Z+jet background;
- pile-up effects are small;
- $H^0 \to ZZ \to \ell\ell\nu\bar{\nu}$ may extend the reach of $H^0 \to ZZ \to \ell\ell\ell\ell$ searches, both at high and low luminosities.

The resulting E_T^{miss} spectrum from the simulation is shown in Figure 43.

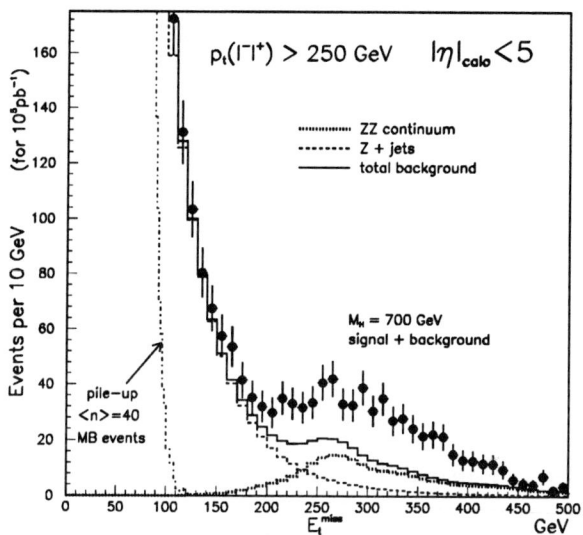

Figure 43. *Expected E_T^{miss} spectrum for the $H^0 \to ZZ \to \ell\ell\nu\bar{\nu}$ signal above various backgrounds. The full signal corresponds to the summed background and the expected signal for one experiment, for $m_H = 700$ GeV and 10^5pb^{-1}, is shown above this background.*

The usefulness of forward jet tagging is demonstrated in the simulation of the search for $H^0 \to WW \to \ell\nu jj$ when considering the signal over $t\bar{t}$ background as shown in Table 15.

The search for Higgs particle(s) in the Minimal Supersymmetric Standard Model (MSSM) has been studied and combines a search for the neutral (h, H^0, A) and charged $H^\pm$ Higgs particles, the remnants of the two Higgs doublets. The main sensitivity beyond LEP 200 which will be achieved by ATLAS at LHC stems from:-

- $H^0 \to \gamma\gamma$
  - this mode needs high luminosity and is usually depressed in rate with respect to SM Higgs, as well as being the most efficient for m_H not far from M_Z;

	Signal	$t\bar{t}$ background
Central cuts and reconstruction	640	240000
Single jet tag $E^j > 600\,\text{GeV}$ $P_T^j > 25\,\text{GeV}$	390	4500
Double jet tag $E^j > 600\,\text{GeV}$ $P_T^j > 25\,\text{GeV}$	90	200

Table 15. *Expected rates for $H^0 \to WW \to \ell\nu jj$ signal ($m_H = 1\,\text{TeV}$) and $t\bar{t}$ background, using jet tagging cuts.*

Particle Type	Signature
gluino/squark	E_T^{miss}+jets same sign dileptons multiple Z+jets+E_T^{miss}
sleptons	isolated high P_T leptons
chargino/neutralino	isolated high P_T leptons

Table 16. *Some supersymmetric channels and their signatures*

- $H^0 \to \tau\tau$ with at least one decay to leptons
  - This needs low luminosity and good calorimeter coverage to reconstruct $\tau\tau$ mass well, but it may be possible to include hadronic τ-decays for increased sensitivity;
- $H^0 \to ZZ^* \to 4\ell$
  - Limited due to depressed $H^0 \to ZZ^*$ rate.

The expected sensitivity of the ATLAS detector is parameterised in terms of m_A and $\tan\beta$, the ratio of the vacuum expectation values of the two Higgs doublets and the discovery curves (5σ) from simulation are shown in Figure 44.

10.3 Supersymmetry

Supersymmetry (SUSY) simulations have progressed furthest for the channels listed in Table 16. The channels that have the largest branching ratios and highest stability against SUSY parameters are E_T^{miss}+jets and same sign dileptons. Therefore these are the most important for a discovery of SUSY. The multiple Z signature is more sensitive to SUSY parameters and is expected to be important as a consistency check.

The simulation for gluinos and squarks in the E_T^{miss}+jets channel has been completed. The case of 5-jet events, where one jet falls in the calorimeter transition region between the barrel and the end-cap has been studied, and the simulated missing energy spectrum observed. The cross section of the signal is plotted against the mass of the gluino for $\tan\beta = 2$, $\mu = -440\,\text{GeV}$, $m_t = 140\,\text{GeV}$, $m_{H^+} = 500\,\text{GeV}$ in Figure 45.

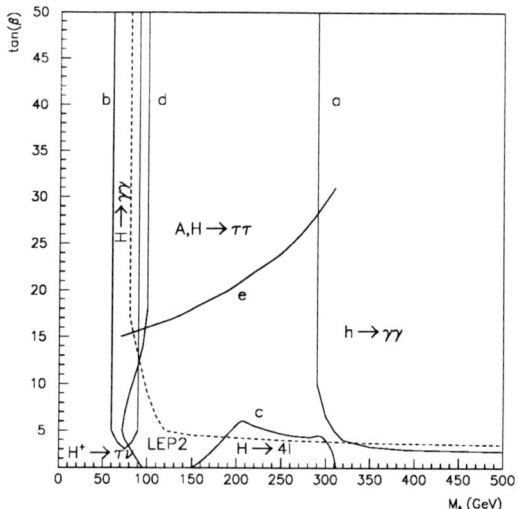

Figure 44. *Discovery contour curves (5σ) in the $m_A, \tan\beta$ plane for various Higgs signals in the MSSM. To the right of line (a) and inside (b) h$\to\gamma\gamma$ may be searched for. Under line (c) $H^0 \to 4\ell$ can be searched for. To the left of line (d) $t\to bH^\pm$ can be searched for. Above line (e) $H^0, A\to\tau\tau$ can be searched for.*

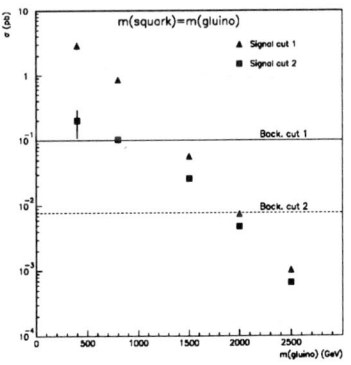

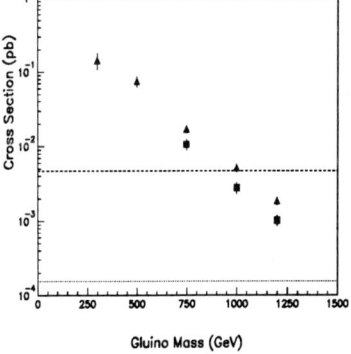

Figure 45. *Cross section of the signal, in the case of $m_{\tilde{q}} = m_{\tilde{g}}$, and the sum of backgrounds. Cut 1 : $E_T^{miss} > 300\,\mathrm{GeV}$. Cut 2 : $E_T^{miss} > 600\,\mathrm{GeV}$.*

Figure 46. *Cross section of the signal for $P_T^\ell > 30\,\mathrm{GeV}$ (triangles) and $50\,\mathrm{GeV}$ (squares). Also shown is the remaining $t\bar{t}$ background, for $P_T^\ell > 30\,\mathrm{GeV}$ (dashed) and $50\,\mathrm{GeV}$ (dotted).*

The mass reach for the search for gluinos and squarks through this channel, with a $S/\sqrt{B}$ ratio > 5 and various values of integrated luminosity is given in the Table 17.

		10^3pb^{-1}	10^4pb^{-1}	10^5pb^{-1}
$m_{\tilde{g}} \sim m_{\tilde{q}}$	$m \leq$	1.6 TeV	2.0 TeV	2.3 TeV
$m_{\tilde{g}} = 2 \times m_{\tilde{q}}$	$m_{\tilde{q}} \leq$	0.75 TeV	1.0 TeV	1.2 TeV
$m_{\tilde{q}} = 2 \times m_{\tilde{g}}$	$m_{\tilde{g}} \leq$	1.0 TeV	1.25 TeV	1.4 TeV

Table 17. *The mass reach for the search for gluinos and squarks*

The search has been performed for gluinos in events with same sign dileptons arising from same sign W's produced in the cascade decays of $\tilde{g}\tilde{g}$. The main background considered is from leptons from the decay $t\bar{t} \to W^+W^-b\bar{b}$ producing real, or fake same sign leptons. The cross section of the signal versus the gluino mass for events that met the following criteria, $P_T^\ell > 30\,\text{GeV}$ and $|\eta| < 2.5$ with $E_T^{\text{miss}} > 100\,\text{GeV}$, is shown in Figure 46.

10.4 Search for new vector bosons

The capability of the ATLAS detector to observe new (neutral or charged) vector bosons that occur naturally in some minimal extensions of the Standard Model, and in models for electroweak symmetry breaking through compositeness has been investigated. The detector is sensitive to $Z' \to e^+e^-$, $\mu^+\mu^-$, jj and $W' \to e\nu$, $\mu\nu$, jj with the best reach achieved through $Z' \to \ell\ell$ and $W' \to e\nu$.

Using the Extended Gauge Model ($\Gamma_{Z'} = \Gamma_Z \times M_{Z'}/M_Z$) the reconstructed mass of a Z' of 4.5 TeV from the $Z' \to \ell\ell$ channel is visible over a small Drell-Yan production background. The result is is shown in Figure 47.

The forward-backward asymmetry is a quantity which is sensitive to the model used. Figure 48 shows the expected forward-backward asymmetries, for $M_{Z'} = 3\,\text{TeV}$. This simulation gave a result of ~5200 events per year and a large observable asymmetry with the a small distortion of the asymmetry due to charge misassignment. This distortion is expected to be negligible in the $\mu\mu$ channel.

A signal for the channel $Z' \to 2$ jets can also be extracted, between 1 TeV (limited by trigger rate) and 3 TeV (limited by signal rate), from the large QCD background. This channel is largely sensitive to the performance of the hadronic calorimetry, in particular the constant term, as shown in Table 18.

A summary of the discovery potential of the ATLAS detector for $Z' \to \ell\ell$, jj is given in Figure 49.

10.5 Gauge boson pair production

Pair production of gauge bosons is an essential test of electroweak theory through the three vector boson coupling. The gauge cancellations predicted by the Standard Model can be studied and possible anomalous couplings detected. Under some general

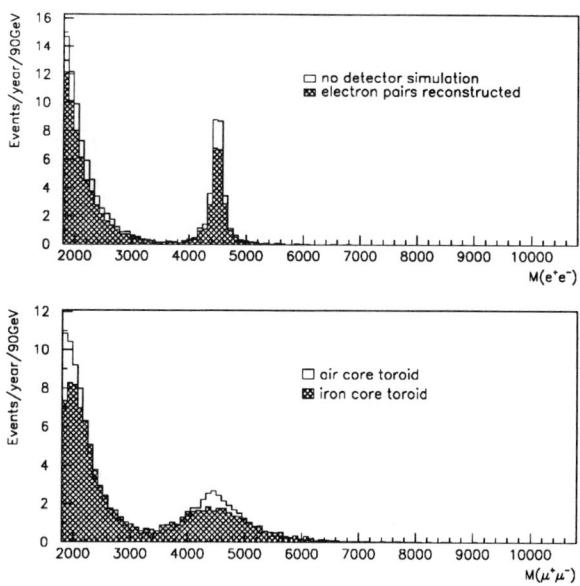

Figure 47. *Reconstructed dilepton mass for $Z' \to e^+e^-$ and $\mu\mu$ decays; $M_{Z'} = 4.5\,\mathrm{TeV}$.*

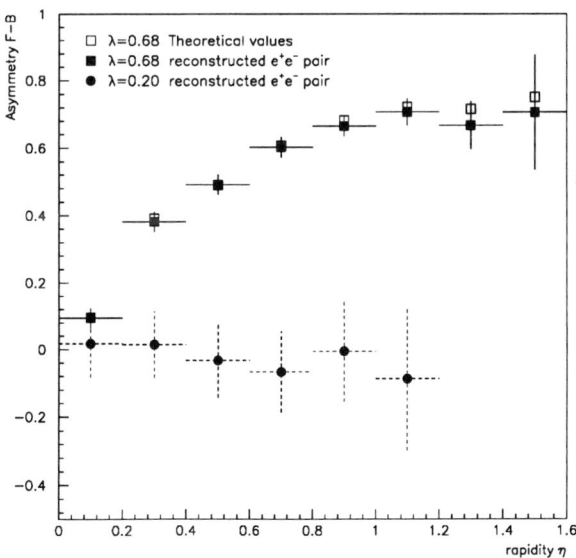

Figure 48. *Rapidity dependence of forward-backward asymmetry for $M_{Z'} = 3\,\mathrm{TeV}$ and two values of λ.*

$\sigma_E/E = a/\sqrt{E} \oplus b$		$\sigma_M/M(\%)$	$S/\sqrt{B}$
a(%)	b(%)		
0	0	2.6	7.6
30	1	3.0	7.5
50	2	3.4	7.4
50	3	4.5	6.8
50	4	5.4	6.1
100	2	3.8	7.2
100	4	5.1	6.0

Table 18. *Expected di-jet mass resolution and statistical significance of signal, for $M_{Z'} = 2\,\mathrm{TeV}$ and various stochastic and constant terms in the calorimeter jet resolution.*

assumptions the anomalies can be described by two parameters, κ and λ (the Standard model values are $\kappa = 1$ and $\lambda = 0$).

The Wγ channel has been simulated using leptonic W decays and observing the photon transverse momentum (P_T^γ), which is very sensitive to any anomalous couplings. The main backgrounds for this channel are $t\bar{t}\gamma$ and $b\bar{b}\gamma$ and a rejection of 10^4 is required against jets.

The simulation concluded that the $t\bar{t}\gamma$ background remained dominant after lepton isolation cuts against $b\bar{b}\gamma$ background. The channel was found to have good sensitivity to κ, see Figure 50. An excess of $\sim$160 events was observed over a total background of $\leq$1000 events for $\kappa = 1.1$.

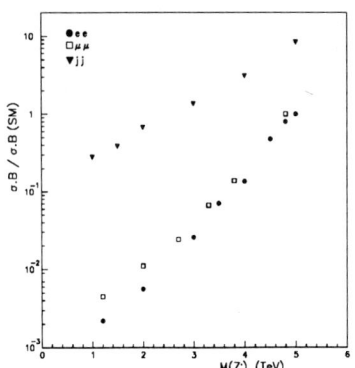

Figure 49. *Discovery mass limits for $Z' \to \ell\ell, jj$.*

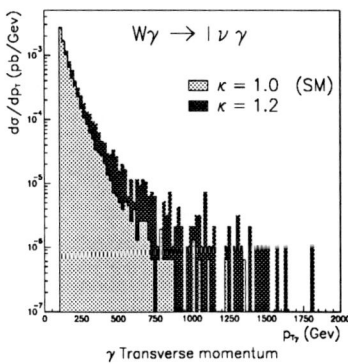

Figure 50. *Transverse momentum of γ in Wγ events for $\kappa = 1.0$ (light grey) and 1.2 (dark grey).*

10.6 $W_L Z_L$ and $Z_L \gamma$ resonances

A study was made of the ability of the ATLAS detector to examine the effects due to alternative symmetry breaking mechanisms such as a techni-rho (ρ_{tc}) that decays to the $W_L Z_L$ resonance, and techni-omega (ω_{tc}) which decays to the $Z_L \gamma$ resonance.

The $W_L Z_L$ resonance has been simulated with a signal cross section of 40 fb, produced from a combination of $W_L Z_L \rightarrow W_L Z_L$ scattering and $q\bar{q}' \rightarrow W_L Z_L$ production. The dominant background is from $t\bar{t}$ decays with three charged leptons in the final state and one dilepton combination close to M_Z and also from continuum $q\bar{q}' \rightarrow WZ$ production. The $W_L Z_L$ resonance is searched for with three charged leptons in the final state (e or μ), with the neutrino 4-momentum reconstructed using E_T^{miss} and kinematics. The reconstructed $W_L Z_L$ mass is shown in upper part of Figure 51.

The $Z_L \gamma$ resonance has been simulated by reconstructing the final states of Z decays to leptons. The dominant sources of background are continuum $Z\gamma$ production and Z+jet production where one jet fakes a γ. It has been found that a rejection of 1000 against jets is sufficient to reduce the second background to negligible levels. The lower part of Figure 51 shows the simulated reconstructed mass distribution of the $Z_L \gamma$ resonance.

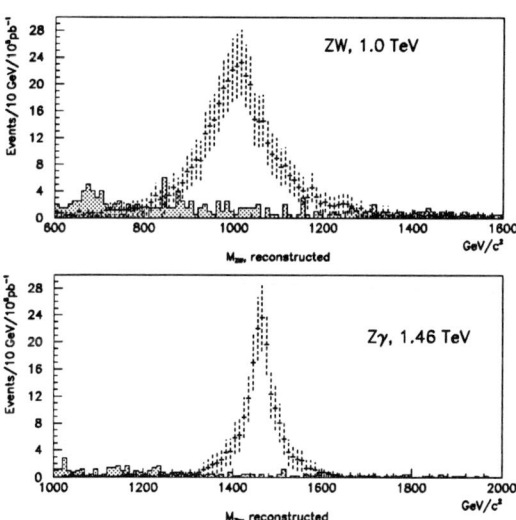

Figure 51. *Reconstructed masses for high mass resonances decaying into gauge boson pairs: (top) a 1 TeV $\rho_{tc} \rightarrow 3\ell$ decays; (bottom) a 1.5 TeV $\omega_{tc} \rightarrow Z\gamma$ with $Z \rightarrow 2\ell$ decays.*

10.7 Search for leptoquarks

The results are presented of a study of the 'gold-plated' pair production channel $pp \rightarrow S\bar{S} \rightarrow eq\bar{e}\bar{q}$. A search for a peak is made in the reconstructed lepton-jet mass for events with two leptons and two jets with $|\eta| < 2.5$ and $P_T > 200$ GeV. The dominant background to this channel is $t\bar{t} \to e^+e^- +$jets.

The results of this simulation show that this channel is sensitive to calorimeter resolution, and a mass reach up to $m_S = 1.5$ TeV is possible. The resulting reconstructed mass for differing calorimeter resolutions is shown in Figure 52.

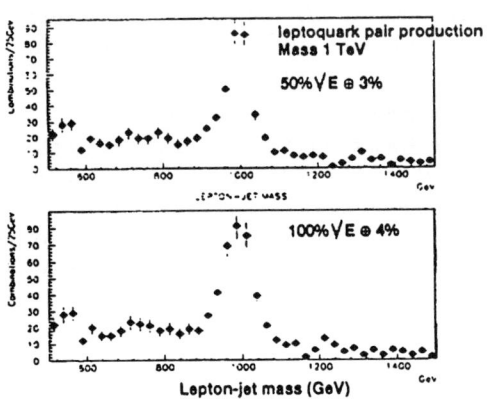

Figure 52. *Reconstructed lepton-jet mass for leptoquark candidate events.*

11 Concluding remarks

The next round of experiments to be performed at the future hadron colliders, SSC and LHC, present the experimentalist with a severe set of problems. Detailed physics simulation has shown that the main physics signals remain secure, with practical assumptions made for granularity and resolution within the detectors. Strong programs of research and development in detector technology are proving that the claimed performance specifications can indeed be achieved. However the very large number of channels, speed of processing, and above all the radiation hardness, present a readout challenge which will probably only be met by close collaboration between the physicists and the electronics industry.

Acknowledgement

The text has been compiled mainly by my young colleague John White of the University of Victoria, to whom I am extremely grateful. I am also very grateful for the help and constructive criticism we have received from Alan Astbury and Michel Lefebvre.

Beyond the Standard Model

Roberto Peccei

Department of Physics, University of California Los Angeles
Los Angeles, California 90024-1547

1 Introduction

In these lectures I describe some of the open questions in the standard model relating to the nature and origin of mass, forces and matter and discuss some of the speculative theoretical ideas put forth in this regard. Some of the topics touched upon include supersymmetry, dynamical symmetry breaking, composite models, grand unified theories and superstrings.

The standard model (SM), the $SU(3) \times SU(2) \times U(1)$ gauge theory of the strong and electroweak interactions, has proven to be extraordinarily robust. To date, all data are in perfect accord with its predictions, with at best very indirect hints for any physics beyond the SM. This is well illustrated by Figure 1 where measurements of the weak mixing angle done in experiments over an energy range from a few eV to 100 GeV are shown. As can be seen, all results are perfectly consistent with each other.

Nevertheless, there are a number of important features in the SM which are poorly understood and some open questions whose answers necessarily can only be found by going beyond the SM. To organise the discussion, it seems useful to divide the issues one wants answers to into two kinds: central issues and side issues.

1.1 Central issues

These are questions whose answers are fundamental for our understanding of the SM and which, most likely, will require going beyond the SM for a proper answer. Broadly speaking there are three such central questions:

1. The Question of Mass;
2. The Question of Forces;
3. The Question of Matter.

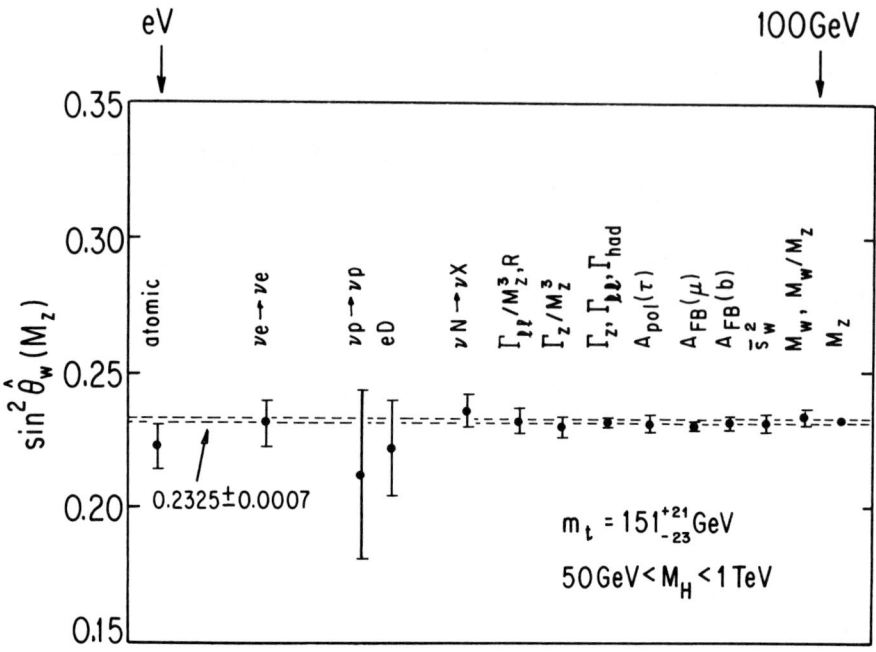

Figure 1. *Measured values of the weak mixing angle, from Langacker 1993.*

I will spend considerable time in what follows defining these questions more clearly and trying to describe suggestive answers to them.

1.2 Side issues

These are issues in the SM which are either open or not addressed, whose answer may involve extensions of the SM or, when better understood, may have an influence on the deeper central issues. Some examples of questions in this class are:

1. Do neutrinos have mass?

2. What is the dark matter in the universe?

3. What is the origin of the matter-antimatter asymmetry in the universe?

4. Why is there no strong CP violation?

5. Is the observed CP violation in the kaon system only a result of quark mixing?

In these lectures, I will mostly focus on the central open questions in the standard model, returning to discuss when appropriate how the answers to these main questions may impinge on these side issues, and vice versa. I will also ignore, for the most part, open calculational problems in the SM. Although problems like calculating the spectrum of hadrons from QCD, or the matrix elements of weak operators between hadronic states, are of great practical importance they are, in some sense, not open problems. Eventually, as our calculational tools get further refined, there is hope that these problems at least will be resolved.

2 The question of mass

One of the fundamental attributes of both the elementary excitations (quarks, leptons, and gauge bosons) and composite states (protons, neutrons, pions,...) present in the SM is that all of these particles have mass. However, the SM possesses only dimensionless coupling constants (g_1, g_2, g_3) and no explicit mass parameters at all! How then are these masses generated?

In natural units, $\hbar = c = 1$, the action S is dimensionless. Since

$$S = \int d^4x \, \mathcal{L},$$

one sees that the Lagrangian density $\mathcal{L}$ scales like (mass)^4. From the kinetic energies of fermions

$$\mathcal{L}_{\text{kinetic}} = -\bar{\psi}\gamma^\mu \frac{1}{i}\partial_\mu \psi$$

and gauge fields

$$\mathcal{L}_{\text{kinetic}} = -\frac{1}{4}(\partial^\mu A^\nu - \partial^\nu A^\mu) \cdot (\partial_\mu A_\nu - \partial_\nu A_\mu) + \cdots$$

one learns that $\psi \sim$ (mass)$^{3/2}$, while $A_\mu \sim$ (mass)^1. Interactions in the SM are introduced via minimal substitution

$$\partial_\mu \to \partial_\mu - ig_i A_{i\mu}$$

and so the coupling constants, g_i, are dimensionless. Obviously no mass scales enter in $\mathcal{L}_{\text{kinetic}}$ or in $\mathcal{L}_{\text{interaction}}$. Furthermore, no explicit mass terms are allowed, both theoretically and experimentally, in the standard model since

(i) the gauge principle, $S \to S$ as $A^\mu \to A^\mu - \frac{1}{g}\partial^\mu \Lambda$, forbids explicit mass terms for the SU(3)×SU(2)×U(1) gauge bosons;

(ii) the observed (V−A) nature of the weak interactions forbids fermion masses to appear.

This last point is readily seen. Recall that, under SU(2)×U(1), for example $\begin{pmatrix} \nu_e \\ e \end{pmatrix}_L$ forms a doublet while e_R is a singlet. Obviously, then, having an explicit electron mass,

$$\mathcal{L}_{\text{mass}} = m\bar{e}e = m\bar{e}_L e_R + m\bar{e}_R e_L,$$

would violate SU(2) symmetry.

Although the asymmetric SU(2) assignments for electrons prevent the above mass term, it is not quite true that all mass terms are forbidden by the nature of the weak interactions. One must really distinguish between two types of mass terms: Dirac masses, and Majorana masses. A Dirac mass is of the form

$$\mathcal{L}_{\text{mass}}^{\text{Dirac}} = m_D(\overline{\psi}_L \psi_R + \overline{\psi}_R \psi_L).$$

Clearly m_D is forbidden by the SU(2) assignments of fermions in the SM, since ψ_L is a doublet, while ψ_R is a singlet. In addition, however, one can have Majorana masses which involve

$$\mathcal{L}_{\text{mass}}^{\text{Majorana}} = \frac{m_R}{2}\left(\psi_R^T C \psi_R + \overline{\psi}_R C \overline{\psi}_R^T\right) + \frac{m_L}{2}\left(\psi_L^T C \psi_L + \overline{\psi}_L C \overline{\psi}_L^T\right),$$

where C is the charge conjugation matrix. In the SM, m_L is forbidden by SU(2)×U(1), for all fermions. However, m_R is allowed by SU(2) and by U(1) for the right-handed neutrinos, ν_R, since right-handed neutrinos have no SU(2)×U(1) quantum numbers at all. Therefore it is possible to have a Majorana mass term in the SM of the form

$$\mathcal{L}_{\text{mass}}^{\text{Majorana}} = \frac{m_R^\nu}{2}\left\{\nu_R^T C \nu_R + \overline{\nu}_R C \overline{\nu}_R^T\right\}.$$

Thus, more correctly, m_R^ν is the only possible explicit mass parameter in the SM. [To be more precise, m_R^ν is a 3×3 mass matrix, not a single mass parameter.] It could very well be zero but, as we shall argue later, is probably very large. At any rate, this parameter does not appear to play a fundamental role except for neutrinos.

Even though $\mathcal{L}_{\text{SM}}$ has no explicit mass scales in it, besides m_R^ν, two dynamical scales enter in the theory whose origins are completely different. These scales are Λ_{QCD}, the dynamical scale associated with the unbroken colour SU(3) theory, and v_F, the Fermi scale—generated as the result of the spontaneous breakdown of the SU(2)×U(1) electroweak group down to U(1)$_{\text{em}}$. Let me briefly discuss, in turn, the origins of these two fundamental scales.

2.1 The origin of Λ_{QCD}

Although coupling constants are dimensionless, they are not really constants but depend on the momentum scale where they are measured. Examining corrections at higher order, one sees that the effective coupling constant in QCD is a running coupling constant. This behaviour arises from quantum corrections to Green's functions in field theory. Green's functions contain, in general, infinities when computed perturbatively. These infinities can be removed by rescaling the fields and couplings,

$$A^\mu = \sqrt{Z_3}\, A_{\text{Ren}}^\mu, \qquad g_3 = Z_g\, g_{3,\text{Ren}},$$

and specifying the value of certain Green's functions at some normalization point. For example, in QCD one fixes the three-gauge vertex of Figure 2 to have the strength $Z_g^{-1} Z_3^{-3/2}$ at $p^2 = q^2 = r^2 = \mu^2$, a convenient normalization point. Physical amplitudes should not depend on the arbitrary normalization point μ^2 which one chooses. Although μ^2 is arbitrary, the response of the theory to μ^2 changes gives physical information.

Figure 2. *Three-gauge vertex in* QCD.

This information is codified in the Renormalization Group Equations (RGE) (see, for example, Zinn Justin 1989).

The RGE arise from studying the μ^2-dependence of the Green's functions in the theory. The relationship between the renormalised (finite) Green's functions and the original Green's functions is μ^2-dependent. For instance, the renormalised gluon propagator in QCD is related to the original propagator by the equation

$$G^{(2)}_{\text{Ren}}(\mu^2) = Z_3^{-1}(\mu^2) G^{(2)}.$$

The dependence of $G^{(2)}_{\text{Ren}}$ on μ^2 enters both explicitly and through the coupling constant dependence on μ^2, $g_{3,\text{Ren}}(\mu^2)$. Since $G^{(2)}$ is independent of μ, it follows that

$$\left[\mu\frac{\partial}{\partial\mu} + \beta(g_{3,\text{Ren}})\frac{\partial}{\partial g_{3,\text{Ren}}} - \mu\frac{\partial}{\partial\mu}\ln Z_3^{-1}(\mu^2)\right] G^{(2)}_{\text{Ren}}(\mu^2) = 0.$$

This is the RGE for the gluon propagator. Here $\beta(g_{3,\text{Ren}})$ describes how the coupling depends on the normalization point μ^2:

$$\beta(g_{3,\text{Ren}}) = \mu\frac{\partial}{\partial\mu}g_{3,\text{Ren}}(\mu^2).$$

Thus, the β-function effectively fixes the way in which the Green's functions depend on μ^2. β can be calculated perturbatively from the defining equation, using

$$g_{3,\text{Ren}}(\mu^2) = \frac{1}{Z_g(\mu^2)} g_3$$

along with a perturbative expansion of Z_g in g_3.

Because QCD has no intrinsic scale, the μ^2 dependence is related to the q^2 dependence of the propagator:

$$G^{(2)}_{\text{Ren}}[\mu^2] \equiv G^{(2)}_{\text{Ren}}[q^2/\mu^2; g_{3,\text{Ren}}(\mu^2)].$$

In turn, the momentum dependence of the Green's functions is related to the scale dependence of the effective coupling $\bar{g}_3(q^2)$. This effective coupling constant—or running coupling constant—is defined through the equation (see Zinn Justin 1989)

$$\left[q^2\frac{\partial}{\partial q^2} - \beta(g_{3,\text{Ren}})\frac{\partial}{\partial g_{3,\text{Ren}}}\right]\bar{g}_3(q^2) = 0,$$

along with the boundary condition that the effective coupling specified at some scale, q_0^2, is $g_{3,\text{Ren}}$. That is,
$$\bar{g}_3(q_0^2) = g_{3,\text{Ren}}.$$
In QCD the β-function $\beta(g_{3,\text{Ren}})$ can be computed perturbatively and one finds (see Gross and Wilczek 1973, Politzer 1973)
$$\beta(g_{3,\text{Ren}}) = -b_3 \frac{g_{3,\text{Ren}}^2}{16\pi^2} + \cdots,$$
where
$$b_3 = \left[11 - \tfrac{2}{3} N_{\text{quarks}}\right].$$
Using this result, the running coupling constant squared for QCD is seen to be
$$\alpha_3(q^2) \equiv \frac{\bar{g}_3^2(q^2)}{4\pi} = \frac{\alpha_3(q_0^2)}{1 + \dfrac{b_3}{4\pi}\alpha_3(q_0^2)\ln\left(\dfrac{q^2}{q_0^2}\right)}.$$
The running coupling constant squared $\alpha_3(q^2)$ given by the above equation is sketched in Figure 3. The running of this coupling constant with q^2 describes qualitatively two important features of the strong interactions:

(i) We do not observe free quarks and gluons but only hadrons (*confinement*), presumably because at small values of q^2 the coupling constant becomes strong.

(ii) At short distances—large q^2—hadrons appear to be composed of essentially free quarks (*asymptotic freedom*).

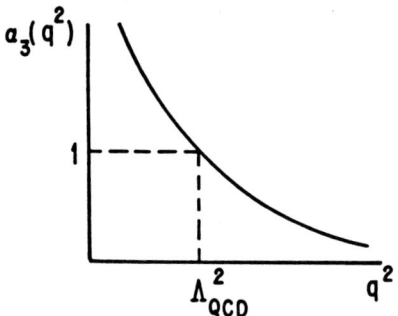

Figure 3. *Behaviour of the QCD running coupling constant with q^2.*

We can define Λ_{QCD} as the scale where $\alpha_3 \approx 1$. Clearly, this is a truly dynamical scale. A measurement of this scale, from deviations at high q^2 of deep inelastic scattering of hadrons from free quark behaviour, gives (for a recent discussion of the QCD running coupling constant, and a technically more precise definition of Λ_{QCD}, see Altarelli 1989)
$$\Lambda_{\text{QCD}} \sim 200\text{--}300 \text{ MeV}.$$
In principle, when the confinement problem is finally calculationally under control, we should be able to compute hadron masses (protons, pions, *etc.*) directly in terms of Λ_{QCD}. [This will not be quite true for bound states involving heavy quarks, since there the masses of the heavy quarks contribute non-negligibly to the corresponding hadronic masses.]

2.2 The origin of the Fermi scale v_F

Although SU(2)×U(1) is a good symmetry of $\mathcal{L}_{SM}$, this symmetry is not respected by the vacuum, leading to the symmetry breakdown:

$$SU(2) \times U(1) \to U(1)_{em}.$$

This breakdown is associated with a physical scale, the Fermi scale v_F. All the masses of elementary excitations in the standard model (quarks, leptons, and gauge bosons) are proportional to v_F, with

$$v_F = \left(\sqrt{2}\, G_F\right)^{-1/2} \approx 250\,\text{GeV}.$$

The role of symmetry breakdown as a generator of mass scales is familiar in superconductivity where a Cooper pairing between opposite spin electrons, $\langle \psi \uparrow \psi \downarrow \rangle$, leads to a mass gap between the normal and the superconducting ground states (see, for example, Anderson 1984). Similarly, here v_F is the scale of the order parameter of the breakdown of SU(2)×U(1)→U(1)$_{em}$. One of the most crucial questions in the electroweak theory is 'what is the origin of this order parameter?'

For the electroweak theory two alternatives have been proposed for the Fermi scale.

(i) The simplest of these is that the Fermi scale v_F is connected to the vacuum expectation value (VEV) of an elementary scalar field.

(ii) Alternatively, it has been proposed that the Fermi scale v_F is associated with the scale of some dynamical condensate of fermions from some deeper theory.

As we shall see in what follows, the physical consequences of v_F being the VEV of some elementary scalar field are quite different from those of having v_F being associated with some underlying fermion-antifermion condensate, $\langle \bar{F} F \rangle$.

For SU(2)×U(1) there is, however, another perhaps deeper mystery. Although both the gauge boson masses and the masses of all fermions are proportional to v_F:

$$M_W \propto v_F; \qquad m_f \propto v_F,$$

there is no real understanding why $M_W \sim 100\,\text{GeV}$, while most fermion masses (except for the top!) have $m_f \ll v_F$. Furthermore, there is also no deep understanding of the hierarchies among fermion masses; why, for example, $m_\tau : m_\mu : m_e$ is approximately in the ratio of 3500 : 200 : 1?

In the simplest version of the SM (Weinberg 1967) the possibility of inducing the breakdown

$$SU(2) \times U(1) \to U(1)_{em}$$

is effected through the introduction of an elementary self-coupled scalar field, Φ, which is a complex SU(2) doublet. This field is assumed to have a potential

$$V = \lambda \left(\Phi^\dagger \Phi - \frac{v_F^2}{2}\right)^2,$$

which leads to the symmetry breakdown. The gauge boson masses, M_W and M_Z, arise from the couplings of this scalar field to the SU(2)×U(1) gauge fields. The replacement of Φ by its VEV, $\langle\Phi\rangle = v_F/\sqrt{2}$, in the seagull graph of Figure 4 gives

$$M_W = \frac{g_2 v_F}{2}.$$

Since the SU(2) coupling constant g_2 is 'known', M_W is fixed directly from a knowledge of the Fermi scale, v_F.

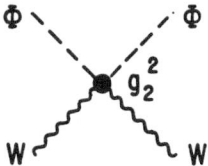

Figure 4. *Seagull graph which gives the W bosons a mass when Φ is replaced by its VEV.*

The fermion masses also originate from Φ, but they require new couplings. These are the so-called Yukawa couplings Γ_f. These couplings connect, by means of the doublet Higgs field Φ, the SU(2) singlet right-handed fermions with the left-handed fermions which are in doublets. When Φ gets replaced by its VEV this leads to a mass term

$$\mathcal{L}_{\text{mass}}^{\text{fermion}} = \frac{(\Gamma_f v_F)}{\sqrt{2}}(\bar{f}_L f_R + \bar{f}_R f_L).$$

Even though all fermion masses are proportional to v_F, they are not determined in the standard model because the Yukawa couplings Γ_f are 'unknown'. Since the Γ_f are arbitrary, any prediction for the fermion masses clearly requires physics beyond the SM.

I will return to the question of fermion masses later, but first I want to ask another question: 'what physics fixes the scale v_F?' By itself, since v_F is the only scale in the standard model, the question makes no sense. However, if we consider the SM within a broader context, as an effective theory up to some cut-off Λ (e.g. $\Lambda = M_{\text{Planck}}$), then it makes sense to ask for the relationship between v_F and Λ. Indeed, because the $\lambda\phi^4$ theory is 'trivial' (Aizenman 1982) [*i.e.* the only consistent version of the theory has $\lambda_{\text{Ren}} = 0$], considering scalar interactions in the SM without some high energy cut-off is itself questionable.

In a $\lambda\phi^4$ theory—unlike what happens in QCD—the effective coupling constant $\lambda(q^2)$ *increases* with increasing q^2. The solution of the RGE

$$\frac{d\lambda}{d\ln q^2} = \frac{3\lambda^2}{4\pi^2} + \cdots,$$

neglecting the higher order contributions, is

$$\lambda(q^2) = \frac{\lambda(\Lambda_0^2)}{1 - \frac{3\lambda(\Lambda_0^2)}{4\pi^2}\ln\left(\frac{q^2}{\Lambda_0^2}\right)},$$

which blows up at the location of the Landau pole (Landau 1955)

$$\Lambda_c^2 = \Lambda_0^2 \, e^{4\pi^2/3\lambda(\Lambda_0^2)}.$$

The behaviour of $\lambda(q^2)$ versus q^2 is sketched in Figure 5.

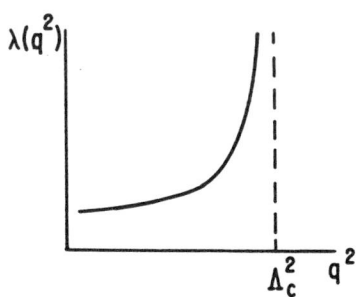

Figure 5. *Behaviour of the scalar effective coupling constant as a function of q^2.*

As one can see from this figure, once the location of the Landau pole is known (in practice, once one knows where the coupling constant λ becomes large) then one knows the value of $\lambda(q^2)$ for all q^2. Furthermore, it is clear that the theory is perfectly sensible for $q^2 \ll \Lambda_c^2$. Note that if one insists that the Landau pole is pushed to infinity, then $\lambda(\Lambda_0^2) \to 0$. This is a qualitative way to see the import of the statement of triviality.

In the electroweak theory, since the scalar doublet Φ is self interacting, one expects to have a very similar situation to the one just described (Luescher and Weisz 1987). Physically, the scalar sector makes perfect sense, as long as we assume that there is some physical high energy cut-off. We can 'measure' where the cut-off is in the scalar theory from the value of the Higgs mass—the remaining physical scalar excitation in the theory with just one scalar doublet Φ. Using the potential V, one finds for this mass the formula

$$M_H^2 = 2\,\lambda(M_H^2)\,v_F^2.$$

From this formula one sees that as long as the Higgs is light, the cut-off is very far away. In fact $\Lambda_c \approx 10^{19}\,\text{GeV} \equiv M_{\text{Planck}}$ would be obtained for $M_H \approx 200\,\text{GeV}$. [For reference, the LEP limit on the Higgs mass is $M_H > 60\,\text{GeV}$ (Mori 1993).] Thus, in this case, the scalar effective field theory is very good and it is sensible to ask whether having $v_F \ll \Lambda_c$ is a stable condition (*naturalness condition*). If, on the other hand, the Higgs is very heavy, so that M_H and Λ_c are comparable, then the effective field theory for the scalar sector no longer makes sense. Numerically, this occurs around (Luescher and Weisz 1987)

$$M_H \sim \Lambda_c \sim 800\text{--}1000\,\text{GeV}.$$

In this situation, $\langle \Phi \rangle$ as an $SU(2) \times U(1)$ order parameter must be replaced by something else.

From the above discussion, one sees that if the Higgs is very light one must worry about the naturalness of the hierarchy

$$v_F \ll \Lambda_c.$$

In fact, radiative effects in a theory with a cut-off destabilise this hierarchy. For instance, the Higgs mass gets a quadratic mass shift and so there is a similar shift in v_F

$$v_F^2 \to v_F^2 + \left[\frac{\alpha}{\lambda}\right]\Lambda_c^2.$$

The existence of quadratic radiative shifts for scalar fields is problematic, for to keep the hierarchy $v_F \ll \Lambda_c$ requires an enormous amount of fine tuning. This fine tuning can only be avoided if:

(i) Instead of $v_F \ll \Lambda_c$, one has that $\Lambda_c \sim v_F$. In this case, however, one abandons the notion of the Fermi scale coming from an elementary scalar field.

(ii) There is some protective symmetry which prevents large radiative shifts.

Remarkably, such a protective symmetry exists; it is called supersymmetry (SUSY)—for an introduction to supersymmetry see, for example, West 1986. This is a boson-fermion symmetry in which bosonic degrees of freedom are paired up with fermionic degrees of freedom, with $m_{\text{Boson}} = m_{\text{Fermion}}$ in the limit of exact SUSY. In a SUSY version of the standard model the quadratic divergences cancel. Even with broken SUSY the divergences are ameliorated. For instance, the quadratic Higgs mass shift due to the presence of a W-loop in a SUSY version of the SM gets cancelled by the equal and opposite shift due to the presence of a Wino-loop ($\widetilde{W}$). Here $\widetilde{W}$ is the spin-1/2 partner of the W. These two graphs are shown in Figure 6.

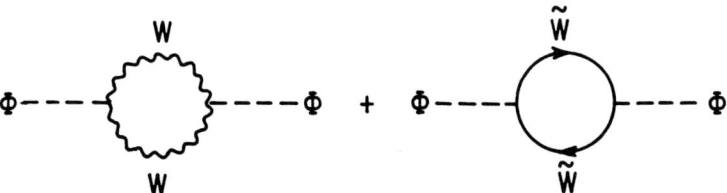

Figure 6. *Graphs contributing to the Higgs mass shift in a SUSY version of the SM. The Wino graph, since it involves fermions, has an overall sign of -1 relative to the W-loop graph.*

Obviously, SUSY is not an exact symmetry of nature, since no superpartners have yet been seen for any of the known particles. However SUSY has no need to be exact to ensure naturalness. Indeed even if $\widetilde{M}_{\widetilde{W}} \neq M_W$, the presence of the Wino-loop now changes the cut-off dependence of the shift to be only logarithmic:

$$v_F^2 \to v_F^2 + \frac{\alpha}{\lambda}(\widetilde{M}_{\widetilde{W}}^2 - M_W^2)\ln\left(\frac{\Lambda_c}{v_F}\right).$$

To guarantee naturalness, therefore, it suffices that the SUSY mass splitting be not too large. That is,

$$(\widetilde{M} - M) \leq O(v_F).$$

Thinking about v_F has brought us to consider two interesting alternatives for the nature of the order parameters for the breaking of $SU(2)\times U(1) \to U(1)_{\text{em}}$. These alternatives lead to quite different physics.

(i) If the breaking is through an elementary scalar VEV $\langle\Phi\rangle$, then naturalness suggests that one should have both a light Higgs and supersymmetry.

(ii) If, however, the breaking is dynamical, then v_F is a scale like Λ_{QCD} and one expects to have a new strong interacting theory in the neighbourhood of 1 TeV.

In the following, I shall discuss these two alternatives in some more detail.

3 The SUSY alternative

It is relatively straightforward to write down a supersymmetric version of the SM (for a discussion see, for example, Zwirner 1992). For these purposes one must:

(i) Associate scalar partners to quarks and leptons, building a chiral supermultiplet

$$\Phi = \{\phi\,;\,\psi\}.$$

In the above, the scalar supermultiplet Φ is composed of a complex scalar ϕ and a Weyl fermion ψ. For example, the left-handed electron supermultiplet is given by

$$\Phi_{e_L} = \{\tilde{e}_L\,;\,e_L\},$$

where $\tilde{e}_L$ is a selectron and e_L is the left-handed electron. Similarly, the Higgs doublet (two are actually needed—see below) is the scalar component of a chiral supermultiplet.

(ii) Associate spin-1/2 partners to the gauge fields, building vector supermultiplets

$$V = \{V^\mu\,;\,\lambda\}.$$

Here V^μ is the gauge field and λ the spin-1/2 gaugino, which is a Weyl fermion.

(iii) Supersymmetrise all interactions; e.g. for the Yukawa coupling one has now the three graphs of Figure 7.

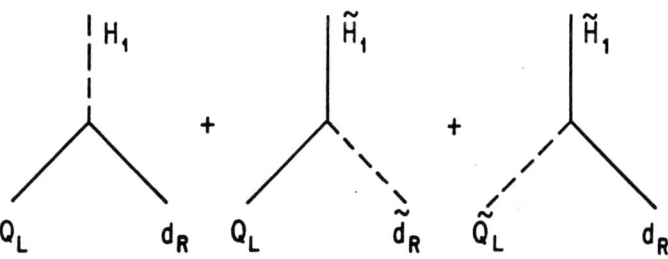

Figure 7. *Yukawa vertices entering in a* SUSY *version of the* SM. *The fields with a tilde are the supersymmetric partners of the appropriate* SM *fields.*

In addition SUSY imposes a few constraints on the allowed interactions:

(i) Interactions involving the chiral supermultiplets should be derivable from a superpotential $W(\Phi_i)$ which involves only Φ_i but not Φ_i^*. This circumstance requires the presence of two different Higgs supermultiplets in the SUSY SM since the vertex diagram of Figure 8 is not allowed. To generate the u-quark Yukawa couplings one replaces H_1^* by a new Higgs doublet H_2, which is itself part of a chiral supermultiplet. [The need for introducing a second Higgs supermultiplet in the SUSY SM can also be motivated by demands that there be no $U(1)$ gauge anomaly. The absence of this anomaly requires $\text{Tr}[Q]_{\text{fermions}} = 0$. Introducing $\widetilde{H}_1$ with $Q_{\widetilde{H}_1} = (0, +1)$ immediately requires a second fermion Higgs, $\widetilde{H}_2$, with $Q_{\widetilde{H}_2} = (0, -1)$ in order to ensure anomaly cancellation.]

(ii) The scalar potential in the theory follows from $W(\Phi_i)$ plus other terms (the so called D-terms) arising from supersymmetrising the gauge interactions. One finds:

$$V(\Phi) = \sum_i \left|\frac{\partial W}{\partial \Phi_i}\right|^2 + \frac{g_a^2}{2}\left|\Phi_i^* T_a \Phi_i\right|^2,$$

where T_a is a group generator.

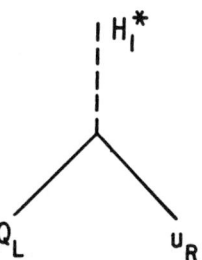

Figure 8. *Vertex diagram forbidden by SUSY.*

These points are of importance for the issue of $SU(2) \times U(1)$ breaking. Ignoring Yukawa couplings, the superpotential only contains a term

$$W(H_1, H_2) = \mu\, H_1 H_2,$$

where μ is a scale parameter, because $(H_1 H_2)^2$ terms in the superpotential would result in a non-renormalisable $V(\Phi)$. Thus, including D-terms but neglecting Yukawa interactions, one has for the potential

$$V(H_1, H_2) = \frac{1}{8}g_2^2 \left(H_1^* \vec{\tau} H_1 + H_2^* \vec{\tau} H_2\right) \cdot \left(H_1^* \vec{\tau} H_1 + H_2^* \vec{\tau} H_2\right)$$
$$+ \frac{1}{2}g_1^2 \left(H_1^* H_1 - H_2^* H_2\right)^2 + \mu^2 \left(H_1^* H_1 + H_2^* H_2\right).$$

Note that in the SUSY SM the quartic coupling constants λ are related to the gauge couplings. Since $\mu^2 > 0$, the above potential cannot break $SU(2) \times U(1)$ if SUSY is exact. However, one must still add SUSY breaking terms and there is the possibility that the mechanism which breaks SUSY might also break $SU(2) \times U(1)$.

A popular scenario is to assume that SUSY is broken at a high scale by a hidden sector, coupled only gravitationally to the real world (Cremmer et al. 1983, Girardello and Grisaru 1984). This scenario introduces only a few parameters ($m_{1/2}$, m_0, A, B) typifying SUSY breaking and is reasonably predictive. One can show that this 'hidden sector breaking' leads to a Lagrangian given by:

$$\mathcal{L} = \mathcal{L}_{SM}^{SUSY} - \sum_i m_0^2 \phi_i^* \phi_i - \sum_i m_{1/2} \lambda_i \lambda_i + A m_0 W_3(\phi) + B m_0 W_2(\phi) + \text{h.c.}$$

Here ϕ_i are all the scalar fields in the theory, while the λ_i are the gauginos. $W_3(\Phi)$ is the tri-linear superpotential and $W_2(\Phi)$ is the bi-linear superpotential.

Ignoring Yukawa couplings again, the potential one derives from this Lagrangian is

$$V(H_1, H_2) = \frac{1}{8} g_2^2 \left| H_1^* \vec{\tau} H_1 + H_2^* \vec{\tau} H_2 \right|^2 + \frac{1}{2} g_1^2 \left| H_1^* H_1 - H_2^* H_2 \right|^2$$
$$+ \mu_1^2 H_1^* H_1 + \mu_2^2 H_2^* H_2 - \mu_3^2 \left(H_1 H_2 + H_1^* H_2^* \right),$$

where $\mu_1^2 = \mu_2^2 = (\mu^2 + m_0^2)$; $\mu_3^2 = B m_0 \mu$. For $SU(2) \times U(1) \to U(1)_{em}$ breaking we need a negative eigenvalue in the Higgs mass matrix

$$M_H^2 = \begin{pmatrix} \mu_1^2 & -\mu_3^2 \\ -\mu_3^2 & \mu_2^2 \end{pmatrix}.$$

Now $\det M_H^2 < 0$ implies that $\mu_1^2 \mu_2^2 < \mu_3^4$. However, if as here, $\mu_1^2 = \mu_2^2$, then one has to worry about the stability of V, when $\mu_1^2 < \mu_3^2$. In fact, there is really no problem at all since the masses also evolve with q^2 and we only expect $\mu_1^2 = \mu_2^2$ at the scale of SUSY breaking. Indeed one can argue that, since the top quark is so heavy, μ_2^2 changes sign as one evolves down from high to low scales (Ibañez and Ross 1982, Alvarez-Gaumé et al. 1982). In effect, SUSY breaking triggers the breaking of $SU(2) \times U(1) \to U(1)_{em}$!

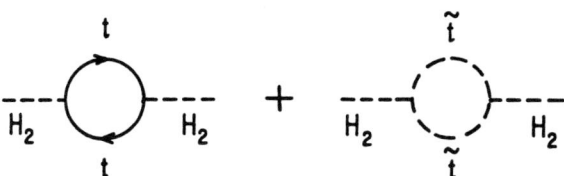

Figure 9. *Top and stop loops contributing to the running of $\mu_2^2(q^2)$.*

Let us try to understand this phenomenon a bit more in detail. The reason why μ_2^2 evolves much more rapidly than μ_1^2 with q^2 is solely due to the presence of a heavy top quark. Roughly speaking, for q^2 near the SUSY breaking scale M_{SB}^2, the loops containing top and stop shown in Figure 9 are dominated predominantly by the stop contribution, since $m_t \simeq 0$ at these large scales while $\tilde{m}_{\tilde{t}} \simeq m_0 \simeq \mu_2$. One finds (Inoue et al. 1982)

$$\delta\mu_2^2 = -\frac{3}{4\pi^2} h_t^2 \mu_t^2 \ln[M_{\rm SB}^2/q^2].$$

These contributions dominate over that of all the other quarks and leptons, because the top Yukawa coupling, h_t, is so much bigger than the rest. Furthermore, the evolution of μ_2^2 is rapid enough so that indeed for $q^2 \sim M_W^2$, $\mu_2^2(q^2)$ becomes negative, triggering the electroweak breakdown. A schematic behaviour of what happens is shown in Figure 10.

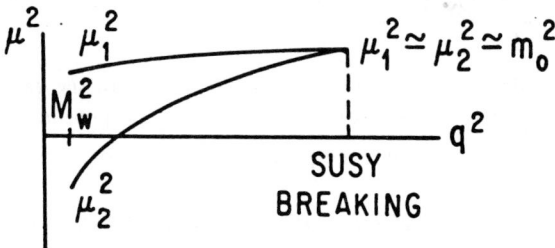

Figure 10. *Different evolution of μ_1^2 and μ_2^2 with q^2.*

Although the above is a very interesting scenario for generating the Fermi scale, it also illustrates that while SUSY solves the naturalness problem, it leaves unexplained a hierarchy problem. Because μ_2^2 evolves only logarithmically with q^2, the Higgs mass is directly related to the SUSY breaking mass parameter m_0^2

$$M_H^2 \simeq -\mu_2^2(M_W^2) \sim \mu_2^2(M_{\rm SB}^2) \simeq m_0^2.$$

That is, in magnitude there cannot be much difference between the value of μ_2^2 at high and low scales. The deep unanswered question is how then at the very large scale where SUSY breaking presumably takes place, $M_{\rm SB}$, can one generate such a small mass as m_0—which is in the 100 GeV–1 TeV range. What is the reason for the hierarchy

$$M_{\rm SB} \gg m_0 ?$$

I should remark that, in effect, the radiative breaking due to stop and top loops evolves even more rapidly than one may think, since the Yukawa coupling h_t is also q^2 dependent and—like the Higgs coupling λ—it is not asymptotically free. In fact, the actual structure of the RGE for h_t is quite interesting in its own right, since the inclusion of strong interaction effects forces h_t to a fixed point (Pendleton and Ross 1981). One has

$$\frac{dh_t}{d\ln q^2} = \frac{h_t}{16\pi^2}\left\{6h_t^2 - \frac{16}{3}g_3^2\right\}.$$

Even though for large h_t the effects of QCD are negligible, the above equation forces h_t to evolve to a value given by the vanishing of the curly bracket. For $q^2 \simeq m_t^2$, this fixed point value

$$h_t^* \simeq \frac{4}{3}\sqrt{2\pi\alpha_s(m_t^2)}$$

is numerically very close to unity and leads to an upper bound for the top mass.

In the minimal SUSY standard model the top quark mass, m_t, is related to $\langle H_2 \rangle = v_2/\sqrt{2}$, while the Fermi scale, v_F, measures both $\langle H_2 \rangle$ and $\langle H_1 \rangle = v_1/\sqrt{2}$. From the Higgs potential $V(H_1, H_2)$, one finds

$$v_F^2 = v_1^2 + v_2^2 \equiv v_F^2 \cos^2 \beta + v_F^2 \sin^2 \beta$$

leading to

$$m_t = \frac{h_t^*}{\sqrt{2}} v_2 = h_t^* \frac{v_F}{\sqrt{2}} \sin \beta \simeq 175 \sin \beta \, \text{GeV}.$$

[A more precise RG evolution gives $m_t(m_t^2) \simeq 190 \sin \beta \, \text{GeV}$; e.g. Barger et al. 1993a.] This value is compatible, for reasonable $\sin \beta$, with the direct limit on the top obtained at Fermilab. [At the time of writing, the best bound on the top mass is $m_t > 131 \, \text{GeV}$, obtained by the D0 Collaboration (1994).]

A recent analysis of the SUSY SM radiative corrections (Barbieri et al. 1992) yields a best value for the top mass of

$$m_t = 131^{+31}_{-23} \, \text{GeV},$$

which, taken in conjunction with the above RGE result, gives [note that, for all practical purposes the result of radiative corrections in the SUSY SM is equivalent to that of the SM with a light Higgs]

$$\sin \beta = 0.68 \pm 0.16 \qquad (0.6 < \tan \beta < 1.5).$$

There are three remarkable results which follow from a SUSY extension of the SM:

(i) SUSY requires a larger Higgs sector and a quite light Higgs boson is expected. The two Higgs doublets necessitated by a SUSY extension H_1 and H_2 contain 8 real fields, 3 of which are absorbed to give masses to the W^+, W^- and Z^0. Thus we are left with 5 physical fields—2 charged (H^+, H^-) and 3 neutral (h^0, H^0, A) Higgs particles. The potential $V(H_1, H_2)$ leads to a number of tree-level relations among these states and the masses of the gauge bosons. One finds (Gunion et al. 1990)

$$M_H^2 = M_W^2 + M_A^2$$
$$M_h^2 + M_H^2 = M_Z^2 + M_A^2$$
$$M_h \leq M_A \leq M_H$$
$$M_h \leq M_Z |\cos 2\beta|.$$

For the value of β given above, the last relation is problematic, since the lightest neutral Higgs, h, has not been observed at LEP (Mori 1993). However, it turns out that radiative corrections to these equations are large (due to the large mass of the top) and can lead to $M_h > M_Z$. One can appreciate the order of magnitude of the effect by focusing again on the radiative mass shift coming from the top and stop loops of Figure 9. The shift is proportional both to the top-stop squared mass difference and to h_t^2. Since the Yukawa coupling h_t itself is proportional to m_t, one sees that the shift due to radiative effects for the lightest Higgs mass actually depends quartically on m_t. An approximate formula for the shift (Ellis et al. 1991; Okada et al. 1991) is

$$\delta M_h^2 \simeq \frac{3 g_2^2 m_t^4}{8 \pi^2 M_W^2 \sin^2 \beta} \ln \left[1 + \frac{m_{\tilde{q}}^2}{m_t^2} \right],$$

where $m_{\tilde{q}}$ is a generic squark mass. For $\cos 2\beta = 1$ and $m_{\tilde{q}} = 1\,\text{TeV}$, this shift is plotted in Figure 11. As can be seen, for large m_t, this shift can be of the order of tens of GeVs.

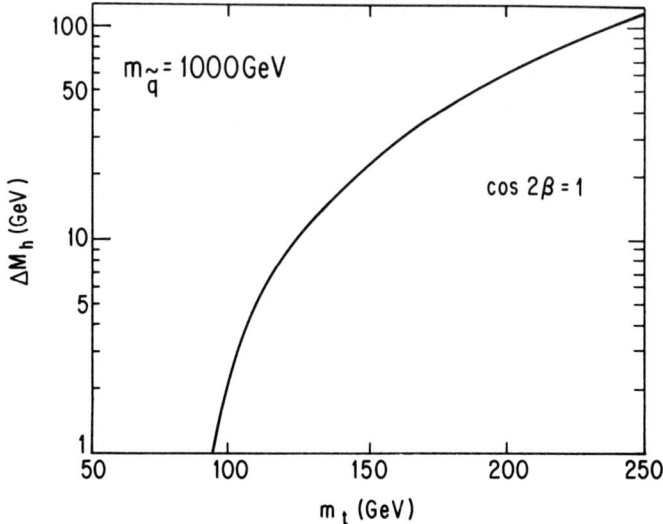

Figure 11. *Lightest Higgs mass shift as a function of m_t. Adapted from Haber and Hempfling 1991.*

A further interesting observation can be made (Barger et al. 1993b). If one takes the range of β indicated from the value of m_t obtained by radiative corrections to the SUSY SM, along with the bounds on M_h which follow from the radiatively corrected version of the $M_h \leq M_Z|\cos 2\beta|$ relation, as well as from the direct search bounds from LEP, one obtains an exclusion plot in the M_h-m_t plane. This plot is shown in Figure 12. As can be seen, it is conceivable that sharp tests of this scenario will emerge soon when bounds on m_t and M_h are sharpened. At LEP-II one should be able to probe values for M_h up to about 80–90 GeV, while the forthcoming run of the Tevatron should either discover top or push the top bound to near 160 GeV.

(ii) SUSY extensions of the SM very naturally provide candidates for the dark matter in the universe. This point is easily understood. Most simple SUSY extension of the SM—like the trivial one we have considered above—preserve a discrete symmetry, known as R-parity (for a discussion, see Nilles 1984). Roughly speaking, one has $R = +1$ for particles and $R = -1$ for sparticles. Thus, as shown in Figure 7, when one supersymmetrises an ordinary vertex, one introduces always *pairs* of sparticles. R-parity of the SUSY interactions ensures that the lightest supersymmetric particle (LSP) is stable. Particularly interesting as dark matter candidates is any one of the 4 'neutralinos'—the SUSY partners of the $(\gamma, Z, H_1^0, H_2^0)$ bosons—$(\tilde{\gamma}, \tilde{Z}, \tilde{H}_1, \tilde{H}_2)$. In general the LSP will be a combination of these states, found by diagonalizing the neutralino mass matrix, and looking for the minimum eigenvalue. It is beyond the purpose of these lectures to discuss this problem in detail. However, I note that the neutralino mass matrix besides depending on the gauge boson masses and β, also depends on the SUSY breaking

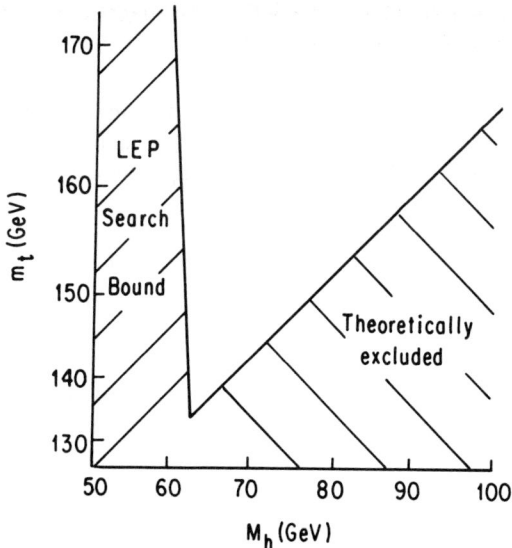

Figure 12. *Exclusion plot in the M_h-m_t plane. Adapted from Barger et al. 1993b.*

gaugino mass $m_{1/2}$ and on μ. Writing the LSP as

$$\tilde{\delta} = a_\gamma \tilde{\gamma} + a_Z \tilde{Z} + a_1 \tilde{H}_1 + a_2 \tilde{H}_2,$$

one finds that

$$\tilde{\delta} = \begin{cases} \tilde{\gamma} & \text{as } m_{1/2} \to 0 \\ \tilde{h} = \sin\beta \tilde{H}_1 + \cos\beta \tilde{H}_2 & \text{as } \mu \to 0. \end{cases}$$

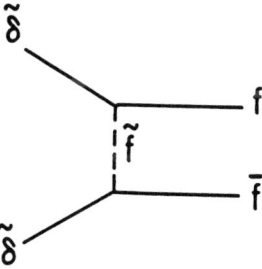

Figure 13. *Neutralino annihilation into fermion-antifermion pairs.*

The reason why neutralinos are excellent candidates for dark matter is because they have an annihilation rate which is typical of the weak interactions. The contribution of weakly interacting massive particles to the universe's relic abundance is of order unity, if these particles have an annihilation rate $\langle \sigma_a v \rangle \sim 10^{-37}$ cm^2 (see, for example, Kolb and Turner 1990). The annihilation rate one computes from the diagram of Figure 13 is roughly of this order of magnitude, since the neutralino's fermion-fermion coupling is an electroweak coupling and, from our naturalness assumption, $m_{\tilde{f}}$ cannot be that different from M_W itself. What is particularly nice is that this diagram not only determines the LSP abundance, but also sets the rate, $\tilde{\delta} + f \to \tilde{\delta} + f$, needed for the detection of neutralinos as dark matter (see Griest 1993).

(iii) The third interesting property of SUSY is the way in which the presence of SUSY matter helps the unification of the SU(3)×SU(2)×U(1) couplings at high scales. Using the RGE one can compute the evolution of the coupling constants in the SM. To one-loop order, one has

$$\frac{1}{\alpha_i(q^2)} = \frac{1}{\alpha_i(M_Z^2)} + \frac{b_i}{4\pi} \ln\left(\frac{q^2}{M_Z^2}\right),$$

where the b_i are the coefficients of the β-function for the particular coupling in question. The detailed evolution of the coupling constants depends on the *matter content* of the theory, since the b_i depend on this content. In particular, in SUSY extensions of the SM there will be a different evolution of the α_i's than in the SM, since there are both new fermions (gauginos and shiggs) and new scalars (sleptons and squarks, as well as now 2 Higgs doublets).

The computation of the contributions of the different particles on the theory to the b_i coefficients is relatively straightforward. One finds (Slansky 1981)

$$b_i = \sum_{\text{particles}} \left[\frac{1}{6}\ell_i^{\text{vector}} - \frac{1}{3}\ell_i^{\text{Weyl fermion}} - \frac{1}{12}\ell_i^{\text{real scalar}}\right]$$

where the ℓ_i are group theoretic factors. For SU(N) groups—which are of interest for our purposes—one has (Slansky 1981)

$$\ell_i = \begin{cases} 2N & \text{adjoint representation} \\ 1 & \text{fundamental representation}. \end{cases}$$

For the U(1) group, ℓ_1 depends on the actual definition of the charge one takes. The value used in the following is

$$\ell_1 = 2Q_1^2 \cdot \frac{3}{5},$$

where Q_1 is the ordinary U(1) charge in the SM (e.g. $Q_{u_R} = 2/3$, $Q_{Q_L} = 1/6$; etc.). As we shall see later on in these lectures, the extra factor of 3/5 identifies among all possible U(1) charges the one that sits in a Grand Unified Theory (GUT).

A straightforward computation leads to the values for the coefficients b_i for the SM and the SUSY SM, shown in Table 1. Note that in the SUSY SM, because of the presence

Coefficient	SM	SUSY SM
b_1	−41/10	−33/5
b_2	19/6	−1
b_3	7	3

Table 1. *Values of b_i in the SM and the SUSY SM.*

of so much matter, the SU(2) coupling is no longer asymptotically free. Using precision values for the coupling constants obtained at LEP one can evolve these values to high q^2 (Amaldi et al. 1991; Langacker and Luo 1991; Ellis et al. 1990). The result of such a study (Langacker 1993), using a SUSY threshold of 10^2–10^3 GeV, is shown in Figure 14. As can be seen, the evolution of the precisely measured couplings at M_Z exclude the possibility of the values of the couplings meeting at some large energy scale in the SM, but including SUSY modifies the running so as to rekindle this possibility. We shall return to this point later on in these lectures.

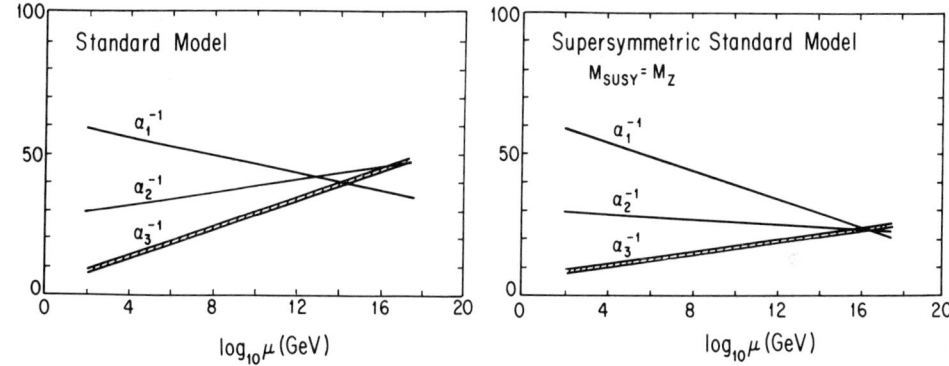

Figure 14. *Evolution of couplings without and with* SUSY *matter. Adapted from Langacker 1993.*

It may be useful to summarise briefly some of the appealing and interesting aspects of supersymmetry, in relation to the question of mass. We have seen that introducing SUSY makes the scalar sector of the SM natural, stabilising the Fermi scale even in the presence of a large cut-off. Furthermore, SUSY suggests an interesting scenario by which v_F itself originates as a radiative effect, triggered by the breaking of supersymmetry through the scalar mass parameter m_0. Why supersymmetry is broken, however, is not explained. Furthermore, in effect, a new 'naturalness' problem arises when SUSY breaking is used to try to trigger $SU(2) \times U(1) \rightarrow U(1)_{em}$. Namely, why do we have

$$M_{SB} \gg m_0 \simeq M_W \simeq v_F.$$

It should be mentioned also that SUSY gives no explanation of the origin of the magnitudes of the Yukawa couplings for quarks and leptons, Γ_f, nor for family replication. In fact, some further unknown mass parameters are introduced into the theory with hidden sector breaking: m_0, $m_{1/2}$, as well as the 3 numerical constants A, B and $\tan \beta$. Thus, in a sense, one has regressed by invoking supersymmetry! Nevertheless, the overall picture remains quite compelling, particularly if one can adduce deeper reasons for introducing supersymmetry than naturalness itself. As we shall see later on, these reasons exist.

4 Dynamical symmetry breaking

The idea behind a dynamical origin for the Fermi scale v_F is rather simple. Rather than having the breakdown of SU(2)×U(1)→U(1)$_{em}$ being caused by an elementary scalar VEV $\langle \Phi \rangle$, one replaces

$$\langle \Phi \rangle \to \langle \bar{F}F \rangle,$$

with $\langle \bar{F}F \rangle$ being a condensate of fermions from some underlying theory. The $\langle \bar{F}F \rangle$ condensate is assumed to form due to an underlying gauge interaction, in much the same way as quark condensates form in QCD. One introduces a new gauge interaction (*Technicolour* (TC); Susskind 1979, Weinberg 1976) which is assumed to become strong at some energy scale Λ_F. It follows then, that

$$\langle \bar{F}F \rangle \sim \Lambda_F^3$$

and that v_F is associated with this dynamical scale. The picture is quite akin to QCD where the colour force becomes strong at scales of order Λ_{QCD} and the quark condensates, which break the global chiral symmetry of QCD, also scale like Λ_{QCD}^3.

The SM value $\rho = 1$, which leads to the mass relation $M_Z^2 \cos^2 \theta_W = M_W^2$, follows from a custodial SU(2) symmetry of the Higgs potential (see Aitchison, these Proceedings). One must resort to a similar symmetry to constrain the technicolour interactions. One assumes that the technicolour interactions have an SU(2)$_L$×SU(2)$_R \approx$ O(4) invariance which breaks to SU(2)$_{L+R} \approx$ O(3) as the custodial symmetry leading to $\rho = 1$. Specifically (Susskind 1979; Weinberg 1976), one takes for technicolour, for example, an SU(N) gauge theory with SU(2)$_L$ doublets

$$\begin{pmatrix} U_i \\ D_i \end{pmatrix}_L$$

$i = 1, \ldots, N$ and SU(2) singlets U$_{iR}$ and D$_{iR}$. The electroweak assignments of U_i and D_i forbid mass terms. In the absence of these mass terms and neglecting the weak electroweak forces, the technicolour theory possess the same kind of chiral symmetry that QCD has: here SU(2)$_L$×SU(2)$_R$. This symmetry is broken down by condensate formation

$$\langle \bar{U}_{Li} U_{Ri} \rangle = \langle \bar{D}_{Li} D_{Ri} \rangle \neq 0$$

to SU(2)$_{L+R}$, providing the required custodial symmetry. The above condensates also break SU(2)×U(1)→U(1)$_{em}$ and thus serve to give mass to the W and Z (Susskind 1979; Weinberg 1976).

Although technicolour accomplishes the same things for the gauge sector as having an elementary Higgs doublet—provided $\langle \bar{U}U \rangle \sim v_F^3$—the results one obtains are not quite the same at the level of radiative corrections. Furthermore, in simple technicolour, there is no way to generate Yukawa couplings, Γ_f, to produce fermion masses. I will consider both these points in turn.

Consider first the radiative corrections. Since the effects of the TC sector are important at scales of order $16\pi^2 v_F^2 \gg M_Z^2$, (see Chanowitz 1993) it suffices to compute the vacuum polarization of the gauge field propagators in a q^2 expansion.

$$\Pi^{AB}(q^2) = \Pi^{AB}(0) + q^2 \Pi^{AB\prime}(0) + \ldots,$$

with $\{AB\} = \{WW; ZZ; \gamma\gamma; Z\gamma\}$, giving 8 parameters, $\Pi^{AB}(0)$ and $\Pi^{AB\prime}(0)$, at order q^2. However, electromagnetic gauge invariance leads to $\Pi^{\gamma\gamma}(0) = \Pi^{Z\gamma}(0) = 0$, leaving only 6 parameters to consider. Three of these are fixed by the 3 independent parameters of SU(2)×U(1), g_1, g_2 and v_F or, more practically, α, M_Z and G_F. Therefore an overall fit of the data, which is independent of whether or not one uses an elementary Higgs, will be characterised by 3 other parameters. These have been called S, T, U (Peskin and Takeuchi 1990, 1992) or ϵ_1, ϵ_2, ϵ_3 (Altarelli et al. 1991/92) in the literature.

In practice, S (or ϵ_3) is the most interesting parameter, since T (or ϵ_1) is dominated by information about m_t and U (or ϵ_2) is not so well determined experimentally, because it depends sensitively on M_W. If the underlying theory conserves an SU(2)$_{L+R}$—as happens in the case of technicolour—then one can write (Peskin and Takeuchi 1990, 1992) a useful spectral representation for S in terms of the functions characterizing the vacuum polarization tensors of the SU(2)$_{L+R}$ and SU(2)$_{R-L}$ currents in the theory. If we denote these vector and axial vector spectral functions by $v(s)$ and $a(s)$, respectively, one finds (Peskin and Takeuchi 1990, 1992)

$$S = \int_0^\infty \frac{ds}{s}[v(s) - a(s)].$$

Using the above formula, in simple technicolour models one can estimate S just by scaling up the spectral functions of QCD from $\Lambda_{\rm QCD}$ to Λ_F. This leads to (Peskin and Takeuchi 1990, 1992)

$$S \simeq a\frac{N}{3} \simeq a\frac{N}{3}\left[\frac{6\pi f_\pi^2}{m_\rho^2}\right] \simeq a\frac{N}{3}0.25,$$

where, in the second step, one assumes that the spectral functions of QCD are dominated by the ρ (vector) and A_1 (axial vector) states. Here a is the number of technicolour doublets in the theory and N is the number of technicolours. Thus in technicolour $S \sim O(1)$ and is positive. Unfortunately, the combined analysis of all the electroweak data shown in Figure 15 gives $S \sim -O(1)$! However, one need not take this as a catastrophe since the fermion mass issue, to be discussed below, already suggests that we need a much more sophisticated technicolour theory than just 'scaled up QCD'.

To include fermion masses, one needs to introduce some communication between the ordinary fermions (quarks and leptons), f, and the condensing fermions, F. The original idea put forth by Dimopoulos and Susskind (1979) and by Eichten and Lane (1980) to generate fermion masses in this context was to introduce *another* underlying theory—Extended Technicolour (ETC)—whose gauge bosons connect both the f and F fermions. Spontaneous breakdown of ETC at a large scale then produces an effective interaction of the form

$$\mathcal{L}_{\rm eff} \simeq \frac{1}{\Lambda_{\rm ETC}^2}(\bar{F}_L F_R)\bar{f}_R f_L.$$

In the presence of the above interaction, the technicolour condensate $\langle \bar{F}_L F_R \rangle \sim v_F^3$ gives a mass to the ordinary fermions, f, of order

$$m_f \simeq \frac{\langle \bar{F}_L F_R \rangle}{\Lambda_{\rm ETC}^2} \sim \frac{v_F^3}{\Lambda_{\rm ETC}^2}.$$

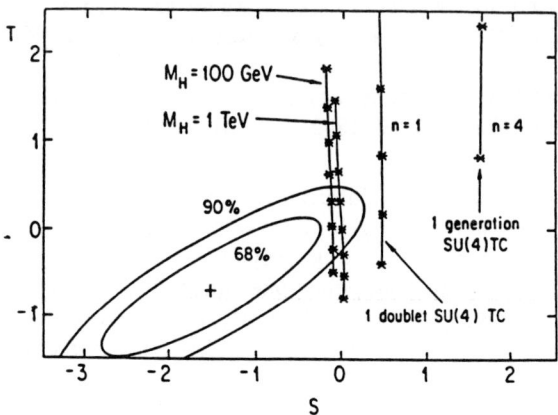

Figure 15. *Fit to all electroweak data in the S-T plane. Adapted from Peskin and Takeuchi 1990, 1992.*

If this formula is to give mass for the top, it is clear that Λ_{ETC} cannot be too large, implying that $\Lambda_{ETC} \sim (1\text{–}10)\,\text{TeV}$. However, this leads to unacceptably large flavour-changing neutral currents (FCNC), since the ETC interaction also relates ordinary fermions to ordinary fermions. For instance, through ETC interactions one can get an effective $(\bar{d}s)^2$ interaction which could directly give $K\text{-}\bar{K}$ mixing of order

$$\mathcal{L}^{eff}_{K\text{-}\bar{K}} \sim \frac{(\cos\theta_c \sin\theta_c)^2}{\Lambda^2_{ETC}} (\bar{d}s)^2 .$$

The experimental absence of such terms requires $\Lambda_{ETC} > 100\,\text{TeV}$ (Dimopoulos and Ellis 1981), leading to a conundrum!

To get round this problem, the concept of 'walking technicolour' was invented (Holdom 1981). There is, in fact, a difference between the Fermi scale, v_F (related to M_W), and the condensate scale $\langle \bar{F}F \rangle$ (which gives m_f). Both are related to the technifermion self-energy, but they probe different scales in this function. The Fermi scale is computed through the contribution of the technifermion bubble, shown in Figure 16a, to the SU(2) current correlation function, while the condensate is proportional to the technifermion self-energy loop of Figure 16b. Thus one has

$$v_F^2 \sim \int \frac{d^4p}{(p^2)^2} \Sigma^2(p^2) \sim \Sigma^2(0)$$

$$\langle \bar{F}F \rangle \sim \int^{\Lambda_{ETC}} \frac{d^4p}{p^2} \Sigma^2(p^2) \sim \Lambda^2_{ETC}\Sigma(\Lambda^2_{ETC}) .$$

The approximate dependences given in the above equations ignore all logarithmic factors. Furthermore, in the case of the technifermion condensate, the integral is effectively cut off by the largest scale in the problem where new physics enters, which here is Λ_{ETC}.

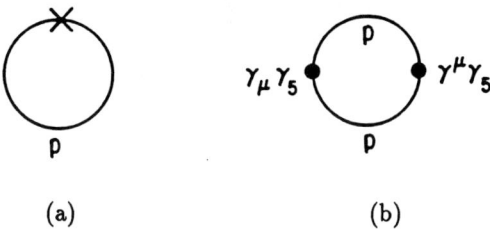

Figure 16. *(a) Graph giving v_F^2; (b) graph determining the $\langle \bar{F}F \rangle$ condensate.*

In Yang Mills theories with broken global chiral symmetries, one can show that (Lane 1974, Politzer 1976), for large q^2

$$\Sigma(q^2) \sim \frac{\Sigma^3(0)}{q^2} \sim \frac{\Lambda^3}{q^2},$$

where Λ is the dynamical scale of theory. Clearly, therefore, if by the time $q^2 \sim \Lambda_{ETC}^2$, $\Sigma(q^2)$ has already reached its asymptotic value, then

$$v_F \sim \Sigma(0) \sim \Lambda_{TC} \quad \text{and} \quad \langle \bar{F}F \rangle \sim \Lambda_{ETC}^2 \Sigma(\Lambda_{ETC}^2) \sim \Lambda_{ETC}^2 \frac{\Lambda_{TC}^3}{\Lambda_{ETC}^2} \sim \Lambda_{TC}^3;$$

whence $\langle \bar{F}F \rangle \sim v_F^3$. However, if $\Sigma(q^2)$ evolves slowly towards its asymptotic value ('walking technicolour') it could well be that

$$\Sigma(\Lambda_{ETC}^2) \gg \frac{\Lambda_{TC}^3}{\Lambda_{ETC}^2}.$$

Then (Holdom 1981)

$$\langle \bar{F}F \rangle \sim \Lambda_{ETC}^2 \Sigma(\Lambda_{ETC}^2) \gg \Lambda_{TC}^3 \sim v_F^3.$$

This behaviour is sketched in Figure 17.

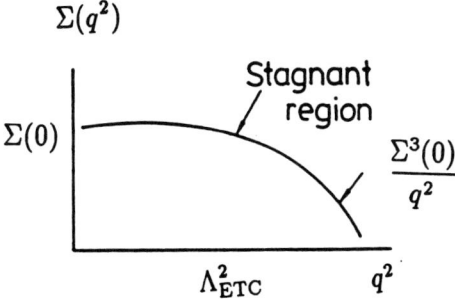

Figure 17. *Behaviour of the technifermion self-energy in a walking technicolour theory.*

In theories where the above behaviour holds, one can get large masses, having relatively large Λ_{ETC}, from a technicolour mechanism

$$m_f \sim \frac{\langle \bar{F}F \rangle}{\Lambda_{\text{ETC}}^2}.$$

However now, because Λ_{ETC} is large, FCNC will be suppressed. One can construct walking technicolour (WTC) models by having enough technifermions so that the WTC coupling is almost not asymptotically free (Yamawaki et al. 1986, Akiba and Yanagida 1986, Appelquist et al. 1986). In effect, this means that the ETC dynamics is not really decoupled from the WTC dynamics. As a result, the 'scaled up QCD' derivation of $S \sim +O(1)$ no longer holds (Appelquist and Triantiphyllou 1992, Hsu and Sundrum 1993). Furthermore, in semi-realistic WTC models one makes use of this non-decoupling of WTC and ETC dynamics to generate fermions mass splittings, through the different dependence of Σ_u and Σ_d on Λ_{ETC} (for an example, see Holdom 1990).

Once one admits that the WTC and the ETC dynamics are coupled—and one introduces sufficient complexity in these theories to try to make them realistic—one has raised the dynamical ante to such an extent that it may be worthwhile considering other possible options. In my view, it appears more sensible instead of introducing an ETC theory to connect the ordinary fermions, f, to the technifermions, F, to appeal to compositeness to make this connection. If the f and F states are made up of common constituents—preons—then the preon dynamics will give rise to effective 4-fermion interactions which play an analogous role to ETC interactions, with the dynamical scale of the preon theory Λ_c substituting for Λ_{ETC}.

I would like to illustrate briefly how this works by discussing a semirealistic model which I developed in collaboration with S. Khlebnikov (Khlebnikov and Peccei 1993). The model is based on 3 replicas of an SU(6) chiral preon theory, augmented by additional vector-like interactions which are felt by some of the preons but not by others. The model gives one generation of quarks (say t and b) which dynamically have $m_t \gg m_b$. [The model has no U(1) anomalies, even in the absence of leptons, due to the presence of technifermions with appropriate U(1) charges.] Two of the SU(6) preon theories have 10 preons in the fundamental representation and 1 preon in the symmetric adjoint representation, while the third preon theory has a doubling of these degrees of freedom. The first two theories will give rise among their bound states to t_R and b_R, while the last preon theory will produce the $\binom{t}{b}_L$ doublet. The preons in this model—besides feeling the SU(2)×U(1) interactions also feel three vector interactions: colour, technicolour and metacolour. The assignments of the preons under these groups are shown below:

$$\text{SU(2) singlets} \begin{cases} \text{SU(6)}_{t_R} : \overline{3}\,\square_t^c + \overline{3}\,\square_t^T + 4\,\square_t^M + \overline{\square\square}_t \\ \text{SU(6)}_{b_R} : \overline{3}\,\square_b^c + \overline{3}\,\square_b^T + 4\,\square_b^M + \overline{\square\square}_b \end{cases}$$

$$\begin{array}{l}\text{SU(2)}\\ \text{doublets}\end{array} \quad \text{SU(6)}_L : \overline{3}\binom{\square^c}{\square^c}_L + \overline{3}\binom{\square^T}{\square^T}_L + 4\,\square_1^M + 4\,\square_2^M + \binom{\overline{\square\square}}{\square\square}_L.$$

Here c labels the SU(3) colour, T the SU(3) technicolour and M the SU(4) metacolour.

Further, to avoid anomalies, the singlet preons are right-handed Weyl fermions, while the doublet preons are left-handed.

If one turns off all the gauge interactions, the preon theories have certain global chiral symmetries. One can argue dynamically—through anomaly matching and complementarity (Khlebnikov and Peccei, 1993)—that each of the SU(6) preon theories preserves a global chiral symmetry

$$H = SU(6) \times SU(4) \times U(1),$$

which is somewhat smaller than the nominal chiral symmetry of the theory. The presence of this chiral symmetry gives rise to cert<u>ain </u>massless bound states which have the quantum numbers of the preon composite ⊡.

Only states involving two preons with SU(6) quantum numbers, or an SU(6) and an SU(4) quantum number bind. If one identifies the SU(4) with metacolour, then the metacolour singlet states—which will be denoted by B—transform under the 15-dimensional representation of SU(6). The other states—denoted by B'—transform according to the fundamental representation of the SU(6) and SU(4). Thus the bound states which are massless in the theory are:

$$B \sim (15,1); \qquad B' \sim (6,4).$$

Since the preons transforming according to the global SU(6) are assigned either to a $\bar{3}$ of colour or a $\bar{3}$ of technicolour, one sees that the B bound states contain the desired quarks, as well as some other states carrying technicolour quantum numbers. Under $SU(3)_c \times SU(3)_T$ one has

$$B = (3,1) + (1,3) + (\bar{3},\bar{3}).$$

Note that since we have 3 distinct preon theories one will find 3 different B and B' massless bound states, which will include, respectively, a t_R, a b_R and a $\binom{t}{b}_L$. When one now includes the gauged vectorial interactions not all of these states will be relevant. For instance, once the metacolour SU(4) is gauged, the B' states will bind into metahadrons with masses of order $M \sim \Lambda_4$—with Λ_4 being the metacolour dynamical scale. Furthermore if, as in QCD, these metacolour interactions cause the formation of condensates, these condensates will tie the left and right sectors of the model together. Indeed, instead of having a separate SU(6) global for each preon theory, these condensates now preserve only a common diagonal SU(6) (Khlebnikov and Peccei 1993).

The formation of the metacolour condensates, in conjunction with the effective 4-fermion interactions arising from each preon theory which tie B and B' bound states of the same helicity together, give rise to additional 4-Fermi interactions of the ETC type. This is shown schematically in Figure 18. The effective ETC scale here, however, is different depending on whether the B_R bound states arise from the $SU(6)_{t_R}$ or the $SU(6)_{b_R}$ preon theory:

$$\Lambda^t_{\text{eff}} \simeq \left(\frac{\Lambda_L}{\Lambda_4}\right)\Lambda_{Rt}; \qquad \Lambda^b_{\text{eff}} \simeq \left(\frac{\Lambda_L}{\Lambda_4}\right)\Lambda_{Rb}.$$

Since the B-states contain both quarks and techniquarks, these effective ETC interactions will give mass to the quarks after one turns on the technicolour gauge interactions.

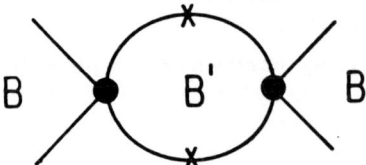

Figure 18. *Formation of an effective* ETC *interaction as a result of the metacolour condensates tying together the effective interactions caused by the common preon constituents of B_L, B'_L and B_R, B'_R.*

One finds in this way

$$m_t \simeq \left[\frac{\Lambda_4^2 \Lambda_{TC}^3}{\Lambda_L^2 \Lambda_{Rt}^2}\right] = K\langle r_t^2 \rangle ; \qquad m_b \simeq \left[\frac{\Lambda_4^2 \Lambda_{TC}^3}{\Lambda_L^2 \Lambda_{Rb}^2}\right] = K\langle r_b^2 \rangle .$$

That is, here the masses of the t and b quark are different since they have different sizes. However, these states are both almost elementary, since mass generation occurs at scales much larger than the binding scale. This is shown schematically in Figure 19.

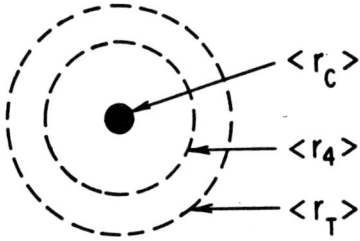

Figure 19. *Scales giving rise to the top-bottom mass hierarchy in the model: $\langle r_C \rangle \sim 1/\Lambda_{Ri}$—binding scale; $\langle r_4 \rangle \sim 1/\Lambda_4$—ties left $\leftrightarrow$ right; $\langle r_T \rangle \sim 1/\Lambda_{TC}$—breaks electroweak symmetry.*

It is easy mechanically to construct family replications in these type of models, obtaining 'by hand' different masses for quarks and leptons. However, these models are unsatisfactory because (Khlebnikov and Peccei 1993):

(i) they give no real understanding of the peculiar hierarchy seen, except mass $\sim$ (size)^2;

(ii) the models have severe technical difficulties, since they have many preons and the vectorial gauge interactions are not asymptotically free;

(iii) the models also have phenomenological difficulties, since they in general have remnant family symmetries leading to a trivial Cabibbo-Kobayashi-Maskawa (CKM) matrix:

$$V_{CKM} = \mathbb{1} .$$

5 The question of forces

Although interactions in the SM are totally defined by specifying the SM gauge group $SU(3) \times SU(2) \times U(1)$ there are a number of open questions which are left unanswered. Some of these questions are:

1. Why are the strong and electroweak forces described by the groups $SU(3)$ and $SU(2) \times U(1)$, respectively? [Theoretically QCD = $SU(5)$ is just as reasonable. Only experiment tells us that the number of colours is 3.]

2. Is there any relation between gravity—the only other force we know—and the SM forces?

3. Are there any other forces in nature besides those of the SM and gravity?

4. Does the remarkable near unification of the SM coupling constants at high energy that we discussed earlier have more than accidental significance?

Let us focus for the moment on this last point. The apparent coupling constant unification at high scales observed could be due to the existence of a Grand Unified Theory (GUT) which suffers spontaneous breakdown of the unifying gauge group G:

$$G \to SU(3) \times SU(2) \times U(1)$$

at the unification scale M_X (Georgi and Glashow 1974; Georgi et al. 1974). This behaviour is shown schematically in Figure 20.

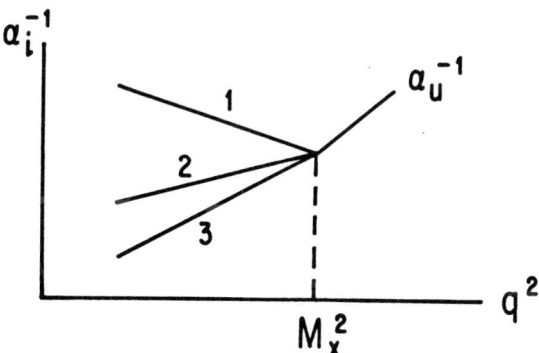

Figure 20. *Schematic plot of the evolution of coupling constants in a GUT spontaneously broken at the scale M_X.*

Let me describe briefly some other nice feature of GUTS, besides coupling constant unification. If G is not a product group, then the idea of grand unification can explain charge quantization, since Q_{em} becomes one of the generators of G. Thus in GUTs the quark charges and the lepton charges are naturally related. Indeed, the known quarks and leptons fit very nicely into representations of simple candidate GUT groups

$$\bar{5} \oplus 10 \subset SU(5); \quad 16 \subset SO(10).$$

It is useful to classify the fermions in terms of left-handed fields only, using $\Psi_L^c = C(\overline{\Psi}_R)^T$. Then for SU(5) (Georgi and Glashow 1974), since Q_{em} is a generator, the $\bar{5}$ must contain d^c and the lepton doublet: [note that $\bar{5} \supset \bar{2}$ of SU(2) and $C\binom{\nu}{e} = \binom{e}{-\nu}$]

$$\bar{5} \; : \; \Psi_L^a \sim \begin{pmatrix} d^c \\ d^c \\ d^c \\ e \\ -\nu \end{pmatrix}_L,$$

while the 10 contains:

$$10 = \Psi_{abL} \sim \frac{1}{\sqrt{2}} \begin{pmatrix} 0 & u^c & -u^c & -u & -d \\ -u^c & 0 & u^c & -u & -d \\ u^c & -u^c & 0 & -u & -d \\ u & u & u & 0 & -e^c \\ d & d & d & e^c & 0 \end{pmatrix}.$$

The group SU(5) has 24 generators. Of these 12 are associated with the generators of SU(3)×SU(2)×U(1) and 12 are additional. Associated with these additional generators are 12 superheavy gauge bosons X, Y with masses of order (M_X), resulting from the spontaneous breakdown

$$\text{SU}(5) \to \text{SU}(3) \times \text{SU}(2) \times \text{U}(1).$$

Schematically, the gauge fields can be organised into a 5×5 matrix

$$\left(\begin{array}{c|c} A_3 - \frac{2}{\sqrt{15}} B & \begin{array}{cc} X & Y \end{array} \\ \hline \begin{array}{c} X \\ Y \end{array} & W + \frac{3}{\sqrt{15}} B \end{array} \right)$$

where A_3 corresponds to the SU(3) bosons, X and Y are the heavy bosons and W, B corresponds to SU(2) and U(1) bosons, respectively.

In SU(5) the electromagnetic charge is $Q_{em} = I_{23} + \sqrt{\frac{5}{3}} I_{24}$. This can be readily checked since, in the fundamental representation, these diagonal generators (normalised to $\text{Tr}\lambda_a \lambda_b = 2\delta_{ab}$) are given by:

$$\frac{\lambda_{23}}{2} = \begin{pmatrix} 0 & & & & \\ & 0 & & & \\ & & 0 & & \\ & & & 1/2 & \\ & & & & -1/2 \end{pmatrix}; \quad \frac{\lambda_{24}}{2} = \frac{1}{\sqrt{15}} \begin{pmatrix} -1 & & & & \\ & -1 & & & \\ & & -1 & & \\ & & & 3/2 & \\ & & & & 3/2 \end{pmatrix}.$$

Comparing with the SU(2)×U(1) definition, $Q_{em} = T_3 + B$, we see that what unifies with g_2 and g_3 is not g_1, but $\sqrt{\frac{5}{3}} g_1$. Thus for an SU(5) GUT one expects

$$g_0(M_X^2) = \sqrt{\tfrac{5}{3}} g_1(M_X^2) = g_2(M_X^2) = g_3(M_X^2),$$

where $g_0(M_X^2)$ is the SU(5) coupling at the breaking scale, M_X. This relationship fixes the weak mixing angle at the GUT scale (Georgi and Glashow 1974):

$$\sin^2 \theta_W(M_X) \equiv \frac{g_1^2(M_X^2)}{g_2^2(M_X^2) + g_1^2(M_X^2)} = \frac{3}{8}.$$

Running this down to M_Z, using as a unification scale $M_X = 10^{16}$ GeV and the 1-loop RGE, gives

$$\sin^2 \theta_W(M_Z) = \begin{cases} 0.194 & \text{GUT} \\ 0.235 & \text{SUSY GUT} \end{cases}$$

to be compared with the experimental value of $\sin^2 \theta_W(M_Z) = 0.2260 \pm 0.0024$. (The data is summarised in the LEP collaborations 1992.) A more refined calculation (see, for example, Langacker and Polonsky, 1993) actually improves the SUSY GUT value to below 0.230, whereas the GUT with ordinary matter still has problems. The GUT value can be improved to $\sin^2 \theta_W(M_Z) = 0.212$ by using $M_X = 10^{14}$ GeV but this leads to other difficulties, predicting too high a rate for proton decay. Clearly the above analysis is the converse of the extrapolation from low energy to high energy discussed earlier and it shows that unification with SUSY works, but otherwise it does not—at least naïvely.

Let me discuss the issue of proton decay further. Since leptons and quarks are in the same SU(5) representations, X or Y boson exchange can change quarks into leptons, mediating proton decay. Figure 21 shows an example of a diagram leading to proton instability, as a result of X exchange.

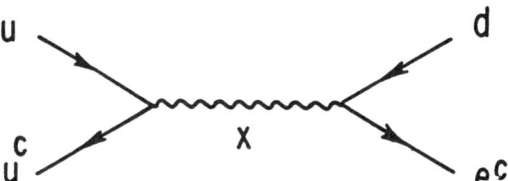

Figure 21. GUT *transitions leading to proton decay.*

Since M_X is very heavy, one gets an effective, 4-fermion Lagrangian for this transition (Buras et al. 1978)

$$\mathcal{L}_{\text{eff}} = \sqrt{2}\pi \frac{\alpha_u(M_X^2)}{M_X^2} \left\{ \varepsilon_{abc}(\overline{u}_{cL}^c \gamma^\mu u_{bL}) \times (2\overline{e}_L^c \gamma^\mu d_{aL} + \overline{e}_R^c \gamma^\mu d_{aR}) \right\}.$$

where $\alpha_u(M_X^2)$ is the grand unified coupling. The above Lagrangian gives a decay rate proportional to M_X^{-4}, or a proton lifetime for the dominant SU(5) $p \to e^+ \pi^0$ mode (Langacker and Polonsky 1993)

$$\tau(p \to e^+ \pi^0) = 10^{31 \pm 0.7} \left\{ \frac{M_X(\text{GeV})}{4.6 \times 10^{14}} \right\}^4 \text{ years},$$

which is in conflict with the experimental bound (Hirata et al. 1988, Totsuka 1992)

$$\tau(p \to e^+ \pi^0)_{\text{exp}} > 10^{33} \text{ years},$$

if $M_X = 10^{14}$ GeV.

SUSY GUTs fare better in this respect, as $M_X \sim 10^{16}$ GeV gives

$$\tau(p \to e^+\pi^0) = (10^{35}\text{–}10^{36}) \text{ years}.$$

However, in these theories, one must be careful since there are other contributions to proton decay besides X and Y exchange (Weinberg 1982, Sakai and Yanagida 1982). In particular, there is another process in SUSY GUTs involving neutralinos, $\tilde{\delta} = \{\tilde{\gamma}, \tilde{Z}, \tilde{H}_1, \tilde{H}_2\}$ and a coloured Higgsino, $\tilde{H}_3$, whose core really involves a dimension-5 operator containing two fermions and two sfermions. The box diagram involving this dimension-5 operator is shown in Figure 22.

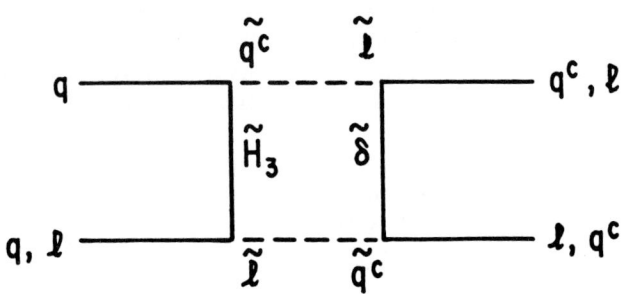

Figure 22. *Dangerous dimension-5 graph contributing to proton decay in SUSY GUTs.*

Due to this dimension-5 operator, the proton lifetime is

$$\tau[\text{dim 5}] \sim \frac{1}{h^4}\frac{1}{\alpha_U^2}\left[M_{\tilde{H}_3}\widetilde{m}_{\text{eff}}\right]^2,$$

where the $1/h^4$ comes from the Yukawa couplings. Because of the appearance of the neutralino mass $\widetilde{m}_{\text{eff}}$, even taking $M_{\tilde{H}_3} \sim M_X$ gives a lifetime only proportional to M_X^2, not M_X^4. To escape troubles with experiment, this requires both $M_{\tilde{H}_3}$ to be big, $M_{\tilde{H}_3} \gg M_X$ and the SUSY breaking parameter, $\widetilde{m}_{\text{eff}}$ not to be too light. The expected range is very uncertain [a recent careful analysis by Arnowitt and Nath 1992, shows that under certain circumstances this lifetime can be above the Kamiokande limits (Hirata et al. 1988, Totsuka 1992)]

$$\tau[\text{dim 5}] \simeq 10^{29\pm4} \text{ years},$$

with the dominant decay modes being $p \to K^+ \nu_\tau$ and $p \to K^0 \mu^+$.

Since the present limits for these modes are already near 10^{32} years, (Hirata et al. 1988, Totsuka 1992) there is some interest in SUSY GUTs models with essentially no dimension-5 operators, such as flipped SU(5) (Antoniadis et al. 1988). Flipped SU(5) is not a true GUT, since the gauge group is SU(5)×U(1), but the extra U(1) allows flipping the assignments of quarks and leptons: $\nu \leftrightarrow e$ and $u \leftrightarrow d$. For instance, one has in this case

$$\bar{5} = \begin{pmatrix} u^c \\ u^c \\ u^c \\ \nu \\ e \end{pmatrix}_L \quad ; \quad 1 = e_L^c.$$

Flipped SU(5) arises quite naturally in various superstring models. Thus, this group is of some intrinsic interest anyway. What makes it more interesting in this context is that, as a result of the flipped assignments, what sits in the 10 of flipped SU(5) is not e_L^c but ν_L^c. Thus in flipped SU(5), one can break the GUT group to the SM using Higgs particles which are in the 10-dimensional representation of SU(5), because a scalar with the quantum numbers of ν_L^c has a VEV which preserves the SM group. The Higgs triplet of Figure 22 in flipped SU(5) can get mass from this VEV and one can well imagine that

$$\langle 10_{\text{Higgs}} \rangle \gg M_X$$

leading to a large mass for $M_{\widetilde{H}_3}$. [The colour triplet Higgsinos in the SUSY SU(5) extension of the SM can get their mass via a superpotential term coupling two 10_{Higgs} with a $\overline{5}_{\text{Higgs}}$.]

I want to, finally, comment briefly about another possible GUT: SO(10) (Fritzsch and Minkowski 1975, Georgi 1975). Perhaps the nicest feature of SO(10) as a GUT is that all fermions sit in a single representation of this group, the 16-dimensional representation. The 16 of SO(10) decomposes in terms of SU(5) as:

$$16 \longrightarrow 10 + \overline{5} + 1.$$

The SU(5) singlet here corresponds to ν_L^c. Because SO(10) is a rank-5 group, it can be broken to the rank-4 SM either directly or in two steps (sequential breaking). In the first case there is only one mass scale

$$\text{SO}(10) \underset{M_X}{\longrightarrow} \text{SU}(3) \times \text{SU}(2) \times \text{U}(1)$$

and one finds the same difficulties with $\sin^2 \theta_W$ and proton decay, in the absence of SUSY, as with an SU(5) GUT. In the second case, there are two mass scales and the GUT breaking scale can be at $M_X = 10^{16}$ GeV, leading to perfectly consistent GUTs without SUSY. The chains

$$\text{SO}(10) \underset{M_X = 10^{16}\text{GeV}}{\longrightarrow} \text{SU}(2)_L \times \text{SU}(2)_R \times \text{SU}(4) \underset{M_I = 10^{12}\text{GeV}}{\longrightarrow} \text{SU}(3) \times \text{SU}(2) \times \text{U}(1)$$

or

$$\text{SO}(10) \underset{M_X = 10^{16}\text{GeV}}{\longrightarrow} \text{SU}(2)_L \times \text{SU}(2)_R + \text{U}(1)_{B-L} \times \text{SU}(3) \underset{M_I = 10^{10}\text{GeV}}{\longrightarrow} \text{SU}(3) \times \text{SU}(2) \times \text{U}(1)$$

give measurable values for the proton lifetime and $\sin^2 \theta_W$ in agreement with experiment, but are not very predictive otherwise. The presence of an intermediate mass scale has beneficial features related to neutrino masses, since it gives a natural mechanism, the see-saw mechanism, (Yanagida 1979, Gell-Mann et al. 1980) for explaining why $m_\nu \ll m_e$ (see below). However, it also introduces another hierarchy problem; why is $M_X \gg M_I \gg v_F$?

It is difficult to be fully convinced of GUTs until the 'smoking gun' of proton decay is found. There may be some hope here, since the new SuperKamiokande detector, being built now in Japan, will push the limits on the proton lifetime to $\tau_p \sim 10^{33}$–10^{34} years.

Nevertheless, GUTs do possess enough attractive features to make their continued investigation very worthwhile. One of these features is connected with the matter-antimatter asymmetry in the universe discussed in Kolb's lectures (these Proceedings). The production of a matter-antimatter asymmetry in the early universe requires that baryon number violating processes should go out of equilibrium in the early universe, and this can happen due to B-violating decays of the heavy X and H_3 states in GUTs. It remains to be seen, however, whether the baryon to photon ratio $n_B/n_\gamma \approx 10^{-10}$ originates from $B - L$ violating processes at the GUT scale (we owe our existence to GUTs) or whether it originates from $B + L$ violating processes at the electroweak scale (we owe our existence to the SM).

6 The question of matter

Perhaps the hardest open problems in the SM are related to the question of matter. Three of these questions are particularly striking:

(i) Why do we see the fermion representations we see in nature?

(ii) Why are the fermions that make up the matter we see chiral?

(iii) Why do quarks and leptons appear in repetitive family structures and why, apparently, are there only three families?

In my opinion (ii) is probably answered by Georgi's survival hypothesis (Georgi H 1979), but the other two are very deep. In fact, ultimately, if we really understood (i) and (iii) I believe we could calculate fermion masses and mixings.

Georgi's survival hypothesis can be succinctly summarised as follows: fermions, unless prohibited by their chiral assignments, will get a mass consistent with the largest scale in the theory. Thus chiral fermions are the only ones which are light. All the examples we discussed are in accord with this hypothesis. Certainly preon models are built this way, since chiral fermions are the ones which are light. This is also true in GUTs. For instance in SO(10) fermions are in the 16 representation only. If they were in both the 16 and the $\overline{16}$, one could write an invariant mass term

$$\mathcal{L}_{\text{mass}} = M\,\overline{16} \cdot 16$$

and there would be no reason for M not to be of $O(M_X)$.

Note that right-handed neutrinos are not in a chiral representation of the SM, since $\nu_R \sim 1$. Thus they are allowed to have a large Majorana mass term

$$\mathcal{L}_{\text{mass}} = \frac{m_R}{2}(\nu_R^T\, C\, \nu_R + \bar\nu_R^T\, C\, \bar\nu_R).$$

Such a term could explain why neutrinos are light via the see-saw mechanism, (Yanagida 1979, Gell-Mann et al. 1980) since including Dirac masses (of order m_ℓ) gives a neutrino mass matrix with vastly different eigenvalues:

$$\begin{pmatrix} 0 & m_D \\ m_D & m_R \end{pmatrix} \iff \begin{matrix} (m_\nu)_{\text{light}} \simeq m_D^2/m_R \ll m_D \\ (m_\nu)_{\text{heavy}} \simeq m_R. \end{matrix}$$

As we saw, GUTs provide a concise way to characterise the fermion matter we know; for instance, by classifying the quarks and leptons in the $\bar{5}$ and 10 representations of SU(5) or the 16 of SO(10). However GUTs do not fix the representation content of the fermions. [There is a mild restriction on which representations are allowed coming from the requirement that there should be no gauge anomalies. In SU(5) having a $\bar{5}$ and 10-dimensional representation guarantees this to be the case. However, a matter content made up of four $\bar{5}$ and a 15 would also be anomaly free, but this is not nature!] Nevertheless, some progress in understanding fermion mass patterns emerges from GUTs, since the larger symmetry can restrict the form of the Yukawa couplings, if only a limited number of Higgs representations are included. In the case of SU(5), with a $\bar{5}$ of Higgs particles, H_1, plus another 5 of Higgs particles, H_2, one has only two distinct Yukawa couplings

$$\mathcal{L}_{\text{Yukawa}} = \Gamma_1 \bar{5}_f 10_f \bar{5} + \Gamma_2 10_f 10_f 5.$$

Then $\langle H_1 \rangle \neq 0$ gives a common mass to the leptons and to the charge $-1/3$ quarks, while $\langle H_2 \rangle \neq 0$ gives a mass to the charge $+2/3$ quarks.

The equality of the lepton and $-1/3$ quark mass matrices at the GUT scale implies relations between masses at low energy via the RGE (Chanowitz et al. 1977)

$$\frac{m_D(q^2)}{m_L(q^2)} = \frac{m_D(M_X^2)}{m_L(M_X^2)} \left(\frac{\alpha_3(q^2)}{\alpha_3(M_X^2)} \right)^{4/b_3}.$$

In SU(5), with just $H_1 = \bar{5}$ as a Higgs, the GUT scale ratio is unity. In this case the above equation gives a good prediction of m_b/m_τ in ordinary SU(5) (where $b_3 = 7$) (Chanowitz et al. 1977)

$$m_b/m_\tau \simeq 2.8,$$

but not such a good prediction in SUSY SU(5), since here $b_3 = 3$ gives $m_b/m_\tau \simeq 4$. However, incorporating two-loop effects one can obtain agreement also for SUSY SU(5) provided a large region in $\tan \beta = v_2/v_1$ is excluded (Langacker and Polonsky, 1994). I shall return to this point below. However, other predictions are not so good. For example, the equality of the lepton and charge $-1/3$ quark matrices at the GUT scale predicts also that

$$\frac{m_s}{m_d} = \frac{m_\mu}{m_e}.$$

But the LHS is around 20 from current algebra estimates (Gasser and Leutwyler 1982), whereas the RHS is 206 from experiment.

This circumstance has generated a lot of interest in having a better set of GUT mass matrix relations. A representative example is the work of Dimopoulos, Hall and Raby (1992; see also Anderson et al. 1993) which uses mass matrices which do not quite have $m_D = m_L$ at M_X, but have a pattern (Georgi-Jarlskog pattern) (Georgi and Jarlskog 1979) and a texture of zeros first suggested by Fritzsch (1977/78):

$$M_D = \begin{pmatrix} 0 & F & 0 \\ F & E & 0 \\ 0 & 0 & D \end{pmatrix} ; \quad M_L = \begin{pmatrix} 0 & F & 0 \\ F & -3E & 0 \\ 0 & 0 & D \end{pmatrix} ; \quad M_U = \begin{pmatrix} 0 & C & 0 \\ C & 0 & B \\ 0 & B & A \end{pmatrix}.$$

The factor of 3 in the (2,2) matrix element of M_L is the Georgi-Jarlskog factor (Georgi and Jarlskog 1979). This factor is related to the number of colours. Instead of a $\bar{5}$,

Georgi and Jarlskog use a $\overline{45}$ Higgs, $H_\gamma^{\alpha\beta}$, with $\{\alpha,\beta,\gamma = 1,\cdots 5\}$, which is antisymmetric in α and β and traceless. What gives mass to M_D and M_L is $\langle H_\gamma^{\alpha\beta}\rangle = \delta_\gamma^\beta \lambda_\alpha$ with λ_α being the diagonal elements of a 4×4 matrix. Since there are 3 colours, and we do not want the VEV to break colour, it follows that $\lambda_\alpha = \{1,1,1,-3\}$. It produces a different pattern of GUT mass relations. Namely, in this case one finds:

$$m_b(M_X) = m_\tau(M_X); \quad m_s(M_X) = \frac{1}{3}m_\mu(M_X); \quad m_d(M_X) = 3\,m_e(M_X),$$

which fixes correctly the m_s/m_d to m_μ/m_e ratio—since this ratio, essentially, does not evolve.

This pattern follows rather naturally in an SO(10) GUT (Harvey et al. 1980/82) as a result of a Higgs 45 breaking. However, in an SO(10) GUT with a Higgs in the 10-dimensional representation to provide the principal Yukawa couplings, one would also expect that $D = A$. That is, that there is a relation between $m_t(M_X)$ and $m_b(M_X)$. Indeed, since the SO(10) 10 representation contains both an SU(5) 5 and a $\bar{5}$, it is easy to see that in SO(10) the two SU(5) Yukawa couplings, Γ_1 and Γ_2, just reduce to a single Yukawa coupling

$$\mathcal{L}_{\text{Yukawa}} = \Gamma\, 16_f\, 16_f\, 10\,.$$

The relation

$$m_b(M_X) = m_t(M_X)$$

is, however, difficult to evolve to low energy to obtain realistic values of m_t. A recent analysis in the context of SUSY GUT theories is shown in Figure 23. Only for large $\tan\beta$ is there a line where the unification of all 3 Yukawa couplings is allowed, while much more of the $\tan\beta$ plane is allowed by not unifying also the top quark Yukawa.

Interesting predictions for mass ratios and mixing angles follow from the mass matrices and textures at M_X assumed by Dimopoulos et al. (1992 and Anderson et al. 1993). Using the freedom of phase redefinition for the quark and lepton fields, the number of parameters in these mass matrices is reduced to seven—6 real parameters and a phase. Thus the effective mass matrices at M_X are given by

$$M_D = \begin{bmatrix} 0 & fe^{i\phi} & 0 \\ fe^{-i\phi} & e & 0 \\ 0 & 0 & d \end{bmatrix}; \quad M_L = \begin{bmatrix} 0 & f & 0 \\ f & -3e & 0 \\ 0 & 0 & d \end{bmatrix}; \quad M_U = \begin{bmatrix} 0 & c & 0 \\ c & 0 & b \\ 0 & b & a \end{bmatrix}.$$

Because Dimopoulos et al. (1992; see also Anderson et al. 1993) assume a SUSY GUT theory, in addition to the above parameters, one also has $\tan\beta$ as a free parameter. These unknowns in the model are fixed by using as input the experimental masses and mixing angles which are best known: m_e, m_μ, m_τ; $m_c(m_c)$; $m_b(m_b)$; m_u/m_d (Gasser and Leutwyler 1982); $|V_{us}|$, $|V_{cb}|$ (PDG 1992). This then allows 6 predictions, one of them being $\sin\beta$. The results are:

$$m_t(m_t) = 176\,\text{GeV} \left[\frac{m_b}{4.25\,\text{GeV}}\right]\left[\frac{m_c}{1.25\,\text{GeV}}\right]\left[\frac{0.053}{V_{cb}}\right]^2$$

$$m_s(1\,\text{GeV}) = 156\,\text{MeV}$$

$$m_d(1\,\text{GeV}) = 6.2\,\text{MeV}$$

$$|V_{ub}|/|V_{cb}| = 0.05\left[\frac{m_u/m_d}{0.6}\right]^{1/2}\left[\frac{1.25\,\text{GeV}}{m_c}\right]^{1/2}$$

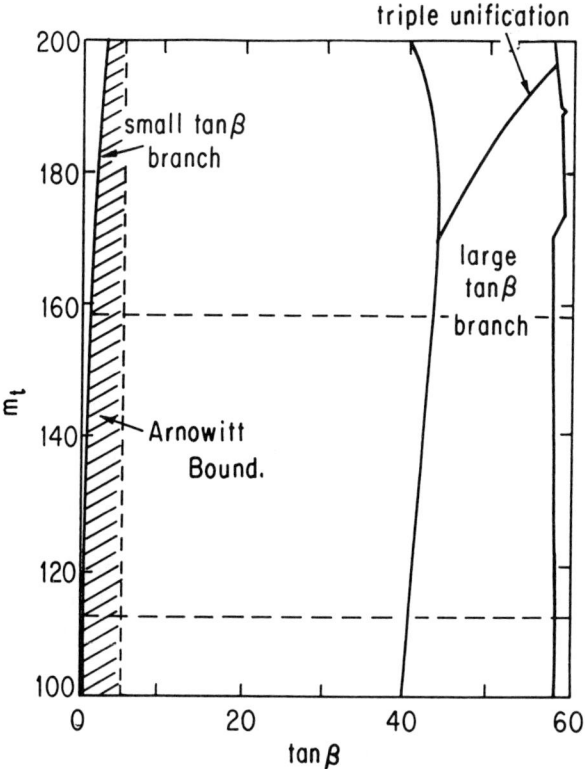

Figure 23. *Allowed values in the m_t-$\tan\beta$ plane assuming that some Yukawa couplings unify at M_X. The triple unification line is shown. The values from $\tan\beta \simeq 1$ to $\tan\beta \simeq 40$ are excluded. The Arnowitt bound (Arnowitt and Nath 1992) comes from an analysis of proton decay in these models. Adapted from Langacker and Polonsky 1994.*

$$\cos\delta = (0.53\pm0.07)\left[\frac{0.6}{m_u/m_d}\right]^{1/2}\left[\frac{m_c}{1.25\,\text{GeV}}\right]^{1/2} - 0.13\left[\frac{m_u/m_d}{0.6}\right]^{1/2}\left[\frac{1.25\,\text{GeV}}{m_c}\right]^{1/2}$$

$$\sin\beta = 0.95\left[\frac{4.25\,\text{GeV}}{m_b}\right]^5\left[\frac{m_c}{1.25\,\text{GeV}}\right]\left[\frac{0.053}{V_{cb}}\right]^2\left[\left[\frac{4.25\,\text{GeV}}{m_b}\right]^{12} - 0.08\right]^{1/2}.$$

These results appear quite reasonable, but one needs real data on m_t, δ and β for a true test. I note here, parenthetically, that the value of V_{cb} more consonant with the latest data on the B lifetime is nearer to 0.043 (Witherell 1993) than the value used here of 0.053.

Although the above attempts may eventually bear fruit, they do not really provide a deep answer for the nature of fermionic matter. Contrast what we know about fermions with the situation regarding gauge bosons. Given G we know immediately that the gauge bosons are in the adjoint representation; therefore the representation of the forces is fixed by G. Why are the fermions in nature in the representations that we see? One could answer this question if there was a SUSY where, instead of having separate chiral and vector multiplets describing matter and forces, one unified matter and forces by having only vector multiplets. Unfortunately, this does not work! The adjoint representation is a real representation and always includes chirally paired representations of fermions upon decomposition.

This problem is nicely illustrated by the following example. Consider a GUT E_6, whose adjoint is a 78. If we decompose the 78 of E_6 into representations of the SO(10) subgroup, we find

$$78 = \underbrace{45 \oplus 1}_{\text{Real}} \oplus \underbrace{16 \oplus \overline{16}}_{\substack{\text{Chirally} \\ \text{paired}}}.$$

The question arises; are there ways in this example where one can get rid of one of the chiral pairs and of the real states? If one could eliminate the real fermions and suppress the wrong-chirality fermions then one could really claim to have achieved a unification between forces and matter!

This is, roughly, what happens in superstring theories. I will close these lectures by briefly indicating how this remarkable fact comes about. These speculations connected to superstrings are, in a sense, in a class of their own. Indeed, they give really beautiful insights into what may be the nature of matter and forces. Other ideas I have discussed here—like preon models—are very much more modest and mechanical. For instance, in the composite model we discussed one got colour triplets of quarks by starting from colour anti-triplets of preons, using the fact that $\overline{3} \times \overline{3} \supset 3$. If this were to be a *real theory*, one would want to eventually understand why the preons in nature come in $\overline{3}$ of colour! These questions are premature in preon models, but are perfectly sensible ones to ask *ab initio* in superstring theories.

String theories were known, almost from the beginning, to require extra dimensions of space-time for their physical consistency (for a comprehensive exposition see, Green et al. 1987). Whether these extra dimensions are actually real or not, however, is not so germane anymore, as one has been able to formulate these theories directly in 4 dimensions by associating the extra dimensions with internal degrees of freedom (see, for example, Kawai et al. 1987 and Lerche et al. 1987). Nevertheless, to understand how some *unwanted* fermions are made to disappear in string theories, it is useful to retain the idea of having these extra dimensions, for it is through the process of *compactifications* of these extra dimensions that chiral fermions are most easily seen to emerge.

It has been known for a long time that physical theories in $D > 4$ can give sensible $D = 4$ theories when the extra dimensions compactify. For example, in the original theory of Kaluza and Klein (1921/26) gravity in D=5, when one of the dimensions was compactified, gave rise to gravity and electromagnetism in D=4. Consider then chiral fermions in $D > 4$ dimensions. Chiral fermions obey the massless Dirac equation

$$\Gamma^\alpha D_\alpha \Psi = 0.$$

where $\alpha = 1, \ldots D$ and D_α is the covariant derivative containing background fields (gravity, Yang-Mills fields, *etc.*). It is easy to see formally that, if the theory compactifies to 4 dimensions, the erstwhile chiral fermions get a mass:

$$(\Gamma^\mu D_\mu + \Gamma^a D_a)\Psi = 0.$$

The first term in the above is the 4-D Dirac kinetic energy operator, while the second term, containing the D−4 dimensional contribution ($a = 5, \ldots, D$), acts precisely as a mass term.

Although, in general, $\Gamma^a D_a$ acts as a mass in four dimensions, this will not be the case if there are 'zero modes' solutions in the D−4 dimensional compactified space

$$\Gamma^a D_a \Psi = 0.$$

If the above holds, there will remain chiral fermions in D=4. This is a constraint equation in the D−4 dimensional compact space, K. In general, solutions to this constraint equation come in repetitive patterns, which depend on the geometry of K. Usually the number of solutions of $\Gamma^a D_a \Psi = 0$ is related to a topological index of K (Wetterich 1983 and Witten 1983). In this scenario not only the type of chiral fermions (matter) but also the number of families is the result of the geometry involved!

A very similar mechanism to the one just outlined obtains in string theories. Before describing a particular realization, it may be useful to very briefly summarise the principle features of string theories (Green et al. 1987):

1. One replaces particles with normal modes of a one-dimensional 'string' ($m = 0$ excitations) and then quantises these excitations.

2. The theories are only consistent if specific gauge groups are attached to the excitations and the dimension of space-time is $D = 10$ for superstrings (or 26 for ordinary strings).

3. The massless spectra contains a spin-2 graviton and the theories give a consistent and finite quantum theory of gravity.

4. In these theories one replaces the point interactions of field theory by string interactions occurring on surfaces.

To illustrate how things could work out, I shall describe a scenario (Candelas et al. 1985) associated with the heterotic string, where $D = 10$ is compactified to $D = 4$ on a 6-dimensional Calabi-Yau manifold (this is only one of many possibilities). The gauge group associated with this theory is (Gross et al. 1985/86)

$$G = E_8 \times E_8,$$

where the first E_8 corresponds to the real world, and the second to the 'hidden' world (ignored here). In 10 dimensions, one can have Weyl fermions (chiral fermions) which obey a Majorana condition and thus are suitable for transforming according to the adjoint representation (Slansky 1981). For the heterotic string these chiral fermions are gauginos (SUSY partners of the E_8 gauge bosons). The SU(3)×E_6 decomposition of the 248 gauginos (248 is the dimension of the adjoint representation of E_8) has a similar pattern of the earlier E_6 example:

$$248 = \underbrace{(1, 78) \oplus (8, 1)}_{\substack{\text{Real} \\ \text{representations}}} \oplus \underbrace{(3, 27) \oplus (\overline{3}, \overline{27})}_{\substack{\text{Complex} \\ \text{conjugate pairs}}}.$$

The 6-dimensional manifold, K, where the compactification is assumed to occur (Candelas et al. 1985) is taken to be a Calabi-Yau manifold which has the special property that the zero modes of the Dirac equation on K have non-trivial SU(3) properties:

$$\Gamma^a D_a \Psi = 0.$$

Thus in this case the zero modes are: $\Psi \sim (3, 27)$ and $(\bar{3}, \overline{27})$. Furthermore, in the compactification the gauge group is reduced to E_6, or one of its subgroups:

$$\begin{array}{ccc} E_8 & \longrightarrow & E_6. \\ D=10 & & D=4 \end{array}$$

In terms of E_6, the chiral matter that appears consists of

$$n_f \cdot 27 + \delta(27 + \overline{27}),$$

with both n_f (the number of families!) and δ related to the topological properties of K. In fact, Candelas et al. (1985) show that $n_f = \frac{1}{2}|\chi|$, where χ is the Euler characteristic of K. Note that the matter here is that of the SM (or of an SU(5) GUT), with some extra matter degrees of freedom which are vectorial:

$$27 = \underbrace{\bar{5} \oplus 10}_{\text{chiral}} \oplus \underbrace{1 \oplus 5 \oplus \bar{5} \oplus 1}_{\text{vectorial}}$$

$\uparrow$
E_6

Two remarks are in order which transcend this simple example:

(i) In general, for reasonably low levels of the associated string Kac-Moody algebra (for a discussion of this point see, for example, Ginsparg 1987), one cannot be left with a GUT group after compactification because matter scalars do not contain the adjoint representation. With scalars in an E_6 78 one can effect the breaking:

$$E_6 \underset{\langle 78 \rangle}{\longrightarrow} SU(3) \times SU(2) \times U(1),$$

but this is not possible with scalars in the 27:

$$E_6 \underset{\langle 27 \rangle}{\not\longrightarrow} SU(3) \times SU(2) \times U(1).$$

Thus superstrings and SUSY GUTs appear to be not truly compatible! There is unification in superstrings, but it occurs at the Planck scale, with (Ginsparg 1987)

$$g_3^2 k_3 = g_2^2 k_2 = g_1^2 k_1 = \frac{4\pi G_N}{\alpha'},$$

where G_N is Newton's gravitational constant and α' is the string slope. Here k_1, k_2 and k_3 are numbers associated with the Kac-Moody levels of the string (e.g. $k_3 = k_2 = 1$, $k_1 = 5/3$). Thus the 'experimental' unification of couplings at

$M_X = 10^{16}$ GeV appears to be a coincidence for superstring theory! What happens in many examples is that the compact manifold K is not simply connected. In this case then non-trivial gauge field configurations (Wilson loops) can be trapped in K, leading to a breakdown of the gauge group, much as if there was an effective scalar in the adjoint (Hosotani 1983, Witten 1985). Typical breakdown chains for E_6 produced by this, so called, Hosotani-Witten mechanism are (for a more detailed exposition see, for example, Ibañez 1990):

$$E_6 \longrightarrow SU(3) \times SU(2) \times U(1)^n$$
$$E_6 \longrightarrow SU(5) \times U(1)$$
$$E_6 \longrightarrow SU(3) \times SU(3) \times SU(3).$$

To get rid of some of the extra gauge interactions—as in the case of the flipped SU(5) chain (Antoniadis et al. 1988)—one needs some intermediate scale breaking, something that is also quite against the simple picture of SUSY GUT unification discussed earlier.

(ii) More on the positive side, compactification (either real or effective) has a very nice advantage since it easily generates Yukawa couplings. The $D > 4$ gauge couplings can be transmuted into Yukawa couplings in four dimensions, since the gauge fields in the extra dimensions become Higgs fields in four dimensions after compactification. This is shown schematically in Figure 24.

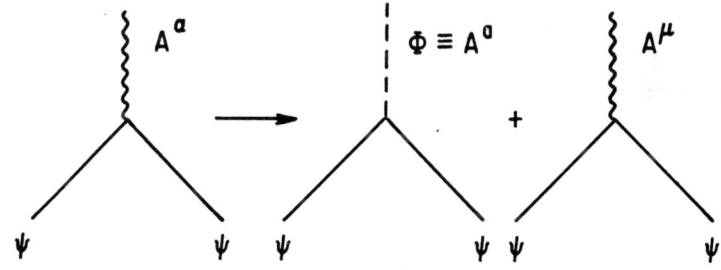

Figure 24. *Generation of Yukawa couplings through compactification.*

Although one cannot really compute the Yukawa couplings (that would require a complete normal mode decomposition of Ψ and A^a on the compact space) one can, in certain circumstances, infer some information for a given manifold K on discrete symmetries which can give the overall texture of the mass matrix (*e.g.* which elements are zero). Some encouraging results exist (Greene et al. 1987), but these mass matrices are not quite ready to be confronted with experiment yet!

Concluding remarks

I hope I have given you in these lectures a panorama of the ideas that have been put forth to understand the deep structures underlying the SM, summarised by the 3 questions:

- the question of mass;
- the question of forces;
- the question of matter.

I do not believe, however, that the answers to these questions will come forth from purely theoretical arguments, although a good theory argument is always nice! But I do believe that there is a good chance that, with the experiments now being planned with present accelerators (before the year 2000) and, certainly, with future accelerators (LHC and an e^+e^- linear collider—before the year 2010) we will get a glimmer of which (if any!) of these ideas correctly describe nature.

Acknowledgements

I am grateful to David Saxon for inviting me to lecture at St. Andrews. The Scottish Summer School had an extremely spirited and very engaged group of students who made the delivery of my lectures a real challenge and a great pleasure. I also owe a great debt of gratitude to Lance Vick, for without his invaluable help the written version of my notes would have remained a chimera. This work was supported in part by the Department of Energy under Grant DE-FG03- 91ER40662.

References

Aizenman M, 1982, *Comm Math Phys* **86** 1; Frölich J, 1982, *Nucl Phys* **B200** [FS4] 281.
Akiba T and Yanagida T, 1986, *Phys Lett* **169B** 432.
Altarelli G, 1989, *Ann Rev Nucl Part Sci* **39** 357.
Altarelli G, Barbieri R, 1991, *Phys Lett* **B253** 161; Altarelli G, Barbieri R and Jadach S, 1992, *Nucl Phys* **B369** 3.
Alvarez-Gaumé L, Claudson M and Wise M, 1982, *Nucl Phys* **B207** 96.
Amaldi U, de Boer W and Fürstenau H, 1991, *Phys Lett* **B260** 447.
Anderson G, Raby S, Dimopoulos S and Hall L, 1993, *Phys Rev* **D47** 3707.
Anderson P W, 1984, *Basic Notions of Condensed Matter Physics* (Addison-Wesley, Reading, Mass.).
Antoniadis A, Ellis J, Hagelin J and Nanopoulos D, 1988, *Phys Lett* **B194** 231; *Phys Lett* **B205** 459; 1989, *Phys Lett* **B208** 209.
Appelquist T, Karabali D and Wijewardhana L, 1986, *Phys Rev Lett* **57** 982.
Appelquist T and Triantiphyllou G, 1992, *Phys Lett* **B278** 345.
Arnowitt R and Nath P, 1992, *Phys Rev Lett* **69** 725; *Phys Lett* **287B** 89; *Phys Rev* **D46** 3981.
Barbieri R, Caravaglios F and Frigeni M, 1992, *Phys Lett* **B279** 169.

Barger V, Ohmann P and Phillips R J N, 1993a, *Phys Rev* **D47** 1093.
Barger V, Berger M S, Ohmann P and Phillips R J N, 1993b, *Phys Lett* **B314** 35.
Buras A J, Ellis J, Gaillard M K and Nanopoulos D V, 1978, *Nucl Phys* **B135** 66.
Candelas P, Horowitz G T, Strominger A and Witten E, 1985, *Nucl Phys* **B258** 46.
Chanowitz M S, Ellis J and Gaillard M K, 1977, *Nucl Phys* **B128** 506.
Chanowitz M, 1993, in *Proceedings of the XXVI International Conference on High Energy Physics, Dallas, Texas*, ed Sanford J (AIP, New York, NY 1993).
Cremmer E, Ferrara S, Girardello L, and Van Proeyen A, 1983, *Nucl Phys* **B212** 413.
Dimopoulos S and Ellis J, 1981, *Nucl Phys* **B182** 505.
Dimopoulos S, Hall L and Raby S, 1992 *Phys Rev Lett* **68** 752; *Phys Rev* **D45** 4192.
Dimopoulos S and Susskind L, 1979, *Nucl Phys* **B155** 237.
D0 Collaboration, 1994, Fermilab Preprint (1994).
Eichten E and Lane K D, 1980,*Phys Lett* **90B** 237.
Ellis J, Kelley S and Nanopoulos D V, 1990, *Phys Lett* **B249** 441.
Ellis J, Ridolfi G and Zwirner F, 1991, *Phys Lett* **B257** 83;
Fritzsch H and Minkowski P, 1975, *Ann Phys (NY)* **93** 173.
Fritzsch H, 1977 *Phys Lett* **70B** 436; 1978 *Phys Lett* **73B** 317.
Gasser J and Leutwyler H, 1982, *Phys Rept* **87C** 77.
Gell-Mann M, Ramond P and Slansky R, 1980, in *Supergravity*, ed Van Nieuwenhuizen P (North Holland, Amsterdam 1980).
Georgi H and Glashow S L, 1974, *Phys Rev Lett* **32** 32.
Georgi H, 1975, in *Particles and Fields 1974*, ed Carlson C E, (AIP, New York, NY 1975).
Georgi H, 1979, *Nucl Phys* **B156** 126.
Georgi H and Jarlskog C, 1979, *Phys Lett* **86B** 297.
Georgi H, Quinn H and Weinberg S, 1974, *Phys Rev Lett* **33** 451.
Ginsparg P, 1987, *Phys Lett* **197B** 139.
Girardello L and Grisaru M, 1984, *Nucl Phys* **B194** 65.
Green M, Schwarz J and Witten E, 1987, *Superstring Theory* (Cambridge University Press, Cambridge, UK 1987).
Greene B, Kirklin K H, Miron P J and Ross G G, 1987, *Nucl Phys* **B292** 606.
Griest K, 1993, *Ann N.Y. Acad Sci* **688** 390.
Gross J D, Harvey J A, Martinec E and Rohm R, 1985, *Nucl Phys* **B256** 257; 1986, *Nucl Phys* **B267** 75.
Gross D and Wilczek F, 1973 *Phys Rev Lett* **30** 1343; *Phys Rev* **D8** 3633.
Gunion J F, Haber H E, Kane G and Dawson S, 1990, *The Higgs Hunter's Guide* (Addison-Wesley, Redwood City, California 1990).
Haber H and Hempfling R, 1991 *Phys Rev Lett* **66** 1815.
Harvey J, Ramond P and Reiss D B, 1980, *Phys Lett* **B92** 309; 1982 *Nucl Phys* **B199** 223.
Hirata K S *et al.* , 1988, *Phys Lett* **205B** 416.
Holdom B, 1981, *Phys Rev* **D24** 1441.
Holdom B, 1990, *Phys Lett* **B246** 169.
Hosotani Y, 1983, *Phys Lett* **126B** 309; *Phys Lett* **129B** 193.
Hsu S D and Sundrum R, 1993, *Nucl Phys* **B391** 127.
Ibañez L and Ross G G, 1982, *Phys Lett* **B110** 215.
Ibañez L, 1990, *Proceedings of the 1990 CERN School of Physics*, Mallorca, Spain 1990.
Inoue K, Kakuto A, Komatsu H and Takashita S, 1982, *Prog Theo Phys* **68** 927; 1986, *ibid* **71** 412.
Kaluza Th, 1921, *Sitz Preuss Akad Wiss Berlin, Math Phys* **K1** 966; 1926, Klein O, *Nature* **118** 516.
Kawai H, Lewellen D and Tye S H, 1987, *Nucl Phys* **B288** 1.

Khlebnikov S and Peccei R D, 1993, *Phys Rev* **D48** 361.
Kolb E W and Turner M, 1990, *The Early Universe* (Addison-Wesley, Redwood City, California 1990).
Landau L, 1955 in *Niels Bohr and The Development of Physics*, ed Pauli W, (McGraw-Hill, NY 1955).
Lane K D, 1974, *Phys Rev* **D10** 2605.
Langacker P, 1993, *Ann N.Y. Acad Sci* **688** 34.
Langacker P and Luo M, 1991, *Phys Rev* **D44** 817.
Langacker P and Polonsky P, 1993, *Phys Rev* **D47** 4028.
Langacker P and Polonsky N, 1994, *Phys Rev* **D49** 1454.
LEP collaborations, 1992, *Phys Lett* **B276** 247.
Lerche W, Lüst D and Schellekens J N, 1987, *Nucl Phys* **B287** 477.
Luescher M and Weisz P, 1987, *Nucl Phys* **B290** 25; 1988 **B295** 65; 1989 **B318** 705.
Mori T, 1993, in *Proceedings of the XXVI International Conference of High Energy Physics, Dallas, Texas*, ed Sanford J, (AIP, New York, NY 1993).
Nilles H P, 1984, *Phys Rept* **110C** 1.
Okada Y, Yamaguchi M and Yanagida T, 1991, *Prog Theo Phys Lett* **85** 1.
PDG 1992, Particle Data Group, Hikasa K et al. , *Phys Rev* **D45** 51.
Pendleton B and Ross G G, 1981, *Phys Lett* **B98** 291; Hill C, 1981, *Phys Rev* **D24** 691.
Peskin M and Takeuchi T, 1990, *Phys Rev Lett* **65** 964; 1992 *Phys Rev* **D46** 381.
Politzer H D, 1973, *Phys Rev Lett* **30** 1346.
Politzer H D, 1976, *Nucl Phys* **B117** 397.
Sakai N and Yanagida T, 1982, *Nucl Phys* **B197** 533.
Slansky R, 1981, *Phys Rept* **79C** 1.
Susskind L, 1979, *Phys Rev* **D20** 2619.
Totsuka Y, 1992, in *Particles and Cosmology*, eds Matveev V A et al. (World Scientific, Singapore 1992).
Weinberg S, 1967 *Phys Rev Lett* **19** 1266.
Weinberg S, 1976, *Phys Rev* **D13** 974.
Weinberg S, 1982, *Phys Rev* **D26** 287.
West P, 1986, *Introduction to Supersymmetry and Supergravity* (World Scientific, Singapore 1986).
Wetterich C, 1983, *Nucl Phys* **B233** 109.
Witherell M, 1993, to appear in the *3rd KEK Workshop on CP Violation*, Tsukuba (Japan).
Witten E, 1983, in *Proceedings of the 1982 Shelter Island Conference II*, (MIT Press, Cambridge, Mass. 1983).
Witten E, 1985, *Nucl Phys* **B258** 75.
Yamawaki K, Bando M and Matsumoto K, 1986, *Phys Rev Lett* **56** 1335.
Yanagida T, 1979, in *Proceedings of the Workshop on the Unified Theory and the Baryon Number of the Universe*, KEK 1979.
Zinn Justin J, 1989, *Quantum Field Theory and Critical Phenomena* (Clarendon Press, Oxford 1989).
Zwirner F, 1992, in *Proceedings of the 1991 Summer School in High Energy Physics and Cosmology, Trieste 1991*, eds Gava E et al. (World Scientific, Singapore 1992).

Particle Physics and Cosmology

Edward W Kolb

Fermi National Accelerator Laboratory
and The University of Chicago

In these lectures on the early Universe I will discuss some recent developments in particle cosmology, taking particular care to highlight the rôle of particle physics in our understanding of cosmology. I will assume that the reader is familiar with basic particle physics, but not necessarily basic astronomy.

Before starting, I would like to discuss the motivation for particle physicists to be interested in cosmology. The aim of modern cosmology is to understand the origin and the large-scale structure of the Universe on the basis of physical law. The modern framework for this effort is the hot big-bang model. With knowledge of the laws of physics, the fundamental forces, and the fundamental particles, in principle the model should be able to explain the gross features of our Universe. It is also possible to 'reverse engineer' this standard approach: by observations of the outcome, we might be able to tell something about the fundamental ingredients that went in. Therefore we might be able to discover something about particle physics by studying cosmology.

Let me also describe my own approach to cosmology. How should one approach the study of the history of the ancient Universe? There are two types of people who study old things, antiquarians and historians. An antiquarian is interested in things that are old simply because they are old. They do not attempt to differentiate between the relative importance of objects from antiquity. In the extreme, an antiquarian would see no difference between a grocery shopping list from 1215 and the Magna Carta. A historian on the other hand is interested in events and objects from the past because they have a bearing upon the present. It is the job of the historian to sort through the past to find the objects and events that had an impact upon the future development of history. I consider it the job of the cosmologist to be a historian of the Universe. The cosmologist should not be interested in the early Universe because it was very old, or very hot, or very dense. Rather a cosmologist studies the early Universe because he or she has the faith that events in the early Universe are responsible for shaping the present Universe, and that it is impossible to understand the Universe today without an understanding of the early Universe.

Therefore in these lectures I will concentrate on events in the early Universe that have the potential to explain the present state of the Universe. In a very real sense the job of cosmology is to provide a canvas upon which other fields of science, including particle physics, can weave their individual threads into the tapestry of our understanding of the Universe. Nowhere is the inherent unity of science better illustrated than in the interplay between cosmology, the study of the largest things in the Universe, and particle physics, the study of the smallest things.

1 A quick look at the Universe

Before concentrating on the particle physics aspects of cosmology, I will start with a look at the most important observational features of the Universe. I will then discuss the Robertson-Walker metric, and discuss some particle kinematics in the expanding Universe. Then I will develop the dynamics of the Friedmann-Robertson-Walker (FRW) cosmology. The final part of the introductory section will be a brief review of the radiation-dominated era and primordial nucleosynthesis. More details can be found in Kolb and Turner (1989).

1.1 Expansion of the Universe

It was Hubble who discovered a linear relationship between the recessional velocities of nebulae and their distances. The recessional velocity is determined via the Doppler effect. If the relative velocity between a source and observer is v_R, then the measured wavelength of the light, $\lambda_{\rm obs}$, will differ from the wavelength of the emitted light, $\lambda_{\rm emitted}$. This difference is expressed in terms of a redshift z, defined as

$$z \equiv \frac{\lambda_{\rm obs} - \lambda_{\rm emitted}}{\lambda_{\rm emitted}}. \tag{1}$$

If one interprets the observation of a redshift of light from distant galaxies as a Doppler effect, then $z = v_R/c$. (Of course this is a non-relativistic expression. The special relativistic expression relating v_R and z is $v_R/c = [(1+z)^2 - 1]/[(1+z)^2 + 1]$.) If the relative distance is increasing, then z is positive.

The linear relationship between the distance and the redshift, Hubble's law, can be written in several equivalent forms:

$$\begin{aligned} cz &= H_0 d_L \\ v_R &= H_0 d_L \\ d_L &= (3000 h^{-1}) z \text{ Mpc} = 10^{-2} h^{-1} cz \text{ Mpc}, \end{aligned} \tag{2}$$

where a megaparsec (Mpc) is 3.1×10^{24} cm. As to the meaning of these symbols: c is the speed of light (1 unless you are an astronomer), the redshift z and recessional velocity v_R have already been defined, d_L is the (luminosity) distance, and H_0 is Hubble's constant. Let us postpone for a moment questions about what exactly the luminosity distance is, and just think of it as the distance to the object without worrying whether it is the distance when the light was emitted, the distance when the light was detected, *etc.*

Sixty-four years after the discovery of Hubble's law, Hubble's constant H_0 is still not well known. It is traditional to express the uncertainty in Hubble's law in terms of a dimensionless parameter h:

$$H_0 = 100\, h\, \frac{\text{km}}{\text{s Mpc}} \qquad (1 \gtrsim h \gtrsim 0.4). \tag{3}$$

The Hubble constant is *the* fundamental parameter in cosmology, and it is not known to better than a factor of two! This uncertainty traces to the oldest and most fundamental problem of astronomy—the distance scale (for a review see Rowan-Robinson, 1985). The uncertainty in H_0 will result in a proliferation of factors of h in many of the equations in subsequent sections.

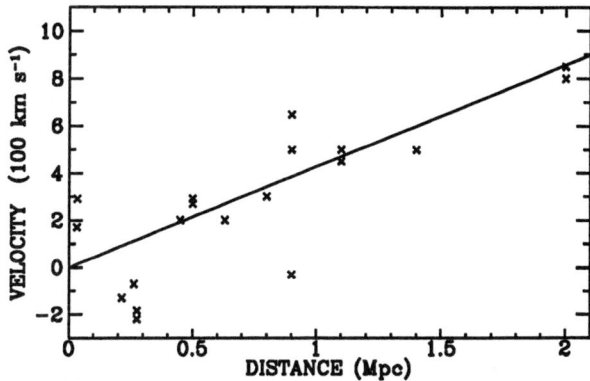

Figure 1. *Hubble's 1929 data. The solid line is a guide to the eye.*

It is somewhat amusing to look at the original data upon which Hubble based his claim, shown in Figure 1. Clearly it took a leap of imagination, intuition, and genius to see a linear relationship in the data. After all, some of the nearby nebulae are approaching, rather than receding. With modern (hopefully more reliable) methods for determining distances, astronomers are able to extend Hubble's program to much greater distances. All agree on the linear nature of the relationship, but do not agree on the value of H_0.

To orient the particle physicist I have included a small table of extragalactic distances. The distance to nearby objects in our local group of galaxies, like Andromeda, can be determined by direct means. Hence the distance is independent of h. For more distant objects such as the Virgo cluster of galaxies we can accurately determine the red shift (or equivalently v_R) but not the distance. Using the measured z, Hubble's law will give the distance in terms of the annoying factor of h^{-1}. Depending upon your favourite value of h, the distance to Virgo is somewhere between 12 and 30 Mpc.

OBJECT	z	d_L	v_R
M31 (Andromeda)	−0.0009	0.65 Mpc	−270 km s^{-1}
Virgo Cluster	0.004	12 h^{-1} Mpc	1150 km s^{-1}
Coma Cluster	0.02	67 h^{-1} Mpc	6700 km s^{-1}
Hydra	0.2	600 h^{-1} Mpc	60600 km s^{-1}

Clearly Hubble's law as given in Equation 2 must break down for $z \to 1$. Even if one adopts the special relativistic Doppler formula, we will see in the section on kinematics that there are important corrections for $z \to 1$.

The expansion of the Universe and Hubble's law will be discussed further, but for our first quick view of the Universe, it will suffice to note that the Universe is expanding, and furthermore the expansion seems to be isotropic about us.

1.2 The cosmic background radiation

The cosmic background radiation (CBR) provides fundamental evidence that the Universe began from a hot big bang. The surface of last scattering for the CBR was the Universe at an age of about 300,000 years. The first thing to learn about the CBR is its spectrum. It is a blackbody to a remarkable accuracy. The best measurement of the spectrum of the CBR was made with the Cosmic Background Explorer (COBE) satellite (Mather et al. 1993). The measurements are summarized in Figure 2. Note that the true error bars for the measurements are a factor of 100 times smaller than shown in the figure. Clearly the CBR is a blackbody, with the present temperature of the Universe $T_0 = 2.726 \pm 0.01$ K, and deviations from a blackbody shape over the wavelength interval 0.05 cm to 0.5 cm less than 0.03%.

Once the temperature of the CBR is known, the number density and energy density of the background photons are also known. For a temperature of $T_0 = 2.726$ K $= 2.36 \times 10^{-4}$ eV, the number density and energy density of the CBR is given by

$$n_\gamma = (2\zeta(3)/\pi^2)T_0^3 = 411 \text{ cm}^{-3}$$
$$\rho_\gamma = (\pi^2/15)T_0^4 = 4.71 \times 10^{-34} \text{ g cm}^{-3}. \qquad (4)$$

After the spectrum, the next most important feature of the CBR is its isotropy. Anisotropy is expected due to several effects. For instance, a dipole moment of the CBR is expected as a result of the motion of our local reference frame with respect to the CBR rest frame. Motion with velocity $\vec{\beta} = \vec{v}/c$ through an isotropic blackbody radiation field of temperature T results in a frequency-independent formula for the temperature distribution: $T(\theta) = T\sqrt{1-\beta^2}/(1-|\beta|\cos\theta)$.

The most accurate measurement of the CBR dipole anisotropy was by COBE: an amplitude of 3.336 mK, corresponding to a velocity of 627 ± 22 km s^{-1} in the general direction of Hydra for our local group of galaxies. COBE has also determined that the dipole anisotropy has a thermal spectrum.

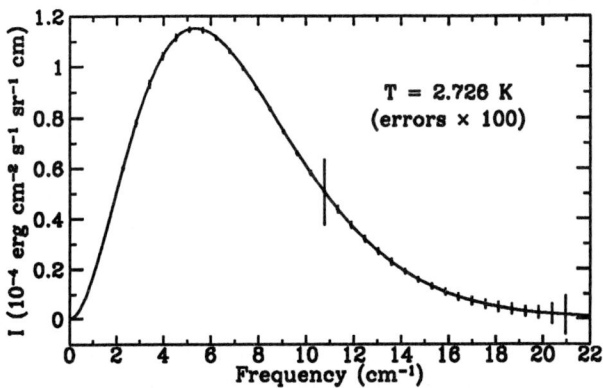

Figure 2. *The spectrum of the cosmic background radiation.*

Additional fluctuations in the CBR temperature are also expected due to the presence of density inhomogeneities presumed to have triggered structure formation. The search for anisotropies in the CBR beyond the dipole anisotropy has occupied physicists since the discovery of the CBR itself in 1965. Finally, in 1992 the long search was rewarded when the COBE collaboration announced the discovery of anisotropy on angular scales from about 7° to 90° at a magnitude of about 1 part in 10^5 (Smoot et al. 1992). There are several methods to analyze the anisotropy. The cleanest and most reliable indication of anisotropy is an *rms* temperature variation of $30 \pm 5\,\mu K$ on the sky averaged over a beam of FWHM 10°. COBE also reported a quadrupole anisotropy of $11 \pm 3\,\mu K$.

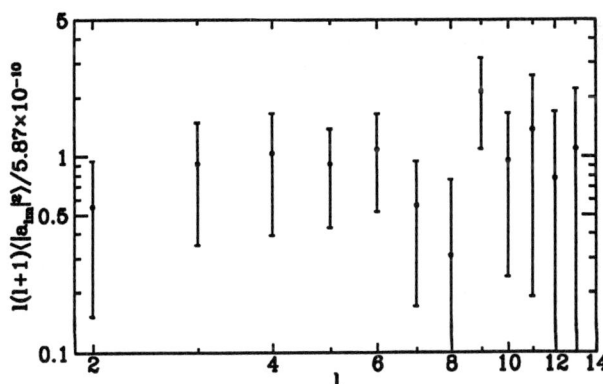

Figure 3. *Anisotropy multipoles of the cosmic background radiation.*

If one expands the observed temperature fluctuation as a function of angles θ and

ϕ in spherical harmonics,

$$\frac{\delta T}{T} = \sum_{l=2}^{\infty} \sum_{m=-l}^{+l} a_{lm} Y_{lm}(\theta, \phi), \qquad (5)$$

then measurement of fluctuation can be expressed in terms of the multipole amplitudes a_{lm}. In this expansion the dipole moment has been left out since it arises to our peculiar velocity with respect to the CBR rest frame. The inferred multipole amplitudes from $l = 2$ to $l = 13$ as measured by COBE are shown in Figure 3.

In this first discussion of the CBR, the most important feature is that the temperature of the Universe is well determined: $T_0 = 2.726\,\mathrm{K}$. A dipole moment of the CBR is also well known. In addition there is now evidence for higher multipole moments in the CBR anisotropy. However, in the excitement and publicity of the discovery of CBR anisotropy one should not loose sight of the most important aspect of the CBR: its remarkable isotropy. The CBR is isotropic about us to better than one part in 10^5.

1.3 Homogeneity and isotropy

We live in a hot, expanding Universe. We also live in a Universe that on 'large' scales is homogeneous, the same at every point, and isotropic, the same in every direction. There is an ample (and growing) body of evidence for homogeneity and isotropy. The homogeneity and isotropy of the Universe is the most fundamental principle in modern cosmology. In fact, it is called the *Cosmological Principle*.

The assumption of the isotropy and homogeneity of the Universe in modern cosmology dates back to the work of Einstein, who made the assumption not based upon observational evidence, but to simplify the mathematical analysis. Today there is ample evidence for the isotropy and homogeneity for the part of the Universe we can observe, our present Hubble volume, characterized by a length $H_0^{-1} \simeq 3000 h^{-1}$ Mpc $\simeq 10^{28} h^{-1}$ cm.

The best evidence for the isotropy of the observed Universe is the uniformity of the temperature of the CBR as discussed above. If the expansion of the Universe were not isotropic, the expansion anisotropy would lead to a temperature anisotropy of similar magnitude. Likewise, inhomogeneities in the density of the Universe on the surface of last scattering would lead to temperature anisotropies. In this regard, the CBR is a very powerful probe.

Additional evidence for the isotropy of the Universe is the isotropy of the x-ray background radiation. Some large fraction of the x-ray background is believed to be from unresolved sources (*e.g.* QSO's) at high redshift. Likewise, a substantial fraction of faint radio sources are radio galaxies at high redshift ($z \simeq 1$), and their distribution is also isotropic about us.

Evidence for homogeneity and isotropy from the distribution of galaxies is somewhat less certain, mostly because we are only now mapping the distribution of galaxies on scales large enough to see homogeneity. One can find some measure of homogeneity of the Universe by taking a sphere of radius R which contains on average N galaxies, and placing it down at all points in the Universe, counting the number of galaxies inside it, and computing the root-mean-square (*rms*) number fluctuations. One finds that the rms number fluctuations, $\delta N/N$, decrease with increasing scale, and drop below unity for radius $R_0 = 8h^{-1}$ Mpc. This indicates that on scales less than R_0 the Universe is lumpy, and for scales greater than R_0 the Universe becomes smooth.

This is not to say that the Universe is structureless on scales greater than R_0. The best known example of structures on larger scales comes from the Center for Astrophysics (CFA) slices of the universe (de Laupparent, Geller, and Huchra, 1986), shown in Figure 4 (recall from Equation 2 that $d_L = 10^{-2} h^{-1} cz$ Mpc). Some of the structure is the result of the presentation of the data in redshift space (which stretches out things in the radial direction), but clearly there is structure on scales much larger than $8h^{-1}$ Mpc.

Clear evidence from the distribution of galaxies for homogeneity of the Universe must await surveys on scales much larger than the CFA survey. These should be completed before the end of this century. Until that time, we can only look at larger regions of the Universe with 'sparse' samples of the location of galaxies, *i.e.* only a fraction of galaxies in the sample volume are mapped. Such a sparse sample for the Automatic Plate Measuring (APM) survey (Loveday *et al.* 1992) is shown in Figure 5. Note that the APM survey is nearly three times as deep as the CFA survey. It does not seen to show structures on the size of the survey as the CFA survey does. Just by comparing the two figures one concludes that the distribution of matter in the Universe is not a fractal, but rather approaches homogeneity on large scales.

In conclusion, the Universe is lumpy on small scales, containing people, planets, stars, galaxies, galaxy clusters, superclusters, *etc.* However on large scales the distribution of matter and radiation in the Universe is smooth. We are only now probing the transition region from lumpy to smooth in the distribution of galaxies. It is exciting to be a cosmologist at the time when the largest structures in the Universe are being discovered.

The large degree of spatial symmetry in a spatially homogeneous and isotropic Universe will greatly simplify the dynamics of the expansion of the Universe. Through the action of cosmic inflation, we will be able to understand why the Universe is homogeneous and isotropic on observable scales, as well as understanding why there is structure in the Universe.

1.4 The present Universe

I will conclude this quick look at the Universe today by presenting the parameters that describe the present Universe.

Expansion: The Universe is expanding at a present rate given by Hubble's constant, expressed in terms of a dimensionless parameter h: $H_0 = 100h$ km s^{-1} Mpc^{-1}, with $0.4 < h < 1$.

Temperature: The present temperature of the Universe is $T_0 = 2.726 \pm 0.01$ K, with a dipole moment of 3.336 mK, and *rms* temperature fluctuations on a scale of 10° of about 30 μK.

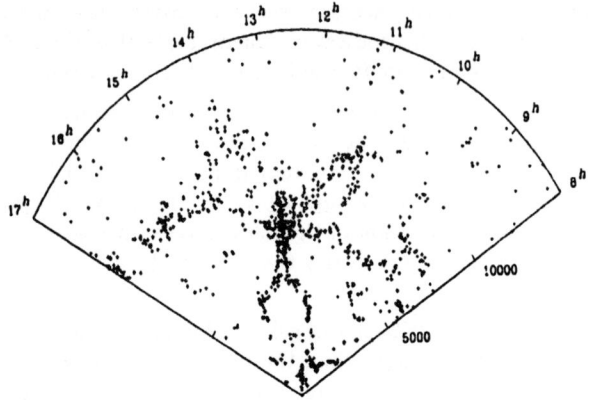

Figure 4. *One of the CFA slices of the Universe containing 1074 galaxies.*

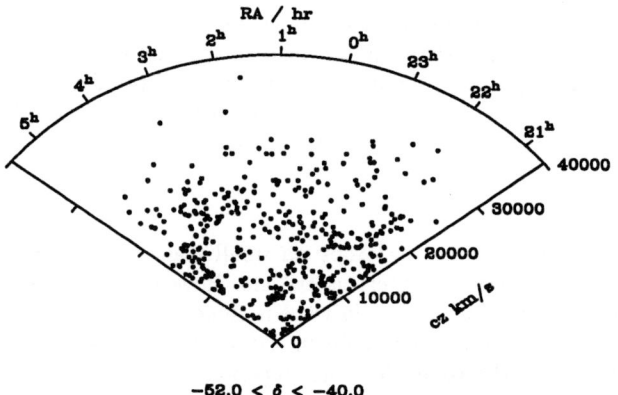

Figure 5. *A sparse sample of the APM survey.*

Homogeneity and Isotropy: The Universe is homogeneous and isotropic on large scales and clumpy on small scales. The transition region is about $R_0 = 10h^{-1}$ Mpc.

Mass and Energy Density: The mass density of the Universe is poorly known. It is convenient to express mass densities in terms of a critical density, ρ_C, formed by Hubble's constant and Newton's constant:

$$\rho_C = \frac{3H_0^2}{8\pi G} = 1.88h^2 \times 10^{-29} \text{g cm}^3. \tag{6}$$

Expressed as a fraction of the critical density, the matter (M), photon (γ), and radiation (R) energy densities of the Universe are

$$\begin{aligned}
\Omega_M &\equiv \rho_M/\rho_C = 0.01 \text{ to } 1 \\
\Omega_\gamma &\equiv \rho_\gamma/\rho_C = 2.6 \times 10^{-5} h^2 \\
\Omega_R &\equiv \rho_R/\rho_C = 4.3 \times 10^{-5} h^2,
\end{aligned} \tag{7}$$

where I have included 3 massless neutrino species at a temperature of 1.96 K in addition to the photons in determining the radiation energy density. Clearly today we live in a 'matter-dominated' Universe since $\Omega_M \gg \Omega_R$.

1.5 The Robertson-Walker metric

The metric for a space with homogeneous and isotropic spatial sections is the Robertson-Walker (RW) metric, which can be written in the form

$$ds^2 = dt^2 - a^2(t) \left\{ \frac{dr^2}{1 - kr^2} + r^2 d\theta^2 + r^2 \sin^2\theta d\phi^2 \right\} \tag{8}$$

where (r, θ, ϕ) are spatial coordinates (referred to as comoving coordinates), $a(t)$ is the cosmic scale factor, and with an appropriate rescaling of the coordinates k can be chosen to be $+1$, -1, or 0 for spaces of constant positive, negative, or zero spatial curvature, respectively. The coordinate r in Equation 8 is dimensionless, i.e. $a(t)$ has dimensions of length, and r ranges from 0 to 1 for $k = +1$. Notice that for $k = +1$ the circumference of a one-sphere of coordinate radius r in the $\phi = $ const plane is just $2\pi a(t)r$, and that the area of a two-sphere of coordinate radius r is just $4\pi a^2(t)r^2$; however, the physical radius of such one- and two-spheres is $a(t)\int_0^r dr/(1-kr^2)^{1/2}$, and not $a(t)r$.

The time coordinate in Equation 8 is the proper time measured by an observer at rest in the comoving frame, i.e. (r, θ, ϕ)=const. Observers at rest in the comoving frame remain at rest, i.e. (r, θ, ϕ) remain unchanged, and observers initially moving with respect to this frame will eventually come to rest in it. Thus, if one introduces a homogeneous, isotropic fluid initially at rest in this frame, the $t = $ const hypersurfaces will always be orthogonal to the fluid flow, and will always coincide with the hypersurfaces of both spatial homogeneity and constant fluid density.

The dynamical equations that describe the evolution of the scale factor $a(t)$ follow from the Einstein field equations, $R_{\mu\nu} - \frac{1}{2} R g_{\mu\nu} = 8\pi G T_{\mu\nu}$. Before proceeding we must specify the stress-energy tensor. To be consistent with the symmetries of the metric, the total stress-energy tensor $T_{\mu\nu}$ must be diagonal, and by isotropy the spatial components must be equal. The simplest realization of such a stress-energy tensor is that of a perfect fluid characterized by a time-dependent energy density $\rho(t)$ and pressure $p(t)$: $T^\mu_\nu = \mathrm{diag}(\rho, -p, -p, -p)$. The $\mu = 0$ component of the conservation of stress energy equation $(T^{\mu\nu}_{;\nu} = 0)$ gives the 1st law of thermodynamics: $d(\rho a^3) = -p d(a^3)$. For the simple equation of state $p = w\rho$, where w is independent of time, ρ evolves as $\rho \propto a^{-3(1+w)}$. Examples of this simple equation of state we will employ include:

$$\begin{aligned}
\text{RADIATION} \quad & (p = \tfrac{1}{3}\rho) \implies \rho \propto a^{-4} \\
\text{MATTER} \quad & (p = 0) \implies \rho \propto a^{-3} \\
\text{VACUUM ENERGY} \quad & (p = -\rho) \implies \rho \propto \text{const.}
\end{aligned} \tag{9}$$

The 0-0 component of the Einstein equation gives the Friedmann equation

$$\left(\frac{\dot{a}}{a}\right)^2 + \frac{k}{a^2} = \frac{8\pi G}{3}\rho. \tag{10}$$

A combination of the i-i component with the Friedmann equation gives an equation for the deceleration of the expansion:

$$\frac{\ddot{a}}{a} = -\frac{4\pi G}{3}(\rho + 3p). \tag{11}$$

The expansion rate of the Universe is determined by the Hubble *parameter* $H \equiv \dot{a}/a$. The Hubble parameter is not constant, and in general varies as t^{-1}. The Hubble time (or Hubble radius) H^{-1} sets the time scale for the expansion: a roughly doubles in a Hubble time. The Hubble *constant*, H_0, is the present value of the expansion rate. The Friedmann equation can be recast as

$$\frac{k}{H^2 a^2} = \frac{\rho}{3H^2/8\pi G} - 1 \equiv \Omega - 1. \tag{12}$$

Since $H^2 a^2 \geq 0$, there is a correspondence between the sign of k, and the sign of $\Omega - 1$

$$\begin{aligned} k = +1 &\implies \Omega > 1 \quad \text{CLOSED} \\ k = 0 &\implies \Omega = 1 \quad \text{FLAT} \\ k = -1 &\implies \Omega < 1 \quad \text{OPEN}. \end{aligned} \tag{13}$$

1.6 Particle kinematics

The first application of particle kinematics with the Robertson-Walker metric is a calculation of the proper distance to the horizon, *i.e.* for a comoving observer with coordinates (r_0, θ_0, ϕ_0), for what values of (r, θ, ϕ) would a light signal emitted at $t = 0$ reach the observer at, or before, time t? A light signal satisfies the geodesic equation $ds^2 = 0$. Because of the homogeneity of space, without loss of generality we may choose $r_0 = 0$. A light signal emitted from coordinate position (r_H, θ_0, ϕ_0) at time $t = 0$ will reach $r_0 = 0$ in a time t determined by

$$\int_0^t \frac{dt'}{a(t')} = \int_0^{r_H} \frac{dr}{\sqrt{1 - kr^2}}, \tag{14}$$

and the proper distance to the horizon measured at time t, $d_H(t) = \int_0^{r_H} \sqrt{g_{rr}} dr$, is simply $a(t)$ times the above integral:

$$d_H(t) = a(t) \int_0^t \frac{dt'}{a(t')} = a(t) \int_0^{a(t)} \frac{da(t')}{\dot{a}(t')a(t')}. \tag{15}$$

We know the behaviour of $a(t)$ from the Friedmann equation. For the early Universe we can ignore the curvature term. For a radiation-dominated Universe, $a \propto t^{1/2}$, and $d_H(t) = 2t$, while for a matter-dominated Universe $a \propto t^{2/3}$, and $d_H(t) = 3t$. If $d_H(t)$ is finite, then our past light cone is limited by a particle horizon, which is the boundary between the visible Universe and the part of the Universe from which light signals have not reached us. The behaviour of $a(t)$ near the singularity will determine whether or not d_H is finite.

The next application of particle kinematics is the redshift. The four-velocity u^μ of a particle with respect to the comoving frame is referred to as its *peculiar* velocity. The equation of geodesic motion for u^μ is

$$\frac{du^\mu}{d\lambda} + \Gamma^\mu_{\nu\alpha} u^\nu \frac{dx^\alpha}{d\lambda} = 0, \tag{16}$$

where $u^\mu \equiv dx^\mu/ds$, and λ is some affine parameter, which we will choose to be the proper length ds. The $\mu=0$ component of the geodesic equation is $du^0/ds + \Gamma^0_{\nu\alpha} u^\nu u^\alpha = 0$. Using the fact that for the Robertson-Walker metric, the only non-vanishing component of $\Gamma^0_{\nu\alpha}$ is $\Gamma^0_{ij} = (\dot a/a) h_{ij}$ (where h_{ij} is the spatial metric), the geodesic equation gives $|\dot{\vec u}|/|\vec u| = -\dot a/a$, which implies that $|\vec u| \propto a^{-1}$. In an expanding Universe, a freely-falling observer is destined to come to rest in the comoving frame even if he has some initial peculiar velocity. Recalling that the four-momentum is $p^\mu = mu^\mu$, we see that the magnitude of the three-momentum of a freely-propagating particle also 'redshifts' as a^{-1}. The wavelength of light is inversely proportional to the photon momentum ($\lambda = 2\pi\hbar/p$). If the momentum changes, the wavelength of the light also changes. The wavelength at time t_0, denoted as λ_0, will differ from that at time t_1, denoted as λ_1, by $\lambda_0/\lambda_1 = 1 + z = a(t_0)/a(t_1)$. This means that the redshift of the wavelength of a photon is due to the fact that the Universe was smaller when the photon was emitted!

Our final foray into particle kinematics will be Hubble's law. Suppose a source, e.g. a galaxy, has an absolute luminosity $\mathcal{L}$. Its luminosity distance is *defined* in terms of the measured flux $\mathcal{F}$ by $d_L^2 \equiv \mathcal{L}/4\pi\mathcal{F}$. If a source at comoving coordinate $r = r_1$ emits light at time t_1, and a detector at comoving coordinate $r = 0$ detects the light at $t = t_0$, conservation of energy implies $\mathcal{F} = \mathcal{L}/4\pi a^2(t_0) r_1^2 (1+z)^2$, which implies $d_L^2 = a^2(t_0) r_1^2 (1+z)^2$. In order to express d_L in terms of the redshift z, the explicit dependence upon r_1 must be removed. Since light travels on geodesics,

$$\int_0^{r_1} \frac{dr}{(1-kr^2)^{1/2}} = \int_{a_1}^{a_0} \frac{da(t')}{\dot a(t') a(t')}. \tag{17}$$

By use of the Friedmann equation, for zero pressure solution is easily found to be

$$r_1 = \frac{2\Omega_0 z + (2\Omega_0 - 4)(\sqrt{\Omega_0 z + 1} - 1)}{H_0 R_0 \Omega_0^2 (1+z)}. \tag{18}$$

Therefore Hubble's law becomes

$$H_0 d_L = 2\Omega_0^{-2} \left[2\Omega_0 z + (2\Omega_0 - 1)\left(\sqrt{2\Omega_0 z + 1} - 1\right)\right] \simeq z + \frac{1}{2}(1-q_0) z^2 + \cdots, \tag{19}$$

where q_0 is the deceleration parameter, $q_0 = -\ddot a(t_0)/\dot a^2(t_0) a(t_0) = 2\Omega_0$. Clearly for $z \to 1$ departures from the linear relationship are expected. In principle these departures would lead to a value for Ω_0. But in practice, evolutionary effects in the brightness of galaxies for large z have prevented realization of the promise.

1.7 The radiation-dominated era

In a radiation-dominated Universe, the energy density and pressure can be expressed in terms of the photon temperature T as

$$\rho_R = \frac{\pi^2}{30} g_* T^4, \qquad p_R = \rho_R/3 = \frac{\pi^2}{90} g_* T^4, \tag{20}$$

where g_* counts the total number of effectively massless degrees of freedom (those species with mass $m_i \ll T$), given by

$$g_* = \sum_{i=\text{bosons}} g_i \left(\frac{T_i}{T}\right)^4 + \frac{7}{8} \sum_{i=\text{fermions}} g_i \left(\frac{T_i}{T}\right)^4. \qquad (21)$$

The relative factor of 7/8 accounts for the difference in Fermi and Bose statistics. Note that g_* is a function of T since the sum runs over only those species with mass $m_i \ll T$. For $T \gtrsim 300\,\text{GeV}$, all the species in the standard model—8 gluons, $W^\pm Z^0$, 3 generations of quarks and leptons, and 1 complex Higgs doublet—should have been relativistic, giving $g_* = 106.75$.

During the early radiation-dominated epoch ($t \lesssim 4 \times 10^{10}$ sec) $\rho \simeq \rho_R$; and further, when $g_* \simeq$ const, $p_R = \rho_R/3$ (i.e. $w = 1/3$) and $a(t) \propto t^{1/2}$. From this it follows that the expansion rate and expansion age is

$$H = 1.66 g_*^{1/2} \frac{T^2}{m_{Pl}}\,; \qquad t = 0.301 g_*^{-1/2} \frac{m_{Pl}}{T^2} \sim \left(\frac{T}{\text{Mev}}\right)^{-2} \text{sec}. \qquad (22)$$

1.8 Primordial nucleosynthesis

Primordial nucleosynthesis is a most useful probe of the early Universe and the consistency of the big bang model. The basic idea is that at very high temperatures ($T \gg 1\,\text{MeV}$) there were no nuclei, but as the Universe expanded and cooled, conditions became hospitable for the formation of nuclei. The outcome of primordial nucleosynthesis, the relative abundances of the various nuclei, depend upon the interplay of the expansion rate of the Universe and the nuclear reactions. This has to occur in a setting of enormous specific entropy. In other words, primordial nucleosynthesis occurs with the ratio of photons to nucleons of about 10^9.

The outcome of primordial nucleosynthesis is very sensitive to the baryon-to-photon ratio, usually denoted by η. The reason for this is simple. In nuclear statistical equilibrium (NSE), the fraction of the total baryon mass contributed by a nucleus with A protons (p) and neutrons (n) and binding energy B_A is

$$X_A = g_A [\zeta(3)^{A-1} \pi^{(1-A)/2} 2^{(3A-5)/2}] A^{5/2} (T/m_N)^{3(A-1)/2} \eta^{A-1} X_p^Z X_n^{A-Z} \exp(B_A/T), \qquad (23)$$

where as usual,

$$\eta \equiv \frac{n_N}{n_\gamma} = 2.68 \times 10^{-8}\,(\Omega_B h^2) \qquad (24)$$

is the present baryon-to-photon ratio, g_A is the number of spin degrees of freedom of the nucleus, and m_N is the mass of a nucleon. The sensitive dependence upon of primordial nucleosynthesis upon η arises from the factor of η^{A-1} in this equation. Of course primordial nucleosynthesis represents a departure from nuclear statistical equilibrium and the NSE abundance is not always the actual abundance, but nevertheless the NSE abundance sets the value of what the abundance 'wants to be'.

Recent analysis (Walker et al. 1991) of the outcome of primordial nucleosynthesis suggest that η must be in the range of 3 to 5 times 10^{-10} to agree with the inferred primordial abundances.

2 The formation of structure

Before discussing the physical processes important in the theory of structure formation through gravitational instability, I will briefly review some preliminaries related to a Fourier analysis of the density field of the Universe.

It is convenient to discuss the density field of the Universe in terms of the density contrast, where

$$\delta(\vec{x}) \equiv \frac{\delta\rho(\vec{x})}{\bar{\rho}} = \frac{\rho(\vec{x}) - \bar{\rho}}{\bar{\rho}}, \tag{25}$$

and to express the density contrast $\delta(\vec{x})$ in a Fourier expansion:

$$\delta(\vec{x}) = \frac{V}{(2\pi)^3} \int \delta_k \exp(-i\vec{k} \cdot \vec{x}) d^3k. \tag{26}$$

Here $\bar{\rho}$ is the average density of the Universe, periodic boundary conditions have been imposed, and V is a normalization volume. Since $\delta(\vec{x})$ is a scalar quantity, one can use either comoving or physical coordinates in the Fourier expansion; unless otherwise specified, I will use comoving coordinates.

A particular Fourier component is characterized by its amplitude $|\delta_k|$ and its comoving wavenumber k. Since $\vec{x}$ and $\vec{k}$ are (comoving) coordinate quantities, the physical distance and physical wavenumber are related to the comoving distance and wavenumber by $dx_{\text{phys}} = a(t)dx$, $k_{\text{phys}} = k/a(t)$. The wavelength of a perturbation is related to the wavenumber by $\lambda \equiv 2\pi/k$, $\lambda_{\text{phys}} = a(t)\lambda$.

All statistical quantities for *gaussian* random fluctuations can be specified in terms of the *power spectrum* $|\delta_k|^2$. In the absence of a better idea it is assumed that $|\delta_k|^2 \propto k^n$, that is, a featureless power law.

The *rms* density fluctuation is defined by,

$$\frac{\delta\rho}{\rho} = \langle \delta(\vec{x})\delta(\vec{x}) \rangle^{1/2}, \tag{27}$$

where $\langle \cdots \rangle$ indicates the average over all space. Some manipulation yields:

$$\left(\frac{\delta\rho}{\rho}\right)^2 = V^{-1} \int_0^\infty \frac{k^3 |\delta_k|^2}{2\pi^2} \frac{dk}{k}. \tag{28}$$

The contribution to $(\delta\rho/\rho)^2$ from a given logarithmic interval in k is given by

$$\left(\frac{\delta\rho}{\rho}\right)_k^2 \approx \Delta^2(k) \equiv V^{-1} \frac{k^3 |\delta_k|^2}{2\pi^2}. \tag{29}$$

The fluctuation power per logarithmic interval, denoted by $\Delta^2(k)$, will appear often.

Now consider $(\delta M/M)$, the *rms* mass fluctuation on a given mass scale. This is what most people mean when they refer to the density contrast on a given mass scale. Mechanically, one would measure $(\delta M)_{rms}$ as follows: take a volume V_W, which on average contains mass M, place it at all points throughout space, measure the mass within it, and then compute the *rms* mass fluctuation. Although it is simplest to choose a spherical volume V_W with a sharp surface, to avoid surface effects one often wishes to smear the surface. This is done by using a *window function* $W(r)$, which smoothly defines a volume V_W and mass $M = \bar{\rho} V_W$, where

$$V_W = 4\pi \int_0^\infty r^2 W(r) dr. \tag{30}$$

The *rms mass* fluctuation on the mass scale $M \equiv \bar{\rho} V_W$ is given in terms of the density contrast and window function by

$$\left(\frac{\delta M}{M}\right)^2 = \frac{1}{V_W^2} \int \Delta^2(k) |W(k)|^2 \frac{dk}{k}. \tag{31}$$

Notice that the *rms* mass fluctuation is given in terms of an integral over $\Delta^2(k)$.

Taking $|\delta_k|^2 = A k^n$, and $W(r) = 1$ for $r \leq r_0$ and zero otherwise, one finds

$$\left(\frac{\delta M}{M}\right)_{r_0} \simeq \Delta(k = r_0^{-1}) \tag{32}$$

for $n > -3$. $(\delta M/M)$ is given by an integral over all wavelengths longer than about r_0 and is roughly equal to the *rms* value of $\Delta(k = r_0^{-1})$. Using this 'top hat' window function, Davis and Peebles (1983) find for the CFA-I redshift survey that $(\delta M/M) = 1$ for a sphere of radius of $r_0 = 8h^{-1}$ Mpc. This value of r_0 separates the linear from the non-linear regime.

It is commonly assumed that the observed structure is a result of the growth of small seed inhomogeneities. In the next section on inflation and in Section 4 on phase transitions I will discuss possible origins for the seed perturbations. But before doing that, here I will discuss the theory of gravitational instability in an expanding Universe, and discuss the physical, non-gravitational, processes that might affect the perturbation spectrum. First, consider the linear theory of gravitational instability.

2.1 Gravitational instability—linear theory

We will start by considering the simplest possible form of gravitational instability, the Jeans instability in a non-expanding, perfect fluid. The Eulerian equations of Newtonian motion describing a perfect fluid are

$$\frac{\partial \rho}{\partial t} + \vec{\nabla} \cdot (\rho \vec{v}) = 0,$$
$$\frac{\partial \vec{v}}{\partial t} + (\vec{v} \cdot \vec{\nabla})\vec{v} + \frac{1}{\rho}\vec{\nabla}p + \vec{\nabla}\phi = 0,$$
$$\nabla^2 \phi = 4\pi G \rho. \tag{33}$$

Here ρ is the matter density, p the matter pressure, $\vec{v}$ the local fluid velocity, and ϕ the gravitational potential. The simplest solution is the static one where the matter is at rest ($\vec{v}_0 = 0$) and uniformly distributed in space ($\rho_0 = $ const, $p_0 = $ const). Throughout, we will denote unperturbed quantities with the subscript 0. Now consider perturbations about this static solution, expanded as

$$\rho = \rho_0 + \rho_1; \quad p = p_0 + p_1; \quad \vec{v} = \vec{v}_0 + \vec{v}_1; \quad \phi = \phi_0 + \phi_1. \tag{34}$$

We will consider adiabatic perturbations, that is, perturbations for which there are no spatial variations in the equation of state. The (adiabatic) sound speed, v_S^2, is defined as $v_S^2 \equiv \partial p/\partial \rho$, and by assumption there are no spatial variations in the equation of state, $v_S^2 = p_1/\rho_1$. To first order, the small perturbation ρ_1 satisfies the second-order differential equation:

$$\frac{\partial^2 \rho_1}{\partial t^2} - v_S^2 \nabla^2 \rho_1 = 4\pi G \rho_0 \rho_1. \tag{35}$$

Solutions to Equation 35 are of the form

$$\rho_1(\vec{r}, t) = \delta(\vec{r}, t) \rho_0 = A \exp\left[-i\vec{k} \cdot \vec{r} + i\omega t\right] \rho_0, \tag{36}$$

and ω and $\vec{k}$ satisfy the dispersion relation $\omega^2 = v_S^2 k^2 - 4\pi G \rho_0$, with $k \equiv |\vec{k}|$.

If ω is imaginary, there will be exponentially growing modes; if ω is real, the perturbations will simply oscillate as sound waves. It is clear that for k less than some critical value, ω will be imaginary. This critical value is called the Jeans wavenumber, k_J, and is given by

$$k_J = \left(\frac{4\pi G \rho_0}{v_S^2}\right)^{1/2}. \tag{37}$$

For $k^2 \ll k_J^2$, ρ_1 grows exponentially on the dynamical timescale $\tau_{\rm dyn} \simeq (4\pi G \rho_0)^{-1/2}$.

It is useful to define the Jeans mass, the total mass contained within a sphere of radius $\lambda_J/2 = \pi/k_J$:

$$M_J \equiv \frac{4\pi}{3}\left(\frac{\pi}{k_J}\right)^3 \rho_0 = \frac{\pi^{5/2}}{6} \frac{v_S^3}{G^{3/2}\rho_0^{1/2}}. \tag{38}$$

Perturbations of mass less than M_J are stable against gravitational collapse, while those of mass greater than M_J are unstable.

The classical Jeans analysis is not directly applicable to cosmology because the expansion of the Universe is not taken into account, and because the analysis is Newtonian. For modes of wavelength less than that of the horizon, i.e. $\lambda_{\rm phys} \ll H^{-1}$, a Newtonian analysis suffices so long as the expansion is taken into account.

When the expansion of the Universe is taken into account, the wave equation becomes

$$\ddot{\delta}_k + 2\frac{\dot{a}}{a}\dot{\delta}_k + \left(\frac{v_S^2 k^2}{a^2} - 4\pi G\rho_0\right)\delta_k = 0. \tag{39}$$

Again, for $k \ll k_J$ there are unstable (growing mode) solutions. In the limit $k \ll k_J$ for a spatially flat, matter-dominated FRW model where $\dot{a}/a = (2/3)t^{-1}$ and $\rho_0 = (6\pi G t^2)^{-1}$,

$$\ddot{\delta} + \frac{4}{3t}\dot{\delta} - \frac{2}{3t^2}\delta = 0, \qquad k \ll k_J. \tag{40}$$

This equation has two independent solutions, a growing mode, δ_+, and a decaying mode, δ_-, with time dependence given by

$$\delta_+(t) = \delta_+(t_i)\left(\frac{t}{t_i}\right)^{2/3}; \quad \delta_-(t) = \delta_-(t_i)\left(\frac{t}{t_i}\right)^{-1}. \tag{41}$$

Here we see the key difference between the Jeans instability in the static regime and in the expanding Universe: the expansion of the Universe slows the exponential growth of the instability and results in power-law growth for unstable modes.

Consider a two-component model with non-relativistic (NR) species i (e.g. baryons or a WIMP—weakly interacting massive particle) and photons during the radiation-dominated era. In this case $\dot{a}/a = 1/2t$ and $\rho_0 \ll \rho_{TOTAL}$. Consider the evolution of perturbations that are Jeans unstable, $k \ll k_J$. If the photons have no perturbations then the solution is

$$\ddot{\delta}_i + \frac{1}{t}\dot{\delta}_i = 0. \tag{42}$$

In this case the solution is $\delta_i(t) = \delta_i(t_i)[1 + a \ln(t/t_i)]$, so only a perturbation with an 'initial velocity,' $\dot{\delta}(t_i) \neq 0$, can actually grow, and only logarithmically at that.

The growth of linear perturbations in a radiation-dominated Universe is inhibited compared to the static situation. The physical reason behind this fact is easy to see. The classical Jeans instability with exponential growth is moderated by the expansion of the Universe. In a matter-dominated epoch, perturbations grow as a power law. In the radiation-dominated epoch, the expansion rate is faster than what it would be if there were only matter present, and so the growth of perturbations is further slowed.

2.2 Damping processes

The theory of gravitational instability discussed so far assumes that the Universe is filled with a perfect fluid. However there are important departures from this ideal situation.

Collisionless damping occurs during the radiation-dominated era when (linear) perturbations do not grow. If the particle species is collisionless the perfect-fluid approximation is clearly invalid. In this case collisionless phase mixing, or Landau damping, will occur. Perturbations will be damped on length scales smaller than the distance the particle will travel while decoupled. If the species becomes non-relativistic at time t_{NR}, then at time t_{EQ} when the Universe becomes matter-dominated and perturbations can start to grow the species would have free-streamed a distance

$$\lambda_{FS} = a(t) \int_0^{t_{NR}} \frac{1}{a(t')} dt' + a(t) \int_{t_{NR}}^{t_{EQ}} \frac{v(t')}{a(t')} dt', \tag{43}$$

where the integral has been split into two pieces: the relativistic regime, with $v \simeq 1$; and the non-relativistic regime, when $v \lesssim 1$. Assuming the Universe is radiation dominated at t_{NR},

$$\lambda_{FS} \simeq t_{NR}(1 + z_{NR})[2 + \ln(t_{EQ}/t_{NR})]. \tag{44}$$

For a light neutrino species

$$\lambda_{FS-\nu} \simeq 20 \text{ Mpc} \left(\frac{m_\nu}{30 \text{ eV}}\right)^{-1}. \tag{45}$$

Perturbations on scales less than $\lambda_{FS-\nu}$ are damped by free streaming. Note that this length scales is much larger than the length scale associated with galaxies, containing a mass (in neutrinos) of $4 \times 10^{14} (m_\nu/30 \text{ eV})^{-2} M_\odot$, where $M_\odot = 2 \times 10^{33}$g is a solar mass.

The perfect-fluid approximation also breaks down for baryons. Before recombination the photons mean free path is small, but as the matter particles in Universe become electrically neutral during recombination the photon mean free path becomes longer, and photons can diffuse out of dense regions. To the degree that the photons are not completely decoupled from the baryons, they can drag the baryons along, also damping perturbations in the baryons. This effect is known as Silk damping. A careful calculation of the damping requires solving the Boltzmann equation, but to a good approximation the scale for Silk damping is about an order of magnitude smaller than the horizon scale at decoupling. Thus, baryon perturbations should be damped on scales less than $\lambda_S \simeq (\Omega_0/\Omega_B)^{1/2}(\Omega_0 h^2)^{-3/4}$ Mpc, corresponding to a mass of $M_S = 6.2 \times 10^{12}(\Omega_0/\Omega_B)^{3/2}(\Omega_0 h^2)^{-5/4} M_\odot$.

2.3 Super-horizon-size perturbations

So far the analysis of the evolution of density perturbations has been Newtonian. For modes that are well within the horizon, $\lambda_{\text{phys}} \ll H^{-1}$, the Newtonian analysis is adequate. To treat the evolution of modes outside the horizon one needs a full, general-relativistic analysis. Clearly this is beyond the scope of these lectures. However I will illustrate the idea in a simple, geometric way. Consider perturbations of a spatially flat ($k = 0$) FRW model. The Friedmann equation for the unperturbed $k = 0$ model is

$$H^2 = 8\pi G \rho_0/3 \qquad (k=0). \qquad (46)$$

Now consider a similar FRW model, one with the same expansion rate H, but one that has higher density, $\rho = \rho_1$, and is therefore positively curved. The expansion rate is

$$H^2 = \frac{8\pi G \rho_1}{3} - \frac{k}{a^2} \qquad (k>0). \qquad (47)$$

If we compare the models when their expansion rates are equal, we have made a choice of gauge; in this case the uniform Hubble gauge. We immediately see that the density contrast between the two models is given in terms of the curvature of the closed model by

$$\delta \equiv \frac{\rho_1 - \rho_0}{\rho_0} = \frac{k/a^2}{8\pi G \rho_0/3}. \qquad (48)$$

The evolution of δ has been reduced to that of the curvature k/a^2 relative to the energy density ρ_0. As long as δ is small, equivalently $k/a^2 \lesssim 8\pi G \rho_0$, the scale factors for the two models are essentially equal (fractional difference of order δ). In a matter-dominated Universe, $\rho \propto a^{-3}$, while in a radiation-dominated Universe $\rho \propto a^{-4}$, so

$$\delta \propto \frac{a^{-2}}{\rho_0} \propto \begin{cases} a^2 & \text{RADIATION DOMINATED} \\ a & \text{MATTER DOMINATED.} \end{cases} \qquad (49)$$

Recall that in a matter-dominated Universe $a \propto t^{2/3}$, while in a radiation-dominated Universe $a \propto t^{1/2}$, so

$$\delta = \delta_i \begin{cases} t/t_i & \text{RADIATION DOMINATED} \\ (t/t_i)^{2/3} & \text{MATTER DOMINATED.} \end{cases} \qquad (50)$$

This simple model illustrates several important points about the evolution of super-horizon-sized perturbations. (1) The geometric character of a density perturbation, which is why density perturbations are referred to as curvature fluctuations. (2) What is actually relevant is the difference in the evolution of the perturbed model as compared to some unperturbed, reference model. (3) The freedom in the choice of the reference model is equivalent to a gauge choice, so that in general, δ will be gauge dependent.

As in particle physics, when confronted with gauge ambiguity, the correct thing to do is to ask a physical question, one whose answer cannot depend upon the gauge. Here, the perturbed and reference models are compared by matching their expansion rates. A very useful quantity is $\zeta \equiv \delta\rho/(\rho_0 + p_0)$. The evolution of ζ is particularly simple and *independent* of the background space-time. For super-horizon-sized modes the evolution is $\zeta = $ const for $\lambda_{\rm phys} \gtrsim H^{-1}$.

2.4 Summary of the evolution of perturbations

For sub-horizon-sized perturbations one can perform a Newtonian treatment of the evolution of perturbations. During the radiation-dominated epoch, perturbations do not grow. During the matter-dominated epoch, perturbations on scales larger than the Jeans length grow as $\delta \propto a(t) \propto t^{2/3}$. Perturbations on scales smaller than the Jeans length oscillate as acoustic waves. Before recombination, the baryon Jeans mass is larger than the horizon mass, by a factor of about 30; after recombination the baryon Jeans mass drops to about $10^5 M_\odot$.

Collisionless phase mixing damps perturbations on scales smaller than the free-streaming scale, λ_{FS} is a few times (t_{NR}/a_{NR}), where the subscript NR denotes the value of the quantity at the epoch when the species became non relativistic. Taking the particle's mass to be m_X and the ratio of its temperature to the photon temperature to be T_X/T, the free-streaming scale is roughly $\lambda_{FS} \simeq 1$ Mpc $(m_X/{\rm keV})^{-1}(T_X/T)$. For cold dark matter (CDM) models this damping scale is smaller than any cosmologically interesting scale. For hot dark matter (HDM) models the damping scale is larger than the galactic scale, so HDM must be augmented with some other seeds to grow galaxies.

Due to photon diffusion, adiabatic fluctuations in the baryons are strongly damped on scales smaller than the Silk scale. This damping occurs primarily just as the photons and baryons decouple. The Silk scale is given by $\lambda_S \simeq 3.5(\Omega_0/\Omega_B)^{1/2}(\Omega_0 h^2)^{-3/4}$ Mpc.

For modes that are super-horizon sized the subtleties of the gauge non-invariance of $\delta\rho$ are important, and a full general relativistic treatment is required. There are two physical modes, a decaying mode and a growing mode, as well as pure gauge modes. In the synchronous gauge, the growing mode evolves as $\delta \propto a^2 \propto t$ (radiation dominated) and $\delta \propto a(t) \propto t^{2/3}$ (matter dominated). Alternatively, the evolution of super-horizon-sized modes can be described by the quantity $\zeta = \delta\rho/(\rho_0 + p_0)$, which is constant.

The primeval spectrum is modified by the above physical processes. The processing of the initial spectrum by the damping processes depend upon the mix of matter: cold, hot, and baryons. It is useful to quantify the processing of the spectrum by specifying a 'transfer function'. Since matter fluctuations start to grow when the Universe becomes matter dominated, it is convenient to specify the spectrum at this time, $t_{\rm EQ} = 4 \times 10^{10} (\Omega_0 h^2)^{-2}$ sec. Again, in the absence of a better idea, it is convenient to

OBJECT	MASS	LENGTH SCALE	ANGULAR SCALE
Stars	$1 M_\odot$	$0.0004 h^{-2/3}$ Mpc	$0.0065'' h^{1/3}$
Globular Clusters	$10^6 M_\odot$	$0.04 h^{-2/3}$ Mpc	$0.65'' h^{1/3}$
Galaxies	$10^{12} M_\odot$	$2 h^{-2/3}$ Mpc	$65'' h^{1/3}$
Groups of Galaxies	$10^{13} M_\odot$	$4 h^{-2/3}$ Mpc	$2.2' h^{1/3}$
Non-Linear Scale	$10^{14} h^{-1} M_\odot$	$8 h^{-1}$ Mpc	$4.3'$
Thickness of LSS	$5 \times 10^{14} h^{-1} M_\odot$	$15 h^{-1}$ Mpc	$8.5'$
Clusters	$10^{15} M_\odot$	$20 h^{-2/3}$ Mpc	$11' h^{1/3}$
Superclusters	$10^{16} M_\odot$	$40 h^{-2/3}$ Mpc	$23' h^{1/3}$
Horizon at LSS	$10^{18} h^{-1} M_\odot$	$200 h^{-1}$ Mpc	$2°$

Table 1. *Length and angular scales in an $\Omega_0 = 1$ Universe. The mass is related to scale by $M(\lambda) = 1.45 \times 10^{11} h^2 \lambda_{\rm Mpc}^3 M_\odot$, and the angle is related to scale by $\Delta\theta = 34'' h \lambda_{\rm Mpc}$. LSS stands for last scattering surface, $z = 1100$.*

specify the *unprocessed* power spectrum as a power law, $|\delta_k|^2 \propto k^n$.

The processed spectrum for hot dark matter is given by

$$|\delta_k|^2 = A k^n 10^{-2(k/k_\nu)^{1.5}} = A k^n \exp[-4.61(k/k_\nu)^{1.5}], \quad (51)$$

where the neutrino damping scale is $k_\nu = 0.16(m_\nu/30{\rm eV})\,{\rm Mpc}^{-1}$ (which is equivalent to $\lambda_\nu = 40(m_\nu/30{\rm eV})^{-1}$ Mpc), and A provides the overall normalization of the power spectrum. For cold dark matter the processed spectrum is given by

$$|\delta_k|^2 = \frac{A k^n}{(1 + \beta k + \omega k^{1.5} + \gamma k^2)^2}, \quad (52)$$

with $\beta = 1.7(\Omega_0 h^2)^{-1}$ Mpc, $\omega = 9.0(\Omega_0 h^2)^{-1.5}$ Mpc$^{1.5}$, $\gamma = 1.0(\Omega_0 h^2)^{-2}$ Mpc^2. Of course there is an intermediate case known as warm dark matter.

2.5 Confronting the spectrum

In the next two sections we will discuss the generation of perturbations in inflation and due to defects produced in phase transitions. He we discuss how we can probe the spectrum by present-day observations. For more details, see Liddle and Lyth (1993) or Copeland, Kolb, Liddle, and Lidsey (1993b).

The range of scales of interest stretches from the present horizon scale, $6000 h^{-1}$ Mpc, down to about $1 h^{-1}$ Mpc, the scale which contains roughly enough matter to form a typical galaxy. On the microwave sky, an angle of θ (for small enough θ) samples linear scales of $100 h^{-1}(\theta/1°)$Mpc. For purposes of discussion, it is convenient to split this range into three separate regions.

- **A: Large scales: $6000 h^{-1}$ Mpc $\longrightarrow \sim 200 h^{-1}$ Mpc**
  These scales entered the horizon after the decoupling of the microwave background. Except in models with peculiar matter contents, perturbations on these

scales have not been affected by any physical processes, and the spectrum retains its original form. At present the perturbations are still very small, growing in the linear regime without mode coupling. Here, we are still seeing the primeval spectrum.

- **B: Intermediate scales:** $\sim 200 h^{-1}\,\mathrm{Mpc} \longrightarrow 8 h^{-1}\,\mathrm{Mpc}$
  These scales remain in the linear regime, and their gravitational growth is easily calculable. However, they have been seriously influenced by the matter content of the Universe, in a way normally specified by a *transfer function*, which measures the decrease in the density contrast relative to the value it would have had if the primeval spectrum had been unaffected. Even in CDM models, where the only effect is the suppression of growth due to the Universe not being completely matter dominated at the time of horizon entry, this effect is at the 25% level at $200 h^{-1}$ Mpc. To reconstruct the primeval spectrum on these scales, it is thus essential to know the matter content of the Universe, including dark matter, and of its influence on the growth of density perturbations.

- **C: Small scales:** $8 h^{-1}\,\mathrm{Mpc} \longrightarrow 1 h^{-1}\,\mathrm{Mpc}$
  On these scales the density contrast has reached the nonlinear regime, coupling together modes at different wavenumbers, and it is no longer easy to calculate the evolution of the density contrast. This can be attempted either by an approximation scheme such as the Zel'dovich approximation (Efstathiou, 1990), or more practically via N-body simulations (for example, see Davis *et al.* 1992, and references therein). Further, hydrodynamic effects associated with the nonlinear behaviour can come into play, giving rise to an extremely complex problem with important non-gravitational effects. Again, the transfer function plays a crucial role on these scales. In hot dark matter models, perturbations on these scales are most likely almost completely erased by free streaming, and hence no information can be expected to be available.

Let us now consider each range of scales in turn, starting with the largest scales and working down to the smallest scales.

A. Large scales ($6000 h^{-1}\,\mathrm{Mpc} \longrightarrow \sim 200 h^{-1}\,\mathrm{Mpc}$)

Without doubt the most important form of observation on large scales for the near future is large-angle microwave background anisotropies. Scales of a couple of degrees or more fall into our definition of large scales. Such measurements are of the purest form available—anisotropy experiments directly measure the gravitational potential at different parts of the sky, on scales where the spectrum retains its primeval form. Such measurements also are of interest in that the tensor modes may contribute.

In addition to perturbations from the scalar density perturbations, the presence of gravitational waves will lead to temperature fluctuations. One can think of gravitational waves as propagating modes associated with transverse, traceless tensor metric perturbations of $g_{\mu\nu} \to g_{\mu\nu}^{\mathrm{FRW}} + h_{\mu\nu}$.

Tensor modes do not participate in structure formation and most measurements we shall discuss are oblivious to them. Further, tensor modes inside the horizon redshift away relative to matter, and so tensor modes also fail to participate in small-angle microwave background anisotropies.

Nevertheless, these large-scale measurements still exhibit one crucial and ultimately uncircumventable problem. On the largest scales, the number of statistically independent sample measurements that can be made is small. Given that the underlying inflationary fluctuations are stochastic, one obtains only a limited set of realizations from the complete probability distribution function. Such a subset may insufficiently specify the underlying distribution. This effect, which has come to be known as the *cosmic variance*, is an important matter of principle, being a source of uncertainty which remains even if perfectly accurate experiments could be carried out. At any stage in the history of the Universe, it is impossible to accurately specify the properties (most significantly the mean, which is what the spectrum specifies assuming gaussian statistics) of the probability distribution function pertaining to perturbations on scales close to that of the observable Universe.

Observations other than microwave background anisotropies appear confined to the long term future. Even such an ambitious project as the Sloan Digital Sky Survey (SDSS) (Gunn and Knap 1992; Kron 1992) can only reach out to perhaps $500h^{-1}$ Mpc, which can only touch the lower end of our specified large scales. However, in order to specify the fluctuations accurately, one needs many statistically independent regions (100 seems an optimistic lower estimate) which means that the SDSS may not specify the spectrum with sufficient accuracy above perhaps $100h^{-1}$ Mpc.

A much more crucial issue is that the SDSS will measure the galaxy distribution power spectrum, not the mass distribution power spectrum. In modern work it is taken almost completely for granted that these are not the same, and it seems likely too that a bias parameter (relating the two by a multiplicative constant) which remains scale independent over a wide range of scales may be hopelessly unrealistic. Consequently, converting from the galaxy power spectrum back to that of the matter may require a detailed knowledge of the process of galaxy formation and the environmental factors around distant galaxies. Once one attempts to reach yet further galaxies with a long look-back time, one must also understand something about evolutionary effects on galaxies. As we shall discuss in the section on intermediate scales, it seems likely that peculiar velocity data may be rather more informative than the statistics of the galaxy distribution.

A more useful tool for large scales is microwave background anisotropies on large angular scales. Our formalism closely follows that of Scaramella and Vittorio (1990). On large angular scales, the most convenient tool for studying microwave background anisotropies is the expansion into spherical harmonics

$$\frac{\Delta T}{T}(\vec{x}, \theta, \phi) = \sum_{l=2}^{\infty} \sum_{m=-l}^{l} a_{lm}(\vec{x}) Y_m^l(\theta, \phi), \qquad (53)$$

where θ and ϕ are the usual spherical polar angles and $\vec{x}$ is the observer position. With spherical harmonics defined as in Press *et al.* 1986, the reality condition is $a_{l,-m} = (-1)^m a_{l,m}^*$. In the expansion, the unobservable monopole term has been removed. The dipole term has also been completely subtracted; the intrinsic dipole on the sky cannot be separated from that induced by our peculiar velocity relative to the comoving frame.

With gaussian statistics for the density perturbations, the coefficients $a_{lm}(\vec{x})$ are gaussian distributed stochastic random variables of position, with zero mean and rotationally invariant variance depending only on l: $\langle a_{lm}(\vec{x})\rangle = 0$; $\langle |a_{lm}(\vec{x})|^2\rangle \equiv \Sigma_l^2$.

It is crucial to note that a single observer sees a single realization from the probability distribution for the a_{lm}. The observed multipoles as measured from a single point are defined as

$$Q_l^2 = \frac{1}{4\pi}\sum_{m=-l}^{l}|a_{lm}|^2, \qquad (54)$$

and indeed the temperature autocorrelation function can be written in terms of these

$$C(\alpha) \equiv \left\langle \frac{\Delta T}{T}(\theta_1,\phi_1)\frac{\Delta T}{T}(\theta_2,\phi_2)\right\rangle_\alpha = \sum_{l=2}^{\infty} Q_l^2 P_l(\cos\alpha), \qquad (55)$$

where the average is over all directions on a single observer sky separated by an angle α, and $P_l(\cos\alpha)$ is a Legendre polynomial. The expectation for the Q_l^2, averaged over all observer positions, is just $4\pi\langle Q_l^2\rangle = (2l+1)\Sigma_l^2$.

A given model predicts values for the averaged quantities $\langle Q_l^2\rangle$. On large angular scales, corresponding to the lowest harmonics, only the Sachs-Wolfe effect operates. One has two terms corresponding to the scalar and tensor modes—we denote these contributions by square brackets. The scalar term is given in terms of the amplitude of the scalar density perturbation when it crosses the Hubble radius:

$$\left(\frac{\delta\rho}{\rho}\right)_\lambda^{\text{HOR}} \equiv A_S \qquad (56)$$

by the integral

$$\Sigma_l^2[S] = \frac{8\pi^2}{m^2}\int_0^\infty \frac{dk}{k} j_l^2(2k/aH)A_S^2 T^2(k), \qquad (57)$$

where j_l is a spherical Bessel function and the transfer function $T(k)$ is normalized to one on large scales.

The amplitude of a given Fourier mode of the dimensionless strain on scale λ when it crosses inside the Hubble radius is given by

$$\left|k^{3/2}h_\mathbf{k}\right|_\lambda^{\text{HOR}} \equiv A_G. \qquad (58)$$

The expression equivalent to $\Sigma_l^2[S]$ for the tensor modes contribution to temperature fluctuations is a rather complicated multiple integral which usually must be calculated numerically (Abbot and Wise, 1984; Starobinsky, 1985; Lucchin et al. 1992). Under many circumstances (Lucchin, Matarrese and Mollerach suggest $0.5 < n < 1$ for power-law inflation) there is a helpful approximation which is that the ratio $\Sigma_l^2[S]/\Sigma_l^2[T]$ is independent of l and given by

$$\frac{\Sigma_l^2[S]}{\Sigma_l^2[T]} = \frac{A_S^2}{A_G^2}. \qquad (59)$$

On the sky, one does not observe each contribution to the multipoles separately. As uncorrelated stochastic variables, the expectations add in quadrature to give

$$\Sigma_l^2 = \Sigma_l^2[S] + \Sigma_l^2[T]. \qquad (60)$$

There are two obstructions of principle. These are

- Even if one could measure the Σ_l^2 exactly, the last scattering surface being closed means one obtains only a discrete set of information—a finite number of the Σ_l covering some effective range of scales. There will thus be an uncountably infinite set of possible spectra which predict exactly the same set of Σ_l.

- One cannot measure the Σ_l^2 exactly. What one can measure is a single realization, the Q_l^2. As a sum of $2l + 1$ gaussian random variables, Q_l^2 has a probability distribution which is a χ^2 distribution with $2l + 1$ degrees of freedom, χ^2_{2l+1}. The variance of this distribution is given by

$$\text{Var}[Q_l^2] = \frac{2}{2l+1} \langle Q_l^2 \rangle^2, \tag{61}$$

though one should remember that the distribution is not symmetric. Each Q_l^2 is a single realization from that distribution, when we really want to know the mean. From a single observer point, there is no way of obtaining that mean, and one can only draw statistical conclusions based on what can be measured. Thus, a larger set of spectra which give different sets of Σ_l^2 can still give statistically indistinguishable sets of Q_l^2. The variance falls with increasing l but is significant right across the range of large scales.

B. Intermediate scales ($\sim 200 h^{-1}$ Mpc $\longrightarrow 8 h^{-1}$ Mpc)

It is on intermediate scales that determination of the primeval spectrum is most promising. Here a range of promising observations are available, particularly towards the small end of the range of scales. In terms of technical difficulties in interpreting measurements, a trade-off has been made compared to large scales; on the plus side, the cosmic variance is a much less important player as far more independent samples are available, while on the minus side the spectrum has been severely affected by physical processes and thus has moved a step away from its primeval form.

1. Intermediate-scale microwave background anisotropies

In the absence of reionization, the relevant angular scales are from about $2°$ down to about 5 arcminutes. (Should reionization occur, a lot of the information on these scales could be erased or amended in difficult to calculate ways.) Several experiments are active in this range, including the South Pole and MAX experiments.

Unlike the large-scale anisotropy, one cannot write down a simple expression for the intermediate-scale anisotropies, even if it is assumed that one has already incorporated the effect of dark matter on the growth of perturbations via a transfer function. The reason is due to the complexity of physical processes operating. A case in point is the expected anisotropy (specified by the Σ_l^2, but now for larger l) in CDM models ($n = 1$), as calculated in detail by Bond and Efstathiou (Bond and Efstathiou, 1987).

On large scales, $l^2 \Sigma_l^2$ is approximately independent of l. Once we get into the intermediate regime, $l^2 \Sigma_l^2$ exhibits a much more complicated form, which is dominated by a strong peak at around $l = 200$. This is induced by Thomson scattering from moving electrons at the time of recombination. Bond and Efstathiou's calculation gives a peak height around 6 times as high as the extrapolated Sachs-Wolfe effect. Beyond the first peak is a smaller subsidiary peak at $l \sim 800$.

In their calculation, Bond and Efstathiou assumed both the primeval spectrum and the form of the dark matter. Of course, given the number of active and proposed dark matter search experiments, one should be optimistic that this information will be obtained in the not too distant future. However, even with this information, the complexity of the calculation makes it hard to conceive of a way of inverting it, should a good experimental knowledge of the Σ_l^2 ($l \in [30, 750]$) be obtained. Once again, it's much easier to compare a given theory with observation than to extract a theory from observation.

One of the interesting applications of these results might be a combination with the large-scale measurements. The peak on intermediate scales is due only to processes affecting the scalar modes, whereas we have pointed out that the large-scale Sachs-Wolfe effect is a combination of scalar and tensor modes. On large scales, one cannot immediately discover the relative normalizations of the two contributions. However, if the dark matter is sufficiently well understood, the height of the peak in the intermediate regime gives this information. Should it prove that the tensors do play a significant role, then this would be a very interesting result as it immediately excludes slow-roll potentials for the regime corresponding to the largest scales. Should the tensors prove negligible, then although the conclusions are less dramatic one has an easier inversion problem on large scales as one can concentrate solely on scalar modes.

2. Galaxy clustering in the optical and infrared

A. Redshift surveys in the optical.

Over the last decade, enormous leaps have been made in our understanding of the distribution of galaxies in the Universe from various redshift surveys. Most prominent is doubtless the ongoing Center for Astrophysics (CFA) survey (Ramella, Geller, and Huchra, 1992), which aims to form a complete catalogue of galaxy redshifts out to around $100h^{-1}$ Mpc. Other surveys of optical galaxies, often trading incompleteness for greater survey depth, are also in progress. On the horizon is the Sloan Digital Sky Survey which aims to find the redshifts of one million galaxies, occupying one quarter of the sky, with an overall depth of $500h^{-1}$ Mpc and completeness out to $100h^{-1}$ Mpc.

The redshifts of galaxies are relatively easy (though time consuming) to measure and interpret, and so provide one of the more observationally simple means of determining the distribution of matter in the Universe. The main technical problem is to correct the distribution for redshift distortions (which gives rise to the famous 'fingers of God' effect). However, the distribution of galaxies, specified by the galaxy power spectrum (or correlation function) is two steps away from telling us about the primeval mass spectrum.

- We have already discussed that the primeval spectrum on intermediate scales has been distorted by a combination of matter dynamics and amendments to the perturbation growth rate when the Universe is not completely matter dominated. If we know what the dark matter is, then this need not be a serious problem.

- Galaxies need not trace mass, and in modern cosmology it is almost always assumed they do not. This makes the process of getting from the galaxy power spectrum to the mass power spectrum extremely non-trivial. Models such as biased CDM rely on the notion of a scale-independent ratio between the two, but this too can only be an approximation to reality. In recent work, authors have emphasised the possible influence of environmental effects on galaxy formation (for instance, a nearby quasar might inhibit galaxy formation, and indeed it has been demonstrated that only very modest effects are required in order to profoundly affect the shapes of measured quantities such as the galaxy angular correlation function.

Despite this, attempts have been made to reconstruct the power spectrum from various surveys. In particular, this has been done for the CFA survey (Vogeley et al. 1992), and for the Southern Sky Redshift Survey (Park et al. 1992). These reconstructions remain very noisy, especially at both large scales (poor sampling) and small scales (shot noise and redshift distortions), and at present the best one could do would be to try and fit simple functional forms such as power-laws or parametrized power spectra to them. Even then, the constraints one would get on the slope of say a tilted CDM spectrum are very weak indeed. However, these reconstructions go along with the usual claim that standard CDM is excluded at high confidence due to inadequate large-scale clustering, without providing any particular constraints on the choice of methods of resolving this conflict.

Nevertheless, with larger sampling volumes such as those which the SDSS will possess, one should be able to get a good determination of the *galaxy* power spectrum across a reasonable range of scales, perhaps $10h^{-1}$ to $100h^{-1}$ Mpc.

B. Redshift surveys in the infrared.

A rival to redshifts of optical galaxies is those of infra-red galaxies, based on galaxy positions catalogued by the Infra-Red Astronomical Satellite (IRAS) project in the mid-eighties. The aim here is to sparse-sample these galaxies and redshift the subset. This is being done by two groups, giving rise to the QDOT survey (Saunders et al. 1991) and the 1.2 Jansky survey (Fisher et al. 1992). Taking advantage of the pre-existing data-base of galaxy positions has allowed these surveys to achieve great depth with even sampling and reach some interesting conclusions.

The main obstacle to comparison with optical surveys is due to the selection method. Infra-red galaxies are generally young, and appear to possess a distribution notably less clustered than their optically selected counterparts. They are thus usually attributed their own bias parameter which differs from the optical bias. The mechanics of proceeding to the power spectrum are basically the same as for optical galaxies.

The most interesting and relevant results here are obtained in combination with peculiar velocity information, as discussed below.

C. Projected catalogues.

As well as redshift surveys, one also has surveys which plot the positions of galaxies on the celestial sphere. At present the most dramatic is the APM survey, encompassing several million galaxies. The measured quantity is the projected counterpart of the correlation function, the angular correlation function usually denoted $w(\theta)$ where θ is the angular separation. Though arguments remain as to the presence of systematics, one in principle has accurate determinations of the galaxy angular correlation function. The first aim is to reconstruct the full three dimensional correlation function from this (proceeding thence to the galaxy power spectrum). Unfortunately, present methods of carrying out this inversion [based on inverting Limber's equation which gives $w(\theta)$ from $\xi(r)$] have proven to be very unstable, and a satisfactory recovery of the full correlation function has not been achieved.

In its preliminary galaxy identification stage, the SDSS will provide a huge projected catalogue on which further work can be carried out.

3. Peculiar velocity flows

Potentially the most important measurements in large-scale structure are those of the peculiar velocity field. Because all matter participates gravitationally, peculiar velocities directly sample the mass spectrum, not the galaxy spectrum. Were one to know the peculiar velocity field, this information is therefore as close to the primeval spectrum as is microwave background information.

Perhaps the most exciting recent development in peculiar velocity observations is the development of the POTENT method by Bertschinger, Dekel and collaborators (1989). Using only the assumption that the velocity can be written as the divergence of a scalar (in gravitational instability theories in the linear regime this is naturally associated with the peculiar gravitational potential), they demonstrate that the radial velocity towards/away from our galaxy (which is all that can be measured by the methods available) can be used to reconstruct the scalar, which can then be used to obtain the full three dimensional velocity field. This has been shown to work very well in simulated data sets, where one mimics observations and then can compare the reconstruction from those measurements with the original data set. So far, the noisiness and sparseness of available real radial velocity data has meant that attempts to reconstruct the fields in the neighbourhood of our galaxy have not yet met with great success; however, once better and more extensive observational data are obtained one can expect this method to yield excellent results.

At present, POTENT appears at its most powerful in combination with a substantial redshift survey such as the IRAS/QDOT survey. As POTENT supplies information as to the density field and the redshift survey to the galaxy distribution, the two in combination can be used in an attempt to measure quantities such as the bias parameter and the density parameter Ω_0 of the Universe. Reconstructions of the power spectrum have also been attempted. At present, the error bars (due to cosmic variance because of small sampling volume, due to the sparseness of the data in some regions of the sky, and due to iterative instabilities) are large enough that a broad range of spectra (including standard CDM) are compatible with the reconstructed present-day spectrum.

With larger data sets and technical developments in the theoretical analysis tools, POTENT (and indeed velocity data in general) appears to be a very powerful means of investigating the present-day power spectrum. To that, one need only add a knowledge of the dark matter.

C. Small scales ($8h^{-1}$ Mpc $\longrightarrow$ $1h^{-1}$ Mpc)

It is worth saying immediately that this promises to be the least useful range of scales. For many choices of dark matter, including the standard hot dark matter scenario, perturbations on these scales are almost completely erased by dark matter free-streaming to leave no information as to the primeval spectrum. Only if the dark matter is cold does it seem likely that any useful information can be obtained.

There are several types of measurement which can be made. Quite a bit is known about galaxy clustering on small scales, such as the two-point galaxy correlation function. However, the strong nonlinearity of the density distribution on these scales erases information about the original linear-regime structure, and the requirement of N-body simulations to make theoretical predictions makes this an unpromising avenue for reconstruction even should nature have chosen to leave significant spectral power on these scales. There exist very small-scale (arcsecond–arcminute) microwave background anisotropy measurements, though these are susceptible to a number of line of sight effects, and further the anisotropies are suppressed (exponentially) on short scales because the finite thickness (about $7h^{-1}$ Mpc) of the last scattering surface comes into play.

Up to now, the most useful constraints on small scales have come from the pairwise velocity dispersion (the dispersion of line-of-sight velocities between galaxies). These are sensitive to the normalization of the spectrum at small scales, though unfortunately susceptible to power feeding down from higher scales as well. There are certainly noteworthy constraints—for instance it is generally accepted that unbiased standard CDM generates excessively large dispersions. However, the calculations required involve N-body simulations and because a wide range of wavelengths contribute, obtaining knowledge of any structure in the power spectrum on these scales is likely to prove impossible, even if the amplitude can be determined to reasonable accuracy.

Already we are able to test various models for the power spectrum. For example a galaxy survey gives the power spectrum on 'small' scales, while CBR anisotropies probe the power spectrum on large scales. For preliminary models we can take a primordial power-law spectrum $|\delta_k|^2 \propto k^n$ ($n = 1$ is the Harrison-Zel'dovich spectrum) and some choice for dark matter, hot, cold, or mixed (fraction hot plus a fraction cold). Processing the primordial spectrum through the transfer function gives the curves of Figure 6. The simplest model is $n = 1$ CDM. Clearly the general shape is correct, but the normalization for the galaxy points does not match the normalization for the COBE points.

To better fit the observed spectrum several variations on the theme of CDM has been proposed. In the mixed dark matter (MDM) variant, a small amount of hot dark matter is added: $\Omega_{\text{HDM}} \sim 0.3$, $\Omega_{\text{CDM}} \sim 0.65$, and $\Omega_B \sim 0.05$. The 'pinch' of hot dark matter, e.g. in the form of neutrinos of mass 7 eV or so, leads to the suppression of fluctuations on small scales because of the free streaming of neutrinos. Another variant involves a modification of the spectrum of perturbations. When the spectrum of perturbations is normalized to the COBE DMR result, which fixes the spectrum on a very-large scale, a modest amount of 'tilt', say $n \simeq 0.8$, can reduce fluctuations on small scales. Another variant involves introducing a cosmological constant; a possibility too unpalatable to consider further.

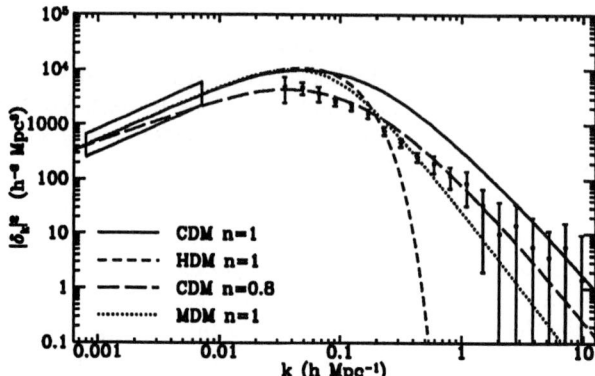

Figure 6. *Comparison of the measured power spectrum of density perturbations and the predictions of several models. The power spectrum of density perturbations from galaxies (the points on the right-hand side) is from Fisher et al. 1992, and the power spectrum from the COBE DMR measurements (the box on the left-hand side) is from Smoot et al. 1992. The models shown are cold dark matter; hot dark matter; tilted cold dark matter; and mixed dark matter.*

Data on large-scale structure is accumulating rapidly. In a few years we will have much better information about the power spectrum, and we will see if any of the simple models is correct. Let us turn now to possibilities for generating the perturbations.

3 Inflation

The basic idea of inflation is that there was an epoch early in the history of the Universe when potential, or vacuum, energy was the dominant component of the energy density of the Universe. During that epoch the scale factor grew exponentially. During this phase (known as the de Sitter phase), a small, smooth, and causally coherent patch of size less than the Hubble radius H^{-1} can grow to such a size that it easily encompasses the comoving volume that becomes the entire observable Universe today.

In the original proposal, inflation occurred in the process of a strongly first-order phase transition. This model was soon demonstrated to be fatally flawed. Subsequent models for inflation involved phase transitions that were second-order, or perhaps weakly first-order; some even involved no phase transition at all. Recently the possibility of inflation during a strongly first-order phase transition has been revived. Before discussing the latest developments in inflation, I will briefly review some of the history of inflation.

3.1 The art of inflation

Pre-history

The years before the birth of the inflationary Universe contained a rich pre-history of work in cosmology investigating the cosmological consequences of a Universe dominated by vacuum energy. Vacuum energy is interesting in cosmology because it acts as a cosmological constant, and will drive the Universe in exponential expansion. Recall that the expansion of the Universe is determined by the Friedmann equation:

$$\left(\frac{\dot{a}}{a}\right)^2 + \frac{k}{a^2} = \frac{8\pi G_N}{3}\rho, \tag{62}$$

where $a(t)$ is the cosmic scale factor, ρ is the energy density of the Universe, and the constant k is ± 1 or 0 depending upon the spatial curvature. If the contribution of the vacuum energy density ρ_V dominates, then ρ is a constant (it does not decrease with a), and for $k = 0$ the solution to the Friedmann equation is

$$a(t) = a(0)\exp(Ht); \qquad H \equiv \frac{\dot{a}}{a} = \left(\frac{8\pi G_N}{3}\rho_V\right)^{1/2} = \text{const}. \tag{63}$$

Such a rapid expansion may solve several cosmological problems, including the flatness/age problem, the homogeneity/isotropy problem, the problem of the origin of density inhomogeneities, and the monopole problem.

The possibility of a Universe dominated by vacuum energy became much more relevant with the realization that the Universe may have undergone a series of phase transitions associated with spontaneous symmetry breaking. The work of Kirzhnits and Linde (1972) showed that symmetries that are spontaneously broken today should have been restored at temperatures above the energy scale of spontaneous symmetry breaking, and as the Universe cooled below some critical temperature, denoted as T_C, there should have been a phase transition in which the symmetry was broken. Thus, phase transitions associated with spontaneous symmetry breaking might offer a mechanism whereby the early Universe may be dominated by vacuum energy for some period of time (Kolb and Wolfram, 1980).

The Classical Era of Old Inflation

Although there was a rich pre-history, the classical era of inflation crystallized with the paper of Guth (1980). In this classical picture, the Universe underwent a strongly first-order phase transition associated with spontaneous symmetry breaking of some Grand Unified Theory (GUT). Whether the phase transition is first order or higher order depends upon the details of the 'Higgs' potential for the scalar field whose vacuum expectation value is responsible for symmetry breaking. This theory is now usually referred to as 'old' inflation.

In old inflation the crucial feature of the potential was the barrier separating the symmetric high-temperature minimum, say located at $\phi = 0$, from the low-temperature true vacuum located at $\phi \neq 0$. If the transition is strongly first order, the transition from the high-temperature to the low-temperature minimum occurs by the quantum-mechanical process of nucleation of bubbles of true vacuum. These bubbles of true vacuum expand at the velocity of light, converting false vacuum to true.

The bubble nucleation rate ('per volume' will always be understood when discussing the bubble nucleation rate) depends upon the shape of the potential, but in general, it is written as $\Gamma = A\exp(-B)$, where A is a parameter with mass dimension 4, and B, the bounce action, is dimensionless. Let us simply assume that $A \sim \phi_0^4$, where ϕ_0 is the mass scale of spontaneous symmetry breaking (SSB).

In the classical picture, the energy density of the Universe became dominated by the false-vacuum energy of the Higgs field and the Universe expanded exponentially. Sufficient inflation was never a real concern; the problem with the classical picture is in the termination of the false-vacuum phase; usually referred to as the graceful exit problem.

Inside the true vacuum bubble is just what one expects—vacuum. For successful inflation it is necessary to convert the vacuum energy to radiation. The way this is accomplished in a first-order phase transition is through the process of collision of vacuum bubbles. In bubble collisions the energy density tied up in the bubble walls may be converted to entropy. Thus, if a first-order phase transition is to have a graceful exit, there must be many bubble collisions. The decline of the classical era began with the realization that bubbles of true vacuum do not percolate and fill the Universe; *i.e.* there is no graceful exit. The basic reason is that the exponential expansion of the background space overwhelms the bubble growth. To see this, consider the expression for the *coordinate* (or *comoving*) radius of the bubble. Assume that the bubble is nucleated at time t_0 with zero radius, and expands outward at the speed of light. At some time $t > t_0$ after nucleation, the comoving bubble radius is

$$r(t,t_0) = \int_{t_0}^{t} dt'\, a^{-1}(t') = \frac{\exp(-Ht_0) - \exp(-Ht)}{Ha(0)}. \tag{64}$$

The *physical* size of the bubble of course is simply $R(t,t_0) = a(t)r(t,t_0)$. Notice that as $t \to \infty$, the comoving bubble size approaches a *finite* value:

$$r(\infty, t_0) = \frac{\exp(-Ht_0)}{Ha(0)}. \tag{65}$$

Bubbles nucleated at larger t_0 reach a smaller comoving size than bubbles nucleated earlier in the transition. If a bubble is nucleated at time t_0, at some later time t the bubble has comoving volume $v(t,t_0)$ and physical volume $V(t,t_0)$ given by

$$\begin{aligned} v(t,t_0) &= \frac{4\pi}{3}r^3(t,t_0) \longrightarrow \frac{4\pi}{3}\frac{\exp(-3Ht_0)}{[Ha(0)]^3} \\ V(t,t_0) &= \frac{4\pi}{3}R^3(t,t_0) \longrightarrow \frac{4\pi}{3}\frac{\exp[3H(t-t_0)]}{H^3}, \end{aligned} \tag{66}$$

where here arrows indicate the asymptotic values as $t \to \infty$.

The probability that a point remains in the old (false-vacuum) phase at time t is simply

$$p(t) = \exp\left[-\int_0^t dt_0\, \Gamma V(t,t_0)\right] \longrightarrow \exp\left[-\frac{4\pi}{3}\left(\frac{\Gamma}{H^4}\right)Ht\right]. \tag{67}$$

Thus, the probability that a point remains in the false-vacuum phase decreases exponentially in time, just as expected.

Although $p(t)$ decreases exponentially, the volume of space in the false vacuum is increasing exponentially. A measure of whether true vacuum regions will percolate the space is the fraction of physical space in false vacuum:

$$f(t) = p(t)a^3(t) \longrightarrow \exp\left[-\frac{4\pi}{3}\left(\frac{\Gamma}{H^4}\right)Ht\right]\exp[3Ht]. \tag{68}$$

Clearly whether this fraction increases or decreases in time depends upon the competition between the decreasing probability for a point to be in the false vacuum and the increasing volume of space in the false vacuum. A rough estimate of whether $f(t)$ will increase or decrease is the criteria that $\epsilon = \Gamma/H^4$ is much greater or much less than unity. If ϵ is much less than one the transition will never be completed, while if ϵ is much greater than one the transition will be completed, but there won't be a sufficient period of inflation. So if ϵ is small enough to guarantee sufficient inflation, it will be too small for percolation to result.

This graceful exit problem led to the decline of the classical era of inflation and the dawn of the inflationary dark ages.

Slow-Rollover Renaissance of New Inflation

Soon after the demise of the original model, inflation was revived by the realization that it was possible to have an inflationary scenario without recourse to a strongly first-order phase transition. Linde, and Albrecht (1982) and Steinhardt (1982) proposed that the Universe inflates in the process of the classical evolution of the vacuum. In the classical evolution of the field to its true minimum the field has 'kinetic' energy and 'potential' energy. If one has a region of the scalar Higgs potential that is 'flat,' then the velocity of the Higgs field in the evolution to the ground state will be slow, and the potential energy of the Higgs field might dominate the kinetic energy. This can be made more quantitative by writing the classical equation of motion for a spatially homogeneous scalar field ϕ (called the *inflaton*) in an expanding Universe under the influence of a potential $V(\phi)$:

$$\ddot{\phi} + 3\frac{\dot{a}}{a}\dot{\phi} + \frac{dV(\phi)}{d\phi} = 0. \tag{69}$$

If the potential is flat enough that the $\ddot{\phi}$ term can be neglected, the scalar field will undergo a period of 'slow roll.' The energy density contributed by the scalar field is $\rho_\phi = \dot{\phi}^2/2 + V(\phi)$, and in the slow-roll region $V(\phi) \gg \dot{\phi}^2$, so the expansion closely approximates the exponential solution. This theory is sometimes referred to as 'new' inflation.

The original proposal of slow-rollover inflation was also based upon an SU(5) GUT phase transition. The potential was 'flattened' by assuming that it took the Coleman-Weinberg form. However it was soon realized that even this potential was not flat enough. If the scalar potential is approximated by a simple potential of the form $V(\phi) = \lambda(\phi^2 - \phi_0^2)^2$, in order for density fluctuations produced in inflation to be small enough required $\lambda \lesssim \mathcal{O}(10^{-15})$. Clearly such small numbers did not arise naturally in simple unified models, and a successful slow-rollover inflation model must be somewhat more complicated. Unfortunately, it was soon discovered that there is no cosmological upper bound on the complexity of models.

It was soon realized that the requirement of a small coupling constant could not easily be accommodated in simple particle physics models. Of course the usual temptation is to modify the Higgs sector by adding more representations than required in the minimal model. In fact a successful model was constructed along these lines by Pi (1984) and by Shafi and Vilenkin (1984).

For a while it was thought that supersymmetric GUTS could hold the key, but they were soon abandoned for a variety of reasons. After supersymmetric models, some very interesting supergravity models emerged. Although many supergravity models raised new problems of their own, some supergravity models were quite successful, and (at least) gave a proof of existence that the inflationary scenario might be implemented in particle models.

All of these Baroque models suffered from a low re-heat temperature as a result of a weakly coupled inflaton. This made baryogenesis problematical, although not impossible. All post-renaissance inflation models involved second-order transitions, and because inflation occurred in a smooth patch of the Universe that originally contained a single correlation region, the observable Universe should contain less than one topological defect produced in the transition. This is good news for the monopole problem, but bad news for cosmic strings and texture.

Rococo Inflation

The complexity of inflationary models was again increased as people started modifying the gravitational sector of the theory. In Rococo inflation the identity of the inflaton is up for grabs. There are models where the inflaton is associated with the radius of internal dimensions, with the extra degree of freedom in fourth-order gravity, with the scalar field of induced gravity, *etc.* Some of these models can be made to work; it might be said that none work naturally.

Perhaps somewhere along the line as more and more detail was added to make the models satisfy all of the constraints, the message, or at least the spirit, of inflation was lost.

Impressionism

In response to the excesses of Baroque and Rococo inflation, there grew up around Andrei Linde a Russian school of 'Impressionist' inflation. In the impressionist style no serious attempt is made to connect the details of inflaton with any specific particle physics models. In this way the true essence and beauty of the inflationary Universe is realized without any of the cluttering details. The best example of this the the 'chaotic' inflation model. In this model the scalar potential is assumed to be simply $V(\phi) = \lambda \phi^4$. What a perfect example of impressionism! This potential embodies features common to all scalar potentials without any of the details. Of course it is not 'realistic' in the sense that no one would accept the existence of a scalar field whose sole purpose is to make inflation simple, but it can be taken to represent the impressions of every scalar field, while at the same time representing no scalar field.

As Linde has repeatedly emphasized, it is not even necessary to connect inflation with a phase transition. In the $\lambda\phi^4$ chaotic model the ϕ field is expected to start away from its minimum (at $\phi = 0$) due to 'chaotic' initial conditions. From there, inflation can be analyzed as in slow rollover models.

Despite the seductive beauty of the impressionist approach we must demand more realism. Eventually we want a description of the Universe that has the fine details of the Baroque or Rococo but with the simplicity and spirit of impressionism.

The Postmodern Era

One of the most interesting recent developments is postmodernism. The postmodern movement is characterized by an eclectic mixture of classical tradition with some aspect of the recent past. With this definition, it may be said that first-order inflationary cosmologies represent a postmodern trend. The classical tradition is a first-order transition, while the aspect of the recent past will be embodied by the slow rolling of a second scalar field.

The key to first-order inflation is the relaxation of the assumption that $\epsilon \equiv \Gamma/H^4$ is constant in time. There are two ways one might imagine a time dependence for ϵ. The first way is for H to change. Since $H = \sqrt{G\rho_V}$, either the effective gravitational constant G must change or the vacuum energy ρ_V must change. (We will see that in many cases the two possibilities are equivalent representations of the same physics.) The second way is for Γ to change. Of course, in general, both H and Γ might change.

If ϵ starts small, much less than one, then there might be a sufficient amount of inflation. If ϵ grows and eventually becomes much greater than one, then the bubbles of true vacuum will percolate and collisions between the bubble walls might convert the false-vacuum energy into entropy. This is the hope of first-order inflation.

The best example of a first-order inflation model is extended inflation (La and Steinhardt, 1989). The difference between extended inflation and Guth's model is the theory of gravity: Jordan-Brans-Dicke (JBD) in extended inflation rather than GR in Guth's model.

In JBD the gravitational 'constant' is set by the value of a scalar field. During inflation this scalar field evolves and gravity becomes weaker; as a result the cosmic-scale factor grows as a large power of time rather than exponentially. This means that in extended inflation the physical volume of space remaining in the false vacuum grows only as power of time and not exponentially, and unlike Guth's original model, bubble nucleation can convert all of space to the true vacuum.

3.2 Scalar field dynamics

Regardless of the particular model of inflation, scalar field dynamics plays an important role in the cosmology, so let us study the scalar field dynamics in more detail. Consider a minimally coupled, spatially homogeneous scalar field ϕ, with Lagrangian density

$$\mathcal{L} = \frac{1}{2}\partial^\mu\phi\,\partial_\mu\phi - V(\phi) = \frac{1}{2}\dot\phi^2 - V(\phi). \tag{70}$$

With the assumption that ϕ is spatially homogeneous, the stress-energy tensor takes the form of a perfect fluid, with energy density and pressure given by $\rho_\phi = \dot\phi^2/2 + V(\phi)$, and $p_\phi = \dot\phi^2/2 - V(\phi)$. The classical equation of motion for ϕ is given in Equation 69. All minimal slow-roll models are examples of sub-inflationary behaviour, which is defined by the condition $\dot H < 0$. Super-inflation, where $\dot H > 0$, cannot occur here, though it is possible in more complex scenarios. This allows us to eliminate the time-dependence in the Friedmann equation and derive the first-order, non-linear differential equations

$$(H')^2 - \frac{3}{2}\kappa^2 H^2 = -\frac{1}{2}\kappa^4 V(\phi) \tag{71}$$

$$\kappa^2 \dot\phi = -2H', \tag{72}$$

where $\kappa^2 = 8\pi G$.

We can define two parameters, which we will denote as slow-roll parameters, by

$$\epsilon \equiv \frac{3\dot\phi^2}{2}\left(V + \frac{\dot\phi^2}{2}\right)^{-1} = \frac{2}{\kappa^2}\left(\frac{H'}{H}\right)^2$$

$$\eta \equiv \frac{\ddot\phi}{H\dot\phi} = \frac{2}{\kappa^2}\frac{H''}{H}. \tag{73}$$

Slow-roll corresponds to $\{\epsilon, |\eta|\} \ll 1$. With these definitions, the end of inflation is given *exactly* by $\epsilon = 1$. A small value of η guarantees $3H\dot\phi \simeq -V'(\phi)$, which is often called the slow-roll equation.

Density perturbations arise as the result of quantum-mechanical fluctuations of fields in de Sitter space. First, let's consider scalar density fluctuations. To a good approximation we may treat the inflaton field ϕ as a massless, minimally coupled field. (Of course the inflaton does have a mass, but inflation operates when the field is evolving through a 'flat' region of the potential.) Just as fluctuations in the density field may be expanded in a Fourier series the fluctuations in the inflaton field may be expanded in terms of its Fourier coefficients $\delta\phi_\mathbf{k}$: $\delta\phi(\mathbf{x}) \propto \int \delta\phi_\mathbf{k}\exp(-i\mathbf{k}\cdot\mathbf{x})d^3k$. During inflation there is an event horizon as in de Sitter space, and quantum-mechanical fluctuations in the Fourier components of the inflaton field are given by

$$k^3|\delta\phi_\mathbf{k}|^2/2\pi^2 = (H/2\pi)^2, \tag{74}$$

where $H/2\pi$ plays a role similar to the Hawking temperature of black holes. Thus, when a given mode of the inflaton field leaves the Hubble radius during inflation, it has impressed upon it quantum mechanical fluctuations. In analogy to the discussion of the density perturbations of the previous section, what is called the fluctuations in the inflaton field on scale k is proportional to $k^{3/2}|\delta\phi_\mathbf{k}|$, which by Equation 74 is proportional to $H/2\pi$. Fluctuations in ϕ lead to perturbations in the energy density $\delta\rho_\phi = \delta\phi(\partial V/\partial\phi)$.

Now considering the fluctuations as a particular mode leaves the Hubble radius during inflation, we may construct the gauge invariant quantity ζ using the fact that during inflation $\rho_0 + p_0 = \dot{\phi}^2$:

$$\zeta = \delta\phi \left(\frac{\partial V}{\partial \phi}\right) \frac{1}{\dot{\phi}^2}. \tag{75}$$

Now using Equation 71 and Equation 72, the amplitude of the density perturbation when it crosses the Hubble radius *after* inflation is

$$\left(\frac{\delta\rho}{\rho}\right)_\lambda^{\text{HOR}} \equiv \frac{m}{\sqrt{2}} A_S(\phi) = \frac{m\kappa^2}{8\pi^{3/2}} \frac{H^2(\phi)}{|H'(\phi)|} \propto \frac{V^{3/2}(\phi)}{m_{Pl}^3 V'(\phi)}, \tag{76}$$

where $H(\phi)$ and $H'(\phi)$ are to be evaluated when the scale λ crossed the Hubble radius *during* inflation. The constant m equals 2/5 or 4 if the perturbation re-enters during the matter or radiation dominated eras respectively. The 4 for radiation is appropriate to the uniform Hubble constant gauge. One occasionally sees a value 4/9 instead which is appropriate to the synchronous gauge. The matter domination factor is the same in either case. Note also that it is exact for matter domination, but for radiation domination it is only strictly true for modes much larger than the Hubble radius, and there will be corrections in the extrapolation down to the size of the Hubble radius.

Now we wish to know the λ-dependence of $(\delta\rho/\rho)_\lambda$, while the right-hand side of the equation is a function of ϕ when λ crossed the Hubble radius during inflation. We may find the value of the scalar field when the scale λ goes outside the Hubble radius in terms of the number of e-foldings of growth in the scale factor between Hubble radius crossing and the end of inflation.

It is quite a simple matter to calculate the number of e-foldings of growth in the scale factor that occur as the scalar field rolls from a particular value ϕ to the end of inflation ϕ_e:

$$N(\phi) \equiv \int_{te}^{t} H(t')dt' = -\frac{\kappa^2}{2}\int_\phi^{\phi_e} \frac{H(\phi')}{H'(\phi')} d\phi'. \tag{77}$$

The slow-roll conditions guarantee a large number of e-foldings. The total amount of inflation is given by $N_{\text{TOT}} \equiv N(\phi_i)$, where ϕ_i is the initial value of ϕ at the start of inflation (when $\ddot{a}$ first becomes positive). In general, the number of e-folds between when a length scale λ crossed the Hubble radius during inflation and the end of inflation is given by

$$N(\lambda) = 45 + \ln(\lambda/\text{Mpc}) + \frac{2}{3}\ln(M/10^{14}\text{ GeV}) + \frac{1}{3}\ln(T_{\text{RH}}/10^{10}\text{ GeV}), \tag{78}$$

where M is the mass scale associated with the potential and T_{RH} is the 're-heat' temperature. Relating $N(\lambda)$ and $N(\phi)$ from Equation 77 results in an expression between ϕ and λ. Hopefully this dry formalism will become clear in the example discussed below.

In addition to the scalar density perturbations caused by de Sitter fluctuations in the inflaton field, there are gravitational mode perturbations, $g_{\mu\nu} \to g_{\mu\nu}^{\text{FRW}} + h_{\mu\nu}$, caused by de Sitter fluctuations in the metric tensor. Here, $g_{\mu\nu}^{\text{FRW}}$ is the Friedmann-Robertson-Walker metric and $h_{\mu\nu}$ are the metric perturbations. That de Sitter space fluctuations should lead to fluctuations in the metric tensor is not surprising, since after all, gravitons are the propagating modes associated with transverse, traceless metric perturbations, and they too behave as minimally coupled scalar fields. The dimensionless tensor metric perturbations can be expressed in terms of two graviton modes we will denote as h. Performing a Fourier decomposition of h, $h(\vec{x}) \propto \int \delta h_k \exp(-i\vec{k}\cdot\vec{x})d^3k$, we can use the formalism for scalar field perturbations simply by the identification $\delta\phi_k \to h_k/\kappa\sqrt{2}$, with resulting quantum fluctuations [cf. Equation 74]

$$k^3|h_k|^2/2\pi^2 = 2\kappa^2(H/2\pi)^2. \tag{79}$$

While outside the Hubble radius, the amplitude of a given mode remains constant, so the amplitude of the dimensionless strain on scale λ when it crosses the Hubble radius after inflation is given by

$$\left|k^{3/2}h_k\right|^{\text{HOR}}_\lambda \equiv A_G(\phi) = \frac{\kappa}{4\pi^{3/2}}H(\phi) \sim \frac{V^{1/2}(\phi)}{m_{Pl}^2}, \tag{80}$$

where once again $H(\phi)$ is to be evaluated when the scale λ crossed the Hubble radius *during* inflation.

As usual, it is convenient to illustrate the general features of inflation in the context of the simplest model, chaotic inflation, which is to inflationary cosmology what *drosophila* is to genetics. In chaotic inflation the inflaton potential is usually taken to have a simple polynomial form such as $V(\phi) = \lambda\phi^4$, or $V(\phi) = \mu^2\phi^2$. For a concrete example, let us consider the simplest chaotic inflation model, with potential $V(\phi) = \mu^2\phi^2$. This model can be adequately solved in the slow-roll approximation, yielding

$$\begin{aligned}
\phi(t) &= \phi_i - \frac{2}{\sqrt{3}}\frac{\mu}{\kappa}t \\
a(t) &= a_i \exp\left[\frac{\kappa\mu}{\sqrt{3}}\left(\phi_i t - \frac{\mu}{\sqrt{3}\kappa}t^2\right)\right] \\
H &= \frac{\kappa\mu}{\sqrt{3}}\left(\phi_i - \frac{2}{\sqrt{3}}\frac{\mu}{\kappa}t\right) = \frac{\kappa\mu}{\sqrt{3}}\phi,
\end{aligned} \tag{81}$$

with inflation ending at $\kappa\phi_e = \sqrt{2}$ as determined by $\epsilon = 1$, where ϵ was defined in Equation 73. The number of e-foldings between a scalar field value ϕ, and the end of inflation is just

$$N(\phi) = -\frac{\kappa^2}{2}\int_\phi^{\sqrt{2}/\kappa}\frac{H(\phi')}{H'(\phi')}d\phi' = \frac{\kappa^2\phi^2}{4} - \frac{1}{2}. \tag{82}$$

Equating Equation 82 and Equation 78 relates ϕ and λ in this model for inflation:

$$\kappa^2\phi^2/4 = [45.5 + \ln(\lambda/\text{Mpc})]. \tag{83}$$

Using Equation 76 and Equation 80, A_S and A_G are found to be

$$\begin{aligned}
A_S(\lambda) &= \left(\sqrt{2}\kappa\mu/\sqrt{12\pi^3}\right)[45.5 + \ln(\lambda/\text{Mpc})] \\
A_G(\lambda) &= \left(\kappa\mu/\sqrt{12\pi^3}\right)[45.5 + \ln(\lambda/\text{Mpc})]^{1/2}.
\end{aligned} \tag{84}$$

We can note three features that are common to a large number of (but not all) inflationary models. First, A_S and A_G have different functional dependences upon λ.

Second, A_G and A_S increase with λ. Finally, $A_S > A_G$, for scales of interest, although not by an enormous factor.

The basic picture of the generation of scalar and tensor perturbations is illustrated in Figure 7. The main observational information from the cosmic microwave background arises through the Cosmic Background Explorer (COBE) satellite, and the Tenerife (TEN) and South Pole (SP) collaborations. Galaxy surveys (APM, CFA, and IRAS) may provide useful information up to $100h^{-1}$ Mpc, while the Sloan Digital Sky Survey (SDSS) should extend to the lowest scales measured by COBE. Peculiar velocity measurements using the POTENT (P) methods are important on intermediate scales. The angle θ measures angular scales on the CBR in degrees, and length scales λ are in units of h^{-1} Mpc. d_H refers to the horizon size today and at recombination and $d_{NL} \approx 8h^{-1}$ Mpc is the scale of non-linearity. (See the text for details).

3.3 Reconstructing the Inflaton Potential

Figure 7 illustrates how a knowledge of the potential allows a prediction of the scalar and tensor spectra. Now let's consider the possibility of reverse engineering this approach and try to reconstruct the inflaton potential from knowledge of the scalar and tensor spectra.

Reconstruction of the inflaton potential in this manner was first considered by Hodges and Blumenthal (1990). Recently this question has been studied by Copeland, Kolb, Liddle, and Lidsey (CKLL) (1993a, 1993b, 1993c), and also by Turner (1993). CKLL improved upon the Hodges and Blumenthal (HB) results in two important ways. Firstly, they considered both scalar and tensor modes, whereas HB restricted their study to the scalars alone. This is a vital improvement, because, as HB realized, the scalar spectrum alone is insufficient to uniquely determine the inflaton potential— reconstruction is possible only up to an undetermined constant, and as the reconstruction equations are nonlinear, this leads to functionally different potentials giving rise to the same scalar spectrum. The tensors (even just the tensor amplitude at a single scale) provide just the extra information needed to lift this degeneracy. Secondly, the HB analysis made explicit use of the slow-roll approximation. It is well known that this approximation breaks down unless both the scalar spectrum is nearly flat and the tensor amplitude is negligible. CKLL considered the inflation dynamics in full generality. However, general expressions for the perturbation spectra were studied in a slow-roll expansion.

In CKLL (1993a), analytic expressions were derived for functionally reconstructing the potential in terms of A_S and A_G. Although more complete expressions were given (see especially CKLL 1993c), here I will simply give the expressions to lowest order in A_G/A_S. The first result is a consistency equation relating the slope of the tensor spectrum to A_G and A_S:

$$\frac{\lambda}{A_G(\lambda)} \frac{dA_G(\lambda)}{d\lambda} = \frac{A_G^2(\lambda)}{A_S^2(\lambda)}. \tag{85}$$

This highlights the asymmetry in the correspondence between the scalar and tensor

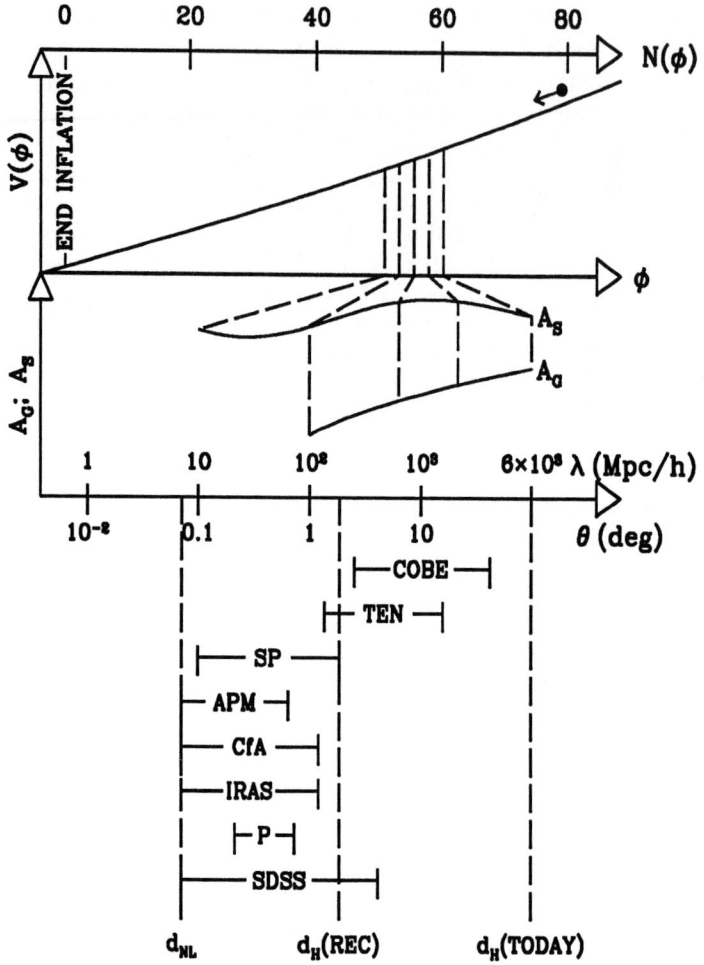

Figure 7. *A schematic figure illustrating the main concepts behind the generation of scalar and tensor perturbations in inflation.*

spectra. If one were given the tensor spectrum, then a simple differentiation supplies the unique scalar spectrum. However, if a scalar spectrum is supplied, then this first-order differential equation must be solved to find the form of $A_G(\lambda)$. This leaves an undetermined constant in the tensor spectrum and, as the consistency equation is nonlinear, this implies that the scalar spectrum alone does not uniquely specify the functional form of the tensors. However, knowledge of the amplitude of the tensor spectrum at one scale is sufficient to determine this constant and lift the degeneracy.

It is the tensor spectrum one requires to proceed with reconstruction. Once the form of the tensor spectrum has been obtained, either directly from observation or by integrating the consistency equation, the potential, as parametrized by λ, may be derived:

$$V[\phi(\lambda)] = \frac{48\pi^3 A_G^2(\lambda)}{\kappa^4}, \qquad (86)$$

where again this is true to lowest order in A_G/A_S. The reconstruction equations allow a functional reconstruction of the inflaton potential. For suitably simple spectra, this can be done analytically, and in CKLL 1993b we illustrated this for well-known cases of scalar spectra which are exactly scale-invariant, logarithmically corrected from scale-invariance, and exact power-laws. An alternative approach, useful for obtaining mass scales, is to concentrate on data around a given length scale λ_0, and perturbatively derive the potential around its corresponding scalar field value $\phi_0 \equiv \phi(\lambda_0)$. If we know $A_G(\lambda_0)$ and $A_S(\lambda_0)$ separately, then $V(\phi_0)$ follows immediately. The derivatives of the potential can also be obtained.

Let me illustrate the idea by example. Within a few years a combination of microwave background anisotropy measurements should give us some information about the scalar and tensor amplitudes at a particular length scale λ_0 (corresponding to an angular scale θ_0). A hypothetical, but plausible, data set that this might provide would be $A_S(\lambda_0) = 1 \times 10^{-5}$; $A_G(\lambda_0) = 2 \times 10^{-6}$; $n_0 = 0.9$. This would lead to (see CKLL 1993a, 1993b)

$$V(\phi_0) = (2\times 10^{16}\,\text{GeV})^4; \quad \pm V'(\phi_0) = (3\times 10^{15}\,\text{GeV})^3; \quad V''_{\text{sr}}(\phi_0) = (5\times 10^{13}\,\text{GeV})^2. \qquad (87)$$

In this way cosmology might be first to get a 'piece of the action' of GUT-scale physics.

4 Cosmological phase transitions

Perhaps the most important concept in modern particle theory is that of spontaneous symmetry breaking (SSB). The idea that there are underlying symmetries of nature that are not manifest in the structure of the vacuum appears to play a crucial role in the unification of the forces. In all unified gauge theories—including the standard electroweak model—the underlying gauge symmetry is larger than the unbroken $SU(3)_C \otimes U(1)_{EM}$. Of particular interest for cosmology is the theoretical expectation that at high temperatures, symmetries that are spontaneously broken today were restored, and that during the evolution of the Universe there were phase transitions associated with spontaneous breakdown of gauge (and perhaps global) symmetries. For example, we can be reasonably confident that there was such a phase transition at a temperature of order 300 GeV and a time of order 10^{-11} sec, associated with the breakdown of $SU(2)_L \otimes U(1)_Y \to U(1)_{EM}$. Moreover, the vacuum structure in many spontaneously broken gauge theories is very rich: topologically stable configurations of gauge and Higgs fields exist as domain walls, cosmic strings, and monopoles. In addition, classical configurations that are not topologically stable, so-called non-topological solitons, may exist and be stable for dynamical reasons. Interesting examples include soliton stars, Q-balls, non-topological cosmic strings, sphalerons, and so on.

Before discussing the cosmological implications, it is useful to review what is meant by the finite-temperature potential.

4.1 Finite-Temperature Potential

Let's start with a simple model of a real scalar field ϕ with Lagrangian

$$\mathcal{L} = \frac{1}{2}\partial_\mu \phi \, \partial^\mu \phi - V_0(\phi); \qquad V_0(\phi) = -\frac{1}{2}m^2\phi^2 + \frac{1}{4}\lambda\phi^4. \tag{88}$$

The Lagrangian is invariant under the discrete symmetry transformation $\phi \leftrightarrow -\phi$. The minima of the *classical* potential of Equation 88 are not at zero but at $\sigma_\pm = \pm\sqrt{m^2/\lambda}$. The origin, $\phi = 0$, is an unstable extremum of the potential because $V_0''(0) < 0$, where prime denotes $d/d\phi_c$. Since the quantum theory must be constructed about a stable extremum of the classical potential, the ground state of the system is either σ_+ or σ_-, and the reflection symmetry $\phi \leftrightarrow -\phi$ present in the Lagrangian is broken by the choice of a vacuum state, as $\phi = 0$ is the only possible vacuum invariant under $\phi \leftrightarrow -\phi$.

The potential of Equation 88 is the classical potential, and it is necessary to consider the effect of quantum corrections. Here I will follow the classic paper of Coleman and Weinberg (1973). For a general Lagrangian $\mathcal{L}(\phi)$ in the presence of a c-number source $J(x)$, the vacuum-to-vacuum amplitude is

$$\langle 0^+|0^-\rangle_J \equiv Z[J] = \int \mathcal{D}\phi \exp\left(i\int d^4x\,[\mathcal{L}(\phi) + J(x)\phi]\right), \tag{89}$$

where of course $Z[J]$ is the generating functional of the Green's functions. It is more useful to consider the generating functional of the *connected* Green's functions, $W[J]$, related to $Z[J]$ by $Z[J] = \exp(iW[J])$. $W[J]$ can be expanded in terms of powers of J, with the coefficients being the connected Green's functions. The classical field ϕ_c is defined as $\phi_c \equiv \delta W/\delta J$. Finally, the effective action is $\Gamma[\phi_c] = W[J] - \int d^4x J(x)\phi_c(x)$.

Now the effective action can be expanded in terms of $\Gamma^{(n)}$, the one-particle irreducible (1PI) Feynman diagrams with n external lines:

$$\Gamma[\phi_c] = \sum_{n=1}^{\infty} \frac{1}{n!} \int \Gamma^{(n)}(x_1 \ldots x_n)\, \phi_c(x_1)\ldots\phi_c(x_n)\, d^4x_1 \ldots d^4x_n. \tag{90}$$

Rather than an expansion in powers of the classical field, one can expand the effective action in powers of derivatives of the classical field:

$$\Gamma[\phi_c] = \int d^4x \left[-V(\phi_c) + \frac{1}{2}(\partial_\mu \phi_c)^2 Z(\phi_c) + \ldots\right]. \tag{91}$$

Now the constant term, $V(\phi_c)$, appearing in this expansion is known as the *effective potential*. By means of a Fourier transform of Equation 90, it is easy to show that the effective potential can also be expressed in terms of a sum of all 1PI Feynman graphs with zero momenta:

$$V(\phi_c) = -\sum_{n=1}^{\infty} \frac{1}{n!} \phi_c^n \Gamma^{(n)}(p_i = 0). \tag{92}$$

A simple example will illustrate the above dry formalism. In this example we follow Lee and Sciaccaluga (1975), and expand the effective potential about $\phi_c = \omega$, rather than about $\phi = 0$:

$$\begin{aligned}\Gamma^{(n)}[\phi_c] &= \sum_{n=1}^{\infty} \frac{1}{n!} \int \Gamma^{(n)}(x_1 \ldots x_n)\, [\phi_c(x_1) - \omega]\ldots[\phi_c(x_n) - \omega]\, d^4x_1\ldots d^4x_n, \\ V(\phi_c) &= -\sum_{n=1}^{\infty} \frac{1}{n!} [\phi_c - \omega]^n\, \Gamma^{(n)}(p_i = 0).\end{aligned} \tag{93}$$

where the coefficients in the expansion are now the generators of the 1PI diagrams in the shifted theory where ϕ_c is replaced by $\phi_c - \omega$. Now $dV/d\omega|_{\phi_c=\omega} = \Gamma^{(1)}$, which is simply the tadpole diagram in the shifted theory (up to a factor of i). So evaluating the tadpole, integrating over ω then setting $\omega = \phi_c$ gives the effective potential.

The shifted theory of Equation 88 gives a potential with mass squared of $-m^2 + 3\lambda\omega^2$, and a ϕ^3 term with coupling $3!i\lambda\omega$. Therefore the tadpole diagram is

$$\Gamma^{(1)} = \bigcirc\!\!\!\!\!\!\!\!\! \bigg| = -\frac{i}{2} \int \frac{d^4k}{(2\pi)^4} \frac{3!\lambda\omega}{k^2 - m^2 + 3\lambda\omega^2} .$$

The total potential to one-loop is the sum of the classical potential of Equation 88 and the one-loop correction:

$$V(\phi_c) = V_0(\phi_c) + \int^{\phi_c} \Gamma^{(1)} d\omega = V_0(\phi_c) + \frac{1}{2} \int \frac{d^4k}{(2\pi)^4} \ln\left(k^2 - m^2 + 3\lambda\phi_c^2\right). \quad (94)$$

A generalization to arbitrary potentials would be to replace the last two terms in the logarithm by $V_0''(\phi_c)$, where V_0 is the classical potential.

A few comments are in order before proceeding.

1. The integral in Equation 94 is infinite. This shouldn't scare us. We introduce a cutoff Λ, and if the theory is renormalizable, all infinities can be absorbed via some renormalization prescription.

2. The physical meaning of the one-loop potential is clear if we integrate over dk_0 to find

$$V(\phi_c) = V_0(\phi_c) + \int \frac{d^3k}{(2\pi)^3} \sqrt{k^2 + V_0''(\phi_c)}. \quad (95)$$

Since $\sqrt{k^2 + V_0''(\phi_c)}$ corresponds to the total energy of a fluctuation of momentum k, the one-loop correction is clearly the sum of zero-point energy fluctuations about the point $\phi = \phi_c$.

3. If the model is generalized to include couplings of ϕ to vectors and fermions, then the one-loop potential will include additional tadpoles where the particle in the loop is the vector or the fermion.

Now the integral in Equation 94 can be done by introducing a cutoff Λ. For the simple model that we are studying, the potential to one-loop is

$$V(\phi_c) = -\frac{1}{2}m^2\phi_c^2 + \frac{1}{4}\lambda\phi_c^4 + \frac{(-m^2 + 3\lambda\phi_c^2)^2}{64\pi^2} \ln(-m^2 + 3\lambda\phi_c^2) + a_1(\Lambda)\phi_c^2 + a_2(\Lambda)\phi_c^4, \quad (96)$$

where $a_i(\Lambda)$ are cutoff-dependent constants that will be determined by renormalization of the mass and coupling constants.

Of course it is the behaviour of the theory at finite temperature that is of interest to us. A simple, heuristic derivation of the effects of the thermal bath is to adopt the 'real-time' formalism in which the propagator, $D(k)$ includes the possibility of emission and absorption from the thermal bath:

$$D_T(k) = \frac{1}{k^2 - m^2 + i\epsilon} + \frac{2\pi}{\exp(E/T) - 1} \delta(k^2 - m^2). \tag{97}$$

Then this propagator is used in the evaluation of the propagator of the tadpole diagram. The additional temperature-dependent part of the propagator leads to an additional, temperature-dependent part of the one-loop potential:

$$V_T(\phi_c) = \frac{T^4}{2\pi^2} \int_0^\infty dx\, x^2 \ln\left[1 - \exp\left(x^2 + V''(\phi_c)/T^2\right)^{1/2}\right]. \tag{98}$$

At high temperature, $T \gg |m|$, it is possible to expand the logarithm and perform the integration: $V_T(\phi_c) = -(\pi^2/90)T^4 + (1/24)V''(\phi_c)T^2 + \cdots$. The first term is simply the free energy of a massless spin-0 boson. The second term in the expansion is ϕ_c-dependent, with a *positive* coefficient. For instance in the simple ϕ^4 theory we have been following, $V''(\phi_c) = -m^2 + 3\lambda\phi_c^2$. Adding $V_T(\phi_c)$ to the classical potential gives a total potential with a coefficient of the term quadratic in ϕ_c of $-m^2/2 + \lambda T^2/8$. Clearly above some critical temperature $T_C = 2m/\lambda^{1/2}$ the coefficient of the quadratic term will be positive, and below T_C the coefficient will be negative. This is a signal that for $T > T_C$ the symmetry will be restored, $\phi = 0$ will be a stable minimum of the potential. In the evolution from the high-temperature phase to the low-temperature phase there is a phase transition.

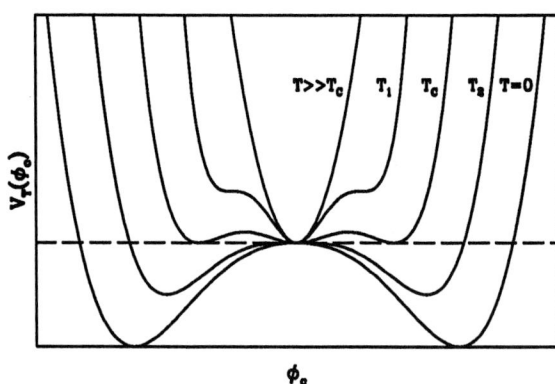

Figure 8. *The temperature dependence of $V_T(\phi_c)$ for a first-order phase transition. Only the ϕ_c-dependent terms in $V_T(\phi_c)$ are shown.*

If ϕ also couples to fermions or gauge bosons, there will be additional terms in the temperature-dependent potential. The $\lambda\phi^4$ theory has a second-order transition. The additional terms in V_T from gauge boson contributions can drive the transition to first order. In general, a symmetry-breaking phase transition can be first or second order.

The temperature dependence of $V_T(\phi_c)$ for a first-order phase transition is shown in Figure 8. For $T \gg T_C$ the potential is quadratic, with only one minimum at $\phi_c = 0$. When $T = T_1$, a local minimum develops at $\phi_c \neq 0$. For $T = T_C$, the two minima become degenerate, and below T_C, the $\phi_c \neq 0$ minimum becomes the global minimum. If for $T \leq T_C$ the extremum at $\phi_c = 0$ remains a local minimum, there must be a barrier between the minima at $\phi_c = 0$ and $\phi_c \neq 0$. Therefore, the change in ϕ_c in going from one phase to the other must be discontinuous, indicating a first-order phase transition. Moreover, the transition cannot take place classically, but must proceed either through quantum or thermal tunnelling. Finally, when $T = T_2$ the barrier disappears and the transition may proceed classically. For a second-order transition there is no barrier at the critical temperature, and the transition occurs smoothly.

As a final illustration let's consider the electroweak phase transition. In the minimal electroweak model there is a complex SU(2) doublet Φ, with potential $V(\Phi) = -m^2 \Phi^\dagger \Phi + \lambda_0 \left(\Phi^\dagger \Phi\right)^2$. Now the complex doublet field Φ can be expressed in terms of 4 real fields:

$$\Phi = \frac{1}{\sqrt{2}} \begin{pmatrix} \phi_1 + i\phi_2 \\ \phi + i\phi_3 \end{pmatrix}. \tag{99}$$

In the standard convention the vacuum expectation value of Φ is chosen to lie in the ϕ direction, $\langle \phi \rangle = \sigma$, and the real field ϕ has a potential like Equation 88. At tree level, the Higgs mass is $M_H^2 = 2\lambda_0 \sigma^2$. The Higgs couples to gauge bosons, and lead to a mass for the W and Z of $M_W^2 = g^2 \sigma^2/4$, $M_Z^2 = (g^2 + g'^2)\sigma^2$. The relationship between the mass of the W and the Fermi constant gives $\sigma = 246$ GeV, $g = 0.66$ and $g' = 0.35$. The Higgs also couples to fermions, with Yukawa coupling $h_i = 0.57(M_i/100\,\text{GeV})$. As the Yukawa coupling is proportional to the fermion mass, the dominant effect is from the top quark, the most massive fermion. Therefore the 4 important coupling constants are $g = 0.66$, $g' = 0.35$, $h_T = 0.57(M_T/100\,\text{GeV})$, and $\lambda_0 = 0.08(M_H/100\,\text{GeV})^2$. Once the top quark and Higgs masses are determined, all the couplings in the minimal electroweak model will be known.

To the classical potential must be added the one-loop corrections from the Higgs, the W, the Z, and all the fermions (of course it is a good approximation to assume the fermion contributions are dominated by the top quark). At zero temperature, the one-loop effective potential is

$$V(\phi_c) = -\frac{1}{2}m^2\phi_c^2 + \frac{1}{4}\lambda\phi_c^4 + \frac{1}{64\pi^2}\left(-m^2 + 3\lambda\phi_c^2\right)^2 \ln\left(\frac{-m^2 + 3\lambda\phi_c^2}{\mu^2}\right)$$
$$+ \frac{3}{1024\pi^2}[2g^4 + (g^2 + g'^2)^2]\phi_c^4 \ln\left(\frac{\phi_c^2}{\mu^2}\right) - \frac{3}{64\pi^2}h_T^4\phi_c^4 \ln\left(\frac{\phi_c^2}{\mu^2}\right), \tag{100}$$

where μ is an arbitrary mass scale which can be related to the renormalized coupling constants. The gauge-boson contribution to the $\phi_c^4 \ln(\phi_c^2)$ term is 1.75×10^{-4}; the top-quark contribution is $-5.19 \times 10^{-4}(M_T/100\,\text{GeV})^4$; and the Higgs boson contribution is $9.73 \times 10^{-5}(M_H/100\,\text{GeV})^4$. A priori, all three contributions could be comparable. Let us consider the case where the Higgs mass is small ($M_H \lesssim 200$ GeV) and the Higgs contribution to the one-loop potential can be ignored. We can then write the potential of Equation 100 as

$$V(\phi_c) = -\frac{1}{2}m^2\phi_c^2 + \frac{1}{4}\lambda\phi_c^4 + B\phi_c^4 \ln\left(\frac{\phi_c^2}{\mu^2}\right)$$
$$= -\frac{1}{2}(2B+\lambda)\sigma^2\phi_c^2 + \frac{1}{4}\lambda\phi_c^4 + B\phi_c^4 \ln\left(\frac{\phi_c^2}{\sigma^2}\right). \tag{101}$$

Here we have used the fact that $V'(\sigma) = 0$ implies that $m^2 = (\lambda + 2B)\sigma^2$, and $B = 1.75\text{–}5.19 \times 10^{-4}(M_T/100\,\text{GeV})^4$. The Higgs mass is $M_H^2 = V''(\sigma) = 2(\lambda + 6B)\sigma^2$.

Now consider the potential at finite temperature. As in the previous example, the finite-temperature potential will have a temperature-dependent piece in addition to the zero-temperature part. The temperature-dependent part receives a contribution from all particles that couple to the scalar field, including the scalar field itself. The one-loop potential at finite temperature can be written as a sum of integrals similar to the one in Equation 98, of the form

$$F_\pm[X(\phi_c)] \equiv \pm \int_0^\infty dx\, x^2 \ln\left[1 \mp \exp[-(x^2 + X(\phi_c)/T^2)^{1/2}]\right] \tag{102}$$

(F_+ applies to boson loops and F_- to fermion loops). For the electroweak model, $V_T(\phi_c)$ is given by

$$V_T(\phi_c) = V(\phi_c) + \frac{T^4}{2\pi^2}\left\{6F_+[g^2\phi_c^2/4] + 3F_+[(g^2+g'^2)\phi_c^2/4]\right.$$
$$\left.+ F_+[M_H^2(\phi_c)] + 12F_-[h_T^2\phi_c^2/2]\right\}, \tag{103}$$

where, as before, $V(\phi_c)$ is the one-loop potential at zero temperature, and for simplicity we have included only the ϕ_c-dependent terms. For $T \gg \sigma$, the terms proportional to T^4 are just given by minus the pressure of a gas of the massless fermions and bosons that couple to ϕ.

Anderson and Hall (1992) showed that a high temperature expansion of the one-loop potential closely approximates the full one-loop potential for $M_H \lesssim 150\,\text{GeV}$ and $M_T \lesssim 200\,\text{GeV}$. (It is important to differentiate between the finite temperature Higgs mass, $M_H(T)$ and the zero-temperature Higgs mass, M_H.) They obtained for the potential

$$V(\phi) = D\left(T^2 - T_2^2\right)\phi^2 - ET\phi^3 + \frac{1}{4}\lambda_T\phi^4, \tag{104}$$

where D and E are given by

$$D = \left[6(M_W/\sigma)^2 + 3(M_Z/\sigma)^2 + 6(M_T/\sigma)^2\right]/24; \quad E = \left[6(M_W/\sigma)^3 + 3(M_Z/\sigma)^3\right]/12\pi.$$

Here T_2 is given by

$$T_2 = \sqrt{(M_H^2 - 8B\sigma^2)/4D}, \tag{105}$$

where the physical Higgs mass is given in terms of the one-loop corrected λ as

$$M_H^2 = (2\lambda + 12B)\sigma^2, \text{ with } B = \left(6M_W^4 + 3M_Z^4 - 12M_T^4\right)/64\pi^2\sigma^4.$$

We use $M_W = 80.6\,\text{GeV}$, $M_Z = 91.2\,\text{GeV}$, and $\sigma = 246\,\text{GeV}$. The temperature-corrected Higgs self-coupling is

$$\lambda_T = \lambda - \frac{1}{16\pi^2}\left[\sum_B g_B \left(\frac{M_B}{\sigma}\right)^4 \ln\left(M_B^2/c_B T^2\right) - \sum_F g_F \left(\frac{M_F}{\sigma}\right)^4 \ln\left(M_F^2/c_F T^2\right)\right] \quad (106)$$

where the sum is performed over bosons and fermions (in our case only the top quark) with their respective degrees of freedom $g_{B(F)}$, and $\ln c_B = 5.41$ and $\ln c_F = 2.64$.

The gauge interactions result in an effective attractive ϕ^3 term in the potential, so *to one-loop* the theory predicts a first-order electroweak phase transition. However the fact that the phase transition is so weakly first order implies that the finite-temperature loop expansion is not reliable, and the one-loop results cannot be trusted for the electroweak transition. There is presently a lot of work in finding an improved potential to describe the electroweak transition.

4.2 Generation of defects

SSB is an intergal part of modern particle physics, and provided that temperatures in the early Universe exceeded the energy scale of a broken symmetry, that symmetry should have been restored. How can we tell if the Universe underwent a series of SSB phase transitions? One possibility is that symmetry-breaking transitions were not 'perfect', and that false vacuum remnants were left behind, frozen in the form of topological defects: domain walls, strings, and monopoles.

As the first example of a topological defect associated with spontaneous symmetry breaking, consider the domain wall. The simple scalar model of the previous section can be used to illustrate domain walls. The Z_2 reflection symmetry, *i.e.* invariance under $\phi \to -\phi$, of the Lagrangian of Equation 88 is spontaneously broken when ϕ takes on a non-zero vacuum expectation. So far we have assumed that all of space is in the same ground state, but this need not be the case! Imagine that space is divided into two regions. In one region of space $\langle\phi\rangle = +\sigma$, and in the other region of space $\langle\phi\rangle = -\sigma$. Since the scalar field must make the transition from $\phi = -\sigma$ to $\phi = +\sigma$ smoothly, there must be a region where $\phi = 0$, *i.e.* a region of false vacuum. This transition region between the two vacua is called a domain wall. Domain walls can arise whenever a Z_2 (or any discrete) symmetry is broken.

The solution to the equation of motion, subject to the boundary conditions that describe a domain wall, is $\phi_W(z) = \sigma \tanh(z/\Delta)$, where the 'thickness' of the wall is characterized by $\Delta = (\lambda/2)^{-1/2}\sigma^{-1}$. It should be clear that the domain wall is topologically stable; the 'kink' at $z = 0$ can move around or wiggle, but it cannot disappear (except by meeting up with an antikink and annihilating). The stress tensor for the domain wall is obtained by substitution of the wall solution into the expression for the stress-energy tensor for a scalar field $T^\mu_\nu = (\lambda/2)\sigma^4 \cosh^{-4}(z/\Delta)\text{diag}(1,1,1,0)$. Note that the z-component of the pressure vanishes, and that the x- and y-components of the pressure are equal to minus the energy density. The surface energy density associated with the wall, given by $\eta \equiv \int T^0_0 dz = (2\sqrt{2}/3)\lambda^{1/2}\sigma^3$, is identical to the integrated, transverse components of the stress, $\int T^i_i dz$. That is, the surface tension in the wall is precisely equal to the surface energy density. Because of this fact walls are inherently relativistic, and their gravitational effects are inherently non-Newtonian (and very interesting).

The existence of large-scale domain walls in the Universe today are ruled out simply based upon their contribution to the total mass density. A domain wall of size $H_0^{-1} \simeq 10^{28} h^{-1}$ cm would have a mass of order $M_{\text{wall}} \sim \eta H_0^{-2} \sim 4 \times 10^{65} \lambda^{1/2} (\sigma/100\,\text{GeV})^3$ grams, or about a factor of $10^{10} \lambda^{1/2} (\sigma/100\,\text{GeV})^3$ times that of the total mass within the present Hubble volume. Walls would also lead to large fluctuations in temperature of the CBR unless σ is very small: $\delta T/T \simeq G\eta H_0^{-1} \simeq 10^{10} \lambda^{1/2} (\sigma/100\,\text{GeV})^3$. Apparently, domain walls are cosmological bad news unless the energy scale and/or coupling constant associated with them are very small.

The existence of domain wall solutions for this simple model traces to the existence of the disconnected vacuum states: $\langle \phi \rangle = \pm \sigma$. The general mathematical criterion for the existence of topologically stable domain walls for the symmetry-breaking pattern $\mathcal{G} \to \mathcal{H}$ is that $\Pi_0(\mathcal{M}) \neq \mathcal{I}$, where $\mathcal{M}$ is the manifold of equivalent vacuum states $\mathcal{M} \equiv \mathcal{G}/\mathcal{H}$, and Π_0 is the homotopy group that counts disconnected components. In the above example, $\mathcal{G} = \mathcal{Z}_2$, $\mathcal{H} = \mathcal{I}$, $\mathcal{M} = \mathcal{Z}_2$, and $\Pi_0(\mathcal{M}) = \mathcal{Z}_2 \neq \mathcal{I}$.

The next example of a topological defect is the cosmic string, a one-dimensional structure. As we shall see, cosmic strings are much more palatable to a cosmologist than domain walls. A simple model that illustrates the cosmic string is the Abelian Higgs model, a spontaneously broken U(1) gauge theory. The Lagrangian of the model contains a U(1) gauge field, A_μ, and a *complex* Higgs field, Φ, which carries U(1) charge e,

$$\mathcal{L} = D_\mu \Phi D^\mu \Phi^\dagger - \frac{1}{4} F_{\mu\nu} F^{\mu\nu} - \lambda(\Phi^\dagger \Phi - \sigma^2/2)^2, \qquad (107)$$

where $F_{\mu\nu} = \partial_\mu A_\nu - \partial_\nu A_\mu$, and $D_\mu \Phi = \partial_\mu \Phi - ieA_\mu \Phi$. We immediately recognize that the theory is spontaneously broken as $V(\Phi)$ is minimized for $\langle |\Phi| \rangle^2 = \sigma^2/2$. The physical states after SSB are a scalar boson of mass $M_S^2 = 2\lambda \sigma^2$ and a massive vector boson of mass $M_V^2 = e^2 \sigma^2$.

The complex field Φ can be written in terms of two real fields: $\Phi = (\phi + i\phi_1)/\sqrt{2}$. *If* the vacuum expectation value is chosen to lie in the real direction, then the potential becomes
$V(\phi) = (\lambda/4)(\phi^2 - \sigma^2)^2$, where $\langle |\Phi| \rangle = \langle \phi \rangle / \sqrt{2}$. However, energetics do not determine the phase of $\langle \Phi \rangle$ since the vacuum energy depends *only* upon $|\Phi|$; this fact follows from the U(1) gauge symmetry. Defining the phase of the vacuum expectation value by $\langle \Phi \rangle = (\sigma/\sqrt{2}) \exp(i\theta)$, we see that $\theta = \theta(\vec{x})$ can be position dependent. However, Φ must be single valued; i.e. the total change in θ, $\Delta \theta$, around any closed path must be an integer multiple of 2π. Imagine a closed path with $\Delta \theta = 2\pi$. As the path is shrunk to a point (assuming no singularity is encountered), $\Delta \theta$ cannot change continuously from $\Delta \theta = 2\pi$ to $\Delta \theta = 0$. There must, therefore, be one point contained within the path where the phase θ is undefined, i.e. $\langle \Phi \rangle = 0$. The region of false vacuum within the path is part of a tube of false vacuum. Such tubes of false vacuum must either be closed or infinite in length, otherwise it would be possible to deform the path around the tube and contract it to a point without encountering the tube of false vacuum. In most instances, these tubes of false vacuum have a characteristic transverse dimension far smaller than their length, so they can be treated as one-dimensional objects and are called 'strings'.

The string solution to the equations of motion was first found by Nielsen and Olesen (1973). At large distances from an infinite string in the z-direction, their solution is

$$\Phi \longrightarrow (\sigma/\sqrt{2})\exp(iN\theta); \qquad A_\mu \longrightarrow -ie^{-1}\partial_\mu\left[\ln(\sqrt{2}\Phi/\sigma)\right], \qquad (108)$$

where θ is the polar angle in the x-y plane, and N is the winding number of the string. The stress-energy tensor associated with a long, thin, straight string is given by $T^\mu_\nu = \mu\delta(x)\delta(y)\mathrm{diag}(1,0,0,1)$, where, μ is the mass per unit length of the string which depends upon the ratio $e^2/2\lambda$ but generally is of order $\pi\sigma^2$. Note that the pressure is negative—i.e. it is a string tension—and equal to $-\mu$. Like domain walls, strings are intrinsically relativistic.

Far from a circular string loop of radius R, the gravitational field is that of a point particle of mass $M_{\mathrm{string}} = 2\pi R\mu$. For a loop of size about that of the present horizon, $M_{\mathrm{string}} \simeq 10^{18}(\sigma/\mathrm{GeV})^2$ grams. As with domain walls, there are non-Newtonian gravitational effects associated with strings. Recall that for a stress tensor of the form $T^\mu_\nu = \mathrm{diag}(\rho,-p_1,-p_2,-p_3)$, the Newtonian limit of Poisson's equation is $\nabla^2\phi = 4\pi G(\rho+p_1+p_2+p_3)$. For an infinite string in the z direction $p_3 = -\rho$ and $p_1 = p_2 = 0$, and Poisson's equation becomes $\nabla^2\phi = 0$, which suggests that space is flat outside of an infinite straight string. Indeed this is so. Vilenkin (1981) has solved Einstein's equations for the metric outside an infinite, straight cosmic string in the limit that $G\mu \ll 1$. In terms of the cylindrical coordinates (r, θ, z) the metric is $ds^2 = dt^2 - dz^2 - dr^2 - (1-4G\mu)^2 r^2 d\theta^2$. By a transformation of the polar angle, $\theta \to (1-4G\mu)\theta$, the metric becomes the flat-space Minkowski metric: as expected, space-time around a cosmic string is that of empty space. However, the range of the flat-space polar angle θ is only $0 \le \theta \le 2\pi(1-4G\mu)$ rather than $0 \le \theta \le 2\pi$. This is referred to as a conical singularity.

The conical nature of space around a string leads to several striking effects: double images of objects located behind the string, fluctuations in the microwave background, and the formation of wakes. To understand the formation of double images, consider the simplified situation of an infinite string normal to the plane containing the source and the observer. The conical space is flat space with a wedge of angular size $\Delta\theta$ removed and points along the cuts identified. Due to this, the observer will see two images of the source, with the angular separation, $\delta\alpha$, between the two images determined by

$$\sin(\delta\alpha/2) = \sin(\Delta\theta/2)\frac{l}{d+l}; \qquad \delta\alpha \simeq \Delta\theta\frac{l}{d+l} = 8\pi G\mu\frac{l}{d+l}. \qquad (109)$$

Here l and d are the distances from the string to the source and observer respectively, and the second equation is a small-angle approximation. The conical metric also leads to discontinuities in the temperature of the microwave background. Imagine as the source, a point on the last scattering surface for the microwave background radiation. An observer at rest with respect to the string will see two images of the same point on the last scattering surface, separated by an angle $\delta\alpha \simeq \Delta\theta$ (for $d \ll l$). Now if the string and observer are not at rest with respect to each other, but instead have a relative velocity v which is perpendicular to the line of sight, the momentum vector of one image will have a small component (order $\Delta\theta$) parallel to the direction of $\vec{v}$, and the other, a small component antiparallel to the direction of $\vec{v}$. The net effect is a small Doppler shift of the radiation temperature $\delta T/T \simeq 8\pi G\mu v$ across the string. Based upon this effect and the observed isotropy of the CBR, we can conclude any strings that exist at present must be characterized by $G\mu \lesssim 10^{-5}$. A third interesting effect of cosmic strings are string wakes. Consider a long, straight string moving through the Universe with velocity v. As the string moves past particles in the Universe the particles will be deflected and will acquire a 'wake' velocity $v_W \simeq 4\pi G\mu v$, transverse to the direction of motion of the string. If the particles have a very small internal velocity dispersion, $e.g.$ cold-dark matter particles, or baryons after decoupling, then matter on both sides of the passing string will move toward the plane defined by the motion of the string. In a Hubble time, a wedge-shaped sheet of matter, with overdensity of order unity, opening angle $\simeq 8\pi G\mu$, and width vH^{-1}, will form in the wake of the string. The mass of the material within the wake-produced sheet can be considerable, about $8\pi G\mu v^3$ of that in the horizon; likewise, the scale of the thin (thickness/width $\sim 8\pi G\mu$) sheets that are formed is comparable to that of the horizon scale. It has been suggested that the sheets that form in the wakes of long, straight cosmic strings play an important role in structure formation.

A final imprint of primordial cosmic strings is gravitational radiation from shrinking string loops. While an infinite straight string is stable, string loops are not. A curved string will move so as to minimize its length. The motion of a small, closed loop is particularly simple: a loop of radius R oscillates relativistically, with a period $\tau \sim R$. As it oscillates it will radiate gravitational waves due to its time-varying quadrupole moment (dimensionally $Q \sim \mu R^3$). The power radiated in gravitational waves is given by $P_{GW} \simeq G(\dddot{Q})^2 \simeq \gamma_{GW} G\mu^2$ where γ_{GW} is a numerical constant of order 100. In a characteristic time τ_{GW} the loop will radiate away its mass-energy, shrink to a point, and vanish. We expect $\tau_{GW} \sim \mu R/P_{GW} \sim (\gamma_{GW} G\mu)^{-1} R$. That is, a loop will undergo about $10^{-2}(G\mu)^{-1}$ oscillations before it disappears.

As cosmic strings stand, they are cosmologically safe and have several potentially interesting consequences: (1) they leave behind a background of relic gravitational waves; (2) relic string present today can lead to temperature fluctuations in the CBR; (3) relic string present today can act as gravitational lenses; and (4) string loops, or flattened structures formed in the wakes of strings, can possibly serve as seeds to initiate structure formation in the Universe.

In our discussion of cosmic strings, we have used the simplest example of a spontaneously broken gauge theory for which string solutions exist. In general, there will be string solutions associated with the symmetry breaking $\mathcal{G} \to \mathcal{H}$, if the manifold of degenerate vacuum states, $\mathcal{M} = \mathcal{G}/\mathcal{H}$, contains unshrinkable loops, $i.e.$ if the mapping of $\mathcal{M}$ onto the circle is non-trivial. This is formally expressed by the statement that topologically stable string solutions exist if $\Pi_1(\mathcal{M}) \neq \mathcal{I}$. In the above example $\mathcal{G} =$U(1), $\mathcal{H} = \mathcal{I}$, and $\mathcal{M} =$U(1). The group U(1) can be represented by the points on a circle, and so $\Pi_1[$U(1)$]$ is the mapping of the circle onto itself. Such a mapping is characterized by the winding number of the mapping, $i.e.$ $\theta \to N\theta$ $(N = 0, 1, \cdots)$, so that $\Pi_1(\mathcal{M}) = \mathcal{Z}$, the set of integers.

Domain walls are two-dimensional topological defects, strings are one-dimensional defects. Point-like defects also arise in some theories which undergo SSB, and remarkably, they appear as magnetic monopoles. A simple model that illustrates the magnetic monopole solution is an SO(3) gauge theory, in which SO(3) is spontaneously broken to U(1) by a Higgs triplet Φ^a, where a is the group space index. The Lagrangian density for this theory is

$$\mathcal{L} = \frac{1}{2}D_\mu \Phi^a D^\mu \Phi^a - \frac{1}{4}F^a_{\mu\nu}F^{a\mu\nu} - \frac{1}{8}\lambda(\Phi^a\Phi^a - \sigma^2)^2,$$
$$F^a_{\mu\nu} = \partial_\mu A^a_\nu - \partial_\nu A^a_\mu - e\varepsilon_{abc}A^b_\mu A^c_\nu,$$
$$D_\mu \Phi^a = \partial_\mu \Phi^a - e\varepsilon_{abc}A^b_\mu \Phi^c. \tag{110}$$

Once again, we encounter a theory that undergoes SSB. In this model, two of the three gauge bosons in the theory acquire a mass through the Higgs mechanism. There is also a physical Higgs particle. The masses of the vector and Higgs bosons are $M_V^2 = e^2\sigma^2$, $M_S^2 = \lambda\sigma^2$.

The magnitude of $\langle \Phi^a \rangle$ is fixed by the minimization of the potential: $|\Phi| = \sigma$. However, the direction of $\langle \Phi^a \rangle$ in group space is not. This is just a manifestation of the SO(3) gauge symmetry. It should be clear that the lowest energy solution is the one where $\Phi^a(\vec{x}) = $ const ($\vec{x} = $ spatial coordinate) since this also minimizes the kinetic energy (spatial gradient term). Even if $\Phi^a(\vec{x}) \neq $ const, the spatial dependence of the direction of Φ^a can often be gauged away, i.e. $D_\mu \Phi^a$ made equal to zero by an appropriate gauge configuration $A^a_\mu(\vec{x})$, with finite energy. However, there are Higgs field configurations that cannot be deformed into a configuration of constant Φ^a by a finite-energy gauge transformation.

An example of a configuration that cannot be gauged away is the 'hedgehog' configuration, in which the direction of Φ^a in group space is proportional to $\hat{r}$, where $\hat{r}$ is the unit vector in ordinary space. This solution is spherically symmetric, and as r tends to infinity, we find that $\Phi^a(r,t) \to \sigma\hat{r}$, and $A^a_\mu(r,t) \to \varepsilon_{\mu ab}\hat{r}_b/er$. Like the domain wall and the cosmic string solutions, continuity requires that the Higgs field vanish as $r \to 0$. The vanishing of the Higgs field at the origin accounts for the topological stability of the hedgehog: There is no way to smoothly deform the hedgehog into a configuration where $\langle|\Phi^a|\rangle = \sigma$ everywhere. The size of the monopole, i.e. the region over which $\langle|\Phi^a|\rangle \neq \sigma$, is of order σ^{-1}. The energy of the hedgehog configuration receives contributions from both the vacuum energy associated with $\langle|\Phi^a|\rangle \neq \sigma$ and spatial gradient energy associated with the variation of $\langle\Phi^a\rangle$.

Gauge and Higgs field configurations corresponding to a magnetic monopole exist if the vacuum manifold ($\mathcal{M} = \mathcal{G}/\mathcal{H}$) associated with the symmetry-breaking pattern $\mathcal{G} \to \mathcal{H}$ contains non-shrinkable surfaces, i.e. if the mapping of $\mathcal{M}$ onto the two-sphere is non-trivial. Mathematically, this is expressed by the statement that monopoles solutions arise in the theory if $\Pi_2(\mathcal{M}) \neq \mathcal{I}$. If $\mathcal{G}$ is simply connected, then $\Pi_2(\mathcal{G}/\mathcal{H}) = \Pi_1(\mathcal{H})$. If $\mathcal{G}$ is not simply connected, then the generalization of the above expression is $\Pi_2(\mathcal{G}/\mathcal{H}) = \Pi_1(\mathcal{H})/\Pi_1(\mathcal{G})$. In the example above, $\mathcal{G} = $SO(3), $\mathcal{H} = $U(1) (SO(3) is not simply connected—it is equivalent to the three-sphere with antipodal points identified), and $\Pi_2[\text{SO(3)}/\text{U(1)}] = \Pi_1[\text{U(1)}]/\Pi_1[\text{SO(3)}] = \mathcal{Z}/\mathcal{Z}_2$, the integers mod 2.

We have discussed the three kinds of topological defects associated with spontaneously broken symmetries: the monopole; the string; and the domain wall. The existence and stability of these objects is dictated by topological considerations.

Many spontaneously broken gauge theories predict the existence of one or more of the above topological defects. These objects are inherently non-perturbative and probably cannot be produced in high energy collisions at terrestrial accelerators. It is very likely that the only place they can be produced is in phase transitions in the early Universe. Although monopoles, strings, and domain walls are topologically stable, they are not the minimum energy configurations. However, their production in cosmological phase transitions seems unavoidable. The 'unavoidable' cosmological production mechanism is known as the Kibble mechanism.

The Kibble mechanism hinges upon the fact that during a cosmological phase transition any correlation length is always limited by the particle horizon. The particle horizon is the maximum distance over which a massless particle could have propagated since the time of the bang. It was given in Equation 15. The correlation length associated with the phase transition sets the maximum distance over which the Higgs field can be correlated. The correlation length depends upon the details of the phase transition and is temperature-dependent. It is related to the temperature-dependent Higgs mass: $\xi \sim M_H^{-1}(T) \sim T^{-1}$. In any case, the fact that the horizon distance is finite in the standard cosmology implies that at the time of the phase transition ($t = t_C$, $T = T_C$), the Higgs field must be uncorrelated on scales greater than d_H, and thus the horizon distance sets an absolute maximum for the correlation length.

During a SSB phase transition, some Higgs field acquires a vacuum expectation value. Because of the existence of the particle horizon in the standard cosmology, when this occurs $\langle \phi \rangle$ cannot be correlated on scales larger than $d_H \sim H^{-1} \sim m_{Pl}/T^2$. Therefore, it should be clear that the non-trivial vacuum configurations must necessarily be produced, with an abundance of order one per horizon volume. While these topological creatures are not the minimum-energy configurations of the Higgs field, they arise as 'topological defects' because of the finite particle horizon. Since they are stable, they are 'frozen in' as permanent defects when they form.

So far we have concentrated on the analysis of topological defects that can arise in gauge theories. However defects can also arise in the spontaneous breaking of global symmetries. The analogies of the defects discussed above are global strings and global monopoles. The global field configurations look like their local counterparts for the scalar field, but of course there is no vector field. This means that formally the string and monopole solutions have infinite energy (recall for the local defects the energy in the gauge fields cancels the energy in the Higgs field far from the defect.) This is really not a problem, because there the divergence in the energy is only logarithmic, and there are many physical effects to cut it off (such as the inter-defect separation). There are just two main differences in the behaviour of gauge and global defects: (1) the energy of the global defects are slightly more spread out, (2) the global strings can radiate energy by the emission of Nambu-Goldstone bosons.

However there are new types of defects in global symmetry breaking that do not appear in the breaking of gauge symmetries. For example, in the spontaneous breaking of a global O(N) model to O($N-1$), for $N = 1$ walls appear, for $N = 2$ global strings result, for $N = 3$ global monopoles are produced. These all have counterparts in local theories. However for $N > 3$ global defects also exist: for $N = 4$ the defect is called global texture, and for $N > 4$ they are called Kibble gradients. Texture corresponds to knots in the Higgs field that arise when the field winds around the three sphere. These knots are generally formed by misalignment of the field on scales larger than the horizon at the symmetry-breaking phase transition because of the Kibble mechanism. As the knots enter the horizon, they collapse at roughly the speed of light, giving rise to nearly spherical energy density perturbations. New knots are constantly coming into the horizon and collapsing, leading to a scale invariant spectrum of density perturbations. The magnitude of the perturbations is set by the scale of the symmetry breaking, and for scenarios of structure formation involving texture, the scale of symmetry breaking must be about 10^{16} GeV.

A theory of texture or Kibble gradients being responsible for the seeds of large-scale structure has been formulated by Turok, Spergel and collaborators (Pen et al. 1993). Texture would provide a very promising alternative to conventional inflation scenarios for generating the primordial density fluctuations if indeed they are ubiquitous in particle physics models. In fact, texture arises in a variety of theories with non-Abelian global symmetries that are spontaneously broken. However, even an extremely small amount of explicit symmetry breaking will spoil the texture scenario. I would like to close these lectures by discussing how sensitive this theory is to Planck-scale effects. This idea was recently discussed by Holman et al. (1992) and Kamionkowski et al. (1992).

To illustrate these possibilities, consider a theory with a global $O(N)$ symmetry spontaneously broken to $O(N-1)$ by an N-vector. The theory is described by the scalar potential $V(\Phi) = \lambda \left(\Phi^a \Phi^a - \sigma^2 \right)^2$. As mentioned above, texture arises for $N = 4$. There are many arguments suggesting that *all* global symmetries are violated at some level by gravity. For example, both wormholes and black holes can swallow global charge. 'Virtual' black holes or wormholes, which should, in principle, arise in a theory of quantum gravity, will lead to higher dimension operators which violate the global symmetry. There are two possible assumptions one might make about the fate of global symmetries in a Universe that includes gravity. The *strong* assumption is that, despite all indications from low-energy, semi-classical gravitational physics (black holes, wormholes, *etc.*), it is possible to have exact global symmetries in the presence of gravity. This is the assumption made in the standard texture scenario. The *weak* assumption is that the global symmetry is not a feature of the full theory. There are two possible realizations of the weak assumption. Either the global symmetry is approximate, in which case one must include the effects of higher-dimensional, non-renormalizable, symmetry-breaking operators, or, consistent with indications from semi-classical quantum gravity, the global symmetry is never even an approximate symmetry unless protected by gauge symmetries.

If one makes the weak assumption, then one must include explicit symmetry breaking terms. If one assumes that gravity does not respect global symmetries at all, then renormalizable operators like $m_{Pl}^2 \lambda_{ab} \Phi^a \Phi^b$, which explicitly break the global symmetry, should be included. These terms are expected, for instance, by the action of wormholes swallowing global charge. If virtual wormholes of size smaller than the Planck length are included, then we expect λ_{ab} to be of order unity. In this case it is wrong to consider an effective low-energy theory with a global symmetry. If one makes the assumption either that wormholes do not dominate the functional integral, or that the global charge is protected by gauge symmetries, then it may be possible to suppress the renormalizable operators. But even in this case higher dimension operators should be included. An example would be a dimension-5 operator, which would add to $V(\Phi)$ terms like $(\lambda_{abcde}/m_{Pl})\Phi^a\Phi^b\Phi^c\Phi^d\Phi^e$. Such terms explicitly break the global symmetry and lead to a mass for the pseudo-Nambu-Goldstone mode of $m^2 \propto \lambda \sigma^3 / m_{Pl}$. Of course the mass is suppressed by m_{Pl}, but we will show below that it still has a drastic effect on the texture scenario.

The implications of the strong and weak assumptions for texture are as follows: With the strong assumption, the texture scenario is unaffected. If one allows unsuppressed wormhole contributions, global symmetries (and hence texture) are a non-starter. If all effects of gravitational physics in the low-energy theory are contained in non-renormalizable terms, a more careful analysis is required. This is the possibility we explore now. In this approach we are then required to include all higher dimension operators consistent with the gauge symmetries of the model and suppressed by appropriate powers of m_{Pl}.

We now consider the effects of the higher dimension operators. These terms will break the symmetry explicitly, generating a complicated potential for the Nambu-Goldstone modes. In general, the vacuum manifold will be reduced to a point, though the potential will likely have many local minima. To see how this works, consider the theory discussed above with $N = 3$. Here, the vacuum manifold is the two sphere and the model, in two spatial dimensions, will have texture. (In three spatial dimensions, the model admits both global monopoles and texture, although the texture in this case is not spherically symmetric. We express the field as

$$\Phi = \sigma(\sin\theta\cos\phi, \sin\theta\sin\phi, \cos\theta), \qquad (111)$$

where θ and ϕ are the angular variables on the two-sphere which represent the Nambu-Goldstone modes of the problem.

The effect of the dimension 5 operators is to introduce 21 terms to the potential for the field which depend explicitly on θ and ϕ. (These are nothing more than the Y_{1m}, Y_{3m}, and Y_{5m} spherical harmonics.) Note that in general, the mass of the Nambu-Goldstone boson in this potential is roughly $\sigma(\sigma/m_{Pl})^{1/2}$.

As long as the mass of the Nambu-Goldstone mode is small compared to the Hubble parameter, the field will evolve essentially as in the original texture scenario. However, once the Compton wavelength of the Nambu-Goldstone mode enters the horizon, the field will begin to oscillate about the minimum (or rather the closest local minimum) of its potential. The field will then align itself on scales larger than the horizon and texture on all scales quickly disappear. For texture to be important for structure formation, they must persist at least until matter-radiation decoupling when $H \simeq 10^{-28}$ eV.

The contribution of a dimension $4 + d$ operator to the Nambu-Goldstone boson mass is $m \sim \sigma(\sigma/m_{Pl})^{d/2}$. Given that the texture scenario requires $\sigma \sim 10^{16}$ GeV, the requirement that $m \lesssim 10^{-28}$ eV implies that $d \gtrsim 35$; i.e. we must be able to suppress all operators up to dimension 40. It is rather difficult to see how this might occur; even the use of additional gauge quantum numbers could not prevent the occurrence of dimension 6 operators which break a non-Abelian symmetry (although they could protect an Abelian symmetry). We note that if we consider dimension-5 operators, then the mass becomes dynamically important immediately after the phase transition: texture therefore never exists.

In conclusion, any model which depends on the dynamics of Nambu-Goldstone modes will be extremely sensitive to physics at very high energies. Texture can by no means be considered a robust prediction of unified theories. This is most discouraging for the texture scenario. On the other hand, if texture is discovered, then this will have profound implications not only for theories of structure formation, but for Planck-scale physics. What better way to close lectures on the implications of cosmology for particle physics.

Finally there are other creatures that might be produced in cosmological phase transitions. Non-topological solitons, or Q-balls, (Frieman *et al.* 1988), and electroweak strings (Achucarro and Vachaspati, 1991).

The lesson for cosmological phase transitions is that even with unlimited energy, accelerators are the wrong tool to probe the non-perturbative sector of field theories. Early-Universe phase transitions continue to provide the best arena for the study of aspects of particle-physics theories related to coherent, soliton-like objects. The only plausible site for the production of objects such as monopoles, strings, walls, sphalerons, and the like is an early-Universe phase transition. All of these can have very significant implications for the evolution of the Universe. Sphalerons, as well as other solitons produced in the electroweak transition, have some promise of a cosmological payoff. Of course there is an enormous difference between finding a soliton-like solution to the field equations and finding solitons in the Universe. However, even if they are not are found, the techniques developed for their study will be useful additions to the theorist's toolbox.

Acknowledgements

I am very grateful to several collaborators who greatly influenced the work recounted here: Ed Copeland, Marcelo Gleiser, Rich Holman, Andrew Liddle, Jim Lidsey, Mike Turner, Sharon Vadas, Yun Wang, and Erick Weinberg.

My work is supported in part by the Department of Energy and the National Aerounatics and Space Administration (Grant NAGW-2381) at at Fermilab.

In times of uncertainty about the future of our field of physics because of political and financial trouble, in times where the advances of our understanding of the Universe seemingly meet with hostility from an increasingly large segment of the public, in times when the prospects for young scientists seem grim, in times when the future of our field seems beyond our powers to influence, perhaps it is worthwhile to recall the words of Johannas Kepler in the last years of his life when his scientific career was caught in personal and political turmoil caused by the Thirty Years War:

> "When the storms rage around us, and the state is threatened by shipwreck, we can do nothing more noble than to lower the anchor of our peaceful studies into the ground of eternity."

References

Achucarro A and Vachaspati T, 1991, *Phys Rev D* **44** 3067.
Anderson G and Hall L J, 1992, *Phys. Rev. D* **45** 2685.
Abbott L F and Wise M B, 1984, *Nucl Phys B* **244** 541.
Albrecht A and Steinhardt P J, 1982, *Phys Rev Lett* **48** 1220.
Anderson G and Hall L J, 1992, *Phys. Rev. D* **45** 2685.
Bertschinger E and Dekel A, 1989 *Ap J Lett* **336** L5.
Bond J R and Efstathiou G, 1987, *Mon Not R astr Soc* **226** 665.
Coleman S and Weinberg E J, 1973, *Phys. Rev. D* **7** 1888.
Copeland E J, Kolb E W, Liddle A R and Lidsey J E, 1993a, *Phys Rev Lett* **71**, 219.
Copeland E J, Kolb E W, Liddle A R and Lidsey J E, 1993b, *Phys Rev D* **48**, 2529.
Copeland E J, Kolb E W, Liddle A R and Lidsey J E, 1993c, *Reconstructing the inflaton potential—perturbative reconstruction to second order*, FNAL preprint.
Davis M and Peebles P J E, 1983, *Ap J* **267** 465.
Davis M, Efstathiou G , Frenk C S and White S D M, 1992 *Nature* **356**, 489.
de Laupparent V, Geller M J, and Huchra J, 1986, *Ap J* **302** L1.
Efstathiou G, 1990, *The Physics of the Early Universe* (SUSSP, Edinburgh).
Fisher K B et al. 1992, *Ap J* **389** 188.
Frieman J A, Gelmini G B, Gleiser M, and Kolb E W, 1988, *Phys. Rev. Lett.* **60** 2101.
Gunn J E and Knapp G R, 1992, 'The Sloan Digital Sky Survey,' Princeton preprint POP-488.
Guth A H, 1980, *Phys Rev D* **20** 30.
Hodges H M and Blumenthal G R, 1990, *Phys. Rev. D* **42** 3329.
Holman R, Hsu S H, Kolb E W, Watkins R and Widrow L M, 1992, *Phys Rev Lett* **69** 1489.
Hubble E P, 1929, *Proc Nat Acad* **15** 168.
Kamionkowski M and March-Russell J, 1992, *Phys Rev Lett* **69** 1485.
Kirzhnits D and Linde A D, 1972), *Phys Lett* **42B** 421.
Kolb E W and Turner M S, 1989, *The early Universe* (Addison-Wesley, Redwood City, CA.)
Kolb E W and Wolfram S, 1980, *Ap J* **239** 428.
Kron R G, 1992, *ESO Conference on Progress in Telescope and Instrumentation Technologies*, ESO conference and workshop proceeding no. 42, p635, ed. Ulrich M-H.
La D and Steinhardt P J, 1989, *Phys Rev Lett* **62** 376.
Lee S Y and Sciaccaluga A M, 1975, *Nucl. Phys. B* **96** 435.
Liddle A R and Lyth D, 1993, *Phys Rep* **231**, 1
Linde A D, 1982, *Phys Lett* **108B** 389.
Loveday J, Peterson B A, Efstathiou G, and Maddox S J, 1992, *Ap J* **390**, 338.
Lucchin F, Matarrese S and Mollerach S, 1992, *Ap J Lett* **401** 49.
Mather J C et al. 1993, *Ap J* in press.
Nielsen H B and Olesen P, 1973, *Nucl Phys* **B61** 45.
Park C, Gott J R and da Costa L N, *Ap J Lett* **392** L51.
Pen U-L, Spergel D N and Turok N *Cosmic structure formation and microwave anisotropies from global field ordering*, Princeton preprint.
Pi S-Y, 1984, *Phys Rev Lett* **52** 1725.
Press W H et al. 1986, *Numerical Recipes* (Cambridge University Press, Cambridge).
Ramella M, Geller M J and Huchra J P, 1992, *Ap J* **384** 396.
Rowan-Robinson M, 1985, *The Cosmological Distance Ladder* (Freeman, San Francisco).
Shafi Q and Vilenkin A, 1984, *Phys Rev Lett* **52** 691.
Saunders W et al. 1991, *Nature* **349** 32.
Scaramella R and Vittorio N, 1990, *Ap J* **353** 372.
Smoot G et al. 1992, *Ap J* **396** L1.

Starobinsky A A, 1985, *Sov Astron Lett* **11** 133.
Turner M, 1993, *Recovering the inflaton potential*, FNAL preprint.
Vilenkin A, 1981, *Phys Rev D* **23** 852.
Vogeley M S, Park C, Geller M J and Huchra, J P, 1992, *Ap J Lett* **391** L5.
Walker T P *et al. Ap J* **376** 151.

Precision Tests of the Standard Electroweak Model at LEP

Alain Blondel

LPNHE, Ecole Polytechnique
France

1 Introduction

We celebrate in 1993 the 20th anniversary of the discovery of the Weak Neutral Current (NC) by the Gargamelle Collaboration at CERN (Hasert et al. 1973/74). [For a complete historical review, consult NC 1993.]

The existence of neutral currents was a prediction of what was called then the 'Weinberg Model', and now universally accepted as the 'Standard Model' (Glashow 1961, Salam 1968, Weinberg 1967/72). Neutral currents had been searched for earlier, in particular in the decay $K_L^0 \to \mu^+\mu^-$ with very stringent limits below 10^{-9}. Theorists like to recall that neutral currents had actually been observed before 1970 in several neutrino experiments and discarded as 'background', or even rejected at the trigger level; that only when a convincing excuse, the Glashow-Iliopoulos-Maiani (GIM) mechanism (Glashow et al. 1970), accounted for the absence of neutral currents in K-decays while allowing their existence in neutrino scattering, only then, did the experimenters find them. Of course, this opinion should be qualified: establishing a new phenomenon requires a very solid demonstration, which was not necessarily feasible in the early experiments—it took long enough for the Gargamelle team to convince themselves and their American competitors that neutral currents existed.

The mere existence of neutral currents was a clear test of the Model. The next consequence of the GIM mechanism was the existence of the charm quark. With the discovery of charm in 1974-1976, the model soon became *the* Standard Model (SM). The early neutrino experiments also provided the first numerical test of the SM: having one free parameter, the 'Weinberg angle' θ_W, the theory could not predict the rate of neutral currents. However the rates of neutral currents in neutrino and antineutrino beams had to be related and the result sit on a curve, 'the Weinberg's nose', Figure 1. That property was important in disentangling the model from its competitors, such as a model with Heavy lepton production (Eichten et al. 1973, Glashow and Georgi 1972), supported by Glashow himself; the lepton died but Glashow survived! LEP has not made discoveries yet, but performed many such consistency tests with high accuracy.

The construction of the Large Electron Positron (LEP) Collider (LEP 1984) at CERN was decided in the wake of spectacular successes from e^+e^- machines in the mid-70's. LEP was then conceived as the machine for physics at the Fermi scale, where weak interaction phenomena appear in the purest form, the direct production of Z bosons, and, at a later stage, of WW pairs.

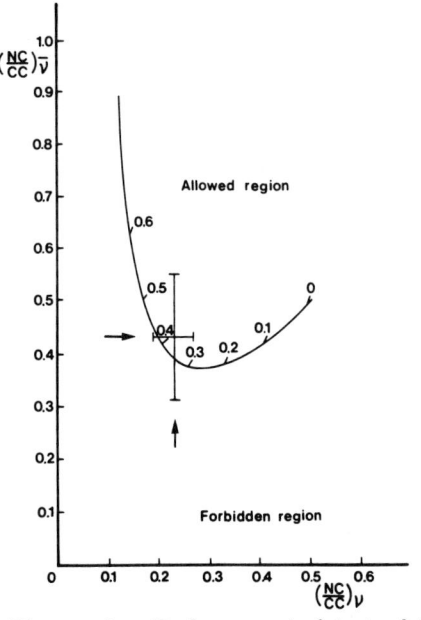

Figure 1. *Early numerical test of the electroweak theory in 1974: 'Comparison of the experimental values with the theoretical bounds'. The rates for NC events in neutrino and antineutrino exposures are used to test the model (from Hasert et al. 1973/74)).*

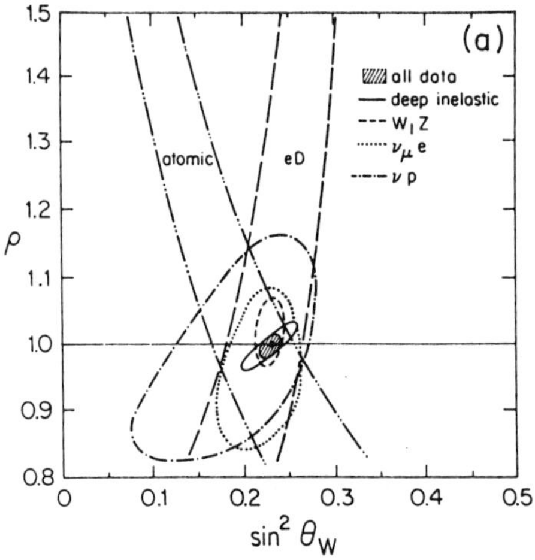

Figure 2. *Before LEP; the allowed region in the ρ-$\sin^2 \theta_W$ plane at the 90% confidence level for various reactions (from Amaldi et al. 1987).*

The plan of these lectures will be as follows: first, I will summarise what we knew before LEP started and then I will describe LEP and the four experiments. A more detailed description of the apparently simple measurement of hadronic cross-sections will follow, as it provided the early and important results on the Z mass and width and on the number of neutrinos. A digression on the beam energy measurements will be made. The very important new measurement of the Z partial width in b quarks will be emphasised. Helicity plays a very important role in LEP physics, providing the source of forward-backward and polarisation asymmetries, and will be described next. Finally the electroweak measurements will be summarised by a discussion of electroweak radiative effects and what can data tell us about them.

The LEP-II programme, due to start in 1996, will bring about the measurement of the W boson mass, a first look at the triple boson couplings and, mostly, another window for finding the Higgs Boson. This important physics programme offers much room for ingenuity and will be described briefly.

2 What we knew before LEP started

2.1 The number of light neutrinos

In a review (Denegri et al. 1989), published in summer 1989, just before SLC and LEP started, limits on the number of light neutrinos were given from the following sources:

- The CERN $p\bar{p}$ experiments, comparing the rates of visible Z and W decays;

- e^+e^- experiments looking for the signal in $\nu\bar{\nu}\gamma$;

- astrophysical limits based on supernova SN1987A;

- cosmological constraints from, for example, the observed He/H ratio;

The review concludes with $N_\nu = 2.1^{+0.6}_{-0.4}$, and the following statement:

> "$N_\nu = 3$ is perfectly compatible with all data. Although the consistency is significantly worse, four families still provide a reasonable fit".

2.2 Electroweak measurements

A status of neutral current data before LEP can be found in Amaldi et al. 1987. The knowledge in 1987 came from 3 sets of measurements.

- Measurements of M_Z and M_W from the CERN $p\bar{p}$ experiments, yielding:

$$M_W = 80.9 \pm 1.4 \,\text{GeV}, \qquad M_Z = 91.9 \pm 1.8 \,\text{GeV},$$

while in 1989, CDF at the Tevatron (Abe et al. 1989) had measured:

$$M_Z = 90.9 \pm 0.3 \,\text{GeV},$$

just before SLC/LEP. As we will see, LEP has improved substantially on the Z mass, but not yet on the W mass. Currently the best measurements of the W mass from $p\bar{p}$ experiments (Abe et al. 1990, Alitti et al. 1991) yield:

$$1 - \frac{M_W^2}{M_Z^2} = 0.2275 \pm 0.0052$$

- Measurements of neutral current over charged current cross-section ratios (NC/CC) of high-energy neutrinos off isoscalar target. This measurement, assuming the $SU(2)_L \times U(1)$ structure of the Standard Model, is nearly equivalent (Blondel 1989) to a measure of the weak boson mass ratio. The present results (Abramowicz et al. 1986, Blondel et al. 1990, Allaby et al. 1986/87, King et al. 1993) lead to:

$$1 - \frac{M_W^2}{M_Z^2} = 0.2256 \pm 0.0047.$$

- Measurements of forward-backward asymmetries at e^+e^- machines, parity violation experiments in atoms or in lepton-nucleon scattering, analysis of neutral current neutrino and antineutrino scattering data on proton, deuterium and isoscalar targets. These data yielded measurements of the weak neutral-current couplings of quarks and leptons, with a precision of typically ± 0.05.

No experiment was found to disagree significantly with the Standard Model, provided the weak radiative effects were taken into account *and* the mass of the top quark were smaller than 200 GeV, in the minimal version of the Standard Model, with only one Higgs boson of mass smaller than a theory motivated upper limit of 1000 GeV.

A global fit to the data yielded a value of the weak mixing angle and of the ρ parameter (Ross and Veltman 1975, Veltman 1977, Marciano and Sirlin 1984) gave

$$\sin^2 \theta_W \equiv 1 - \frac{M_Z^2}{\rho M_Z^2} = 0.229 \pm 0.0064$$

$$\rho = 0.998 \pm 0.0086$$

as shown on Figure 2. The value of ρ could be interpreted either as a measure of isospin-breaking radiative effects, equivalent to the top mass limit given above, or as a limit on isospin-breaking vacuum expectation values in a more complicated model of the Higgs sector, but of course one could not disentangle the effects of these two phenomena, as both would influence that same number. We will see that LEP has now the tool to disentangle these two effects.

The values of the QED coupling constant α, of the weak coupling constant $\alpha_{weak} = \alpha/\sin^2 \theta_W$ and the strong coupling constant α_{strong} were determined with enough precision to rule out the minimal SU(5) Grand Unified Theory, and to favour strongly the supersymmetric version of this model. This could be seen (Amaldi et al. 1987) as 'perhaps the first harbinger of supersymmetry'(!). This observation has now been refined considerably. Among a very large number of publications, see for instance: Langacker and Luo 1991, Amaldi et al. 1991, Langacker and Polonoski 1993. The convergence of running couplings towards a common value at very high energies, the 'grand unification', can be obtained if one introduces new particles, the supersymmetric family for instance, at the energy scale of a TeV or so. This of course only proves supersymmetry if one believes in grand unification and vice-versa, making supersymmetric GUT's appealing, but by no means necessary.

2.3 Searches

The most efficient machine to search for the top quark was and remains the Tevatron. The present published limit (Abe et al. 1991, Barbaro-Galtieri 1993) is

$$M_t > 91 \,\text{GeV at 95\% C.L.},$$

with preliminary limits from the D0 collaboration (Abachi et al. 1994) as high as 131 GeV. This is an area where future progress should come from the Tevatron itself, and I will not mention it further.

There were very few constraints on the mass of the standard model Higgs boson before LEP started. This topic is reviewed in Franzini and Taxil 1982/89. A massless Higgs, as well as Higgs with mass between 10 and 110 MeV were excluded, and a Higgs with mass less than 5–6 GeV was very unlikely. However, *a theoretical loophole existed to every one of the experimental limits.*

3 The LEP storage ring

LEP is a 26.659 km long e^+e^- storage ring running underground in the Pays de Gex in France and Geneva county in Switzerland. Electrons and positrons are injected from a rather complex set of accelerators into LEP at 20 GeV, where they are accelerated up to 47 GeV, providing centre-of-mass energies around the Z pole. So far, no data has been recorded above 95 GeV centre-of-mass energy. The cross-sections decreasing very quickly, a very large step in energy is required to make this operation worthwhile.

Four (eight since 1992) bunches of electrons collide with four (eight) bunches of positrons in four experimental intersections. At the other four points where collisions should occur, the beams are normally separated to avoid unnecessary beam-beam interactions. With eight bunches, other parasitic interaction points would take place in the middle of the arcs. Since 1992, horizontal electrostatic separators are used to generate opposite sign oscillations of the beams in the arcs, in such a way that the beams miss each other. The 'Pretzel Scheme' has allowed an increase of the luminosity by a factor up to 1.5.

The interaction point has an extension of $\sigma_x = 130$–$200\,\mu$m horizontally, $\sigma_y < 10\,\mu$m vertically, $\sigma_z = 1.3$ mm longitudinally. The best peak luminosity achieved in LEP is $1.6 \times 10^{31}/\text{cm}^2/\text{s}$, above the design luminosity of $1.3 \times 10^{31}/\text{cm}^2/\text{s}$ at 45 GeV per beam. [The luminosity $\mathcal{L}$ gives the rate of events for a process with cross-section σ such that N(events per second) $= \mathcal{L} \cdot \sigma$. It is useful to remember that 1 nb$= 10^{-33}$ cm^2.] The average luminosity is reduced by the multiplication of several factors: luminosity decrease during the 10 hour long fills, (factor 0.6), filling time (0.8), scheduled interruptions for machine development (0.75), overall efficiency of the accelerator complex (0.75), and experiments data taking efficiency (0.8–0.9), giving an overall efficiency factor of 0.23.

The typical integrated luminosity recorded by the experiments has been: 1.2 pb^{-1} in 1989, 7 pb^{-1} in 1990, 13 pb^{-1} in 1991, 23 pb^{-1} in 1992 and 35 pb^{-1} in 1993. The results that I shall give are based on the 1989–1993 data as presented at the 1994 Moriond conference (Clarke et al. 1994).

4 LEP experiments

The four LEP detectors are constructed and operated by very large collaborations involving in total more than one thousand physicists and hundreds of institutes around the world. They all include

(i) a measurement of charged particles in *tracking chambers* enhanced by *microvertex* detectors giving a very precise point close (< 10 cm) to the origin;

(ii) detection of photons in *electromagnetic calorimeters*, followed by

(iii) detection of hadronic energy from neutral as well as charged hadrons in *hadronic calorimeters*.

(iv) The hadron calorimeter is usually followed by a *muon detector*.

(v) The tracking is done inside a magnetic field, always oriented parallel to the beam direction and generated by solenoidal coils, the iron yoke of the magnet being used as return for the magnetic flux.

(vi) At small angles, *luminosity detectors* are placed to catch the electrons from the reaction $e^+e^- \to e^+e^-$, of well-known cross-section, which is used to monitor the luminosity.

A complete description of the detectors is given in Decamp *et al.* 1990, Aarnio *et al.* 1991, Adeva *et al.* 1990 and Ahmet *et al.* 1991. A hadronic Z decay making explicit the functions of the various sub-detectors at large angles is shown in Figure 3.

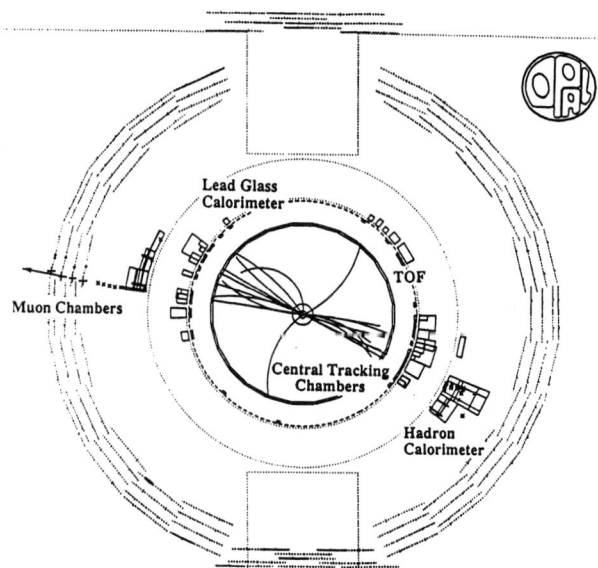

Figure 3. *Hadronic event in the* OPAL *detector.*

5 A synopsis of the measured quantities

5.1 The neutral current matrix element

The general matrix element for $e^+e^- \to f\bar{f}$ of producing a pair of fermions f via photon or Z exchange in the s-channel is:

$$\mathcal{M}_{NC}^{HH'} = -e^2 \frac{QQ'}{s} + \frac{e^2}{\sin^2\theta_W \cos^2\theta_W} \frac{g_{He} g_{H'f}}{M_Z^2 - s + is\Gamma_Z/M_Z}, \tag{1}$$

where each fermion is qualified by its charge Q, helicity $H = L, R$ (for left and right-handed), and weak isospin I_{fH}^3 ($I_{fL}^3 = \pm\frac{1}{2}$, $I_{fR}^3 = 0$). The prime superscript qualifies the final state. The electric charge is $e^2 = 4\pi\alpha$ and θ_W is the weak mixing angle. The first term represents the photon exchange, the second the Z exchange. The Z chiral couplings g_{Hf} are given by

$$g_{Hf} = I_{Hf}^3 - Q_f \sin^2\theta_W. \tag{2}$$

One also commonly uses the vector and axial vector couplings

$$g_{Vf} = (g_{Lf} + g_{Rf}) = I_{Lf}^3 - 2Q_f \sin^2\theta_W; \tag{3}$$

$$g_{Af} = (g_{Lf} - g_{Rf}) = I_{Lf}^3. \tag{4}$$

The cross-section is obtained by summing the matrix elements squared over the final state helicities and averaging over the initial helicity according to the appropriate polarisations. The unpolarised differential cross-section as a function of the scattering angle θ is then

$$\frac{d\sigma}{d\cos\theta} = \frac{(\hbar c)^2 s}{128\pi} \frac{3}{8} \Big\{ \left(|\mathcal{M}^{LL}|^2 + |\mathcal{M}^{RR}|^2 \right)(1+\cos\theta)^2 \\ + \left(|\mathcal{M}^{LR}|^2 + |\mathcal{M}^{RL}|^2 \right)(1-\cos\theta)^2 \Big\}. \tag{5}$$

At the Z pole the Z-channel dominates the cross-section and the photon exchange is only a correction. It can the be written as

$$\sigma = 12\pi(\hbar c)^2 \frac{s\Gamma_e \Gamma_f}{\left(s - M_Z^2\right)^2 + s^2\Gamma_Z^2/M_Z^2} \tag{6}$$

by identifying the Z partial width

$$\Gamma_f = \frac{e^2}{\sin^2\theta_W \cos^2\theta_W} \frac{M_Z}{24\pi} \left(g_{Lf}^2 + g_{Rf}^2 \right), \tag{7}$$

the total width being given by the sum over all open channels. Using Equation 5, one can find that forward-backward asymmetries or polarisation asymmetries are sensitive the following asymmetry of couplings

$$\mathcal{A}_f \equiv \frac{g_{Lf}^2 - g_{Rf}^2}{g_{Lf}^2 + g_{Rf}^2} = \frac{2g_{Vf}g_{Af}}{g_{Vf}^2 + g_{Af}^2}. \tag{8}$$

By measuring all partial widths and asymmetries, LEP can determine all fermion neutral current couplings.

5.2 A strategy of tests and radiative effects

One can organise the measurements at LEP in two broad classes: (i) the measurements providing tests of the basic assumptions that enter the construction of the model, in particular the $SU(2)_L \times U(1)$ gauge structure, and (ii) the measurements which probe electroweak radiative effects.

The main consequence of $SU(2)_L \times U(1)$ invariance is universality: the couplings of all particles with the same quantum numbers, such as e, μ, τ, are the same. Also the chiral couplings of the Z to the fermion f obey the formula of Equation 2. Up to radiative effects, the same value of $\sin^2 \theta_W$ should match all measured couplings. Thus the first programme of LEP is to determine the weak couplings of all quarks and leptons in the least model-dependent fashion.

Within the model, the electroweak radiative effects modify slightly the above predictions. LEP Physics is sensitive to QED radiative effects (emission of real or virtual photons) which are conceptually straightforward, and to electroweak (propagator or vertex) radiative effects, which contain the interesting physics information. It can be shown (Kennedy and Lynn 1986/89, Consoli and Hollik 1989) that electroweak effects can be absorbed in four main radiative effects at the Z pole, plus one when including the W mass.

- The running of the QED coupling constant α from $q^2 = 0$ to $q^2 = M_Z^2$.

- The isospin-breaking loop corrections to the W and Z propagators. They are absorbed conveniently in the ρ parameter, $\rho = 1 + \Delta \rho$.

- The running of the Z and W self-energies, absorbed in the parameters Δ_{3Q} and Δr^{ew}.

- The vertex correction to the $Z \to b\bar{b}$ partial width into a pair of b quarks.

An important feature of the weak corrections is that they introduce a dependence of the LEP observables upon the heavy, yet undiscovered, particles, such as the top quark and the Higgs boson. Other heavy particles which do not belong to the minimal version of the model with three families and one Higgs boson, but that would respect the $SU(2)_L \times U(1)$ structure, would also contribute to the same three blocks.

However, under very general assumptions, these large and heavy physics sensitive radiative corrections modify Equation 2 by only an overall scaling factor ρ and a global change of $\sin^2 \theta_W$, in a universal way. Non-universal corrections are small and—with the notable exception of the $Z \to b\bar{b}$ width—insensitive to heavy physics. As a consequence, it is convenient to express all asymmetry measurements at LEP in terms of the value of $\sin^2 \theta_W$ that one can extract from the charged lepton couplings

$$\sin^2 \theta_W^{eff} \equiv \frac{1}{4}\Big(1 - \frac{g_{V\ell}}{g_{A\ell}}\Big). \tag{9}$$

The relations between the observables, the Fermi constant G_F and the QED running constant $\alpha(M_Z^2)^{-1} = 128.8 \pm 0.1$ (Burkhardt et al. 1989), can be written in terms of these universal electroweak corrections $\Delta \rho$, Δ_{3Q}, Δr^{ew} as (Blondel and Verzegnassi

1993, Altarelli and Barbieri 1991, Blondel et al. 1991, Blondel 1992, Blondel et al. 1992, Altarelli et al. 1993):

$$M_Z^2 = \frac{\pi\alpha(M_Z^2)}{\sqrt{2}G_F(1+\Delta\rho)(1+\Delta_{3Q})\sin^2\theta_W^{\text{eff}}\cos^2\theta_W^{\text{eff}}}; \quad (10)$$

$$\Gamma_\ell = \frac{G_F M_Z^3}{24\sqrt{2}\pi}[1+\Delta\rho]\left[1+\left(\frac{g_{V\ell}}{g_{A\ell}}\right)^2\right]\left(1+\frac{3}{4}\frac{\alpha}{\pi}\right); \quad (11)$$

$$\Gamma_b = \Gamma_d(1+\delta_{vb}); \quad (12)$$

$$M_W^2 = \frac{\pi\alpha(M_Z^2)}{\sqrt{2}G_F(1-\Delta r^{\text{ew}})\left(1-\frac{M_W^2}{M_Z^2}\right)}. \quad (13)$$

Table 1 summarises the main observables and their physics output in terms of radiative effect or specific tests. The most critical technique involved in each of these various observables is also indicated. It turns out that, thanks to this specific choice, these observables are almost uncorrelated, from both points of view of statistical and systematic errors.

Quantity	Main Technologies	Physics Outputs	Relative Precision
line shape			
M_Z	**Absolute energy scale** relative cross-sections line shape fit (**QED** rad. corr.)	input	5×10^{-5}
Γ_Z	**Relative energy scale** relative cross-sections line shape fit (**QED** rad. corr.)	$\Delta\rho$	1.5×10^{-3}
$\sigma_{\text{had}}^{\text{peak},0}$	**Absolute cross-sections**	$N_\nu \cdot \frac{\Gamma_{\text{inv}}}{\Gamma_\ell}$ test $SU(2)_L \times U(1)$	3×10^{-3}
$R_\ell \equiv \frac{\Gamma_{\text{had}}}{\Gamma_\ell}$	lepton, hadron event selection	test universality $f(\alpha_s, \sin^2\theta_W^{\text{eff}}, \delta_{vb})$	4×10^{-3} 2×10^{-3}
$R_b \equiv \frac{\Gamma_b}{\Gamma_{\text{had}}}$	b-tagging	δ_{vb}	10^{-2}
asymmetries		$\sin^2\theta_W^{\text{eff}}$	2×10^{-3}

Table 1. *Synopsis of precision neutral current observables at the Z pole.*

6 The Z line-shape

The major advantage of an e^+e^- collider is the good knowledge of the centre-of-mass energy. The first task is thus to measure the cross-section for the various channels as a function of $\sqrt{s}$. The luck and glory of LEP is the Z resonance, which offers great physics and great event rates.

6.1 Fitting the Z line-shape

The first four parameters of Table 1 are obtained from measurements of the cross-sections for hadrons and leptons, as a function of $\sqrt{s}$. These four quantities constitute a natural choice of variables, as they describe the resonance by its position, its width, its height and the ratio of the two types of decays. These quantities are (i) nearly statistically independent, (ii) easily related to basic electroweak parameters, (iii) easily related to specific experimental errors.

The total cross-section for the reaction $e^+e^- \to f\bar{f}$ of Equation 5 is corrected for initial state radiation by convoluting a radiator function $\mathcal{H}(s,s')$ by an 'Improved-Born-approximation' (Kennedy and Lynn 1986, Consoli and Hollik, 1989) cross-section σ_f^0 (Berends et al. 1989, Kennedy et al., 1989):

$$\sigma_f(s) = \int_{s'=s_{\min}}^{s'=s} \mathcal{H}(s,s')\sigma_f^0(s')ds'. \qquad (14)$$

The function $\mathcal{H}(s,s')$ gives the probability that the initial e^+e^- system with centre-of-mass energy s has radiated one or several initial state photons, so that the remaining available energy for annihilation is s'. This function is calculated up to second-order in α with exponentiation of higher orders in Berends et al. 1988 and Burgers 1988. The effect of initial state radiation is very large, more than 30% on the peak cross-section. The function σ_f^0 (Martinez et al. 1991) contains the interesting parameters:

$$\sigma_f^0(s) = \frac{s}{(s-M_Z^2)^2 + s^2\Gamma_Z^2/M_Z^2}\left[\frac{12\pi\Gamma_e\Gamma_{f\bar{f}}}{M_Z^2(1+\delta_{\text{QED}}^e)} + \frac{I_{f\bar{f}}(s-M_Z^2)}{s}\cdot N_c^f(1+\delta_{\text{QCD}}^f)\right]$$
$$+\frac{4\pi}{3}\frac{\alpha^2(s)}{s}\cdot Q_f^2\cdot(1+\delta_{\text{QED}}^f)\cdot N_c^f(1+\delta_{\text{QCD}}^f) \qquad (15)$$

The term $I_{f\bar{f}}$ represents the γ-Z interference term. For total cross-sections its effect is small. The terms N_c^f, δ_{QCD}^f have the values 1, 0, for leptons and the values 3, $\alpha_s(M_Z^2)/\pi + \ldots$ for quarks. The term $\delta_{\text{QED}}^f = 1 + \frac{3}{4}Q_f^2\frac{\alpha}{\pi}$ represents the final state correction. See Figure 4 for the calculated line shape, with and without radiative effects. Electroweak effects are absorbed in a re-definition of the partial widths (Kennedy and Lynn 1989, Consoli and Hollik 1989, Borelli et al. 1990). The results of the fit as thus expressed in terms of M_Z, widths and ratios of partial widths, from which the Standard Model parameters are consequently determined. This procedure is more satisfactory than, and equivalent to, a direct fit to the Minimal Standard Model, where the only parameters would be M_Z, M_t and M_H.

Clearly, there are a few assumptions involved in writing the Equation 15: one assumes that the process involves only the Z and the photon, with vector and axial-vector

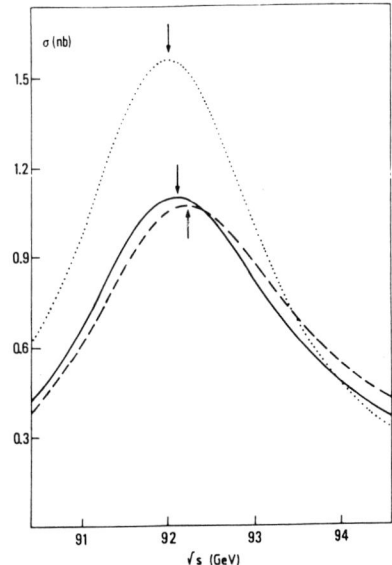
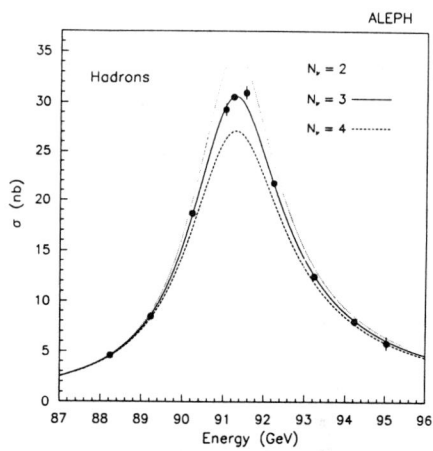

Figure 4. The calculated line shape, at Born level, with first order radiative effects, and exponentiated second order radiative effects.

Figure 5. The Z line shape for hadrons, calculated including exponentiated $\mathcal{O}(\alpha^2)$ initial state radiation. The large sensitivity of the peak cross section to a fourth light neutrino is clearly visible.

couplings. Also, the Standard Model is assumed for the calculation of the interference term. This second assumption can be partly removed if one estimates the interference term from the vector couplings measured in forward-backward asymmetries. However, the hadronic interference term still requires a standard model assumption. The L3 collaboration also presents a fit to the line shape where the interference is left free and find that the Z mass has an error increased by 9 MeV, if one proceeds this way (Adeva et al. 1991). It should be noted, however, that this term is insensitive to the intimate details of the $SU(2)_L \times U(1)$ Standard Model, and in particular to weak radiative effects. Thus as long as one remains within that gauge structure, which is the case for all consequences that will be drawn in the following, this additional error does not apply. Of particular interest are the QED-corrected peak cross-sections:

$$\sigma_f^{\text{peak},0}(s) = \frac{12\pi}{M_Z^2} \frac{\Gamma_e \Gamma_{f\bar{f}}}{\Gamma_Z^2} \tag{16}$$

where the effect of interference and photon exchange have been also corrected for. The hadronic cross-section at the peak, $\sigma_{\text{had}}^{\text{peak},0}$ provides a sensitive measure of the number of light neutrinos. If one writes $\Gamma_Z = \Gamma_{\text{had}} + 3 \cdot \Gamma_\ell + N_\nu \cdot \Gamma_\nu$, it is clear that one additional neutrino will increase Γ_Z by 160 MeV out of 2500 MeV (*i.e.* 6.4%), and *decrease* the peak cross-section by twice as much (ie 13%), as shown on Figure 5. Given the large counting rate at the peak, this is how the SLC/LEP experiments first determined the number of light neutrinos (Abrams et al. 1989, Decamp et al. 1989, Aarnio et al. 1989, Adeva et al. 1989, Akrawy 1989). Indeed, the number of neutrinos can be extracted from quantities that are measured at the peak only:

$$N_\nu = \frac{\Gamma_\ell}{\Gamma_\nu} \cdot \left(\sqrt{\frac{12\pi R_\ell}{M_Z^2 \sigma_{\text{had}}^{\text{peak},0}}} - R_\ell - 3 \right), \qquad (17)$$

where $R_\ell \equiv \Gamma_{\text{had}}/\Gamma_\ell$ is essentially determined by the ratio of cross-sections between leptons and hadrons. The experimental error on R_ℓ almost cancels in this formula, the main error coming from the uncertainty in the measurement of $\sigma_{\text{had}}^{\text{peak},0}$. A precise determination of cross-sections pays off as a precise determination of the number of neutrinos.

6.2 Determination of cross-sections

The measurements of cross-sections by the four LEP detectors are reported in Decamp et al. 1992a/93/94, Aarnio et al. 1991/94, Adeva et al. 1991, Ahmet et al. 1991/93, Akers et al. 1994 and Clarke et al. 1994.

6.2.1 Basics

Measuring an absolute cross-section is difficult. The procedure is the following: signal events, Z decays into hadrons or leptons, are counted in the large angle detectors; at the same time, the normalising events $e^+e^- \to e^+e^-$, subsequently called 'Bhabhas', are counted in the low angle detectors.

The difficulty lies in the fact that many things can go wrong: large angle and small angle detectors must be operational at the same time, with the same dead-time. It is requested for cross-section measurements that all relevant detectors be functioning. Furthermore, the data taking efficiency must be identical for the large angle events, which tend to be very large—over 100 kbytes of memory, and the Bhabhas, which tend to be small. This requires keeping track of all events even when not being processed in the data acquisition system. Finally, a measurement of the trigger efficiency from the data themselves is necessary. The LEP detectors have redundant triggers, yielding efficiencies as high as, e.g. for ALEPH: 99.999% for Bhabhas, 99.98% for hadrons, 99.85% for leptonic channels, measured with better than 10^{-4} precision.

Once these basic requirements are fulfilled, the events must be identified.

6.2.2 Luminosity measurement

The determination of luminosity is based on counting of low-angle Bhabha scattering events $e^+e^- \to e^+e^-$. These events are detected in the small angle luminosity monitors, which consists of two electromagnetic calorimeters, with good spatial resolution, positioned very accurately. A tracking device is often situated in front of the luminosity calorimeters, to improve or verify the spatial resolution. Bhabha events appear as two back-to-back electromagnetic showers, carrying the full beam energy each, as shown on Figure 6. A thorough discussion of the luminosity measurement can be found in the line-shape publications by the experiments, and details of the ALEPH luminosity measurement can be found in Decamp et al. 1992b, Bloch 1993, Bederède et al. 1994.

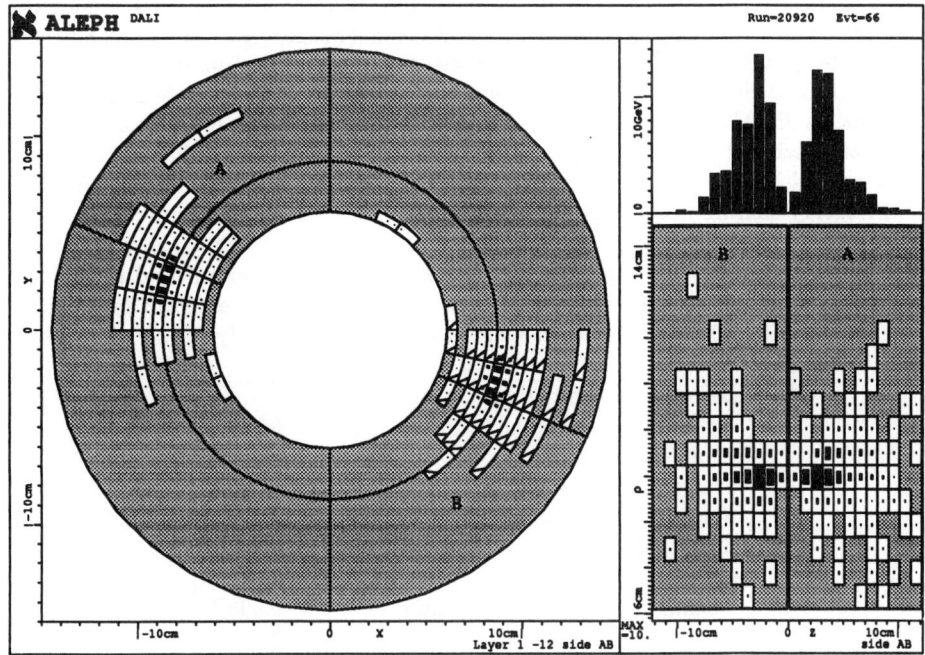

Figure 6. *A luminosity event in the ALEPH SiCAL luminosity monitor.*
Left: front view (x-y): The showers of the electron (side A) and positron (side B) are displayed on the same graph. The Silicium pads are represented with dots of size proportional to the energy collected. The ring shows the radial pad row at which the clusters are reconstructed.
Right: side view (r-z) of the two showers, with the corresponding energy profiles.

Triggering on these events is rather easy in the calorimeter. Redundancy is provided by single arm triggers. The cuts for event selection are specially designed to eliminate uncertainties related to the position of the beam interaction point, as well as to minimise the effect of radiative tails affecting the energy distribution of the outgoing electrons.

The main experimental challenge comes from the very strong angular dependence of the Bhabha cross-section, proportional to $1/\theta^4$, where θ is the scattering angle. As a result, the selected cross-section within the detector acceptance is proportional to $1/\theta_{\min}^2$ where $\theta_{\min}$ is the lowest angle for detection. The value of $\theta_{\min}$ varies from 24 to 55 mrad for the LEP experiments. For a value of $\theta_{\min}$ of 29 mrad, the cross-section for Bhabha scattering exceeds 100 nb. The statistics of these events are therefore very high, providing statistical accuracy of better than 10^{-3}. An uncertainty on the inner radius R of the sensitive region of the luminosity calorimeter induces an uncertainty on the measured cross-section:

$$\frac{\Delta\sigma_{\text{Bhabha}}}{\sigma_{\text{Bhabha}}} = 2 \times \frac{\Delta R}{R}.$$

For the typical value of R (5 cm) a precision well under 25 μm is requested to match the statistical accuracy. The LEP experiments have been upgrading their luminosity monitors to provide high precision knowledge of the inner edge θ_{min}. The ALEPH Silicium calorimeter (SiCAL) has been the first of these new generation of luminosity monitors (Bloch 1993, Bederède et al. 1994). It was operational from the middle of 1992 data onwards. A SiCAL event is shown on Figure 6. It is a silicon-tungsten sandwich with precision machined planes of silicon pads for energy readout. The fiducial cut θ_{min} is made on pad boundaries, where the position resolution, obtained from rapid variation of the energy sharing between pads upon the electron angle, is optimum. The radius of the pad boundary is known with a precision of 18 microns for 1992, and better than 10 microns for 1993. The table showing various sources of experimental systematic errors for the ALEPH SiCAL luminosity measurement in 1992 are shown in Table 2. This error was reduced (by increased Monte-Carlo statistics mostly) to 0.09% in 1993.

Background estimation:	
– off-momentum particles	0.018%
– 'physics' sources	0.010%
Trigger efficiency	0.010%
Reconstruction efficiency	0.001%
Radial fiducial cuts:	
– mechanical precision	0.058%
– beam and module alignments	0.035%
– modules A-B z-separation	0.035%
– asymmetry cuts	0.044%
– shower parametrisation and simulation	0.023%
Energy cuts	0.015%
Acoplanarity cut	0.005%
Simulation statistics	0.120%
TOTAL experimental uncertainty	0.153%

Table 2. *Systematic errors for the ALEPH SiCAL luminosity measurement.*

The other experiments have made similar improvement to their luminosity measurement: OPAL with a detector similar to the ALEPH SiCAL, operational in 1993; L3 with a precision silicon tracker positioned in front of the BGO luminosity calorimeter, operational in 1993; DELPHI-I with a silicium telescope operational in 1993, and a lead/scintillator/silicium sandwich that should be operational in 1994. All offer high statistics and expected systematic errors at the level of 10^{-3} or better. At the low angles considered, the cross-section is dominated by t-channel photon exchange, the Z contribution is less than 5×10^{-4}. The Z-γ interference vanishes at the Z pole, but can be as large as 6.10^{-3} off the pole, leading to a small correction that affects only the Z mass. Other low angle QED processes are also included in the measured cross-section, e.g. $e^+e^- \to \gamma\gamma$, but they are small (10^{-3} or so), and very well calculable. The calculation of radiative corrections to the Bhabha scattering cross-section itself is made delicate by the interplay of experimental cuts with higher order processes, that are not simulated. No single event generator has a complete account of the corrections, so the estimate presently involves a combination of: (i) event generation with BHLUMI (Jadach et al. 1989, 1990/92) a multi-photon $\mathcal{O}(\alpha)$ Monte-Carlo with exclusive exponentiation (many radiative photons are generated, assuming successive occurrence of the first-order process); complete electroweak first-order QED calculations (Böhm et al. 1988, Berends et al. 1988) (ii) estimate of higher order processes by leading-log and second-order calculations (Jadach et al. 1991, Beenhakker and Pietrzyk 1992/93). The present estimate of the theoretical error is ± 0.25% for a minimum angle of 25 mrad. It is slightly higher for larger angles where the correction from Z exchange is larger.

A summary of the present systematic errors on the luminosity measurements is given in Table 3.

	ALEPH		DELPHI-I		L3		OPAL	
	'92	'93 prel.	'92	'93 prel.	'92	'93 prel.	'92	'93 prel.
$\mathcal{L}^{\text{exp.} (b)}$	0.15%	0.09%	0.38%	0.28%	0.5%	0.16%	0.41%	$-^{(a)}$
σ_{had}	0.14%	0.1%	0.13%	0.13%	0.15%	0.25%	0.20%	0.24%
σ_e	0.4%	0.4%	0.59%	$-^{(a)}$	0.3%	$-^{(a)}$	0.22%	0.35%
σ_μ	0.5%	0.5%	0.37%	0.5%	0.5%	$-^{(a)}$	0.19%	0.22%
σ_τ	0.3%	0.3%	0.63%	$-^{(a)}$	0.7%	$-^{(a)}$	0.44%	0.51%
$A_{\text{FB}}^{(e)}$	0.0029	0.0029	0.003	$-^{(a)}$	0.002	$-^{(a)}$	0.002	0.002
$A_{\text{FB}}^{(\mu)}$	0.001	0.001	0.001	0.002	0.002	$-^{(a)}$	0.001	0.001
$A_{\text{FB}}^{(\tau)}$	0.0005	0.0005	0.0017	$-^{(a)}$	0.003	$-^{(a)}$	0.002	0.002

Table 3. *The experimental systematic errors for the analysis of the Z line shape and lepton forward-backward asymmetries. The errors quoted do not include the common uncertainty due to the LEP energy calibration. For the treatment of correlations between the errors for different years see Decamp et al. 1992a/93/94, Aarnio et al. 1991/94, Adeva et al. 1991, Ahmet et al. 1991/93 and Akers et al. 1994.* $^{(a)}$No preliminary result is quoted yet. $^{(b)}$Only the experimental error including the statistics of small angle Bhabha events is quoted.

6.2.3 Selection of hadronic events

Decays of the Z into $q\bar{q}$ pairs are not separately identified, but generically labelled as *hadrons*. The typical hadronic event, Figure 3, is characterised by a large number of charged tracks, on average 20, and a visible energy close to the total centre-of-mass energy. As a result of the enhancement due to the Z pole the non-resonant backgrounds—such as the 'two-photon' events—are reduced by a factor 1000. It is therefore easy to design a set of selection criteria, that keep a large fraction of the signal (99% typically) with very little background contamination (typically 0.3% or less). The typical cuts involve a cut on the number of charged or neutral tracks, and on the visible energy. In the case of ALEPH, two independent selections are used, one using only charged tracks, one using only calorimeter information, providing a cross-check at the level of 0.1%. The errors on the selections performed by the LEP experiments are summarised in Table 3.

A source of energy-dependent systematic error comes from the subtraction of the 'two-photon' background. The models for this process are largely uncertain. The cross-section at the peak of the Z is so large that this uncertainty does not contribute much to its determination. The energy dependence of the background, which is nearly constant, is very different from that of the signal. This causes an uncertainty of a few MeV on the total width determination.

6.2.4 Selection of leptonic events

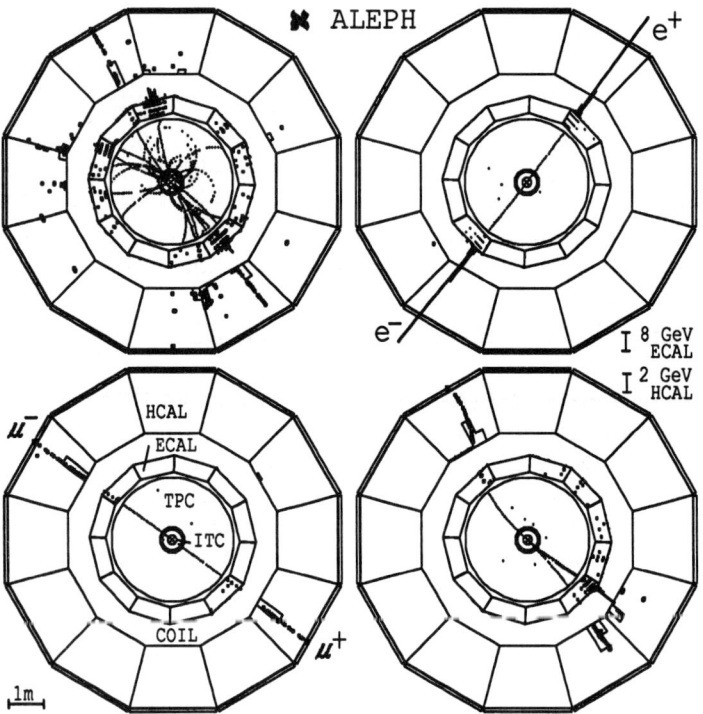

Figure 7. *A hadronic Z decay and leptonic events in the ALEPH detector: a) e^+e^- event, characterised by the large showers in the electromagnetic calorimeter; b) $\mu^+\mu^-$ event, characterised by the two muons penetrating through the hadron calorimeter to the muon chambers; c) $\tau^+\tau^-$ event, characterised by different topologies in the two hemispheres, and missing energy due to the final state neutrinos.*

Leptonic decays of the Z, $e^+e^-\to e^+e^-$, $e^+e^-\to \mu^+\mu^-$ and $e^+e^-\to \tau^+\tau^-$, offer much simpler topologies than hadronic decays, as can be seen in Figure 7. They are however less frequent (1:20), and, since they have fewer tracks, are easier to miss. The selection procedures for individual lepton channels are more involved than for hadronic events, and depend on the specific properties of each detector. Good hermeticity and tracking down to low angles are determinant. A summary of the experimental uncertainties in the selection is given in Table 3. ALEPH, DELPHI-I and OPAL also select leptons inclusively, thereby avoiding the set of cuts that differentiate one lepton type from another and reducing the systematic errors.

An additional delicacy happens for the e^+e^- decays of the Z: the addition of, and interference with, the t-channel Bhabha scattering process, as seen in Figure 11 in Section 6.4. The difficulty is of theoretical nature: while Z production and decay is well adapted to the structure function formalism of Equations 14 and 15, this is not quite the case when the t-channel is included. Therefore, more complete higher order calculations have to be used (Beenakker et al. 1991). The experiments give the 'Z-exchange cross-sections' where the t-channel effect has been subtracted, a procedure which introduces negligibly small systematic errors.

6.3 The beam energy

The LEP data were taken in 1990–1991 at seven different centre-of-mass energies interspaced by 1 GeV from 88.25 GeV to 94.25 GeV. In 1992 all data were taken at the Z pole. In 1993 a scan of the Z line shape was performed again at the energies of 89.4, 91.2 and 93 GeV. This choice of the scan points optimises the resolution on Γ_Z, at energies where beam polarisation can be obtained, allowing be precise energy calibration by resonant depolarisation.

Possible uncertainties on the beam energy result in systematic errors on M_Z, and Γ_Z. The Z mass error is dominated by the knowledge of the absolute energy scale, while the Z width error stems from uncertainties in the differences between the various centre-of-mass energies (Blondel 1992).

The central piece of the LEP beam energy calibration is the resonant depolarisation. For a summary of the methods used for 1990 data, see Hatton 1990, Albert 1991. Transverse spin polarisation of the beam builds up in a storage ring by the Sokolov-Ternov effect (1964): this phenomenon was observed in LEP for the first time in 1990 (Knudsen et al. 1991). The resonant depolarisation method has been used previously in e^+e^- machines, providing accurate measurements of the masses of the J/ψ, ψ', Υ, Υ', at VEPP4 in Novosibirsk (Zholentz et al. 1980, Artamonov et al. 1982), at DORIS in Hamburg (Barber et al. 1984), at CESR in Cornell (McKay et al. 1984). It is performed by the LEP polarisation collaboration since 1991 (Arnaudon et al. 1992).

The spin precession frequency is determined by resonant depolarisation: a fast kicker generates a periodic perturbation to the beam, and the particle's spin. If the perturbation is in tune with the spin precession, a spin resonance condition is obtained, and the polarisation can be destroyed, or even reversed. Figure 8 shows an example of a polarisation reversal, allowing instantaneous determination of the beam energy to 10^{-5}. The spin precession frequency is equal to the product of the revolution frequency, known to one part in 10^9, by the spin tune ν, directly related to the beam energy by the

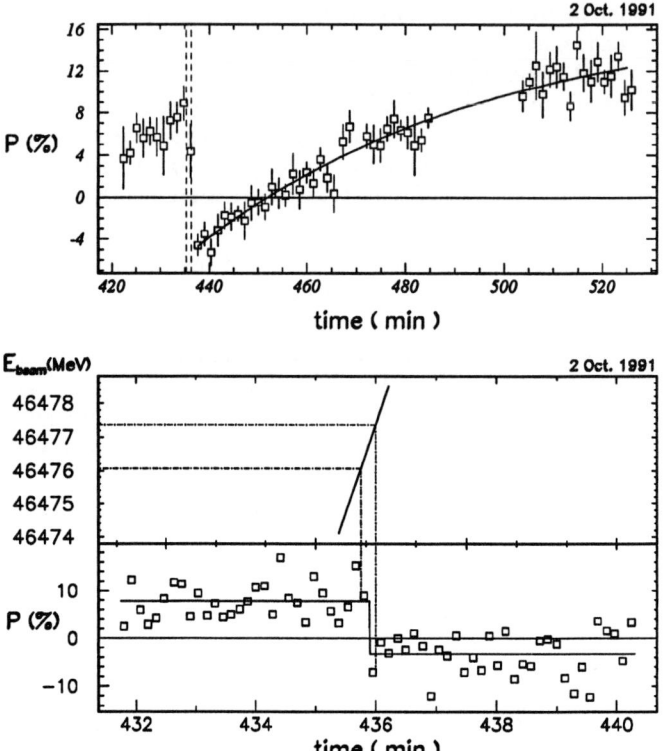

Figure 8. *Polarisation signal on 2nd October 1991, showing the localisation of the depolarizing frequency within the sweep. Top: display of data points, with the frequency sweep indicated with vertical dashed lines. The full line represents the result of a fit with starting polarisation $(-4.9\pm1.0)\%$, polarisation rise-time (60 ± 13) minutes, asymptotic polarisation $(18.4\pm4.1)\%$. Bottom: expanded view of the sweep period, with the individual data sets displayed (there are 10 sets per point); The frequency sweep lasted 7 data sets. The corresponding beam energy is shown in the upper box. Spin flip occurred between the two vertical dash-dotted lines.*

anomalous magnetic moment $a_e = (g_e - 2)/2 = 1.1596521884(43) \times 10^{-3}$ (Van Dyck Jr et al. 1987) and the mass $m_e = 0.51099906(15)$ MeV of the electron:

$$\nu = a_e\gamma = \frac{(g_e - 2)}{2}\frac{E_{\text{beam}}}{m_e} = \frac{E_{\text{beam}}}{0.4406486(1)}. \tag{18}$$

A spin tune $\nu = 105.5$ corresponds to the beam energy of 46.5 GeV.

The resonant depolarisation determines the beam energy with very small errors (Arnaudon et al. 1993, Wenninger 1994). This is shown in Figure 9, where the amplitude of the spin reversal is shown as a function of the exciting frequency. The precision on each energy calibration is 0.3 MeV with systematic effects of less than 0.2 MeV.

However, polarisation runs are performed only seldom, four times in 1991, more often in 1993, where they were made compatible with normal physics operation, but

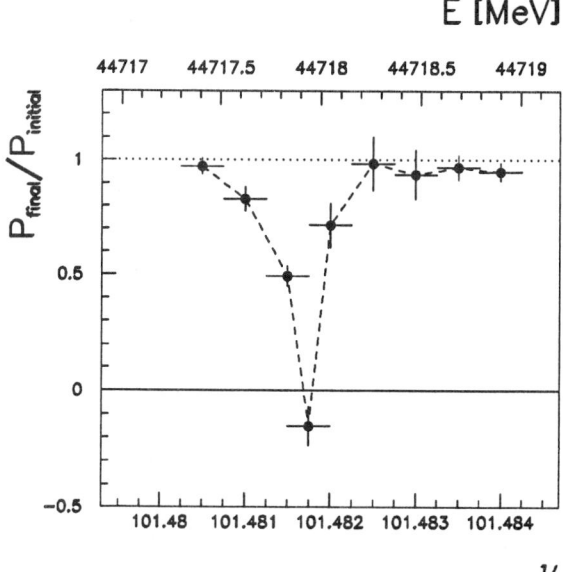

Figure 9. *Measurement of the width of the artificial depolarizing resonance, showing a width of 200 keV.*

only at the end of physics fills, and performed once or twice a week. The extrapolation to the scan data uses several tools to trace in time the properties of the magnets, current, field, temperature, as well as the geometrical properties of the ring. The analysis of the accumulated data is performed in collaboration by the accelerator physicists and members of the LEP collaborations within the LEP Energy working group (Arnaudon et al. 1992/93/94).

Maybe the most spectacular source of energy variations in time is the effect of geological motion of the ground on the geometry of the ring. Because of the strong focusing of LEP, these movements are amplified by a factor of nearly 10^4, so that a small expansion by $\pm 10^{-8}$ leads to a potential error on the Z mass and width of 10 MeV. Terrestrial tides cause such variations on a daily basis. They can be predicted (Fischer and Hofmann 1992, Melchior 1983, Berger and Lovberg 1970) and have indeed been observed (Arnaudon et al. 1994), and their effect measured with an accuracy of 5%, as shown in Figure 10.

All known sources of fluctuations being removed, the LEP energy calibrations of 1993 still showed a scatter of 4 MeV of the beam energy. Careful investigations of the beam orbit measurements (Jacobsen 1992, Wenninger 1994b) indicate that the observed energy jumps are correlated with orbit movement, especially visible in September 1993 when a record rainfall for the century took place!

Neither of these methods has yet reduced the uncertainty in the differences between the various centre-of-mass energies. The present estimate of this uncertainties is 10 MeV, resulting in 5 MeV error on the Z width. To reduce significantly this error, a measurement of two different energies by resonant depolarisation is made.

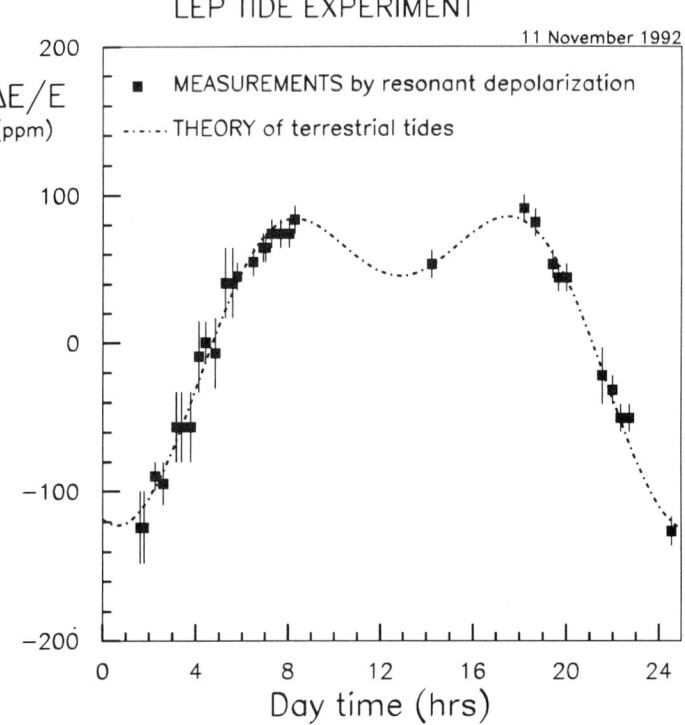

Figure 10. *Beam energy variations measured over 24 hours compared to the expectation from the tides.*

These results allow knowledge of the beam energy with fluctuations of around 4 MeV, leading to systematic errors of 4 MeV on the Z mass, and 3 MeV on the Z width, before correction for the orbit measurement. It is hoped to reduce these to less than two MeV when the analysis is complete.

6.4 Results on the Z line-shape

Once the cross-sections and energies are determined, a fit is performed according to the prescriptions of Equations 14 and 15. Because the forward-backward asymmetry for leptons has a steep dependence on centre-of-mass energy, it receives a small systematic error from the beam energy uncertainty, which correlates it with the Z mass. In addition the lepton forward-backward asymmetry can be used to constrain the Z-photon interference term. For these two reasons, the line shape fit (see Figure 11) usually includes the leptonic forward-backward asymmetries as well.

Two different fits are usually performed:

1. A fit with no assumption of lepton universality: in this case, a set of nine parameters (M_Z, Γ_Z, $\sigma_{\text{had}}^{\text{peak},0}$, R_e, R_μ, R_τ, $A_{\text{FB}}^{(e)}$, $A_{\text{FB}}^{(\mu)}$, $A_{\text{FB}}^{(\tau)}$) is required. The results of such

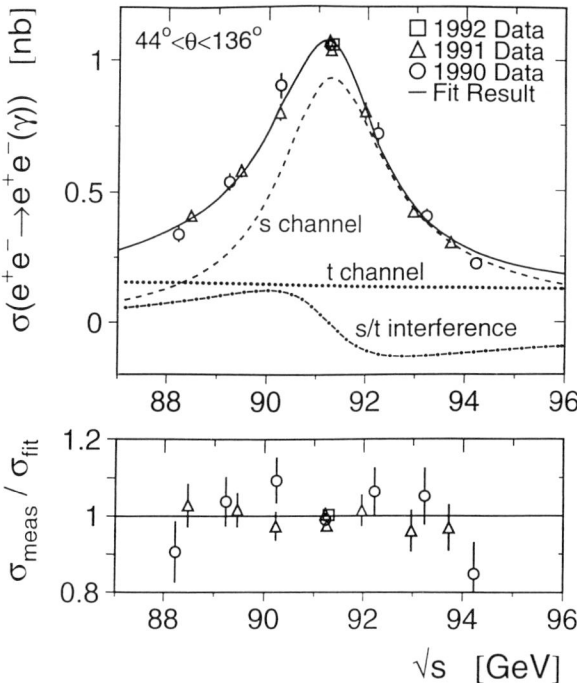

Figure 11. *The* L3 *cross-section for electron final state. In the top diagram are shown the t-channel and s-channel contributions to the e^+e^- final state. The t-channel is subtracted to give the Z exchange cross-section, similar to the muon and lepton channels. Superimposed is the best fit to the model independent Z line shape.*

fits can be found in LEP 1993. This analysis allows a test of lepton universality by comparing the values of R and of the asymmetry for each lepton, as shown in Figure 12

2. The three leptonic widths give consistent values and support lepton universality. One can thus make this assumption and fit for one leptonic width Γ_ℓ defined as the partial Z decay width into a pair of massless leptons, and one asymmetry $A_{FB}^{(\ell)}$. The result is shown in Table 4.

The results given in Table 4 contain all the necessary information. However, one must be careful that the parameters could be strongly correlated. The correlations for the parameters given in Table 4 are given in Table 5. The averages in the last column are performed assuming the following common systematic errors.

- The errors on the beam energy.
- The theoretical uncertainty on σ_{Bhabha}, since all experiments use essentially the same calculations. A common error of 0.25% is assumed.
- The uncertainty in the t-channel subtraction for large angle e^+e^- events, which is also based on the same theoretical calculations.

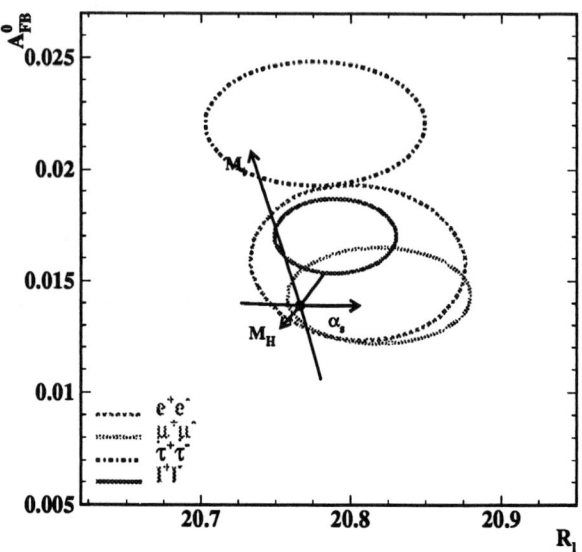

Figure 12. One standard deviation contours (39% probability) in the R_ℓ-$A_{FB}^{(\ell)}$ plane. Also shown as dotted symbol is the SM prediction for $M_Z = 91.1895$ GeV, $M_t = 150$ GeV, $M_H = 300$ GeV, $\alpha_s = 0.123$. The lines with arrows correspond to the variation of the SM prediction when M_t, M_H or α_s are varied in the intervals $50 < M_t(\text{GeV}) < 250$, $60 < M_H(\text{GeV}) < 1000$ and $\alpha_s(M_Z^2) = 0.123 \pm 0.006$, respectively. The arrows point in the direction of increasing values for M_t, M_H and α_s.

	ALEPH	DELPHI-I	L3	OPAL
M_Z(MeV)	$91,191.5 \pm 3.3$	$91,187.0 \pm 3.3$	$91,195.1 \pm 3.6$	$91,186.0 \pm 3.6$
Γ_Z(MeV)	$2,495.9 \pm 5.4$	$2,495.3 \pm 5.2$	$2,502.8 \pm 5.0$	$2,494.7 \pm 5.7$
$\sigma_{had}^{peak,0}$(nb)	41.44 ± 0.36	41.84 ± 0.45	41.1 ± 0.4	41.01 ± 0.41
R_ℓ	20.730 ± 0.078	20.682 ± 0.074	20.93 ± 0.10	20.882 ± 0.077
$A_{FB}^{(\ell)}$	0.0216 ± 0.0026	0.0159 ± 0.0031	0.0184 ± 0.0045	0.0127 ± 0.0026

	Average	χ^2
M_Z(MeV)	$91,189.9 \pm 4.4$	4.3
Γ_Z(MeV)	$2,497.1 \pm 3.8$	1.6
$\sigma_{had}^{peak,0}$(nb)	41.33 ± 0.23	5.7
R_ℓ	20.789 ± 0.040	6.0
$A_{FB}^{(\ell)}$	0.0170 ± 0.0016	6.1

Table 4. Line shape parameters from the four LEP experiments as published in references Decamp et al. 1992a/93/94, Aarnio et al. 1991/94, Adeva et al. 1991, Ahmet et al. 1991/93, Akers et al. 1994 and preliminary results from Clarke et al. 1994. The averages are from LEP 1993

	Γ_Z	$\sigma_{had}^{peak,0}$	R_ℓ	$A_{FB}^{(\ell)}$
M_Z	0.031	0.003	0.01	0.04
Γ_Z		-0.091	0.001	0.002
$\sigma_{had}^{peak,0}$			0.114	0.002
R_ℓ				0.004

Table 5. *Average correlation coefficients for the parameter set of Table 4.*

The agreement between experiments is acceptable, as shown by the values of χ^2 for 3 degrees of freedom given in Table 4.

One can extract from these numbers the values of R_ℓ and N_ν, so that the line shape results for LEP can be summarised as:

$$\begin{aligned}
M_Z &= 91.1895 \pm 0.0018 \pm 0.0040_{\text{LEP}} \\
\Gamma_Z &= 2.4969 \pm 0.0027 \pm 0.0027_{\text{LEP}} \\
N_\nu &= 2.985 \pm 0.012 \pm 0.019_{\text{th.}} \\
R_\ell &= 20.789 \pm 0.040
\end{aligned} \quad (19)$$

This clearly is consistent with three species of light neutrinos (whose mass is much smaller than $M_Z/2$), as shown in Figure 13.

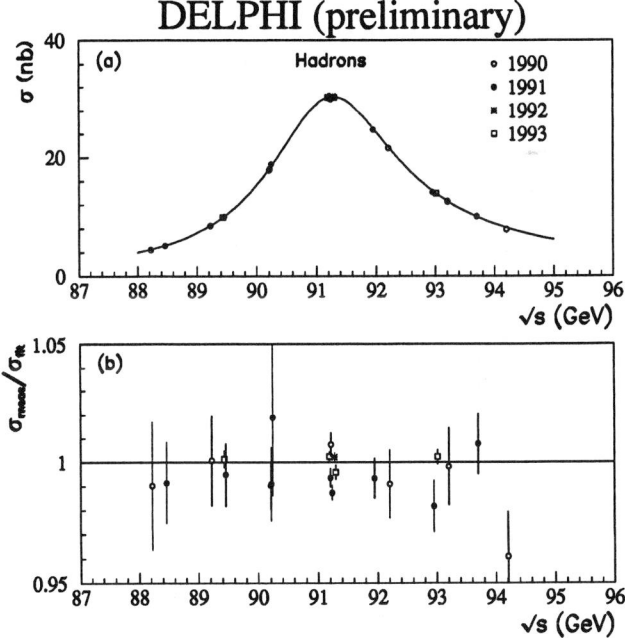

Figure 13. *The ALEPH hadronic cross-sections as a function of centre-of-mass energy. The square points show the considerable statistical improvement from the 1993 scan.*

7 Partial widths into specific flavours

We have seen so far how one can determine the partial widths of inclusive quarks or neutrinos and the three types of charged leptons. Because it belongs to an incomplete multiplet together with the top quark, the $Z \to b\bar{b}$ partial width of the Z into a pair of b quarks is particularly interesting, and receives a specific vertex correction, sensitive uniquely to the top quark mass (Akhundov et al. 1986, Beenakker and Hollik 1988, Lynn and Stuart 1990, Barnabeu et al. 1991). In order to measure the $Z \to b\bar{b}$ partial width, or the b forward backward asymmetry, the first step is identification of b events often called 'b-tagging'. b-tagging is interesting for many reasons: besides allowing electroweak measurements, it is a key tool in selecting clean b samples for study of exclusive b-quark properties, $B^0\overline{B^0}$ mixing, and even as a secondary tool for search for the Higgs boson, which is expected to decay primarily into b-quarks, if it not too heavy. The topic has received the devoted attention of a large fraction of the LEP experimentalists, with many new techniques and refinements. A good review can be found in Brown 1993. The quantity to measure is $R_b = \Gamma_b/\Gamma_{\text{had}}$, where the b-vertex correction is nicely isolated with little theoretical uncertainty (Blondel et al. 1992). The methods group in three categories.

- b-tagging with high P, $P_\perp$ leptons. This is the oldest technique. Leptons are identified among all charged tracks in hadronic events, then selected on the basis of their longitudinal or transverse momentum with respect to the nearest jet. Since charm decay can also produce high momentum leptons, albeit with a different distribution, a statistical separation is obtained from a global fit to the lepton distributions. As a result, the b partial width that is extracted this way is strongly correlated with assumptions made on the charm decays. the efficiency of b-tagging with leptons is reduced to less than 10% by (i) the leptonic branching ratio (40% for either b into electron or muon) and (ii) the high P, $P_\perp$ cuts necessary to isolate a pure sample. Typically, a purity of 80% can be reached with an efficiency of 5%. The heavy flavours analyses using lepton tagging are described in ALEPH 1994, DELPHI 1992a, L3 1991 and OPAL 1993a.

- Global properties of hadronic events (event shapes). This technique has in principle the advantage of using all events. The principle is to try to identify in the various kinematical variables that can be reconstructed in jets those which are affected by the presence of a heavy, fast object decaying isotropically. The most efficient variable is the boosted sphericity product and variations thereof. Such analyses are described in ALEPH 1993a, DELPHI 1992b and L3 1993.

- Vertex tagging. This is the area where most progress has been accomplished in the last year. The tool of identification is now the long (1.5 ps) life time of the b-hadrons, associated with their large decay multiplicity. As a consequence, events containing a b-quark tend to contain several charged tracks originating from a secondary vertex situated several millimeters from the main interaction point. To perform this task, the LEP experiments have equipped themselves with high precision vertex detectors. The vertex detector of ALEPH is described in Batignani et al. 1992. A beautiful example of a fully identified b event is shown on Figure 14. The LEP experiments have used various characterisation of the detached secondary vertex properties: ALEPH (ALEPH 1993b) has used the product of probabilities of the tracks to extrapolate back to the vertex, while OPAL (OPAL 1993b) counts the number of tracks that have a impact parameter significance $\delta/\sigma(\delta)$ above a given threshold. Preliminary DELPHI-I results were given at conferences. The L3 detector is being upgraded to include a vertex detector to be operational in 1994.

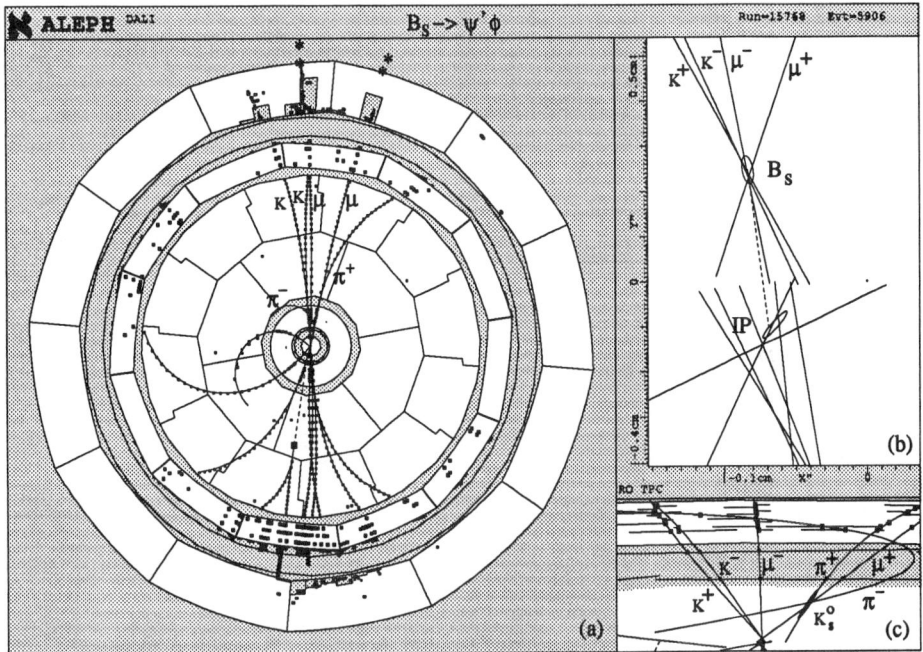

Figure 14. *A fully reconstructed example of a $Z \to b\bar{b}$ event. Here a B_s is identified by its decay $B_s \to \psi'\phi$. The ψ' decays into two muons and the ϕ into two K's. Left: front view (x-y). Right: expanded view of the vertex showing the VDET hits, and the reconstructed primary and seconday vertices.*

These three different methods have one point in common: since the production and decay of b-hadrons is not very well known, the tagging efficiency cannot be calculated with certainty. However, tere are two b's per $Z \to b\bar{b}$ event, and use is made of the *double tag* method to measure the tagging efficiency from the data. In the simple minded situation where the two hemispheres in an event are independent and if the backgrounds are negligible, the fraction of events where both hemispheres are tagged N_2, the fraction with one tagged hemisphere N_1, the tag efficiency ϵ and the b partial width are related by the equations:

$$\varepsilon = \frac{N_2}{N_1} \quad \text{and} \quad \frac{\Gamma_b}{\Gamma_{\text{had}}} = \frac{N_1^2}{N_2}. \tag{20}$$

Of course reality is different and in all methods one has to correct for the backgrounds and hemisphere correlations. The backgrounds come mostly from light quarks which fake the tag. For example, u, d and s quarks can fake a lepton tag because of misidentified hadron, or the vertex tag because of secondary vertices (strange particle decays or secondary interactions). Charm constitutes the more serious background, as it is a source of prompt leptons, and of secondary vertices albeit with lower multiplicities. The charm background estimates require good knowledge of charm production and are presently the dominant source of systematic error.

The hemisphere correlations are mostly of geometrical nature, since the two b quarks in an event are emitted back-to-back, and tend to hit the detector inhomogeneities in a correlated manner. However there are some physical reasons for correlation as well, namely the gluon emission that affects both sides of an event. Better control of systematics can be obtained by calibrating one of the tagging methods against another (mixed tag). The flurry of methods cannot be simply listed and averaged. The most precise methods are those using the vertex tag, and the least precise systematically are those using event shape variables. The results are summarised in Table 6.

Method	Experiment	Γ_b/Γ_{had}		
		value	exp. error	modelling error
High p, $P_\perp$ lepton tag	ALEPH	0.216	±0.006	± 0.005
	L3	0.221	± 0.008	± 0.008
	OPAL	0.224	± 0.011	± 0.007
	DELPHI-I	0.216	± 0.009	± 0.005
Event shape variables:	DELPHI-I	0.232	± 0.005	± 0.017
	ALEPH (mixed)	0.228	± 0.005	± 0.05
	L3	0.222	± 0.003	± 0.007
Microvertex tag:	DELPHI-I	0.2257	± 0.003	± 0.006
	ALEPH	0.2192	± 0.0022	± 0.0026
	OPAL	0.220	± 0.004	± 0.006
	average	0.2208	± 0.0021	

Table 6. Γ_b/Γ_{had} *measurements at* LEP. *The averages have been computed as described in* LEP *1993. Errors stemming from floating the charm partial width are not shown.*

It has been shown that the main background to b-tag is charm. This has for result a potential large correlation between the b and c partial widths. If the measurement of Γ_b/Γ_{had} is to be interpreted within the standard $SU(2)_L \times U(1)$ model, it is easy to see that the charm partial width is essentially fixed to its standard model value, even if one let $\sin^2 \theta_W^{eff}$ and δ_{vb} float independently of each other. On the other hand it is also interesting to measure Γ_b/Γ_{had} and Γ_c/Γ_{had} independently. In this case, the result of Table 6 becomes:

$$\frac{\Gamma_b}{\Gamma_{had}} = 0.2208 \pm 0.0026 \quad \text{and} \quad \frac{\Gamma_c}{\Gamma_{had}} = 0.171 \pm 0.013 \tag{21}$$

with a correlation matrix $\begin{pmatrix} 1. & -0.59 \\ -0.59 & 1. \end{pmatrix}$.

8 Helicity effects in Z production and decays

The Standard Electroweak Model $SU(2)_L \times U(1)$ being left-right asymmetric, helicity effects are expected to play an important role in Z production and decays. Helicity effects in $e^+e^- \to Z \to f\bar{f}$ are sketched in Figure 15.

Initial state helicity	cross-section	Final state helicity
e^- ⇐ Z • e^+ ⇐	$\propto g_{Le}^2$	Forward: $\bar{f}$ ⇐ f ⇐ $\propto g_{Lf}^2$ Backward: f ⇐ $\bar{f}$ ⇐ $\propto g_{Rf}^2$
e^- ⇒ Z • e^+ ⇒	$\propto g_{Re}^2$	Forward: $\bar{f}$ ⇒ f ⇒ $\propto g_{Rf}^2$ Backward: f ⇒ $\bar{f}$ ⇒ $\propto g_{Lf}^2$
e^- ⇐ Z • e^+ ⇒	0	
e^- ⇒ Z • e^+ ⇐	0	

Figure 15. *Helicity effects in $e^+e^- \to Z$ production. The arrows ⇐ and ⇒ indicate the helicity of the particles.*

The fact that the neutral current couplings are different for left-handed and right-handed fermions leads to polarisation and forward-backward asymmetries which can be expressed in terms of the chiral coupling asymmetries:

$$\mathcal{A}_f \equiv \frac{g_{Lf}^2 - g_{Rf}^2}{g_{Lf}^2 + g_{Rf}^2} = \frac{2g_{Vf}g_{Af}}{g_{Vf}^2 + g_{Af}^2} \tag{22}$$

The couplings are related to the weak mixing angle $\sin^2\theta_W$ by the well known relations:

$$g_{Lf} = I_{3f} - Q_f \sin^2\theta_W \tag{23}$$

$$g_{Rf} = -Q_f \sin^2\theta_W \tag{24}$$

$$g_{Af} = (g_{Lf} - g_{Rf}) \tag{25}$$
$$g_{Vf} = (g_{Lf} + g_{Rf}) \tag{26}$$

The values of neutral current couplings and their sensitivity to $\sin^2 \theta_W^{\text{eff}}$ are given in Table 7.

Fermion type	I_{3f}	Q_f	g_{Af}	g_{Vf}	$\mathcal{A}_f$	$\dfrac{\partial \mathcal{A}_f}{\partial \sin^2 \theta_W^{\text{eff}}}$
ν	1/2	0	1/2	1/2	1	0
e^-	$-1/2$	-1	$-1/2$	-0.04	0.16	-7.9
u	1/2	2/3	1/2	0.19	0.69	-3.5
d	$-1/2$	$-1/3$	$-1/2$	-0.35	0.94	-0.6

Table 7. *Numerical values of quantum numbers, Neutral current couplings, chiral coupling asymmetry $\mathcal{A}_f$ and sensitivity of $\mathcal{A}_f$ for the four types of fermions. The value of $\sin^2 \theta_W^{\text{eff}}$ is 0.23.*

On the Z pole, and neglecting the photonic contributions, the observables obtainable if longitudinal beam polarisation is available are:

- the left-right asymmetry of Z production (Prescott 1980, Böhm and Hollik 1982, Lynn and Stuart 1985, Lynn et al. 1985, Lynn and Verzegnassi 1987, Blockus et al. 1986, Alexander et al. 1987, Badier et al. 1987)

$$A_{\text{LR}} = \frac{\sigma_L - \sigma_R}{\sigma_L + \sigma_R} \simeq \mathcal{A}_e, \tag{27}$$

- the forward-backward polarised asymmetries (Blondel et al. 1988):

$$A_{\text{FB}}^{\text{pol}(f)} = \frac{1}{\mathcal{P}} \frac{(N_{\mathcal{P},F} - N_{-\mathcal{P},F}) - (N_{\mathcal{P},B} - N_{-\mathcal{P},B})}{(N_{\mathcal{P},F} + N_{-\mathcal{P},F}) + (N_{\mathcal{P},B} + N_{-\mathcal{P},B})} \simeq \frac{3}{4} \mathcal{A}_f, \tag{28}$$

where $\mathcal{P} = (\mathcal{P}_{e^-} - \mathcal{P}_{e^+})/(1 - \mathcal{P}_{e^-}\mathcal{P}_{e^+})$ is the polarisation of the e^+e^- system. Without polarisation one has at one's disposal the forward-backward asymmetry:

$$A_{\text{FB}}^{(f)} \simeq \frac{3}{4} \mathcal{A}_e \mathcal{A}_f. \tag{29}$$

In either case, for the tau lepton, the polarisation of the final state fermion is measurable. For unpolarised beams, the tau polarisation is a function of polar angle:

$$\mathcal{P}_\tau(\cos\theta) \simeq -\frac{\mathcal{A}_\tau + 2\cos\theta \mathcal{A}_e/(1 + \cos^2\theta)}{1 + 2\cos\theta \mathcal{A}_e \mathcal{A}_\tau/(1 + \cos^2\theta)} \tag{30}$$

from which one can derive the average integrated over polar angle:

$$\langle \mathcal{P}_\tau \rangle \simeq -\mathcal{A}_\tau, \tag{31}$$

or another version of the forward-backward polarisation asymmetry (Jadach and Wąs 1989):

$$A_{\text{FB}}^{\text{pol}(f)} = \frac{(N_{\mathcal{P}_\tau=1,F} - N_{\mathcal{P}_\tau=-1,F}) - (N_{\mathcal{P}_\tau=1,B} - N_{\mathcal{P}_\tau=-1,B})}{(N_{\mathcal{P}_\tau=1,F} + N_{\mathcal{P}_\tau=-1,F}) + (N_{\mathcal{P}_\tau=1,B} + N_{\mathcal{P}_\tau=-1,B})} \simeq \frac{3}{4} \mathcal{A}_e, \tag{32}$$

where $\mathcal{P}_\tau$ is now the polarisation of the tau lepton.

8.1 Definition of $\sin^2 \theta_W$?

Before one starts discussing asymmetries at the Z pole, it is important to define a symbol that will be used as 'monnaie d'échange' throughout the coming pages, $\sin^2 \theta_W$. One can find many different usages and definitions of this mythical parameter in the literature.

- Sirlin's definition (Sirlin 1980):

$$\sin^2 \vartheta_{w,\text{Sirlin}} \equiv 1 - \frac{M_W^2}{M_Z^2}. \tag{33}$$

This definition is perfectly unambiguous. With the advent of LEP, where the Z mass is measured to 10^{-5}, this parameter is clearly redundant with the W mass itself. I will not use this definition, but consistently write $1 - \frac{M_W^2}{M_Z^2}$.

- The $\overline{\text{MS}}$ definition (Sirlin et al. 1983–90):

$$\sin^2_W \hat{\vartheta}_{\overline{\text{MS}}}(q^2) \equiv \frac{\hat{e}^2}{\hat{g}^2}(q^2), \tag{34}$$

where $\hat{e}$ and $\hat{g}$ are the QED and SU(2) running coupling constants in the modified Minimal Subtraction scheme. The advantage of this quantity is quite useful to use in the context of grand unification analysis, see for example Amaldi et al. 1987. It also corresponds to the intuitive picture where $\sin^2 \theta_W$ is related to a ratio of coupling constants. It is however not measurable in any LEP process, even though it is operationally very close to the effective weak mixing angle definition.

- The effective weak mixing angle:

$$\sin^2 \theta_W^{\text{eff}} \equiv \frac{1}{4}\left(1 - \frac{g_{V_e}}{g_{A_e}}\right), \tag{35}$$

where the ratio g_{V_e}/g_{A_e} is extracted from forward-backward asymmetries. There is now an agreement to use this definition that absorbs the vertex corrections for leptons, but not for quarks. See Kennedy and Lynn 1986, Consoli and Hollik 1989, Lynn 1989 and Levinthal et al. 1991 for various avaters of the concept. The numerical values of $\sin^2 \theta_W^{\text{eff}}$ and $\sin^2_W \hat{\vartheta}_{\overline{\text{MS}}}(q^2)$ are numerically very close.

8.2 The left-right cross-section asymmetry

If longitudinal polarisation is available, measurements of A_{LR} and $A_{\text{FB}}^{\text{pol}(f)}$ for the various types of fermions allow a complete determination of the fermion couplings (Blondel et al. 1988). The left-right polarisation asymmetry A_{LR} is a rare example of an observable that reconciles nearly all advantages. A_{LR} is simply obtained by comparing the total cross-section for producing a Z from a left-handed (σ_L) or right-handed (σ_R) e^+e^- system. It is obvious (Figure 15) that A_{LR} at the Z pole will not depend on the way that the Z decays, but only on the electron neutral current couplings (Lynn et al. 1985/89). Detailed calculations show that this is affected only by minor and well understood corrections.

The statistical precision on $A_{\rm LR}$ and $\sin^2\theta_W^{\rm eff}$ from an exposure with alternating polarisation $\mathcal{P}_e$ collecting N Z decays is:

$$\Delta A_{\rm LR} = \frac{1}{\mathcal{P}_e\sqrt{N}}; \qquad \Delta\sin^2\theta_W^{\rm eff} = \frac{\Delta A_{\rm LR}}{7.9}.$$

Given the large cross-section at the top of the resonance a remarkable statistical precision can be envisaged: $\Delta A_{\rm LR} = 0.002$ (statistical) for 10^6 events and 50% polarisation. This corresponds to $\Delta\sin^2\theta_W^{\rm eff} = 0.0003$! The major technological challenge is to obtain largely polarised e^+e^- collisions at the Z pole. For this, we have to cross the Atlantic and go to the SLC at Stanford, California.

8.2.1 Polarised beams at SLC

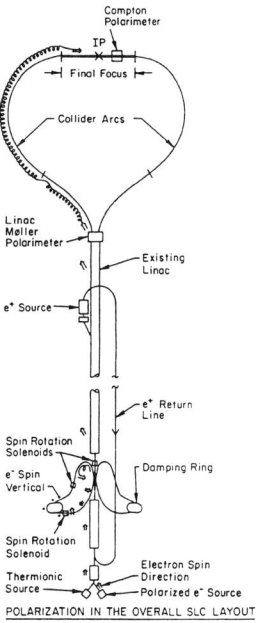

Figure 16.

The overall SLC layout with polarised beams is shown in Figure 16. A more complete description can be found in Blockus *et al.* 1986, Swartz 1988 and Phinney 1992. The orientation of the electron spin is shown along its transport from the electron gun, where polarisation is generated, to the interaction point.

Pulses of longitudinally polarised electrons are produced by photoemission from a polarised cathode (Schultz *et al.* 1992), at a rate of 120 Hz. The positrons are created by the interaction of a second ('scavenger') electron bunch on a radiator situated 2/3 down the linac, and cannot be polarised. In order to reduce the emittance, the beam is transported to damping rings. The polarisation is preserved by combination of spin precession in the transfer line and spin rotation in a superconducting solenoid, so that it is aligned along the damping ring magnetic field. A spin-rotation system based on bumps in the SLC arcs (Limberg et al. 1993) is used to tune the spin orientation so that, taking into account the precession in the arc, the polarisation is longitudinal at the interaction point.

The direction of the polarisation vector is monitored at the end of the linac by a Møller polarimeter, and after the interaction point by a Compton polarimeter (Fero et al. 1994). The performance of SLC has been improving regularly in the last two years, both in luminosity and polarisation: the 1992 analysis (Abe et al. 1993) was based on 10,224 hadronic Z decays, with 22% typical polarisation. In 1993 (Abe et al. 1994), 49,392 hadronic Z decays were used with up to 64% longitudinal polarisation.

8.2.2 The SLC polarised source

The first successful acceleration of polarised electrons to high energy was demonstrated at SLAC in 1974 using an atomic beam source (Cooper et al. 1975). In 1976, Pierce and Meier (1976) observed polarised photo-emission from negative affinity Gallium Arsenide. Most polarised electron sources that have been used in accelerators are based on this technique. Such a photocathode polarised electron source was successfully employed in the neutral current parity violation experiment (Prescott et al. 1978). The SLC polarised source is an improved version of it.

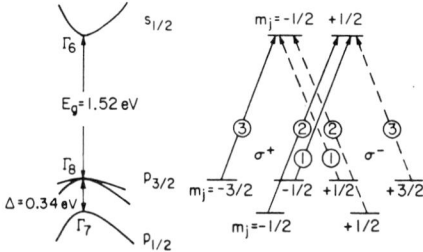

Figure 17. *Left: the band structure of GaAs near the gap minimum. Right: the corresponding energy levels of the $S_{\frac{1}{2}}$, $P_{\frac{1}{2}}$ and $P_{\frac{3}{2}}$ states. The allowed transitions for right-handed (left-handed) photons are marked by full (dashed) lines. The angular momentum state m_j indicates the projection on the incident photon direction. The relative transition rates are indicated by the circled numbers.*

Gallium Arsenide is a semi-conductor with the band structure sketched in Figure 17. The absorption of a photon with pumping of an electron from the valence band into the conduction band obeys the selection rule $\Delta m_j = \pm 1$ for right/left handed photons. If a right-handed photon has an energy that allows the transition ($P_{\frac{3}{2}}$ to $S_{\frac{1}{2}}$), but not ($P_{\frac{1}{2}}$ to $S_{\frac{1}{2}}$), which has an energy gap larger by 0.34 eV, the electron will end up in the $m_j = -\frac{1}{2}$ state with a probability of 3 and in the $m_j = \frac{1}{2}$ state with a probability of 1. m_j is the projection of the angular momentum on the incident photon direction. In the S state, the electron spin carries the total angular momentum, so that $m_j = -\frac{1}{2}$ corresponds to a spin opposite to the photon direction. In a practical set-up where the laser hits a GaAs photocathode at normal incidence, the electron will be accelerated opposite to the incident photon direction. It will therefore be right-handed, with a polarisation:

$$\mathcal{P}_e = \frac{3-1}{3+1} = 0.5. \tag{36}$$

In order to make this a source one must allow the electron to come out of the material. This is not normally possible with pure GaAs, and a coating is necessary to reduce the free electron energy to below the conduction band ('negative affinity'). The best coatings are Cs_2O, and CsF. Keeping good quantum efficiency (1–5%) requires regular cessation (Schultz et al. 1992).

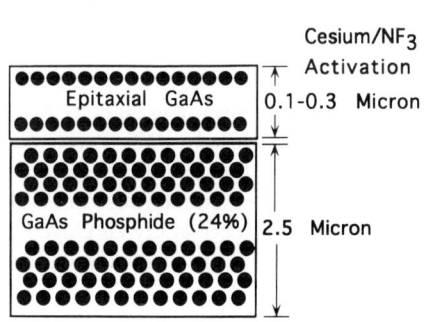

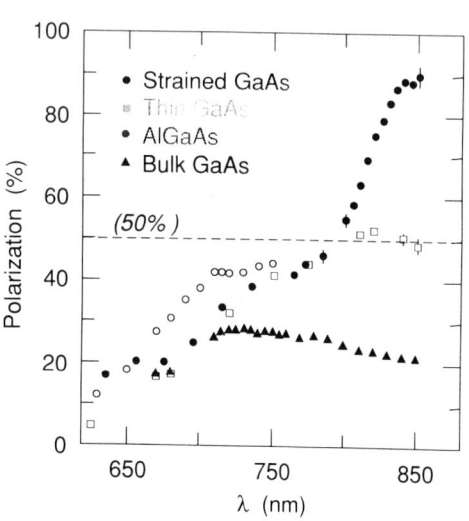

Figure 18. *Sketch of a strained GaAs photocathode. The laser beam hit the upper part, coated with Cesium/NF_3. An epitaxial layer of GaAs is grown on a support of $GaAs_{1-x}P_x$ ($x = 0.24$). The slight mismatch in the lattice structure results in a mechanical strain.*

Figure 19. *Measured dependence of the polarisation at the output of strained GaAs photocathodes on the incident laser wavelength.*

It is possible to improve the polarisation of the electron beam from GaAs if one is able to split the energies of the $P_{\frac{3}{2},\frac{1}{2}}$ and $P_{\frac{3}{2},\frac{3}{2}}$ states so that the $P_{\frac{3}{2},\frac{3}{2}}$ state has higher energy—e.g. the gap is smaller. Then, provided the laser is tuned to a low enough energy, the $m = \frac{3}{2}$ state can be selected and the polarisation can reach a theoretical maximum of 100%. This breakthrough was achieved (Maruyama et al. 1991/92, Nakanishi et al. 1991 and Aoyagi et al. 1992) with strained GaAs photocathodes.

It is a well known fact (in solid state physics) that the band structure of solids is altered by mechanical strain. The band splitting can actually go one way or the other depending on whether the material is under compression or tension. Strain photocathodes with high electron polarisation were first obtained in a thin epitaxial layer of $In_{1-y}Ga_yAs$ grown on a GaAs substrate, and later from epitaxial GaAs grown on a thick GaAsP buffer layer. The strain is due to a small lattice mismatch of the epitaxial layer with respect to the substrate. The strain can be varied by changing the concentration y or the thickness of the layer. Thinner layers lead to optimal polarisation at a lower wavelength. High degrees of electron polarisation have been obtained with an epitaxial layer of 0.15 microns of GaAs on a 2.5 microns layer of $GaAs_{1-x}P_x$ ($x = 0.24$) shown in Figure 18. The wavelength dependence of the polarisation is shown in Figure 19.

Obtaining high intensity beams at 120 Hz with this technique is extremely demanding on the laser. In the absence of commercial lasers meeting the specifications, a new Ti:Sapphire, optically pumped by two Nd:YAG lasers, was specially developed (Frisch et al.). It is currently run at high power at 865 nm wavelength. Both the strained photocathodes and the laser came into operation in 1993, yielding usable polarisation of up to 70%.

The sign of the polarisation of the beam is controlled by the helicity of the laser. This is controlled by Pockels cells and can be reversed on a pulse basis. The SLC polarised source was able, in 1993, to deliver 7×10^{10} electrons per bunch at a repetition rate of 120 Hz with a polarisation of 64%.

8.2.3 Measurement of A_{LR} with SLD

The measurement of A_{LR} is performed with the SLD detector (SLD 1984). The 1993 measurement is based on 49,392 Z decays recorded with an average beam polarisation of $(63.0 \pm 1.1)\%$. The centre-of-mass energy was 91.26 ± 0.02 GeV, measured with magnetic beam spectrometers.

The left-right asymmetry is measured as:

$$A_{LR} = \frac{1}{|\mathcal{P}_e|} \frac{N_L - N_R}{N_L + N_R}, \qquad (37)$$

where $\mathcal{P}_e$ is the average beam polarisation of the e^- beam, and N_L, N_R are the numbers of Z decays selected for left or right beam polarisation.

The helicity of the beam was varied randomly at the source, and recorded by SLD with each event, and by the polarimeter. Hadronic Z decays, with a small content of τ pairs (<1%) were selected with a calorimeter-based selection. The beam related backgrounds, as well as the physics backgrounds, 'two-photon' interactions and e^+e^- final states, were efficiently reduced to a small level. Backgrounds dilute the asymmetry and must be corrected for. Because of its large t-channel content, the e^+e^- final state has a different asymmetry than the other Z decays.

The helicity reversal can potentially change some of the beam parameters, namely the energy and, more critically, the luminosity and polarisation. The energy difference leads to negligible effects. The polarisation asymmetry was directly measured by the polarimeter. The luminosity asymmetry was inferred from a detailed study of possible asymmetries in the relevant beam parameters, intensity, beam position and beam size, to be $(3.8 \pm 5.0) \times 10^{-5}$. This was cross-checked with the SLD luminosity detectors with larger statistical errors. The value of A_{LR} is inferred from the measured asymmetry $A_m = (N_L - N_R)/(N_L + N_R) = (10.24 \pm 0.45)\%$ by the formula

$$A_{LR} = \frac{A_m}{\mathcal{P}_e} + \frac{1}{\mathcal{P}_e}\left[A_m f_b + A_m^2 A_{\mathcal{P}_e} - \frac{E}{\sigma}\frac{d\sigma}{dE}A_E - A_\epsilon - A_\mathcal{L}\right], \qquad (38)$$

where f_b is the background fraction, $A_{\mathcal{P}_e}$ is the asymmetry in the beam polarisation, and A_E, A_ϵ, $A_\mathcal{L}$ are the asymmetries in the beam energy, detector acceptance and luminosity. To obtain the pole asymmetry or $\mathcal{A}_e$ small corrections have to be applied for photon exchange and interference. The result for the 1993 data is:

$$\mathcal{A}_e = A_{\text{LR}}{}^0 = 0.1656 \pm 0.0071_{\text{stat.}} \pm 0.0028_{\text{syst.}}. \tag{39}$$

The systematic error is a quadratic sum of uncertainties stemming predominantly from the beam polarisation measurement itself (1.3%) and from the extrapolation to the beam-beam interaction point of the measured polarisation (1.1%). These errors are given in relative fraction of the measured A_{LR}.

The result, averaged with the 1992 measurement, yields the following value of $\sin^2 \theta_W^{\text{eff}}$:

$$\sin^2 \theta_W^{\text{eff}} = 0.2294 \pm 0.0009_{\text{stat.}} \pm 0.0004_{\text{syst.}}. \tag{40}$$

From the smallness of (i) the corrections, and (ii) the systematic error on the result, one can deduce that the measurement of $\sin^2 \theta_W^{\text{eff}}$ from A_{LR} should soon be considerably improved. Eventually, it is hoped that SLD will record 750,000 Z decays, with a polarisation of 75%, yielding

$$\Delta \sin^2 \theta_W^{\text{eff}} = \pm 0.00025_{\text{stat.}} \pm 0.00012_{\text{syst.}}. \tag{41}$$

8.3 Forward-backward asymmetries for leptons

We now come back to LEP, where longitudinally polarised beams are not (yet) available. Statistics are abundant, and $\sin^2 \theta_W^{\text{eff}}$ is measured from several different asymmetries. The simplest experimentally is the lepton forward-backward asymmetry. The angular distribution for a final state fermion is given by:

$$\frac{d\sigma}{d\cos\theta} = \frac{3}{8}\sigma^{f\bar{f}}\left(1 + \cos^2\theta + \frac{8}{3}A_{\text{FB}}^{(f)}\cos\theta\right) \tag{42}$$

where θ is the scattering angle, defined as the angle between the outgoing fermion f, and the incoming fermion e^-. The resulting forward-backward asymmetry $A_{\text{FB}}^{(f)}$ is a function of centre-of-mass energy: in the vicinity of the pole,

$$A_{\text{FB}}^{(f)}(s) \simeq \frac{3}{4}\mathcal{A}_e\mathcal{A}_f + \frac{s - M_Z^2}{M_Z^2}\frac{3}{2}Q_fQ_e \cdot 4\sin^2\theta_W^{\text{eff}}\cos^2\theta_W^{\text{eff}} \frac{g_{Ae}g_{Af}}{(g_{Ae}^2 + g_{Ve}^2)(g_{Af}^2 + g_{Vf}^2)}. \tag{43}$$

The first term represents the Z exchange, the second the Z-γ interference. For leptons, given the small value of $\mathcal{A}_e = 0.14$, and the larger value of Q_f, the Z-γ interference dominates everywhere but on the pole, where it vanishes. The sensitivity to $\sin^2 \theta_W^{\text{eff}}$ is contained in the Z exchange term. However, from the steep slope of $A_{\text{FB}}^{(\ell)}$ with $\sqrt{s}$, 8% per GeV, results some sensitivity to initial state radiation and centre-of-mass uncertainties.

The initial state radiation effect is treated with great detail in Böhm and Hollik 1989, and implemented in fitting formulae, such as MIZA (Martinez 1991), ZFITTER (Akhundov et al. 1986) and EXPOSTAR (Kennedy et al. 1989). It is believed that the peak asymmetry can be predicted with an accuracy of 0.0008 (Böhm and Hollik 1989).

The uncertainty in the absolute energy scale of LEP does not affect the asymmetry: it is only sensitive to the distance of the scan points from the Z peak, which is determined from the same points. However, a 10 MeV point-to-point error would result in an uncertainty of 0.0008 in $A_{FB}^{(\ell)}$, fully correlated for the three lepton types. Because of the interference with the t-channel, the dependence of the e^+e^- asymmetry on beam energy is of opposite sign to the other two, so the effect is considerably smaller in the average of the three lepton types.

In practice, the experiments extract the forward-backward asymmetry by performing a fit to the angular distribution of lepton pair events, according to Equation 42. The resulting value of $A_{FB}^{(\ell)}$ is shown in Figure 20 for OPAL. Systematic errors from selection cuts and detector imperfections are for the time being much smaller than the experimental statistical error—see Table 3.

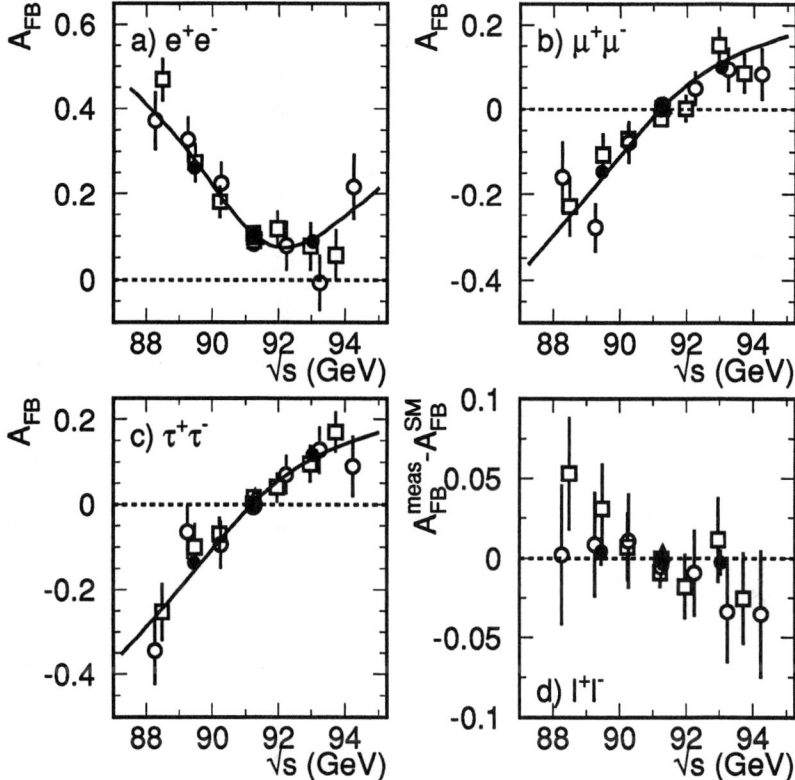

Figure 20. *The OPAL fitted forward-backward asymmetries for electron, muon, tau and inclusive lepton final states. The lines are the results of a global fit to the line shape and lepton asymmetries.*

The main information extracted from the forward backward asymmetry for leptons is the QED-corrected Z pole asymmetry, using Equation 29. The QED corrections include (i) the effect of initial and final state radiation and their interference, (ii) the photon exchange and Z-γ interference terms, including the substantial t-channel contribution for e^+e^- final state. The Z pole asymmetry is so small that small effects such as that of the imaginary part of the photon propagator are important. The summary of results from the four LEP experiments is given in Table 8.

Exp.	ALEPH	DELPHI	L3	OPAL
$A_{FB}^{(e)}$	0.0212 ± 0.0054	0.0247 ± 0.0088	0.0104 ± 0.0092	0.0075 ± 0.0063
$A_{FB}^{(\mu)}$	0.0189 ± 0.0039	0.0128 ± 0.0037	0.0179 ± 0.0061	0.0110 ± 0.0036
$A_{FB}^{(\tau)}$	0.0253 ± 0.0043	0.0215 ± 0.0069	0.0265 ± 0.0088	0.0180 ± 0.0044
$A_{FB}^{(\ell)}$	0.0216 ± 0.0026	0.0159 ± 0.0031	0.0184 ± 0.0045	0.0127 ± 0.0026

Exp.	average
$A_{FB}^{(e)}$	0.0158 ± 0.0035
$A_{FB}^{(\mu)}$	0.0144 ± 0.0021
$A_{FB}^{(\tau)}$	0.0221 ± 0.0027
$A_{FB}^{(\ell)}$	0.0170 ± 0.0016

Table 8. *Results of the forward-backward asymmetry fits for the four LEP experiments. First the results for each of the leptons separately are shown. The χ^2 for the three lepton species to agree is 5.2 for 2 d.o.f. (7% probability). The bottom line shows the result assuming lepton universality.*

From the average value of $A_{FB}^{(\ell)}$ one can derive a value of the effective weak mixing angle:

$$\sin^2 \theta_W^{\text{eff}} = 0.23107 \pm 0.00092. \qquad (44)$$

8.4 τ polarisation

In the case of the τ lepton, the charged decay provides us with a final state polarisation analyser. The ALEPH (Decamp et al. 1991, Buskulic et al. 1993), DELPHI (Abreu et al. 1992), OPAL (Alexander et al. 1991) and L3 (Adeva et al. 1989) collaborations have presented results for the following five decay channels: $\tau \to \pi \nu_\tau$ (B.R. 12%), $\tau \to e \nu_e \nu_\tau$ (B.R. 18%) $\tau \to \mu \nu_\mu \nu_\tau$ (B.R. 18%), $\tau \to \rho \nu_\tau \to \pi^- \pi^0 \nu_\tau$ (B.R. 24%), $\tau \to a_1 \nu_\tau \to \pi^- \pi^+ \pi^0 \nu_\tau$ (B.R. 8%). The analyses do not distinguish here the nature of the charged hadron, channels with kaons are included as well, but have very similar spin properties. Updates of the results for the 1992 data have been given at conferences (Schwartz 1993, Kounine 1994).

Let us consider as an example the decay $\tau \to \pi \nu_\tau$. In this case the only particle that carries the helicity of the τ in the final state is the left-handed neutrino. What happens in the τ rest frame is sketched in Figure 21. The resulting distribution of the decay angle θ of the pion in the τ rest frame relative to the τ line of flight is given by Tsai (1971):

$$W(\cos\theta) = \frac{1}{2}(1 + \alpha \mathcal{P}_\tau \cos\theta), \qquad (45)$$

where $\alpha = 1$. In the laboratory frame, the fractional energy of the pion is $x_\pi = P_\pi / P_{\text{beam}} \simeq \frac{1}{2}(\cos\theta + 1)$. The resulting x_π distribution is $W^+(x) \simeq x$ for $\mathcal{P}_\tau = 1$ and $W^-(x) \simeq 1 - x$ for $\mathcal{P}_\tau = -1$. Thus, the measurement of the τ polarisation is obtained by fitting the momentum spectrum of the pion to a linear combination of W^+ and W^-, with the τ polarisation as fitting parameter.

τ helicity	Decay products in τ rest frame			probability
τ_L^- $\mathcal{P}_\tau = -1$	Forward:	$\nu_\tau L$ $\Rightarrow$ $\leftarrow$	π^- $\rightarrow$	0
	Backward:	π^- $\leftarrow$	$\nu_\tau L$ $\Leftarrow$ $\rightarrow$	1
τ_R^- $\mathcal{P}_\tau = +1$	Forward:	$\nu_\tau L$ $\Rightarrow$ $\leftarrow$	π^- $\rightarrow$	1
	Backward:	π^- $\leftarrow$	$\nu_\tau L$ $\Leftarrow$ $\rightarrow$	0

Figure 21. *The $\tau \to \pi \nu_\tau$ decay as a τ polarisation analyser. The arrows $\Leftarrow$ and $\Rightarrow$ indicate the helicity of the particles.*

For the lepton decays, the resulting momentum spectrum is given in Goggi (1979):

$$W_\ell(x) = \frac{1}{3}\left[4x^3 - 9x^2 + 5 + \mathcal{P}_\tau(8x^3 - 9x^2 + 1)\right]. \tag{46}$$

For ρ and a_1 decays, the decay angle distribution is similar to that of the τ, Equation 45, with a reduced value of α. The information can be retrieved in this case by a full analysis of the ρ or a_1 decay products, as shown by Rougé (1990), and developed in Davier (1992). For the $\tau \to \rho \nu_\tau$ decay, the τ helicity affects the distributions of both τ and ρ decay angles, in a way that depends on the $\pi^- \pi^0$ mass. This set of observables, formally noted $\{\xi\}$, defines the final state. The probability density functions for the ± 1 helicity, $W^\pm(\{\xi\})$, are used to fit the τ polarisation. For the a_1, the decay is defined by six variables. Full use of the density function in this set makes the a_1 channel more sensitive than the leptonic one.

The experimental challenge is the discrimination of the various τ decay modes. Given the high τ momentum, the τ decay products are very collimated. Particle identification and fine granularity both in the tracking detector and the electromagnetic calorimeter play an essential role. Furthermore, since the measurement is based on the analysis of particle spectra, the understanding of the energy dependence of (i) event selection, (ii) particle identification, (iii) backgrounds, set the limits to the experimental accuracy. The typical efficiency of event selection is between 40–70%, with cross-channel contaminations between 3 and 10%. The best sensitivity comes from the π and ρ decay modes. Figure 22 shows the analysis of the π channel in L3. By analysing the polarisation as a function of polar angle one can derive both the average τ polarisation, $\mathcal{P}_\tau$ and the forward-backward polarisation asymmetry $A_{FB}^{pol(\tau)}$, as shown in Figure 23.

The results in the individual channels from the LEP experiments are summarised

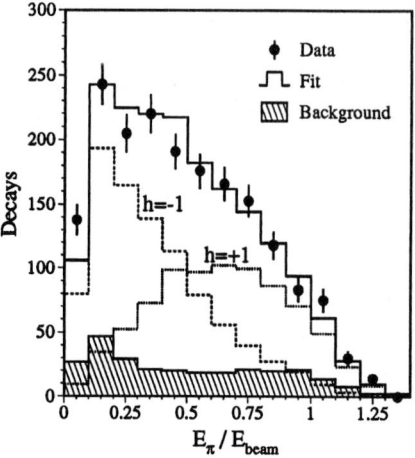 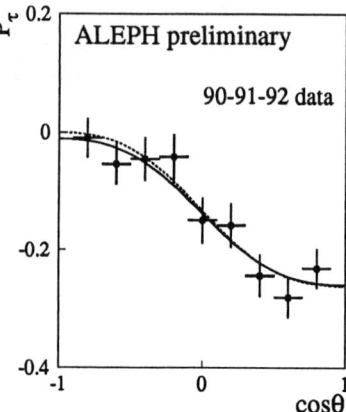

Figure 22. *Momentum distribution of the pion in $\tau \to \pi \nu_\tau$ decays. The full line is the result of the fit, which is a linear sum of the components due to positive (dotted line) or negative (dashed line) τ helicities. From L3 (Adeva et al. 1989).*

Figure 23. *Angular dependence of the τ polarisation in ALEPH. The lines show the result of a fit to Equation 30 both with and without the assumption of lepton universality.*

in Table 9. The measurements are compatible with each other. This table shows the relative sensitivity in the various channels. It can be seen also that experimental systematic errors are now almost as large as the statistical ones. Improvements will require ingenuity!

Channel	ALEPH	DELPHI	OPAL	L3
	$\mathcal{P}_\tau \times 10^3$			
$e\nu_e\nu_\tau$	$-225 \pm 85 \pm 45$	$-92 \pm 115 \pm 80$	$-142 \pm 52 \pm 52$	$+53 \pm 73 \pm 61$
$\mu\nu_\mu\nu_\tau$	$-154 \pm 65 \pm 29$	$-62 \pm 63 \pm 61$	$-107 \pm 58 \pm 45$	$-178 \pm 78 \pm 61$
$\pi\nu_\tau$	$-133 \pm 31 \pm 18$	$-212 \pm 38 \pm 40$	$-150 \pm 28 \pm 31$	$-93 \pm 57 \pm 52$
$\rho\nu_\tau$	$-150 \pm 31 \pm 30$	$-128 \pm 28 \pm 31$	$-123 \pm 19 \pm 25$	$-180 \pm 60 \pm 40$
$a_1\nu_\tau$	$-114 \pm 63 \pm 34$	$-181 \pm 66 \pm 60$	N.A.	$+105 \pm 164 \pm 93$
	$\mathcal{A}_\tau$			
all	0.137 ± 0.014	0.154 ± 0.023	0.174 ± 0.021	0.150 ± 0.050
	$\mathcal{A}_e$			
all	0.127 ± 0.014	0.081 ± 0.023	0.130 ± 0.025	0.117 ± 0.033

Table 9. *Results of the τ polarisation analyses for individual channels from the LEP experiments from Schwartz 1993. Values of $\mathcal{A}_\tau$ and $\mathcal{A}_e$ from Kounine 1994.*

It should be mentioned that the extraction of $\mathcal{P}_\tau$ from the τ decay products assumes that the τ decays through the same V−A maximally parity-violating charged current that governs muon decay. The τ neutrino is *left-handed*. The errors in Table 9 do not allow for possible violation of this assumption. It is possible, however, to place constraints on the τ neutrino helicity ξ by studying the correlation between the helicities of the two τ's in an event. The ARGUS (ARGUS 1993) (at DESY) and ALEPH (1994b) collaborations have performed such analyses, yielding:

$$\xi \nu_\tau = -1.25 \pm 0.23 \pm^{0.08}_{0.15} \text{ (ARGUS)} \quad \text{and} \quad \xi \nu_\tau = -0.99 \pm 0.07 \pm 0.04 \text{ (ALEPH)}.$$

This confirms beautifully that the τ family has the same multiplet structure as the electron and muon. If one uses this empirical value for the ν_τ helicity, an common error of $\Delta \mathcal{P}_\tau = 0.015$ has to be added to the results of Table 9.

The polarisation results can be expressed as a measurement of the τ and electron couplings (Kounine 1994):

$$\mathcal{A}_\tau = 0.150 \pm 0.010, \quad \mathcal{A}_e = 0.120 \pm 0.012. \tag{47}$$

The values of $\mathcal{A}_\tau$ and $\mathcal{A}_e$ are essentially uncorrelated. They are in moderately good agreement with lepton universality, with a χ^2 of 3.6 for 1 d.o.f. (5% probability). If one combines them a measurement of $\sin^2 \theta_W^{\text{eff}}$ is obtained:

$$\sin^2 \theta_W^{\text{eff}} = 0.23270 \pm 0.00097. \tag{48}$$

8.5 Quark asymmetries

In principle the quark asymmetries $A_{\text{FB}}^{(q)}$ offer better sensitivity to the measurement of couplings and $\sin^2 \theta_W^{\text{eff}}$ than the leptonic forward-backward asymmetries, as well as better event statistics. However, it is difficult to tag specific quark final states and to measure their charge. These difficulties limit the potential of quark asymmetries.

A first possibility, carried out by ALEPH (Decamp et al. 1991b, Blondel et al. 1993b), DELPHI (Abreu et al. 1991) and OPAL (Acton et al. 1992), is to measure the inclusive hadronic charge asymmetry. The method is based on the premise, first suggested by Field and Feynman (1978) that the original quark charge is carried out by the resulting jet of particles. This property has been since then verified in several reactions where the original quark flavour is known, in particular (anti)neutrino scattering (Berge et al. 1981).

The difficulty in a hadronic sample is to estimate for each event whether the final state quark was emitted forward or backward. The method used by ALEPH and DELPHI is described in the following. OPAL has used a slightly different one, based on the three highest momentum particles, with somewhat better statistical sensitivity. A charge estimate is constructed as follows: the event is separated in two hemispheres, according to the thrust axis. Then, in each hemisphere, the momentum weighted charge is constructed:

$$Q_{\text{F,B}} = \frac{\sum_{\text{F,B}} p_{\| i}^\kappa q_i}{\sum_{\text{F,B}} p_{\| i}^\kappa}, \tag{49}$$

where q_i is the charge of particle i, $p_{\| i}$ its momentum projected on the thrust axis, and κ is a parameter that is varied to optimise the method. For LEP data the maximum sensitivity is found for $\kappa = 1$. The sum runs for charged particles in the forward hemisphere, for Q_F, or in the backward hemisphere, for Q_B. A good estimate of charge for each event is found to be $Q_{\text{FB}} = Q_\text{F} - Q_\text{B}$.

The charge cannot be recognised event by event, but only in a statistical way. The measured charge distribution has a R.M.S. spread of typically 0.6, and, for a sample of u quarks all in the forward direction, an average value of $\delta_u = 0.42$, which is called *charge separation*. This large dilution down from the parton level value of 4/3, is due, mainly, to the hadronisation process. The value of the average charge separations for the other quarks are shown in Table 10. The good agreement of the distribution of Q_{FB} between data and Monte-Carlo gives confidence in the validity of the modeling of the distribution of the initial quark charge to the final state particles.

	parton level	ALEPH	DELPHI
δ_u	4/3	0.42	0.31
δ_d	−2/3	−0.21	−0.20
δ_s	−2/3	−0.29	−0.26
δ_c	4/3	0.17	0.14
δ_b	−2/3	−0.22	−0.19
$\langle Q_{FB} \rangle$	−0.0127	−0.0084 ± 0.0015	−0.0076 ± 0.0013
$\sin^2 \theta_W^{\text{eff}}$	(0.23)	$0.2308 \pm 0.0035_{\text{exp.}} \pm 0.0038_{\text{frag.}}$	$0.2345 \pm 0.0030_{\text{exp.}} \pm 0.0027_{\text{frag.}}$

Table 10. *Summary of published $\langle Q_{FB} \rangle$ measurements. The first five rows show the charge separations (δ_f) for different quark flavours as expected at parton level and as they appear in the measurement, after hadronisation and detector effects, according to the published ALEPH and DELPHI simulations. The last two rows show the average charge asymmetry for the inclusive sample of quarks and the value of $\sin^2 \theta_W^{\text{eff}}$ that is derived. The parton level value corresponds to $\sin^2 \theta_W^{\text{eff}} = 0.23$.*

A significant average charge asymmetry, $\langle Q_{FB} \rangle$, is observed for the inclusive hadronic event sample. The expected charge asymmetry is given by:

$$\langle Q_{FB} \rangle = \sum_{\text{quark flavours}} \delta_f A_{FB}^{(f)} \frac{\Gamma_f}{\Gamma_{\text{had}}}. \quad (50)$$

The forward backward asymmetries (see Equation 29) are all positive, but the signs of the charge separations are different. This results in a large cancellation, already at the parton level.

The experimental uncertainties are small compared to the statistical error, but the interpretation of $\langle Q_{FB} \rangle$ in terms of $\sin^2 \theta_W^{\text{eff}}$ is affected by the uncertainty in the calculation of the charge separations. This uncertainty is estimated by varying the parameters of the hadronisation models, and by comparing various models. The resulting error in $\sin^2 \theta_W^{\text{eff}}$ is proportional to the value of $\langle Q_{FB} \rangle$ itself. The experiments find different values for the charge separations, which can be traced to a different choice of input parameters in the simulation, and of the decay tables for heavy flavoured particles. The most critical parameters in the simulation of light quark charges, are those controlling pair production of strange particles and baryon pairs. It is hoped that the improved understanding of particle composition and correlations in jets will help in reducing these errors. The decay tables of heavy particles will remain incompletely known; the solution is probably to measure directly the heavy quark jet charges, by means of tagged events.

ALEPH has presented a preliminary analysis (Blondel et al. 1993b) where the fragmentation systematic error is substantially reduced by using a direct measurement of the b jet charge, based on lepton and lifetime tagged b samples, as well as a constraint from the average charge separation between the two hemispheres. This constraint is obtained by measuring the quantity $\langle Q_F \cdot Q_B \rangle$. It can easily be shown that, for a sample consisting only of one type of quarks f, and up to small correlation terms,

$$\langle Q_F \cdot Q_B \rangle \simeq -\frac{(\delta_f)^2}{4}. \qquad (51)$$

This method can be used for a selected b sample to measure δ_b. In the inclusive sample, $\langle Q_F \cdot Q_B \rangle$ measures a weighted sum of the squares of the charge separations, but still constrains usefully some of the fragmentation parameters (unfortunately, not strange particles and baryon pair production!).

The results are expressed in terms of $\sin^2 \theta_W^{\text{eff}}$. It has been shown that QCD corrections (the smearing of the original quark direction due to gluon emission) are automatically taken into account by the method, and no further correction should be necessary. The present status is given in Table 11.

EXPT.	STATUS	s/u	EW Prog.	QCD	cuts
ALEPH	89-90 publ. 89-92 prelim.	0.315 ± 0.045	EXPOSTAR	NO	NO
DELPHI	90-91 publ.	0.315 ± 0.045	ZFITTER	YES	NO
OPAL	90-91 publ.	0.285 ± 0.050	ZFITTER	YES?	S<0.12

EXPT.	$\sin^2 \theta_W^{\text{eff}}$
ALEPH	$0.2317 \pm 0.0013 \pm 0.0011$
DELPHI	$0.2345 \pm 0.0030 \pm 0.0027$
OPAL	$0.2321 \pm 0.0017 \pm 0.0028$
Average	$0.2320 \pm 0.0011 \pm 0.0011$

Table 11. *Summary of the determination of $\sin^2 \theta_W^{\text{eff}}$ from inclusive hadronic charge asymmetries at LEP.*

The present average of $\sin^2 \theta_W^{\text{eff}}$ from inclusive hadronic charge asymmetries is:

$$\sin^2 \theta_W^{\text{eff}} = 0.2320 \pm 0.0016. \qquad (52)$$

8.6 Heavy quark asymmetries

The asymmetry can also be measured for individual quark species if one is able to:

- Tag the specific quark flavour. This can be done for b and c quarks by means of their semi-leptonic decays, which produce prompt leptons. High $P_\perp$ leptons tag b quarks, low $P_\perp$ leptons are enriched in c quarks. Another possibility for b's is the life-time tag and for c the recognition of a high momentum D meson.

- Measure the scattering angle. The thrust axis of the event is usually used as a measure of the original quark-antiquark direction before fragmentation.

- Assign an orientation to the event axis defined above. In the selection with leptons or with D mesons, the sign of the lepton or the flavour of the meson can be used. In case the event is recognised by life-time tag, one can use the jet charge (as in the inclusive sample) to measure in a statistical way the asymmetry.

All LEP collaborations have performed an analysis using lepton tagging (ALEPH 1994, Abreu et al. 1992b, OPAL 1991/93, Adeva et al. 1990b). There the prompt lepton sample is analysed in a global way to extract simultaneously way the b and c partial width, the direct and cascade semi-leptonic branching fractions, the inclusive B^0-$\overline{B^0}$ mixing and the b and c asymmetries. The asymmetry has to be corrected for QCD radiation (Altarelli and Lampe 1993) (3% correction) and for the experimental effects associated with the event axis determination and its orientation with a lepton (a few%).

ALEPH (1993c), DELPHI and OPAL (Clarke et al. 1994) have presented the asymmetry measured from the jet-charge in the lifetime-tagged lepton sample. The weakness of the inclusive jet charge asymmetry described above, namely the dependence upon fragmentation parameters, is avoided by determining the b charge from the data using the charge correlation between opposite hemispheres (Equation 51). As mentioned before, no QCD correction is necessary with this method.

The charm asymmetry is also measured with D mesons by ALEPH (Buskulic et al. 1993b), DELPHI (Abreu et al. 1993), OPAL (Ackers et al. 1993).

The averaging of these results is a very delicate enterprise. A preliminary procedure is attempted by the LEP electroweak working group (LEP 1993). The results are corrected for photonic effects to obtain the pole asymmetries:

$$A_{\text{FB}}^{(b)\,0} = 0.0961 \pm 0.0038_{\text{stat.}} \pm 0.0021_{\text{syst.}}$$
$$A_{\text{FB}}^{(c)\,0} = 0.0703 \pm 0.0080_{\text{stat.}} \pm 0.0072_{\text{syst.}}. \quad (53)$$

The correlation between these numbers is 0.166. They can be expressed as measurements of $\sin^2 \theta_W^{\text{eff}}$:

$$\sin^2 \theta_W^{\text{eff}} = 0.23277 \pm 0.00077 \text{ (from } b\text{)}$$
$$\sin^2 \theta_W^{\text{eff}} = 0.2324 \pm 0.0026 \text{ (from } c\text{)}$$
$$\sin^2 \theta_W^{\text{eff}} = 0.2327 \pm 0.0007 \text{ (combined)}. \quad (54)$$

The b forward-backward polarised asymmetry, defined in Equation 28, can be measured if the beams are polarised. This has recently been done at SLC (Junk 1994), both with lepton tag and the combined vertex-tag/jet charge method. For 5×10^4 Z decays with 63% polarisation, one finds:

$$A_{\text{FB}}^{\text{pol}(b)} = \mathcal{A}_b = 0.99 \pm 0.14. \quad (55)$$

It can be argued that from LEP data, combining the measurement of $A_{\text{FB}}^{(b)\,0} = \frac{3}{4}\mathcal{A}_e\mathcal{A}_b$ and the value $\mathcal{A}_e = 0.143 \pm 0.007$ obtained from purely leptonic asymmetries and τ polarisation, one can derive $\mathcal{A}_b = 0.93 \pm 0.07$. The two values are consistent with each other and with the Standard Model prediction of $\mathcal{A}_b = 0.94$.

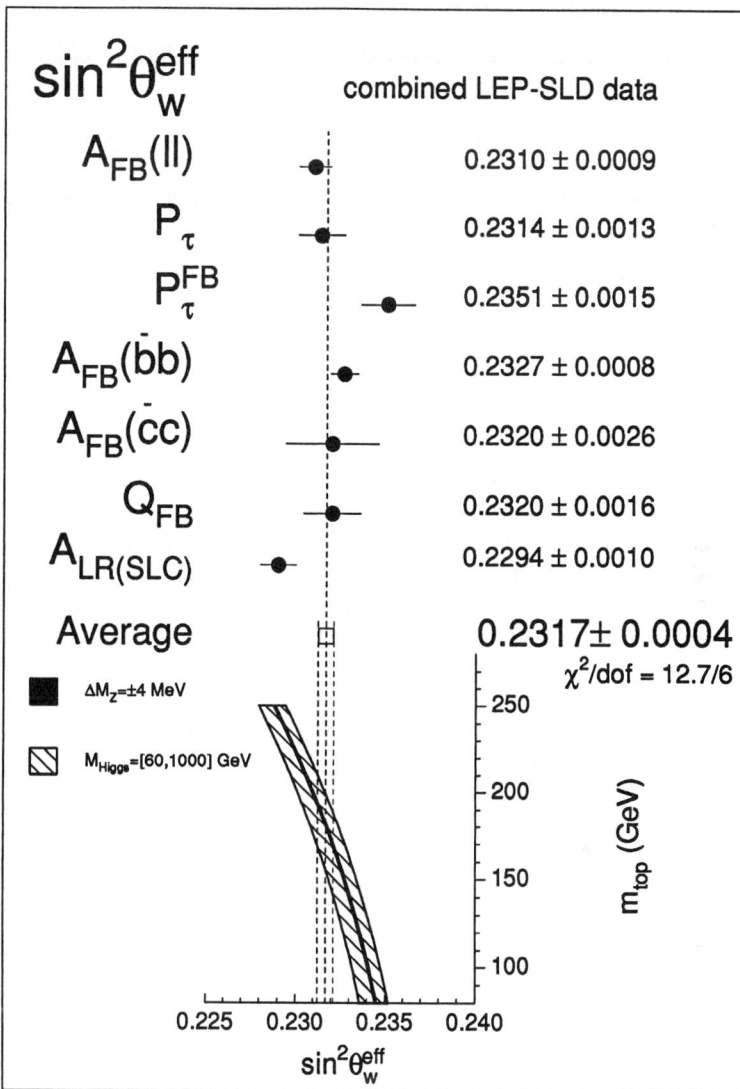

Figure 24. *Summary of measurements of* $\sin^2\theta_W^{\text{eff}}$ *from the forward-backward asymmetries of leptons, τ polarisation, inclusive quarks, heavy quarks asymmetry and the SLC polarisation asymmetry. Also shown is the prediction of the minimal Standard Model as a function of M_t.*

8.7 Summary on $\sin^2\theta_W^{\text{eff}}$

The different values of $\sin^2\theta_W^{\text{eff}}$ measured from asymmetries and τ polarisation at LEP are summarised on Figure 24. Even though it is also possible to extract an equivalent value of $\sin^2\theta_W^{\text{eff}}$ from other measurements in the framework of the minimal $SU(2)_L \times U(1)$ model, and in particular from Γ_e or M_W, asymmetries constitute a direct way of obtaining a value that corresponds exactly to the definition of Equation 35. The measurements presented before average to:

$$\sin^2 \theta_W^{\text{eff}} = 0.23168 \pm 0.00043. \tag{56}$$

This number will play an important role in the determination of electroweak radiative effects.

[Editors' Note: At the time of going to press, the anticipated final sections, on 'top quark prediction' and 'outlook', together with some diagrams, were not available.]

9 Acknowledgements

It is a pleasure to thank my colleagues of ALEPH and of the Polarisation Collaboration for enjoyable company and support during these measurements. The LEP Electroweak working group has performed the averaging of the experimental results. I wish to congratulate the SUSSP organisers for a school which was profitable to everyone, me in particular. I also want to thank the students for lively questions and corrections, in particular on the (non-existent) triple Z vertex. Last, I would like to thank Lance Vick for gentle reminders for these lecture notes.

References

Aarnio P et al. (DELPHI Collaboration)), 1989, *Phys Lett* **B231** 539.
Aarnio P et al. (DELPHI Collaboration), 1991, *Nucl Inst Meth* **A303** 233.
Aarnio P et al. (DELPHI Collaboration)), 1991, *Nucl Phys* **B367** 511; 1994, DELPHI Collaboration, CERN/PPE 94-08, submitted to *Nucl Phys* B .
Abachi S et al. (D0 Collaboration), 1994, FERMILAB-PUB-94-004-E.
Abe F et al. 1989, (CDF Collaboration), *Phys Rev Lett* **63** 720.
Abe F et al. 1990, *Phys Rev Lett* **65** 2243.
Abe F et al. 1991, (CDF Collaboration) *Phys Rev* **D43** 664.
Abe K et al. (SLD Collaboration), 1993, *Phys Rev Lett* **70** 2515.
Abe K et al. 1994, (SLD Collaboration), SLAC-Pub-6456 to appear in *Phys Rev Lett*.
Abramowicz H et al. (CDHS Collaboration), 1986 *Phys Rev Lett* **57** 298.
Abrams G S et al. (MarkII Collaboration)), 1989, *Phys Rev Lett* **63** 724, **63** 2173.
Abreu P et al. 1991, (DELPHI Collaboration), CERN-PPE/91-21, to appear in *Phys Lett B*.
Abreu P et al. 1992, *Z Phys* **C55** 555.
Abreu P et al. 1992b, (DELPHI Collaboration), *Phys Lett* **B276** 536.
Abreu P et al. 1993, (DELPHI Collaboration), *Z Phys* **C59** 533.
Ackers R et al. (OPAL Collaboration), 1993, *Z Phys* **C60** 601.
Acton P D et al. 1992, (OPAL Collaboration), *Phys Lett* **294B** 436.
Adeva B et al. (L3 Collaboration)), 1989, *Phys Lett* **B23189**) 509.
Adeva B et al. (L3 Collaboration), 1990, *Nucl Inst Meth* **A289** 35.
Adeva B et al. (L3 Collaboration), 1990b, *Phys Lett* **B252** 713.
Adeva B et al. (L3 Collaboration)), 1991, *Z Phys* **C51** 179.

Ahmet M et al. (OPAL Collaboration), 1991, Nucl Inst Meth **A305** 275.
Ahmet M et al. (OPAL Collaboration)), 1991, Z Phys **C52** 175; 1993, Z Phys **C58** 219.
Akers R et al. (OPAL Collaboration), 1994, Z Phys **C61** 19.
Akhundov A, Bardin D and Riemann T, 1986, Nucl Phys **B276** 1.
Akrawy M Z et al. (OPAL Collaboration)), 1989, Phys Lett **B231** 530.
ALEPH 1993a, ALEPH Collaboration, Buskulic D et al. Phys Lett **B313** 549.
ALEPH 1993b, ALEPH Collaboration, Buskulic D et al. Phys Lett **B313** 535.
ALEPH 1993c, ALEPH Collaboration, Halley A, results presented at the EPS conference Marseille.
ALEPH 1994, ALEPH Collaboration, CERN-PPE/94-017 to appear in Z Phys C ; CERN-PPE/94-023 to appear in Nucl Inst Meth.
ALEPH 1994b, ALEPH Collaboration, Phys Lett **B321** 168.
Alexander G et al. 1987, Working Group Report CERN/LEPC/87-6 LEPC/M81.
Alexander G et al. (OPAL Collaboration) 1991, Phys Lett **B266** 201.
Alitti J et al. (UA2 Collaboration) 1991, CERN-PPE/91-163.
Allaby J V et al. (CHARM Collaboration), 1986, Phys Lett **B177** 446; 1987, Z Phys **C36** 611.
Altarelli G and Barbieri G, 1991, Phys Lett **B253** 161.
Altarelli G, Barbieri R, Caravaglios F, 1993, CERN-TH 6770/93; CERN-TH 6859/93.
Altarelli G and Lampe B, 1993, Nucl Phys **B391** 3.
Amaldi U et al. 1987, Phys Rev **D36** 1385.
Amaldi U et al. 1991, Phys Lett **B260** 447.
Aoyagi A et al. 1992, Phys Lett **A167** 415.
ARGUS 1993, ARGUS Collaboration, Z Phys **C58** 61.
Arnaudon L et al. (The LEP polarisation Collaboration), 1992, Phys Letters **B284** 431.
Arnaudon L et al. 'The Energy Calibration of LEP in 1991', CERN-PPE/92-125 (1992) and CERN-SL/92-37(DI); LEP Energy Group, ALEPH, DELPHI, L3 and OPAL Collaborations; 1993a, Arnaudon L et al. Phys Lett **B307** 187; 1993b, Arnaudon L et al. 'The Energy Calibration of LEP in 1992', CERN-SL/93-21 (DI), April 1993, Working group on LEP energy: Summary of the 48th, 49th and 50th Energy Meetings (Feb. 1994).
Arnaudon L et al. 1993, 'Energy calibration of LEP in 1993 with resonant depolarisation' CERN-SL/94-YY.
Arnaudon L et al. 1994, CERN-SL/94-07 (BI).
Artamonov A S et al. 1982, Phys Lett **118B** 225.
Badier J et al. 1987, ALEPH Note 87-17.
Barbaro-Galtieri A, 1993, Proceedings of the Europhysics Conference, Marseille (1993).
Barber D P et al. 1984, Phys Lett **135B** 498.
Barnabeu J, Pich A and Santamaria A, 1991, Nucl Phys **B363** 326.
Batignani G et al. 1992, 1991 IEEE Nuclear Science Symp. (Santa Fe), IEEE Trans on Nuclear Science **NS39(4-5)** Vol. 1, 444.
Bederède D et al. 1994, SiCAL – A High precision Luminosity Calorimeter for ALEPH, to appear in Nucl Inst Meth.
Beenakker W and Hollik W, 1988, Z Phys **C40** 141.
Beenhakker W and Pietrzyk B, 1992, Phys Lett **B296** 296; Phys Lett **B304** 366.
Beenakker W, Berends F A, Van der Marck F C, 1991, Nucl Phys **B355** 281.
Berends F A et al. 1989, in Z Physics at LEP-I, eds Altarelli G, Kleiss R and Verzegnassi C, CERN 89-08 Vol I, p 89.
Berends F A, Burgers G, Hollik W, van Neerven W L, 1988, Phys Lett **B203** 177.
Berends F A, Kleiss R, Hollik W, 1988, Nucl Phys **B304** (1988) 712; Computer programme BABAMC by Kleiss R.
Berge J P et al. 1981, Nucl Phys **B184** 13.

Berger G and Lovberg R H, 1970, *Science* **170** (Oct 1970) 296.
Bloch B, 1993, *Proceedings of EPS Conference Marseille*; see also Decamp et al. 1992a/93/94c.
Blockus D et al. 1986, *Proposal for polarisation at* SLC, SLAC-Prop-1, Stanford.
Blondel A et al. 1988, *Nucl Phys* **B304** 438.
Blondel A, 1989, CERN-EP/89-84.
Blondel A et al. 1990, *Z Phys* **C45** 361.
Blondel A, Renard F M and Verzegnassi C, 1991, *Phys Lett* **B269** 419.
Blondel A, 1992 *Electroweak Experiments at* LEP, proceedings of TASI 1991, eds. Ellis, Hill and Lykken (World Scientific) 283.
Blondel A, Djouadi A and Verzegnassi C, 1992, *Phys Lett* **B293** 253.
Blondel A, session on Energy calibration at the second Chamonix Workshop on LEP performance, CERN-SL/92-29 (DI).
Blondel A et al. (ALEPH Collaboration), 1993b ALEPH-Notes 93-041/042/044.
Blondel A and Verzegnassi C, 1993, *Phys Lett* **B311** 346.
Böhm M, Denner A and Hollik W, 1988, *Nucl Phys* **B306** 687.
Böhm M and Hollik W, 1982, *Nucl Phys* **B204** 45.
Böhm M and Hollik W, 1989, *Z Physics at* LEP I, eds Altarelli G et al. CERN 89-08 203.
Borelli A et al. 1990, *Nucl Phys* **B333** 357.
Brown D, 1993, talk given at the 5th international symposium on Heavy Flavours physics, Montreal, July 1993. MPI preprint MPI-PhE 93-25 Munich.
Burgers G, 1988, in *Polarisation at* LEP, CERN 88-06 p 121.
Burkhardt et al. 1989, *Z Phys* **C43** 497.
Buskulic D et al. (ALEPH Collaboration) 1993, *Z Phys* **C59** 369.
Buskulic D et al. (ALEPH Collaboration) 1993b, CERN-PPE/93-208 to appear in *Z Phys C*.
CERN88-06, *Polarisation at* LEP, CERN *88-06*.
Clarke P, Siegrist P, Pietrzyk B, 1994, rapporteurs talks given at XXIX rencontres de Moriond, *Electroweak Interactions and Unified Theories*, Meribel to appear in the proceedings.
Consoli M and Hollik W, 1989, in *Z Physics at* LEP-I, eds. Altarelli G, Kleiss R and Verzegnassi C, CERN 89-08 7.
Cooper P S et al. 1975, *Phys Rev Lett* **34** 1589.
Davier M et al. 1992, Orsay preprint LAL 92-73 or X-LPNHE 92-22.
Decamp D et al. (ALEPH Collaboration) 1989, *Phys Lett* **B231** 519.
Decamp D et al. (ALEPH Collaboration) 1990, *Nucl Inst Meth* **A286** 121.
Decamp D et al. (ALEPH Collaboration) 1991, *Phys Lett* **B265** 430.
Decamp D et al. (ALEPH Collaboration) 1991b, *Phys Lett* **B259** 377.
Decamp D et al. (ALEPH Collaboration) 1992a, *Z Phys* **C53** 1; Decamp D et al. (ALEPH Collaboration)), 1993, *Z Phys* **C60** 71; ALEPH Collaboration, 1994, CERN-PPE/94-030, submitted to *Z Phys C*.
Decamp D et al. (ALEPH Collaboration) 1992b, *Z Phys* **C53** 375.
DELPHI 1992a, DELPHI Collaboration, Abreu P et al. *Z Phys* **C56** 201.
DELPHI 1992b, DELPHI Collaboration, Abreu P et al. *Phys Lett* **B281** 383; **B295** 383.
Denegri D, Sadoulet B and Spiro M, 1989, CERN-EP/89/72.
Eichten T et al. 1973, *Phys Lett* **46B** 274.
Fero M et al. 1994, SLAC-pub-6423.
Field R D and Feynman R P, 1978, *Nucl Phys* **B136** 1.
Fischer G and Hofmann A, 1992, in second workshop on LEP performance, ed. Poole J, CERN-SL/92-29 (DI) 337.
Franzini P J and Taxil P, 1982, *Nucl Phys* **B197** 45.
Franzini P J and Taxil P, 1989, in *Z Physics at* LEP-I, eds. Altarelli G, Kleiss R and Verzegnassi C CERN 89-08 vol. II, 58.

Frisch J, Woods M and Zolotorev Z, 'A New Ti:Sapphire laser for the new polarized electron source', SLAC-Pub-5950.
Glashow S L, 1961, *Nucl Phys* **22** 579.
Glashow S L, Iliopoulos J, Maiani L, 1970, *Phys Rev* **D2** 1285.
Glashow S L and Georgi H, 1972, *Phys Rev Lett* **28** 1494.
Goggi G, 1979, CERN 79-01 Vol II, p 483.
Hasert F J et al. 1973, *Phys Lett* **46B** 121; *Phys Lett* **46B** 138.
Hasert F J et al. 1974, *Nucl Phys* **B73** 1.
Hatton V et al. 1990, CERN LEP performance note 12.
Hofmann A, 1991, in *Proceedings of the First Workshop on* LEP *Performance*, ed. Poole J, CERN-SL/91-23 265.
Jacobsen B, 1992.*'Observation of tidal distortions of the* LEP *orbit'*, SL-MD note 62.
Jadach S and Wąs Z, 1989, in *Z Physics at* LEP I, eds. Altarelli G, Kleiss R and Verzegnassi C, CERN 89-08 Vol I, 235.
Jadach S and Ward B F L, 1989, *Phys Rev* **D40** 3582.
Jadach S, Richter-Wąs E, Wąs Z, Ward B F L, 1991, *Phys Lett* **B253** 469; **B260** 438.
Jadach S, Richter-Wąs E, Wąs Z, Ward B F L, 1991, *Phys Lett* **B268** 253; 1992, *Comput Phys Commun* **70** 305.
Junk T, 1994, talk given at XXIX rencontres de Moriond, *Electroweak Interactions and Unified Theories*, Meribel.
Kennedy D C and Lynn B W, 1986, SLAC-Publication 4039 (1986, revised 1988).
Kennedy D C and Lynn B W, 1989, *Nucl Phys* **B321** 83.
Kennedy D C et al. , 1989, *Nucl Phys* **B321** 83.
King B J et al. (CCFR Collaboration) 1993, *A precise measurement of the Weak Mixing Angle in Neutrino-Nucleon Scattering*, Nevis Preprint R1489, Submitted to the European Physical Society Conference Marseille, 1993.
Knudsen L et al. 1991, *Phys Lett* **B270** 97.
Kounine A 1994, rapporteur talk given at *XXIX rencontres de Moriond, 'Electroweak Interactions and Unified Theories'*, Meribel.
L3 1991, L3 Collaboration, Adeva B et al. *Phys Lett* **B261** 177.
L3 1993, L3 Collaboration, Adriani O et al. *Phys Lett* **B307** 237.
Langacker P and Luo M, 1991, *Phys Rev* **D44** 817.
Langacker P and Polonoski N, 1993, *Phys Rev* **D47** 4028.
LEP 1984, Design Report, CERN-LEP/84-01.
LEP 1993 The LEP Collaborations ALEPH, DELPHI, L3, OPAL and the LEP Electroweak Working Group, 'Updated Parameters of the Z Resonance from Combined Preliminary Data of the LEP Experiments', CERN-PPE/93-157.
Levinthal D et al. 1991, CERN-TH 6094/91.
Limberg T, Emma P and Rossmanith R, 1993, 'The north arc of SLC as a spin rotator', *Proc of the Particle Accelerator Conference*, Washington, USA.
Lynn B W, Peskin M E and Stuart R G, 1985, SLAC-PUB 3725 and *Physics at* LEP CERN 86-02, Geneva,(1986), 90.
Lynn B W and Stuart R G, 1985, *Nucl Phys* **B253** 84
Lynn B W, 1989, SLAC-PUB 5077.
Lynn B W and Stuart R G, 1990, *Phys Lett* **B252** 676.
Lynn B W and Verzegnassi C, 1987, *Phys Rev* **D35** 3326.
Marciano W J and Sirlin A, 1984, *Phys Rev* **D29** 945.
Maruyama T et al. 1991, *Phys Rev Lett* **66** 2376.
Maruyama T et al. 1992, *Phys Rev* **B46** 4261.
Martinez M et al. 1991, *Z Phys* **C49** 645.

McKay W W et al. 1984, *Phys Rev* **D29** 2483.
Melchior P, 1983, 'The tides of planet earth', 2nd edition, Pergammon Press.
Nakanishi T et al. 1991, *Phys Lett* **A158** 345.
NC 1993, *Neutral Currents 20 Years after*, Editions Frontières, Paris 1993.
OPAL 1991/93, OPAL Collaboration, Akrawy M Z et al. 1991, *Phys Lett* **B263** 311; Acton P D et al. 1993, *Z Phys* **C60** 19.
OPAL 1993a, OPAL Collaboration, Acton P D et al. *Z Phys* **C58** 523; Ackers R et al. *Z Phys* **C60** 199.
OPAL 1993b/94, OPAL Collaboration, Acton P D et al. *Z Phys* **C60** 579; Ackers R et al. *Z Phys* **C61** 357.
Phinney N, 1992, *Review of SLC performance*, SLAC-Pub.-5864.
Pierce D T and Meier F, 1976, *Phys Rev* **B13** 5484.
Prescott C Y 1980, *Proc. 1980 Int Symp on High-Energy Physics with Polarised Beams and Targets*, Lausanne, eds. Joseph C and Soffer J, (Birkhäuser Verlag, Basel, 1981), 34.
Prescott C Y et al. 1978, *Phys Lett* **B77** 347.
Ross D A and Veltman M, 1975, *Nucl Phys* **B95** 135.
Rougé A, 1990, *Z Phys* **C48** 45.
Salam A, 1968, *Proc. 8th Nobel Symposium, Aspenäsgården*, eds. Almqvist and Wiskell (Stockholm, 1968) 367.
Schultz D et al. 1992, *Polarized source performance for SLC-SLD*, SLAC-Pub 5768; 1993 ibid. SLAC-Pub 6060.
Schwartz A, 1993, *XVI Int. Symp. on Lepton-Photon interactions, Cornell University*, Ithaca, N.Y. (August 1993).
Sirlin A, 1980, *Phys Rev* **D22** 971; Marciano W and Sirlin A, 1980, *Phys Rev* **D22** 2695.
Sirlin A et al. 1983–90, Sirlin A et al. 1983 *Nucl Phys* **B217** 84; Marciano W and Sirlin A, *Phys Rev* **D29** 75; Sirlin A, *Phys Lett* **B232** 123; Sirlin S, *Nucl Phys* **B332** 20.
SLD 1984, 'The SLD design report', SLAC-Report 273.
Sokolov A A and Ternov I M, 1964, *Sov Phys Doklady* **8** 1203.
Swartz M L, 1988, in CERN 88-06 Vol. II, p. 163.
Tsai Y S, 1971, *Phys Rev* **D4** 2821.
Van Dyck Jr R S et al. 1987, *Phys Rev Lett* **59** 26.
Veltman M, 1977, *Nucl Phys* **B123** 89.
Weinberg S, 1967, *Phys Rev Lett* **19** 1264; 1972, *Phys Rev* **D5** 1412.
Wenninger J, 1994, talk given at XXIX rencontres de Moriond, 'Electroweak Interactions and Unified Theories', Meribel, to appear in the proceedings.
Wenninger J, 1994b, CERN-SL/94-14 (BI).
Zholentz A A et al. 1980, *Phys Lett* **96B** 214.

List of Acronyms

The following acronyms are used in the papers presented in this volume:

AEEC	Asymmetry of EEC (Energy-Energy Correlation)
AGS	Alternating Gradient Synchrotron
ALEPH	Apparatus for LEP Physics
ALLM	Abramowicz et al. parametrization
AMY	A collaboration at KEK
APM	Automatic Plate Measuring
ARIADNE	A computer simulation code
ATLAS	A detector at LHC
BCAL	Barrel Calorimeter
BCDMS	A collaboration
BCS	Bardeen, Cooper & Schrieffer (theory of superconductivity)
BEMC	Barrel Electromagnetic Calorimeter
BFKL	Evolution equations
BGF	Boson Gluon Fusion
BPC	Beam-Pipe Calorimeter
CAL	Uranium-Scintillator Calorimeter
CERN	European Centre for Particle Physics
COBE	Cosmic Background Explorer
CBR	Cosmic Background Radiation
CC	Charged Current
CDF	Collaboration at the Fermilab collider
CDM	Cold Dark Matter *or* Colour Dipole Model
CFA	Center for Astrophysics
CFAI	The name of a Sky Survey
CKM	Cabibbo-Kobayashi-Maskawa (matrix)
CMBR	Cosmic Microwave Background Radiation
CMS	Compact Muon Solenoid (detector) at LHC
COBE	Cosmic Background Explorer satellite
CTD	Central Tracking Detector
CTEQ	A collaboration
DAQ	Data AcQuisition
DELPHI	Detector for Electron, Photon and Hadron Identification
DESY	Deutsches Elektronen-Synchrotron
DG	Drees and Grassie (parametrization)
DMR	Differential Microwave Radiometers
DIS	Deep Inelastic Scattering

EEC	Energy-Energy Correlation
EHLQ	Structure functions parametrisation
EM	ElectroMagnetic (calorimeter)
EMC	European Muon Collaboration at the CERN SPS
ERT	Ellis Ross Terrano (matrix)
ETC	Extended Technicolour
FCAL	Forward Calorimeter
FCNC	Flavour-Changing Neutral Currents
FMUON	Forward Muon detector
FNC	Forward Neutron Calorimeter
FNAL	Fermi National Accelerator Laboratory
FRW	Friedmann-Robertson-Walker (cosmology/metric)
FWHM	Full-width half maximum
GEANT	A simulation program
GEM	Gamma Electron Muon (detector) at SCC
GIM	Glashow Iliopoulos Maiani (mechanism)
GL	Ginzburg-Landau (theory of superconductivity)
GLDAP	evolution equation for the density of partons
GR	General Relativity (theory)
GRV	Glück $et\ al.$ parametrization
GS	Gordon-Storow parametrisation
GUT(S)	Grand Unified Theory/Theories
HAC	Hadron Calorimeter
HDM	Hot Dark Matter
HERA	Hadron Electron Ring Anlage (at DESY)
HERA-B	A collaboration at HERA for the study of B-physics
HERACLES	A computer simulation code
HERMES	Collaboration at Hera for the study of polarised e-p scattering
HERWIG	Hadron Emission Reactions with Interfering Gluons (A computer simulation code)
H1	Collaboration at HERA
HO	Higher Order
HSC	Honeycomb Strip Chamber
IP	Interaction Point
IRAS	Infra-red Astronomical Satellite
ISASUSY	A Monte Carlo simulation including SUSY
ISR	Intersecting Storage Rings (at CERN)
IVB	Intermediate Vector Boson
JADE	Japan, England, Deutschland (detector at PETRA)
JBD	Jordan-Brans-Dicke (theory)
JETSET	A Monte-Carlo Computer Simulation code
JBD	Jordan-Brans-Dicke (theory)
LAC	Liquid-Argon Calorimeter
LAC1,2,3	Parametrizations
LEP	Large Electron Positron (Collider)
LEP-I	First version of LEP
LEP-II	Second version of LEP (also called LEP200)

LEPTO	A computer simulation code
LHC	Large Hadron Collider
LSP	Lightest Supersymmetric Particle
LSS	Last Scattering Surface
LLA	Leading-Log Approximation
LO	Leading Order
LPS	Leading Proton Spectrometer
L3P	L3 plus 1
LSP	Lightest Supersymmetric Particle
LST	Limited Streamer Tube
LUMI	Luminosity Detector
MAX	Milimeter Anisotropy Experiment
MDM	Mixed Dark Matter
ME	Matrix Elements
MIPS	Million Instructions per Second
MIT	Massachusetts Institute of Technology
MRS	Martin, Stirling and Roberts parametrisation
MSGC	MicroStrip Gas Counter
MSSM	Minimal Supersymmetric Standard Model
NC	Neutral Current
NLC	Next Linear Collider
NLLA	Next-to-Leading-Log Approximation
NLO	Next-to-Leading Order
NMC	New Muon Collaboration at the CERN SPS
NR	Non-Relativistic
NSE	Nuclear Statistical Equilibrium
OMEGA	Photon Collaboration
OPAL	Omni-Purpose Apparatus for LEP
PDG	Particle Data Group
PDP	Particle Data Properties (Tables of)
PETRA	e^+e^- storage ring at DESY
POTENT	A method used in calculation
PS	Particle Shower
PYTHIA	Event Generator — computer simulation code
QED	Quantum ElectroDynamics
QCD	Quantum ChromoDynamics
QCDC	QCD Compton
QDOT	Queen Mary, Durham, Oxford, Toronto (survey)
QPM	Quark-Parton Model
QSO	Quasi-Stellar Object
RCAL	Rear Calorimeter
RG	Renormalization Group
RGE	Renormalization Group Equation(s)
RISC	Reduced Instruction Set Chip
RMUON	Rear Muon Detector

/continued

RPP	Review of Particle (Data) Properties (same as PDP)
RW	Robertson-Walker (model or metric)
SDC	Solenoidal Detector Collaborations (detector) at SSC
SDSS	Sloan Digital Sky Survey
SIT	Silicon Tracker (detector)
SIVT	Silicon Tracker/Vertex (detector)
SLAC	Stanford Linear Accelerator Center
SLC	Stanford Linear Collider
SLD	Stanford Linear Detector
SLUG	A Monte Carlo Programme
SM	Standard Model (of electroweak interaction)
SP	South Pole (collaboration)
SSB	Spontaneous Symmetry Breaking
SSC	Superconducting Super Collider
SUSY	Super Symmetric (theories)
SUSY GUT	Super Symmetric Grand Unified Theory
SUSY SM	Super Symmetric Standard Model
TASSO	A collaboration at PETRA
TC	Technicolour
TEN	Tenerife (collaboration)
TERAD	A collaboration
TeVatron	TeV accelerator at FNAL
TGV	Triple Gluon Vertex
TOF-VETO	Time-Of-Flight Veto
TOPAZ	A collaboration
TPC	Time Projection Chamber
TRD	Transition Radiation Detector
TRDT	TRD Tracker
UA1/UA8	Collaborations
VDM	Vector Dominance Model
VEV	Vacuum Expectation Value
VXD	Vertex Detector
WIMP	Weakly Interacting Massive Particle
WLS	Wave Length Shifting (fibre)
WTC	Walking Technicolour
ZEUS	Collaboration at HERA

Participants

•Mr. Giovanni Abbiendi
Dipartimento di Fisica
Padova University
Via Marzolo, 8
Padova 35131
Italy

•Dr. Ian J.R. Aitchison
Department of Theoretical Physics
University of Oxford
1 Keble Road
Oxford OX1 3NP
United Kingdom

•Mr. Sandro Ambrosanio
Dipartimento di Fisica
Università di Roma"La Sapienza"
Piazzale Aldo Moro 2
Roma I-00185
Italy

•Ms. Gulay A. Bahadiroglu
Department of Physics
Istanbul University, Faculty of Sciences
Vezneciler 34459
Istanbul
Turkey

•Dr. Christine J. Beeston
Physics Department
University of Manchester
Manchester, M13 9PL
United Kingdom

•Ms. Angela Benelli
Sezione di Bologna
Istituto Nazionale di Fisica Nucleare
Via Inerio 46
Bologna 40100
Italy

•Mr. Roland Bernet
HPK F8
Institute of Particle Physics
ETH Hönggerberg
Zürich CH-8093
Switzerland

•Ms. Laura M. Bertolotto
Department of Physics
University of Edinburgh
James Clerk Maxwell Building
The King's Buildings
Edinburgh EH9 3JZ
United Kingdom

•Dr. Siegfried Bethke
Physikalisches Institüt
Universität Heidelberg
Philosophenweg 16
Heidelberg D-6900
Germany

•Mr. Jacques Bloch
Department of Physics
University of Durham
South Road
Durham, DH1 3LE
United Kingdom

•Dr. Alain Blondel
Laboratoire de Physique Nucleaire des Hautes Energies
Ecole Polytechnique Palaiseau
IN2P3-CNR3
Palaiseau F-91128
France

•Mr. Harry Blundell
Physics Department
Carleton University
1125 Colonel By Drive
Ottawa ON K1S 5B6
Canada

•Mr. Pierre E. Bourdon
Laboratoire de Physique Nucleaire des
Hautes Energies
Ecole Polytechnique Palaiseau
Route de Saclay
Palaiseau F-91128
France

•Mr. Giancarlo Brugnola
Sezione di Bologna
Istituto Nazionale di Fisica Nucleare
Via Inerio 46
Bologna 40100
Italy

•Mr. Maria Chamizo
Centro de Investigaciones Energeticas
Medioambitales y Tecnologicas
Av. Complutense 22
Madrid 28040
Spain

•Mr. Denis Comelli
Dipartimento di Fisica Teorica
Università di Trieste
Stard acostiera 11
Miramare
Trieste 34014
Italy

•Prof. John F. Cornwell
Department of Physics and Astronomy
University of St Andrews
North Haugh
St Andrews KY16 9SS
United Kingdom

•Mr. Alessandro Culatti
Dipartimento di Fisica "Galileo Galilei"
University of Padova
via F Marzolo 8
Padova 35131
Italy

•Ms. Beatriz De Carlos
Consejo Superior de Inv. Cientificas
Instituto de Estructura de la Materia
Serrano 123
Madrid 28006
Spain

•Dr. Alberto De Min
Dipartimento di Fisica
Università di Milano
Via Celoria 16
Milano 20133
Italy

•Mr. David Delepine
Département de Physique FYMA
Université Catholique de Louvain
Chemin du Cyclotron 2
Louvain-la-Neuve B-1348
Belgium

•Mr. Antonio Di Domenico
Dipart. di Fisica G. Marconi
Univ. di Roma I "La Sapienza"
Piazzale Aldo Moro 2
Roma I-00185
Italy

•Mr. Thomas G. Dignan
High Energy Physics Laboratory
Harvard University
42 Oxford Street
Cambridge MA 02138
USA

•Dr. Brian Dolan
Institut für Theoretische Physik
Universt Hannover
Appelstrasse 2
Hannover W-3000
Germany

•Dr. Tony Doyle
Department of Physics and Astronomy
University of Glasgow
Glasgow G12 8QQ
United Kingdom

•Dr. Sarah J. Durston Johnson
Department of Physics and Astronomy
University of Rochester
1125 Genesee St.
Rochester NY 14627
USA

•Mr. Michael Earnshaw
Department of Applied Mathematics
and Theoretical Physics
University of Cambridge
Silver Street
Cambridge CB3 9EW
United Kingdom

•Mr. Victor D. Elvira
D0 - Mail Station 352
Fermilab PO Box 500
Wilson Road
Batavia IL 605100
USA

•Mr. Marcus Englert
Sektion Physik
University of Munich
Am Coulombwall 1
Garching D-8046
Germany

•Dr. Eda Eskut
Physics Department
Cukurova University
Science and Art Faculty
Bacali-Adana
Turkey

•Mr. Jose R. Espinosa
Consejo Superior de Investigaciones
Cientificas
Instituto de Estructura de la Materia
Calle Serrano 123
Madrid 28006
Spain

•Mr. Federico Farchioni
Dipartimento di Fisica
Università di Pisa
Piazza Toricelli 2
Pisa I-56100
Italy

•Mr. Dimitris Fassouliotis
Physics Department
National Technical University - Athens
Zografou Campus, Zografou
Athens
Greece

•Dr. Roger W. Finlay
Physics Department
Ohio University
Athens OH 45701
USA

•Mr. Roman K. Friedrich
Institut für Theoretische Physik
Universität Tubingen
Auf der Morgenstelle 14
Tübingen D 72076
Germany

•Ms. Cecilia Gerber
D0 - Mail Station 352
Fermilab PO Box 500
Wilson Road
Batavia IL 60510
USA

•Mr. Bostjan Golob
Institut "Jozef Stefan"
Jamova 39
p.p. 100
Ljubljana 61111
Slovenia

•Mr. Aaron K. Grant
Enrico Fermi Institute
University of Chicago
5640 Ellis Avenue
Chicago IL 60637-1433
USA

•Mr. Geza Gyuk
Department of Astronomy and
Astrophysics
University of Chicago, Box 47
5640 S. Ellis Avenue
Chicago IL 60637
USA

•Mr. Thomas Hambye
Institut de Physique Theorique
Université Catholique de Louvain
Unité FYMA, Chemin du Cyclotron 2
Louvain-la-Neuve B-1348
Belgium

•Mr. Jose Illana
Departamento de Fisica Teorica y del
Cosmos
Universidad de Granada
Granada E-18071
Spain

•Dr. Henry R. Jaqaman
Physics Department
Birzeit University
P.O. Box (14), Birzeit
West Bank, Via Israel

•Dr. Peter Jenni
PPE, CERN
Geneva CH1211
Switzerland

•Mr. Yong-Yeon Keum
LPTHE
Université Pierre et Marie Curie
Tour 24 - 5ème étage
2 Place Jussieu
Paris 75251
France

•Prof. Edward W. (Rocky) Kolb
Fermilab
PO Box 500
Batavia IL 60510
USA

•Mr. Castulus Kolo
Sektion Physik
Universität München
Am Coulombwall 1
Garching D-8046
Germany

•Mr. Nikolaos Konstantinidis
Department of Physics
Imperial College
Prince Consort Road
London SW7 2BZ
United Kingdom

•Dr. Ayse Kuzucu
Fen-Edebiyat Fakultesi
Cukurova Universitesi
Fizik Bolumu Balcali
Adana 01330
Turkey

•Miss. Fani Lebessis
"Demokritos"
National Centre for Scientific Research
PO Box 60228
Aghia Parakvesi 15310
Greece

•Mr. Dietrich Lehner
IfH, DESY
Platanenallee 6
Zeuthen D-15738
Germany

•Dr. Sergio Lupia
Dipartimento di Fisica Teorica
Università di Torino
Via Pietro Giuria 1
Torino 10125
Italy

•Mr. Alick Macpherson
Department of Physics
University of Alberta
Edmonton T6G 2J1
Canada

•Dr. Paul Maley
Particle Physics Department
Rutherford Appleton Laboratory
Office 2-90, R1
Chilton, Didcot OX11 0QX
United Kingdom

•Mr. Andre W. Maul
Department of Physics and Astronomy
Michigan State University
Room 251, Physics-Astronomy Bldg.
East Lansing MI 48824-1116
USA

•Mr. Stefano Moretti
Dipartimento di Fisica Teorica
Università di Torino
Via Pietro Giuria 1
Torino 10125
Italy

•Mr. Krzysztof Muchorowski
Institute of Experimental Physics
University of Warsaw
ul. Hoza 69
Warsaw 00-681
Poland

•Dr. Peter Negus
Department of Physics and Astronomy
University of Glasgow
Glasgow G12 8QQ
United Kingdom

- Mr. Remus Nicolaescu
  Institute of Gravity and Space Sciences
  Institute of Atomic Physics
  PO Box MG-6
  Bucharest-Magurele 769900
  Rumania

- Mr. Andreas Nyffeler
  Institute for Theoretical Physics
  University of Bern
  Sidlerstrasse 5
  Bern CH 3012
  Switzerland

- Dr. Marco Paganoni
  Dipartimento di Fisica & INFN
  Università di Milano
  Via Celoria 16
  Milano 20133
  Italy

- Mr. Philip R. Page
  Department of Theoretical Physics
  University of Oxford
  1 Keble Road
  Oxford OX1 3NP
  United Kingdom

- Dr. Ken J. Peach
  Department of Physics and Astronomy
  University of Edinburgh
  James Clerk Maxwell Building
  The King's Buildings
  Edinburgh EH9 3JZ
  United Kingdom

- Prof. Roberto D. Peccei
  Department of Physics
  University of California at Los Angeles
  405 Hilgard Avenue
  Los Angeles CA 90024-1547
  USA

- Mr. Jose R. Pelaez
  Departamento de Fisica Teorica II
  Universidad Complutense de Madrid
  Faculdad de Ciencias Fisicas
  Madrid 28040
  Spain

- Mr. Krzysztof Piotrzkowski
  High Energy Physics Laboratory
  Institute of Nuclear Physics
  ul. Kawiory 26A
  Krakow
  Poland

- Mr. Johan C. Rathsman
  Department of Radiation Sciences
  Uppsala University
  BOX 535
  Uppsala S-751 21
  Sweden

- Mr. Philip Reeves
  Department of Physics and Astronomy
  University of Glasgow
  Glasgow G12 8QQ
  United Kingdom

- Ms. Stefania Ricciardi
  Dipartimento di Fisica & INFN
  Università di Ferrara
  via Paradiso 12
  Ferrara 44100
  Italy

- Mr. Pieter J. Rijken
  Institute Lorentz
  University of Leiden
  PO Box 9506
  Leiden 2300 SB
  Netherlands

- Dr. Antonio Riotto
  ISAS-SISSA
  International School for Advanced Studies
  Via Beirut 2
  Trieste 34014
  Italy

- Ms. Simona Rolli
  Department of Nuclear and Theoretical Physics
  University of Pavia
  Via Bassi 6
  Pavia 27100
  Italy

•Dr. Robert Rylko
Institute of Electronics
Technical University
ul. Partyzantów 17
Koszalin 75-411
Poland

•Prof. David H. Saxon
Department of Physics and Astronomy
University of Glasgow
Glasgow G12 8QQ
United Kingdom

•Mr. Markus Schmidt
Institut für Hochenergiephysik
Universität Heidelberg
Schröderstrasse 90
Heidelberg D-6900
Germany

•Mr. Christoph Schwick
DESY
Notkestrasse 85
Hamburg D-22607
Germany

•Ms. Gabriella Sciolla
Istituto di Fisica
Universita' di Torino
Via Pietro Giuria 1
Torino 10125
Italy

•Mr. Rupert Seidlein
Physics Department
Ohio State University
174 West 18th Avenue
Columbus OH 43210
USA

•Dr. Yakov M. Shnir
B I Stepanov Institute of Physics
Academy of Sciences of Belarus
F. Scaryna Av.70
Minsk 220072
Belarus

•Mr. Mike G. Smith
Department of Physics and Astronomy
University of Glasgow
Glasgow G12 8QQ
United Kingdom

•Mr. Jürgen Steegborn
Institut für Theoretische
Teilchenphysik
Universität Karlsruhe
Karlsruhe D-76128
Germany

•Mr. Johannes Steuerer
Department of Physics and Astronomy
University of Victoria
PO Box 3055
Victoria V8W 3P6
Canada

•Prof. W. James Stirling
Department of Physics
University of Durham
Science Laboratories
South Road
Durham DH1 3LE
United Kingdom

•Mr. Raimund Ströhmer
Physikalisches Institut
University of Heidelberg
Philosophen Weg 12
Heidelberg D-6900
Germany

•Mr. Pedro Teixeira-Dias
Physikalisches Institut
University of Heidelberg
Philosophenweg 12
Heidelberg D-6900
Germany

•Dr. Ravdandorj Togoo
Laboratory of High Energies
Joint Institute for Nuclear Research
Dubna 141980
Russia

•Mr. Orhan Toker
Sektion Physik
University of Munich
Am Coulombwall 1
Garching D-8046
Germany

•Dr. Roberto Ugoccioni
Dipartimento di Fisica Teorica
Università di Torino
Via Pietro Giura 1
Torino 10125
Italy

•Mr. Victor Uros Corrales
Laboratoire de Physique Nucléaire et de
Hautes Energies
LPNHE
4 Place Jussieu, BP 2000
Paris 75252
France

•Mr. Patrick Van Esch
Dienst Elem
Vrije Universiteit Brussel
Fac. Wetenschappen
Pleinlaan 2
Brussel B - 1050
Belgium

•Dr. Lance L.J. Vick
Department of Physics and Astronomy
University of Edinburgh
James Clerk Maxwell Building
The King's Buildings
Edinburgh EH9 3JZ
United Kingdom

•Mr. Christian Völcker
Sektion Physik
Ludwig-Maximilians-Universität
München
Garching D-8046
Germany

•Ms. Doreen Wackeroth
Werner Heisenberg Institut
Max-Planck-Institut für Physik
Föhringer Ring 6
Munich D-8000
Germany

•Mr. Alan Walker
Department of Physics and Astronomy
University of Edinburgh
James Clerk Maxwell Building
The King's Buildings
Edinburgh EH9 3JZ
United Kingdom

•Mr. Nick H. Willis
Department of Applied Mathematics
and Theoretical Physics
University of Cambridge
Silver Street
Cambridge CB3 9EW
United Kingdom

•Dr. Gunter Wolf
F1, DESY
Notkestrasse 85
Hamburg D-2000
Germany

•Mr. Songhoon Yang
Nevis Laboratory
Columbia University
Irvington NY
USA

•Prof. Shigeo Yazaki
College of Liberal Arts & Sciences
Kitasato University
1-15-1 Kitasato
Sagamihara
Kanagawa 228
Japan

Index

adiabatic perturbations, 374
adjoint or regular representation, 31, 39
α_s parameter, 86, 106, 123
anomalous magnetic moment (μ^-, e^-), 24
antiscreening, 39
asymptotic freedom, 37–39, 87, 101, 324
ATLAS detector (LHC), 280–304
ATLAS, higgs simulations, 308
ATLAS, leptoquark simulations, 317
ATLAS, new vector boson simulations, 314
ATLAS, supersymmetry simulations, 312
ATLAS, top quark simulations, 306

β function, 40, 304
beam polarisation, 138
beta-function in QCD, 86
bias, 381, 386, 387
Bjorken variable, 156
bunch configuration of HERA, 147

Cabibbo angle, 64
calorimetry at hadron colliders, see ATLAS, calorimetry, 280
CBR (cosmic background radiation), 361–408
charged current processes, 57, 61, 62, 64
charged Higgs bosons, 254, 262
charginos, 266
chiral preon theory, 342
chiral supermultiplet, 329
chiral symmetries, broken, 341
chiral symmetry, 42, 67
CKM matrix, 64, 344
cluster fragmentation, 92
collisionless damping, 376, 378
colour, 82, 86, 89
comoving frame, 369
Compact Muon Solenoid, 279
compactifications, 354
compositeness, 276
confinement, 87, 324
conservation of stress energy, 369
cosmic string wakes, 409

cosmic strings, electroweak, 413
cosmic variance, 381
cosmological constant term, 66
cosmological principle, 366
counter terms, 16, 28
covariant derivative, SU(2), 32, 33
covariant derivative, SU(3), 37
covariant derivative, U(1), 10
CP-violation, 64
cross sections for DIS, 159
current conservation, 8
curvature fluctuations, 378
custodial SU(2), 74, 338
cut-off, 14, 15
cut-off mass, 326

damping processes, 376
dark matter, 335
dark matter, cold, 378, 380, 383, 385–387
dark matter, hot, 378
dark matter, warm, 379
data acquisition at hadron colliders, 300
decoupling, 22
deep inelastic lepton nucleon scattering, 154
deep inelastic neutral current scattering, 139
deep inelastic scattering (DIS), 154
density field of the Universe, 372
diffractive scattering, 217
dimension-5 operator, 348
Dirac Lagrangian, 3
Dirac masses, 322
DIS cross sections at HERA energies, 164
DIS final states, 201
domain walls, 400, 405, 406
Doppler effect, 362
double-angle method, 156
drosophila, 396
dynamical current, 10, 31

E_6 and E_8, 355
early universe, 46
effective degrees of freedom, 15
effective four-fermion interactions, 61

effective Lagrangian, 26
Einstein field equations, 369
electric quadrupole moment, of W, 36
electron identification in colliders, 289
electron-positron accelerators, 84
electroweak theory, 53–75
energy density, 364, 367
equation of state, 369
evolution, Altarelli-Parisi, 169
evolution, Lipatov, 169
expansion of the Universe, 362, 367
extended technicolour (ETC), 339
extragalactic distances, 363

family, 56
Fermi constant, 25, 53
Fermi scale, 322, 325
Fermi theory, 53, 57
fermion masses, 58, 59, 65, 68, 71
ferromagnet, 41, 44
Feynman gauge, 6, 48
Feynman graphs and rules, 7
fine tuning problems, 66
finite-temperature potential, 400
flavour independence of α_s, 119
flavour-changing neutral currents, 64, 340
flipped SU(5), 348
forward-backward asymmetries for leptons, 450
four fermion interaction, 20
free energy, 44
free streaming, 376, 378
Friedmann equation, 369, 389
FRW cosmology, 362

$\gamma\gamma$-scattering, 25, 173
g-factor for the W, 25, 35
gauge condition, 6
gauge dependence, 378
gauge fields, 321
gauge invariance, 18, 19, 26, 67
gauge transformations, 6
gauge-fixing term, 6, 37
gaugino, 329, 335–336
gaussian fluctuations, 373
GEM detector (SSC), 278
Georgi's survival hypothesis, 350
ghost terms, 38
GIM mechanism, 62
global and local non-Abelian symmetries, 28

global and local symmetries, 1
global SU(2), 29, 37, 72–74
global SU(2)×U(1), 50
global SU(3), 31
global U(1), 9, 43
gluinos, 267
gluinos in QCD β-function, 126
gluinos, search for, 126, 132
gluon, 83
gluon self coupling, 38, 86, 104
gluon self-energy, 39
gluon spin, 104
gluon structure functions, 199
gluon trilinear self-coupling, 39
gluon-tagged jets, properties of, 127
Goldstone boson, 41–50, 254
grand unification, 230, 250–251, 336, 345
grand unified scale, 14, 66
gravitational instability, 374
group constants of QCD, 104

H1 detector, 140, 152, 170
hadron jets, 88
hadronic cross-section in e-p annihilation, 89
hadronic event shape observables, 93
hadronic events at LEP, 431
hadronisation, 40, 87, 90, 93
Harrison-Zel'dovich spectrum, 387
heavy quark asymmetries, 458
heavy vector bosons at colliders, 276
helicity effects in Z interactions, 443
HERA collider, 136–137
HERA kinematics, 154
heterotic string, 355
hidden symmetries, 1
hierarchy of scales, 15
hierarchy problem, 332, 349
Higgs boson, 46–47, 58, 70, 226
Higgs boson mass, 66, 68, 76, 228
Higgs couplings, 60, 61
Higgs decay channels, 233
Higgs doublet, 49, 63, 71, 72, 329, 333, 336
Higgs physics at hadron colliders, 273
Higgs potential, 65, 69, 71
Higgs supermultiplets, 330
Higgs triplet, 349
Higgsino, 348, 349
high energy behaviour of tree graphs, 58
homogeneity, 366, 367, 369, 370

Hubble parameters, 362–363, 370

independent fragmentation, 92
inflation, 66, 388
inflation, chaotic, 392, 396
inflation, extended, 393
inflation, first order, 393
inflation, slow rollover, 391–394
inflaton potential, 398, 391, 399
invariance, global, 8
invariance, local, 8
invariance, local U(1), 10
isospin operators, 30
isotropy, 366, 367, 369

Jacquet-Blondel method, 155
Jeans instability, 376
Jeans parameters, 375–378
jet algorithms, 97
jet formation, 183
jet multiplicity, 206
jet production, 203
jet recombination schemes, 97
jet tagging, 249
jets of hadrons, resolution criteria, 96

Kibble mechanism, 410
kinematics of $e\text{-}p$ scattering, 154

L3P detector (LHC), 279
Lagrangian density, 321
Lagrangian for QED, 3
$\Lambda_{\overline{MS}}$, 86
Landau damping, 376
Landau pole, 230, 327
large rapidity gap events, 179, 210, 215
leading order (LO) QCD predictions, 184
left-right cross-section asymmetry, 445
LEP, Higgs production at, 237, 259, 260
LEP Collider, 418
LEP, leptonic events at, 432
LHC, design parameters, 272
LHC, Higgs production at, 240
lightest supersymmetric particle (LSP), 334
linear colliders, 131, 263
local SU(2) symmetry, 31, 35
local SU(3) symmetry, 37
local U(1) symmetry, 10, 11, 31
logarithmic violations of scaling, 135
luminosity distance, 362

magnetic moment, of W, 35

magnetic moments, anomalous, 11
Majorana masses, 322
manifest symmetries, 1
mass density, 367
mass generation, 344
mass generation, dynamical, 49
mass matrices, 62
mass scale in non-renormalisable theory, 27
mass shift, quadratic, 328
matrix element models in QCD, 91
matter-antimatter asymmetry, 350
Maxwell Lagrangian, 3
Meissner effect, 47
metacolour, 342–343
Micro Gas Strip (MSGC) detectors, 292
minimal substitution, 321
monopoles, magnetic, 400, 405, 409, 411
Muon detectors in colliders, 296

Nambu-Goldstone bosons, 411
naturalness, 67, 251
naturalness condition, 327
NC matrix element, 423
neutral currents, 61–74, 417
neutralinos, 266, 334, 335, 348
neutrino mass, 60, 65
neutrino mixing, 65
Newton's gravitational constant, 66, 356
Noether's theorem, 9
non-Abelian field strength tensor, 34, 37
non-Abelian symmetry, 29
non-decoupling, 74
non-renormalisable interaction, 20–28, 36, 44
non-renormalisable theory, 27, 54
non-topological solitons, 400
number density, 364
number of light neutrinos, 419

parton distributions of the proton, 162
parton showers (PS), 202
parton substructure at colliders, 276
Pauli matrices, 29
peculiar velocity, 370, 381
phase transitions, 374–413
photon mass, and gauge invariance, 11, 18
photon self-energy, 17, 48, 67
photon vacuum polarisation, 18
photoproduction, 171
photoproduction, hard scattering, 180
photoproduction, partial cross section, 177

photoproduction, total cross section, 173
physical charge, 21
Planck length, 14, 66, 411
polarised beams at SLC, 446
polarised source at SLC, 447
pomeron exchange, 176, 217, 218
pomeron structure function, 217
POTENT, 386, 397
power spectrum, 373
electroweak theory tests, 67, 75
preons, 342
primordial nucleosynthesis, 372
propagator, fermion, 5
propagator, massive vector particle, 47
propagator, photon, 6, 48
proper distance, 370
proper time, 369
proton lifetime, 348, 349
pseudorapidity, 181

Q-balls, 400
QCD models, 202
QCD beta-function, 86
QCD group constants, 104
QCD matrix element models, 91
QCD scaling violations, 96, 121
QCD shower models, 91
QCD, basics of, 31, 37, 86
QED, 3
quadratic Casimir operator, 39
quark asymmetries, 455
quark flavours, 62, 86
quark-parton model cross sections, 160

ρ parameter, 61, 71, 74, 338
radiative corrections, 231, 256, 338
radiative corrections to the electroweak theory, 69
radiative effects, 424
rapidity, 181
redshift survey, 374, 384–386
redshift, cosmological, 362, 366, 370
Regge fits, 176, 217
reionization, 383
renormalisation, 2, 13–19, 27, 55
renormalisation group, 22, 69
renormalisation group equation, 40, 323
renormalisation scale dependence of α_s, 112
renormalisation scale in QCD, 86
renormalised charge, 18
resummed QCD calculations, 114

Robertson–Walker metric, 362, 369
running charge, 22
running coupling constant, 13, 322, 323
running of α_s, 38, 102, 124

Sachs-Wolfe effect, 382, 384
scale factor, cosmological, 369
scaling violations in QCD, 96, 121
scaling violations of F_2, 199
screening, 22, 48
screening of Higgs sector, 74
SDC detector (SSC), 277
sea quarks, 163
seagull graph, 326
see-saw mechanism, 349
selectron, 329
self-coupled scalar field, 325
self energy, electron, 13
self-energy, photon, 17
sfermions, 348
shiggs, 336
shower models in QCD, 91
Silicium calorimeter (SiCAL), 430
silicon trackers at colliders, 291
Silk damping, 377
$\sin \theta_W$, 60, 69, 70, 333
$\sin^2 \theta_W$, 445
$\sin^2 \theta'_{\text{eff}}$, 75
sleptons, 266, 336
small x physics, 166
SO(10), 345, 349
Solenoidal Detector Collaboration, 306
soliton stars, 400
sparticle, 251, 265
spatial curvature, 369
sphalerons, 400
spontaneous breakdown of $SU(2) \times U(1)$, 322
spontaneous symmetry breaking, 41, 45, 71, 226, 345, 390, 400–410
spontaneously broken $SU(2)_L \times U(1)$, 49, 56
squarks, 267, 336
SSC, Higgs production at, 240
SSC, design parameters, 272
standard model (SM), 319
stop (squark), 332
string fragmentation, 92
strings, cosmic, 400, 405, 406
strings, cosmic, global, 411
strong coupling, 232

structure formation, 372
structure function, gluon, 168
structure function, longitudinal, 159
structure function, photon, 159, 173, 190
$SU(N)$, 9, 31, 39
$SU(2)$, 29–31, 39
$SU(2)$ gauge coupling constant g, 34
$SU(2) \times U(1)$, 35, 49–51, 56
$SU(2)_{f_5}$, 41–43, 46, 51
$SU(2)_f$, 41, 49
$SU(3)$, 29–31, 37, 39
$SU(3)_c$, 31, 37
$SU(3)_c$ field strength tensor, 37
$SU(3)_{f_5}$, 43
$SU(3)_f$, 41
$SU(5)$, 346
$SU(6)$, 342
subtracted vacuum polarisation, 21
super-horizon-size perturbations, 377
superconductor, 47
superstring, 349, 354
supersymmetric SM, 329
supersymmetry, 68, 250, 328
supersymmetry at hadron colliders, 276
symmetry breaking, 41
symmetry current, 11, 31, 46

$t\bar{t}$ models, 49
τ polarisation, 451
technicolour, 49, 67, 250, 338
technicolour, extended (ETC), 339
technicolour, walking, 340
technifermion condensate, 340
technifermion self-energy, 340
techniquarks, 343
texture, 411
three-family model, 62
tilted cold dark matter, 385
top quark, 230–232, 241, 244, 331
top quark loops, 70
top quark mass, 69, 75, 333
top quark at hadron colliders, 275
topological defects, 405
tracking detectors at hadron colliders, 290
transfer function, 378, 380
TRD trackers at hadron colliders, 291
tree diagrams, 11
trigger selection, 147
triggering at hadron colliders, 300
triple gluon vertex, 104

two-point galaxy correlation function, 387

$U(N)$, 9
$U(1)$, 9, 29, 32, 43, 45, 46
$U(1)$ Higgs model, 18, 46, 55
$U(2)$, 29
unitarity bound, 229
unitary gauge, 55
universal charge renormalisation, 19
unphysical Higgs field, 48, 55

V–A structure, 43
vacua, degenerate, 45
vacuum, 14, 22, 45, 46, 49–51, 53
vacuum polarisation, 18, 20, 49, 70
valence quarks, 162
vector boson, 54–56, 60
vector boson vacuum polarisation, 71
vector dominance model (VDM), 171
vector supermultiplets, 329
vertex correction, 18, 23, 75

W-W interactions, 34
W-W-Z^0 vertex, 36
W-W-γ vertex, 34–36
W-W-W-W vertex, 36
walking technicolour, 340
Ward identity, 19
wave function renormalisation, 16, 18
weak hypercharge, 56
weak isospin, 56
weak mixing angle, 34, 52, 60, 445
Weinberg angle, 417
Weyl fermion, 329, 343
WIMP, 376
window function, 373
wino-loop, 328

Yang Mills theories, 34, 37, 341
Yukawa coupling, 60, 62, 65, 71, 326
Yukawa coupling, top, 332
Yukawa interaction, 228, 230

Z line-shape, 426, 436
ZEUS detector, 143, 149, 174